DATE DUE

Beneficial Effects
of Endotoxins

Beneficial Effects
of Endotoxins

Edited by

Alois Nowotny

Center for Oral Health Research
School of Dental Medicine
University of Pennsylvania
Philadelphia, Pennsylvania

Plenum Press • New York and London

Library of Congress Cataloging in Publication Data

Main entry under title:

Beneficial effects of endotoxins.

Bibliography: p.
Includes index.
1. Endotoxin — Physiological effect. 2. Endotoxin — Therapeutic use. 3. Immune
response — Regulation. I. Nowotny, A. (Alois), 1922—   . [DNLM: 1. Endotoxins — Therapeutic use. QW 630 B464]
QP632.E4B46 1983   615′.373   83-2256
ISBN 0-306-41147-4

QP
632
.E4
B46
1983

© 1983 Plenum Press, New York
A Division of Plenum Publishing Corporation
233 Spring Street, New York, N.Y. 10013

Printed in the United States of America

Contributors

Alexander M. Abdelnoor • Department of Microbiology, Faculty of Medicine, American University of Beirut, Beirut, Lebanon

Ken-ichi Amano • Department of Bacteriology, Hirosaki University School of Medicine, Hirosaki 036, Japan

Ulrich H. Behling • Oyster Creek Nuclear Generating Station, Forked River, New Jersey 08731, and School of Dental Medicine, University of Pennsylvania, Philadelphia, Pennsylvania 19104

L. Joe Berry • Department of Microbiology, University of Texas, Austin, Texas 78712

Lóránd Bertók • "Frèdèric Joliot-Curie" National Research Institute for Radiobiology and Radiohygiene, Budapest, Hungary

Abraham I. Braude • Department of Medicine and Pathology, University of California, San Diego, California 92103

R. Christopher Butler • Departments of Microbiology and Immunology, The Arlington Hospital, Arlington, Virginia

John L. Cantrell • Ribi ImmunoChem Research, Inc., Hamilton, Montana 59840

Herman Friedman • Department of Microbiology and Immunology, University of South Florida College of Medicine, Tampa, Florida 33612

Jeanne E. Gaska • Department of Microbiology, University of Texas, Austin, Texas 78712

Katherine A. Gollahon • Department of Microbiology, Comprehensive Cancer Center and Institute of Dental Research, The University of Alabama in Birmingham, Birmingham, Alabama 35294

T. L. Graham • Monsanto Agricultural Products Co., St. Louis, Missouri 63166

Wolfgang Graninger • Department of Chemotherapy, University of Vienna, Vienna, Austria

Sheldon Edward Greisman • Department of Medicine and Physiology, University of Maryland School of Medicine, Baltimore, Maryland 21201

Wolfgang Hinterberger • First Department of Medicine, University of Vienna, Vienna, Austria

Monto Ho • Department of Microbiology, Graduate School of Public Health, and Department of Medicine, School of Medicine, University of Pittsburgh, Pittsburgh, Pennsylvania 15261

Diane M. Jacobs • Department of Microbiology, Schools of Medicine and Dentistry, State University of New York at Buffalo, Buffalo, New York 14214

Arthur G. Johnson • Department of Medical Microbiology/Immunology, University of Minnesota, Duluth, Minnesota 55812

Lawrence B. Lachman • Immunex Corporation, Seattle, Washington 98101

P. Liacopoulos • Institut d'Immunobiologie (INSERM U.20, CNRS, L.A. 143), Hôpital Broussais, 75674 Paris Cedex 14, France

Jerry R. McGhee • Department of Microbiology, Comprehensive Cancer Center and Institute of Dental Research, The University of Alabama, in Birmingham, Birmingham, Alabama 35294

Suzanne M. Michalek • Department of Microbiology, Comprehensive Cancer Center and Institute of Dental Research, The University of Alabama, in Birmingham, Birmingham, Alabama 35294

Masayasu Nakano • Department of Microbiology, Jichi Medical School, Tochigi 329-04, Japan

Erwin Neter • Departments of Microbiology and Pediatrics, State University of New York at Buffalo, and Laboratory of Bacteriology, Children's Hospital, Buffalo, New York 14222

A. Nowotny • Center for Oral Health Research, School of Dental Medicine, University of Pennsylvania, Philadelphia, Pennsylvania 19104

Monique Parant • Immunothérapie Expérimentale, Institut Pasteur, 75724 Paris Cedex 15, France

Walter R. Paukovits • Institute for Cancer Research, University of Vienna, Vienna, Austria

Dov H. Pluznik • Department of Life Sciences, Bar-Ilan University, Ramat-Gan, Israel; present address: Laboratory of Microbiology and Immunology, National Institute of Dental Research, NIH, Bethesda, Maryland 20205

Edgar Ribi • Ribi ImmunoChem Research, Inc., Hamilton, Montana 59840

Paul H. Saluk • Department of Microbiology and Immunology, Hahnemann Medical College and Hospital, Philadelphia, Pennsylvania 19102

Ewa Skopińska-Różewska • Division of Experimental Immunosuppression, Transplantation Institute, Warsaw Medical School, 02-006 Warsaw, Poland

Steven Specter • Department of Microbiology and Immunology, University of South Florida College of Medicine, Tampa, Florida 33612

John J. Spitzer • Department of Physiology, Louisiana State University Medical Center, New Orleans, Louisiana 70112

Judy A. Spitzer • Department of Physiology, Louisiana State University Medical Center, New Orleans, Louisiana 70112

Barnet M. Sultzer • Department of Microbiology and Immunology, State University of New York Downstate Medical Center, Brooklyn, New York 11203

Kuni Takayama • William S. Middleton Memorial Veterans Hospital, Madison, Wisconsin 53705

Robert L. Truitt • May and Sigmund Winter Research Laboratory, Mount Sinai Medical Center, Milwaukee, Wisconsin 53233

Takehiko Uchiyama • Department of Microbiology, Kitasato University School of Medicine, Kanagawa 228, Japan

Richard I. Walker • Naval Medical Research Institute, Bethesda, Maryland 20014

Michael Wannemuehler • Department of Microbiology, Comprehensive Cancer Center and Institute of Dental Research, The University of Alabama in Birmingham, Birmingham, Alabama 35294

Elizabeth J. Ziegler • Department of Medicine and Pathology, University of California, San Diego, California 92103

Preface

The study of endotoxins has undergone cycles of enthusiastic research followed by neglect and revival again due to newly described biological effects attributed to endotoxins. It is almost a generally applicable rule that both extremes in such fluctuations in the popularity of certain research fields are unjustified. It is obvious that exaggerated expectations will lead to disappointments, and complete neglect of a once-exciting field is to be followed by renewed attention, if the scientific basis of earlier interest was solid and remained so throughout. The healing effect of fever was praised as nature's blessed cure, until it was discovered that it is caused by bacteria. When it became evident that by careful application one can arrive at therapeutic doses without introducing shock and hypotension, the era of fever therapy began. During this time, indiscriminate use of pyrogens was so widespread that the high expectations had to be followed by disillusionment. Since then, interest in endotoxins experienced a few more oscillations, with varying wavelengths and amplitudes, the latest being triggered by the mitotic effect of endotoxin. This led to justified excitement among researchers who study the activation of the cells of the lymphatic system and the consequences thereof, including the release of important mediator molecules.

Fever therapy is passé, but the healing that often followed the application of endotoxins remains a well-documented fact. Endotoxins can be beneficial, and the realization of this survived the upheavals of changing popularity. Endotoxin remained a substance with a fascinating variety of biological effects, unparalleled by any other natural products. Each time that the tide of curiosity has risen, it did not ebb before bringing better insight into a few aspects of the mechanism of endotoxin action, both from biological as well as from molecular points of view. We felt it necessary to review some of these developments, but this time with a sharp focus on the beneficial effects of endotoxins. This is the first monograph that addresses itself exclusively to this feature of the substance. We revisited this field from every available angle, and attempted to use current knowledge to reinvestigate the possible mechanisms involved. Needless to say, we could not offer clear explanations of these multiple effects. What we hope is that the reader who had high expectations when he or she picked up this volume will not put it down with blighted hope, but will join the few of us who hang on doggedly to this topic and endure the consequences of its often changing prominence.

A. Nowotny

Merion, Pennsylvania

Contents

In Search of the Active Sites in Endotoxins

A. Nowotny

> The most acute, strongest, most dangerous, and fatal diseases, occur in the continual fever. The least dangerous of all, and the mildest and most protracted is the quartan, for it is not only such from itself, but it also carries off other great diseases.
>
> —HIPPOCRATES, OF THE EPIDEMICS

I. Introduction

The history of fever, studies of its effects and search for its origin, go back as far as the history of medicine. The beneficial effects of fever were praised by famous healers in Egypt, Greece, and in the Roman and Persian empires several millenia ago. This appreciation continued through the Middle Ages into our modern times, when "fever therapy" was applied to diseases as diverse as rheumatoid arthritis (Barsi, 1947), manic depression (Terry, 1939), headaches (Sutherland and Wolf, 1940), neurosyphilis (Favorite and Morgan, 1946), herpes zoster (Bernstein and Klotz, 1947), various eye and skin diseases (Tucker, 1946), and many others.

In the 19th century, the search for the fever-causing substances led to bacterial products. It was shown that ordinary distilled water could cause the elevation of body temperature when injected into dogs, but only if the water contained live or killed microorganisms (Billroth, 1865). Burdon-Sanderson named the fever-producing product *pyrogen* in 1876, and Pfeiffer coined the name *endotoxin* for such compounds in 1892. Burdon-Sanderson isolated pyrogens from infected and putrefy-

A. Nowotny • Center for Oral Health Research, School of Dental Medicine, University of Pennsylvania, Philadelphia, Pennsylvania 19104.

ing meat, while Centanni in 1894 and Roussy in 1889 attempted to fractionate bacteria in order to isolate their fever-producing component.

Although it is tempting to review the history of all of the long-forgotten experiments that aimed at isolating the pyrogenic molecule(s) from bacteria, the purpose of this chapter would not be served well by such a survey. To fully appreciate the problems relevant to the identification of active components and active sites in endotoxin preparations, a brief survey of the most frequently investigated biological activities followed by a critical review of selected results obtained by the various investigators who attempted to isolate the active substance or substances will be a more helpful introduction to the complexities of the topic discussed in this chapter.

It is advisable to discuss briefly the terms used in this chapter. *Endotoxins* are those bacterial cell wall constituents that elicit characteristic endotoxicity symptoms. The best known of these are pyrogenicity, Sanarelli–Shwartzman reactivity, and changes in blood clotting, microcirculation, and in blood pressure which may lead to shock and death. These are referred to as *toxic effects.*

Immunogenicity, mitogenicity, stimulation of leukopoiesis, protection against lethal irradiation, immune adjuvant effect, enhancement of nonspecific resistance to infections or to toxic endotoxin doses, and induction of antitumor effects are considered to be the *beneficial* effects of endotoxins.

The term *lipopolysaccharide* (LPS) is most frequently used to describe the endotoxins obtained by the phenol–water extraction procedure of Westphal and Lüderitz. I have maintained this distinction.

Endotoxins consist of a polysaccharide moiety covalently bound to a lipid-rich zone. The latter is termed the *lipid moiety,* indicating the lipophilic part of the intact endotoxin molecule. Other authors call this *lipid A* and use the same term to identify a precipitate formed during acidic hydrolysis of the intact endotoxin (or LPS). Herein, the term *lipid A* is restricted to the acid-degraded precipitates cleaved from the native endotoxin during hydrolysis. Some rough mutant gram-negative bacteria do not synthesize the polysaccharide moiety. In this chapter the glycolipid-rich endotoxins are called *mutant endotoxins.*

Finally, although the coverage of this review endeavored to include all major results obtained by others, and particularly the findings of those who made the first observation, I unwittingly put greater emphasis on the achievements made in my laboratory. This review therefore only scans through the outstanding results of my former masters and of my distinguished fellow endotoxicologists, but it attempts to show our views and our research approaches by presenting the results of many trials and tribulations experienced and achieved together with my students and associates in the last 25 years. I wish to assure the reader that any omission is by no means a valid measure of merit.

II. Biological Effects Attributed to Endotoxin

In 1969, I tabulated the most significant activities of endotoxins (Nowotny, 1969) known at that time. This list contained 22 entries. Since then, several excellent reviews have been published, some of which deal with the more recently studied biological effects in great detail (Elin and Wolff, 1976; Morrison and Ulevitch, 1978; Morrison and Ryan, 1979). Therefore, only the highlights of recent achievements will be briefly surveyed here, before starting the discussion of the components of the endotoxin preparations that may be involved in the elicitation of these biological effects.

The effect of endotoxin on the cells of the lymphatic system has been a major focus of recent surveys. These studies were launched by the findings of Takano et al. (1968), who observed an endotoxin-induced enhanced rate of incorporation of ^{32}P into the nucleic acid of mouse spleen cells in vivo. Peavy et al. (1970) reported elevated incorporation of labeled thymidine into spleen cells in vitro following endotoxin exposure. The mitogenic response of spleen cells was further studied to identify the responder cell type. Several laboratories, almost simultaneously and in full accord, concluded that endotoxin activates B-cell mitogenesis (Gery et al., 1972; Andersson et al., 1972).

The practical significance of the mitogenic effect remains to be established. It was assumed that mitogenicity is a prerequisite for the immune adjuvant effect elicited by endotoxin, but this assumption later became questionable since many mitogens have no adjuvant effect, but also because nonmitogenic endotoxin derivatives could be isolated that had an adjuvant effect on T-dependent SRBC antigens both in vivo and in vitro (Frank et al., 1977; Behling and Nowotny, 1977; Nowotny, 1977).

The effect of endotoxin on T cells is much less evident. Earlier reports described no response of T cells to endotoxin (Andersson et al., 1972; Möller et al., 1972; Peavy et al., 1974). Subsequently, it was shown that endotoxin can act indirectly on T cells (Schmidtke and Najarian, 1975; Ozato et al., 1975), one example being the potentiation of proliferation of Con A-activated T cells. Maturation of T cells could be accelerated by endotoxins according to Scheid et al. (1973) and Adorini et

4 A. Nowotny

al. (1976). LaGrange and Mackaness (1975) reported that endotoxin stimulated the efferent arc of cell-mediated immunity. Gately *et al.* (1976) supported this, using different systems. McGhee *et al.* (1979) showed that T-cell activation is part of the adjuvant action of LPS.

Another cell type that responds to endotoxin exposure with demonstrable changes is the macrophage. *In vitro* studies were initiated by DeLaunay and Lasfargues (1945) on the mechanism of endotoxin-induced cell damage. Heilman and Bernton (1961) reported profound morphological changes in splenic macrophages maintained in a medium containing 0.05 μg/ml endotoxin. These observations were greatly extended in recent years and several biochemical changes were found in the macrophages that indicated a highly activated state (Spitznagel and Allison, 1970; Allison *et al.*, 1973). The significance of this became evident through studies of not only the role that macrophages play in infectious diseases but also the enhanced capacity of macrophages to kill tumor target cells in an apparently nonspecific fashion after endotoxin stimulation (Alexander and Evans, 1971; Currie and Basham, 1975; Hibbs *et al.*, 1977; Doe and Henson, 1978; Doe *et al.*, 1978).

Probably the most important new development regarding the biological effect of endotoxin is the realization that a great number of the reactions are not the consequences of direct interaction between endotoxin and the affected cell types, as had been supposed rather naively in the 1960s. An increasing amount of experimental evidence shows that several endotoxic reactions are mediated by factors released by a few direct endotoxin–cell interactions. These mediators act on cells that do not respond to direct exposure to endotoxins. Some of the mediators have been known for many years, such as endogenous pyrogen or interferon, but the possibility that this may be applicable to many reactions was first assumed by Berry in 1971. Later, in 1980, an entire conference was devoted to the role of humoral factors elicited by endotoxin (*Microbiology 1980*).

Our first relevant observation came from experiments comparing the antitumor resistance-enhancing and the colony-stimulating factor (CSF)-inducing potencies of several endotoxin preparations. It was found that those endotoxin samples that were potent CSF inducers were similarly active in anti-tumor resistance (TUR) enhancement (Butler and Nowotny, 1976; Butler *et al.*, 1978). We also reported that there is no correlation between the toxicity and CSF-inducing capacity of various endotoxins and their derivative, indicating that different chemical structures are required for CSF induction and for the elicitation of toxic reactions (Abdelnoor *et al.*, 1980).

Endotoxin-induced CSF as a regulator of granulopoiesis was described by McNeill (1970), Metcalf (1971), and Quesenberry *et al.*

(1972). These studies shed some light on the possible mechanisms involved in endotoxin-induced protection against lethal irradiation, described first by Mefferd *et al.* (1953) and studied in detail by Smith *et al.* (1957, 1959, 1961, 1966). It was only logical to attempt to protect against lethal irradiation by passive transfer of CSF-rich sera. We found (Butler *et al.*, 1978) that particularly high CSF levels can be demonstrated in BCG-infected and endotoxin-induced (post-BCG-endotoxin) sera obtained according to the method of Carswell *et al.* (1975). Using these sera, we could induce a significant resistance to subsequent irradiation (Behling *et al.*, 1980). Addison and Berry (1981) pretreated mice with zymosan before injecting them with LPS. The serum of these mice could protect against irradiation if they were fractionated and injected into normal mice 24 hr before irradiation with 650 R.

The post-BCG-endotoxin serum also contains the so-called tumor-necrotizing factor (TNF), described first by Carswell *et al.* (1975) and investigated in detail by researchers at the Sloan Kettering Institute (Hoffmann *et al.*, 1976; Green *et al.*, 1976) followed by other groups (Männel *et al.*, 1980; Parant *et al.*, 1980). We used such post-BCG-endotoxin sera to enhance resistance of mice to subsequent challenge with viable TA3-Ha transplantable tumor and observed a highly significant protection (Butler and Nowotny, 1976, 1979; Butler *et al.*, 1978).

The immune adjuvant effect of endotoxin may contribute to protection against the damage caused by irradiation, such as immunosuppression (Behling and Nowotny, 1978). Immune adjuvancy may also enhance the immune response to transplantable tumor, both at humoral and cellular levels. Therefore, we wished to see whether postendotoxin and post-BCG-endotoxin sera have mediators with immune adjuvant effects. In collaboration with H. Friedman and his associates, we used *in vitro* immunization of mouse spleen cells with SRBC as immunogen in the presence or absence of post-endotoxin sera. We could show that good enhancement of the immune response (measured by the Jerne plaque assay) was evident if postendotoxin, and particularly if post-BCG-endotoxin sera were added to the *in vitro system* (Butler *et al.*, 1979, 1980; Friedman *et al.*, 1980). We could also identify macrophages as the cellular source of the adjuvant mediator released into the post-BCG-endotoxin sera of mice (Butler *et al.*, 1979). More details of this work are summarized in the chapters by Friedman *et al.* (Chapter 14) and Butler (Chapter 25).

In summary, in the last decade the major emphasis regarding the biological effects of endotoxin has been on its action on white blood cells. We have mentioned only a few of these here, omitting such important cells as thrombocytes or others less intensively studied. What seems to be evident is that some of the beneficial effects of endotoxin involve direct interaction with cells of the lymphatic system resulting in the generation of humoral factors, which can trigger further cellular events, entire chain reactions, leading to the end result which can be observed as enhanced host defenses.

III. Isolation of Active Endotoxin Preparations

1. From Cell-Free Culture Media

The very first studies on the nature of the pyrogenic substances revealed that sterilization by filtration, which produced a cell-free culture supernatant, did not remove pyrogenic materials (Centanni, 1894; Muller, 1911; Hort and Penfold, 1911; Samelson, 1912; Jona, 1916; Seibert, 1925). These very early results indicated that spontaneous release of heat-stable and filterable endotoxin occurs. It was assumed earlier that dead and disintegrating bacteria are the sole source of these pyrogenic materials, although more recent findings clearly show that some gram-negative cells can shed pyrogenic endotoxins during apparently normal growth and such release will not necessarily lead to subsequent death of the microorganisms (Russell, 1976; Nowotny, 1983).

Accordingly, cell-free supernatants of some cultures were used successfully for the isolation of endotoxins. One of the most important findings relevant to this was reported by Shear (1941), who reported in numerous papers that *Serratia marcescens* culture filtrate contained a "polysaccharide" that induced hemorrhagic necrosis of a few experimental tumors (Shear and Turner, 1943; Shear *et al.*, 1943; O'Malley *et al.*, 1963). It was established that this substance was also pyrogenic, and contained spontaneously released endotoxin. Ikawa *et al.* (1952) studied the tumor hemorrhage-causing component in *E. coli* culture supernatants and their chemical analytical data reveal great similarity to endotoxin extracted directly from the same bacterial cells. Work *et al.* (1966; Bishop and Work, 1965) conducted detailed studies not only on the chemical composition of a spontaneously released component found in *E. coli* culture supernatant but also measured a number of characteristic endotoxic activities of these and studied the growth conditions that facilitated their release. Marsh and Crutchley (1967) named this *free endotoxin* and elaborated fractionation procedures that showed that such spontaneous release will produce a mixture of various cell wall products.

Some of the studies cited above reported that the amount of endotoxin present in the culture supernatant depends upon the composition of the medium. Lysine-starvation, for example, can induce considerable shedding in several bacterial species. The best yield, as expected, is obtained from culture filtrates that are allowed to autolyse. Degradation of the cell wall may lead to the release of all endotoxins from it, and in all likelihood the researchers who used cell-free culture supernatants as a starting material harvested significantly more endotoxin deriving from cell decay than shedding from live cells. It is necessary to emphasize here that the heterogeneity of autolyzed cell supernatants will also increase with increased yield of endotoxins, since not only endotoxin will be released in larger quantities during prolonged autolysis, but everything else as well. This means that an active endotoxin preparation obtained by such procedure will be highly contaminated with nonendotoxin cell products. Therefore, to assign the observed biological activities (whatever is being measured) to endotoxin alone in such preparations may not be correct. As will be seen later, deliberate efforts to dislodge endotoxin from the cell wall by heat or by chemical means are prone to the same unwanted consequences.

2. Release by Heating

The first isolation procedure applying elevated temperature was described by Centanni (1894). He obtained such substances (called *pyrotoxina bacteria*) by heating the mixture to 100°C and filtering it through porous ceramic filters. He precipitated the filtrate with alcohol and dried the preparation *in vacuo*. This toxin produced strong leukotaxis when injected subcutaneously. Higher concentrations led to edematous swelling and local necrosis of the tissues.

Heat-enhanced liberation of endotoxin was applied by Roberts (1949, 1966), who dispersed bacterial cells in water and heated the suspension to 80°C for 30 min. The supernatant contained endotoxin mixed with a high percentage of proteins and other components released by the bacteria. Purification of the crude preparation was attempted by repeated precipitation with 75% ethanol. This simple procedure was tested for efficiency in extracting several gram-negative strains (Nowotny *et al.*, 1963). Most of the preparations obtained were highly contaminated but very potent endotoxins.

3. Release and Purification by Enzymes

Besides the poorly identified enzymes involved in autolysis, a number of well-characterized hydrolytic enzymes were used to liberate endotoxin from bacterial cell walls. Douglas and Fleming (1921) digested

acetone-killed microorganisms with trypsin at pH 8.5 for 5 days at 32°
C. Shwartzman (1928) used the same method and isolated a substance
from the cell-free supernatant that induced hemorrhagic necrosis in the
skin of rabbits when the sensitizing i.d. injection was followed by i.v. injection of the same filtrate. Raistrick and Topley (1934) further refined
this procedure, and carried out a few steps of purification. The homogeneity of the preparation was not tested. Chemical analyses indicated
that the material obtained is very similar to the one obtained by Boivin
et al. (1933) by trichloroacetic acid (TCA) extraction. Freeman and Anderson (1941) used tryptic digestion to free endotoxin from *Salmonella
typhi*. The preparation was subjected to partial hydrolysis with 0.1 N
acetic acid, which yielded 50–60% polysaccharides, 20–30% peptides
and soluble N-containing split products, and a few percent lipids. In
our laboratories, Lehrer (1971) attempted to degrade TCA-extracted
endotoxin with a number of commercially available hydrolytic enzymes.
None of the enzymes tested diminished the measured biological activities; on the contrary, proteolytic enzymes reduced the amino acid content of the TCA-extracted endotoxin, and enhanced its toxicity. The
results were more important when whole bacteria were exposed to selected hydrolytic enzymes, applied in a certain sequence. The supernatant of the digested mixture contained the most active endotoxin
preparation ever isolated by us. This material was purified by ethanol
precipitation, and by column chromatography on ion-exchange resins.
Its amino acid content, in spite of the use of proteolytic enzymes, was
27% (Lehrer and Nowotny, 1972).

4. Solubilization with Detergent

Since several detergents were used to dissolve tissues or isolated
cell walls, we applied this approach to elute endotoxin from gram-negative cell walls (Nowotny *et al.*, 1963). Cetyltrimethylammonium bromide solution in water (0.5%) extracted endotoxin from three strains,
E. coli K12, *S. typhi* O901, and *Serratia marcescens* O8. Precipitation with
alcohol removed the detergents and yielded a protein-rich, highly toxic
preparation. An added advantage of the procedure was that this deter-
gent precipitated nucleic acids, and thus the extract did not contain detectable amounts of UV-positive contaminants.

Braun and Rehn (1969) used hot 4% sodium laurylsulfate to dissolve bacterial cell walls. Wu and Heath (1973) applied this method to
extract an LPS–protein complex from *E. coli*. All these results indicate
that detergents can serve as very efficient and very mild solvents to obtain fully active and most probably structurally unchanged starting materials.

5. Acids

The most effective method using acids was developed by Boivin *et al.* (1933a), who used 0.5 N TCA at cold room temperature to elute endotoxin in 3 to 4 hr. The preparation could be purified by dialysis and precipitation with alcohol. This procedure and the chemical and biological studies of the endotoxin obtained by it, signaled the beginning of modern research in this field. The advantages of the method are simplicity, good immunogenicity, and low or no nucleic acid content of the preparations. Since the extraction is carried out at low temperature, it is unlikely that the macromolecular structure of the endotoxin is irreversibly altered. It is important to note that the TCA-extracted endotoxin contains a considerable percentage of amino acid (5–10%), some of which appears to be covalently bound to the lipopolysaccharide (Nowotny, 1961, 1963, 1969, 1971a,b; Wober and Alaupovic, 1971) although Webster *et al.* (1955) could reduce it below 0.1% by NaCl saturation and alcohol precipitation. Until recently, this peptide was considered to be biologically inert, but the results of Sultzer and Goodman (1976), Morrison *et al.*(1976), and others showed that it has mitogenic activity on mouse spleen cells. This will be discussed in more detail below as well as in Chapter 11 by Sultzer.

Several other acids were used to isolate endotoxin, but almost all of these caused simultaneous degradation of the macromolecule. The consequences of this are clearly manifested by rapidly reduced toxicity. Freeman and Anderson (1941) used 0.1 N acetic acid at 100°C to isolate a polysaccharide-rich preparation from whole bacterial cells that was nontoxic, but active serologically. Recent findings in our laboratories (Chang *et al.*, 1974; Nowotny *et al.*, 1975) renewed interest in this preparation, and Urbaschek (1980) reported that it could elicit stimulation of colony formation by bone marrow cells and protect against lethal irradiation.

6. Organic Compounds

The largest group of chemicals used for the extraction of endotoxins belongs to organic chemistry. Besides the above-mentioned detergents, TCA, and acetic acid, many others were tested and the most suitable ones are discussed below in chronological order.

a. The use of diethylene glycol as a solvent for endotoxic somatic antigens was introduced by Morgan (1937) and improved later by Morgan and Partridge (1940). It had to be applied to dry cells, since, in the presence of water, unwanted proteins and nucleic acids were also extracted. Further purification of the extract was obtained by dialysis and the usual precipitation with alcohol or acetone. The applicability of this procedure is limited, as some gram-negative bacteria do not yield diethylene glycol-soluble endotoxin (Davies, 1960).

b. Phenol is probably the most effective solvent for endotoxin extraction. Its use goes back to the work of Miles and Pirie (1939), who used 2% phenol in water and left the bacteria in the solution at 0°C for a few days. The supernatant contained a barely degraded somatic antigen that was further purified using ammonium sulfate precipitation and centrifugatin at 14,000 rpm. The chemical properties of the antigen were studied and these experiments showed that it consists of phospholipids and proteins. The presence of polysaccharides was assumed but not verified by analysis.

Ninety percent phenol used on dried bacteria was introduced by Palmer and Gerlough, who published in 1940 "a simple method for preparing antigenic substances from the typhoid bacillus." These authors found that 90% phenol dissolves bacterial proteins while it renders the polysaccharides soluble in water. The polysaccharides were precipitated with alcohol. Morgan and Partridge (1941) adapted the phenol–water method of Palmer and Gerlough to isolate antigenic material from "smooth" and "rough" strains of *B. dysenteriae* Shiga and found that while the smooth strains yielded a good preparation, the rough strains did not.

Forty-five percent phenol in water is the most widely applicable extraction procedure. It was developed by Westphal *et al.* (1952), who used 45% phenol at 68°C. At this temperature, 45% phenol is soluble in water and the two form a homogeneous system. The bacterial cells are extracted at that temperature for 10–15 min, then cooled down and centrifuged in a refrigerated centrifuge. Here the phenol and the water phases will separate. The proteins will be dissolved in the phenol phase and the lipopolysaccharides, polysaccharides, nucleic acids, and a number of other water-soluble bacterial components will be in the water phase. Westphal and Lüderitz (1954) separated the endotoxic LPS by precipitation in alcohol containing 2% $MgCl_2$, dissolved the precipitate in water, and sedimented the LPS by centrifugation at $110,000g$ for 2 hr. The LPS formed a gellike clear sediment.

c. Pyridine in various concentrations was also used for the extraction. Goebel *et al.* (1945) successfully extracted O antigens from *Shigella flexneri* with 50% pyridine in water, but could not use the same procedure for *Shigella sonnei.*

We used an equimolar mixture of pyridine and formic acid to dissolve the cell walls of three different gram-negative bacteria (Nowotny *et al.*, 1963). This solvent was prepared by mixing the correct quantities of anhydrous pyridine and 90% formic acid. The lyophilized bacteria were suspended in it at boiling water bath temperature and stirred for 30 min. The cooled suspension was diluted with 10 vol of cold distilled water, centrifuged, the supernatant dialyzed, precipitated with methanol, and finally lyophilized.

d. Urea (7 M), applied to the 50% pyridine-extracted residue of *Shigella sonnei*, released O antigens in good yield, according to Baker *et al.* (1949). The extraction was carried out at cold room temperature for 24 hr. Further purification of the antigen as well as studies of its serological properties were described.

e. Diethyl ether-saturated water was introduced by Ribi et al. (1959) as a mild method that avoided potential degradation of the macromolecules by strong chemicals. Ribi and associates used this procedure to dislodge endotoxic LPS from *Salmonella enteritidis*, and the purified products showed high activity in a number of characteristic endotoxicity assays. Several publications from this laboratory followed, which reported the lack of correlation between biological activities and nitrogen or fatty acid contents (Ribi *et al.*, 1961a,b). These results will be discussed later.

f. Butanol saturated with water was introduced by Maddy (1964) to isolate erythrocyte membrane components. Leive and Morrison (1972) used it to isolate endotoxin.

g. The usefulness of EDTA during attempts to solubilize cell walls has been studied by Repaske (1958) among others. It was found that EDTA facilitates lysozyme-induced degradation of bacterial cell walls. Colobert (1958) assumed that EDTA has a detergentlike action in this process. Gray and Wilkinson (1965) applied EDTA for the isolation of LPS-like substances. They tested the efficiency of this chemical for endotoxin extraction in strains that are resistant to EDTA and in strains that are readily killed by it. They found that EDTA-sensitive *Alcaligenes faecalis* and *Pseudomonas aeruginosa* rapidly released cell wall components when exposed to 0.0068 M EDTA dissolved in 0.05 M borate buffer, pH 9.2. Later, biological tests proved that EDTA extracted endotoxin from these bacteria. Leive (1965) reported that 5×10^{-3} final concentration of EDTA releases 35–50% of the total endotoxin content of *E. coli* cells in a 2-min exposure at 37°C.

h. Dimethylsulfoxide (Me_2SO) was used for the isolation of bacterial antigens first by Ormsbee *et al.* (1962). Adams (1967) found that Me_2SO, an excellent solvent for various naturally occurring polysaccharides, is also eminently applicable for LPS isolation. Me_2SO at 60°C extracted about five times more material than 45% phenol when isolated cell walls were used as starting material.

i. Extraction of endotoxin from rough mutants was much more difficult than from the parent smooth strains (Morgan and Partridge, 1941). TCA could not be used at all, and, as we found (Kasai and Nowotny, 1967), only a small portion of the endotoxic glycolipid was partitioned into the water phase after 45% phenol extraction. The majority of the glycolipid appeared in the middle phase between phenol and water (Nowotny *et al.*, 1967). Considering the greatly reduced polarity of the mutant endotoxins (lack of long polysaccharide chains), this is to be expected. Galanos *et al.* (1969) modified the 45% phenol procedure to obtain mutant endotoxins. The new method used 90% phenol/chloroform/petroleum ether in a 2:5:8 ratio.

Further simplification of the extraction procedure was achieved. Kasai *et al.* (1970) eliminated the use of phenol and replaced it with chloroform/methanol 4:1. We extracted lyophilized bacteria in a Soxhlet apparatus with this mixture and precipitated the endotoxic glycolipid product by adjusting the chloroform/methanol ratio to 1:2. Repeating the dissolution and precipitation several times, a fully active endotoxin could be obtained from the Re mutants of *S. minnesota* R595 (Chen *et al.*, 1973) and from similar heptoseless rough mutants of *E. coli* and *S. typhimurium* (Ng *et al.*, 1974, 1976a). Investigating the homogeneity of this preparation, we found that the endotoxic glycolipid was separated by thin-layer chromatography (TLC) into four or five bands. All major bands were isolated by a preparative TLC procedure and both the molar ratios and the biological activities of the fractions were determined (Chen and Nowotny, 1974; Chen *et al.*, 1975). It could be shown that all bands represented fully active endotoxic compounds. These were the first descriptions of the isolation of chromatographically homogeneous endotoxins.

In summarizing the survey of the extraction procedures, a few points should be emphasized. The first is that no procedure can be applied for the extraction of endotoxin from all gram-negative bacteria. We reached this conclusion earlier (Nowotny *et al.*, 1963) and many laboratories have confirmed this. This means that the optimal method and the optimal experimental conditions have to be determined for almost every strain.

The second and even more important point leads us to a discussion of the heterogeneity of such extracts. Knowing the complexities of gram-negative bacterial cell walls (Salton, 1971), no one can expect that any one of the above extraction methods will yield anything but a mixture of cell constituents. It only follows that the composition of the mixture and the proportions of the components in it will vary to a great extent, depending not only upon the chosen extraction method but also upon the bacterial strain selected for study.

IV. Heterogeneity

Martin wrote in 1934: "There is a voluminous literature dealing with attempts to isolate . . . endotoxins of bacteria . . . . But the methods hitherto employed have seldom been such as to allow the isolation of the toxic material in a state approaching chemical purity." Almost half a century later this statement is still applicable.

First the heterogeneity of the "lipid A" precipitates was demonstrated (Nowotny, 1963; Nowotny et al., 1963; Burton and Carter, 1964). The most dramatic heterogeneity was shown by a TLC system elaborated by Kasai (1966). By this procedure, over 40 components were seen in acid-hydrolyzed lipid A samples (Nowotny, 1969; Chang and Nowotny, 1975). The heterogeneity of the lipid A preparations has been confirmed by numerous independent laboratories (Konno, 1974, 1978; Buttke and Ingram, 1975). The homogeneity of intact LPS preparations was investigated by ion-exchange chromatography (Nowotny, 1966; Nowotny et al., 1966). It was found that all preparations, including phenol–water-extracted and purified LPS, are heterogeneous. As already stated above, when we separated mutant endotoxins by TLC and determined the biological activities of the visible bands, we saw that at least four endotoxic glycolipids are present in these preparations also (Kasai and Nowotny, 1967; Nowotny et al., 1967; Chen et al., 1973; Ng et al., 1974, 1976a).

We have discussed the problem of heterogeneity repeatedly (Nowotny, 1966, 1971a,b; Nowotny et al., 1966) but it remained largely ignored by the scientific community. Later it was reemphasized by Jann et al., (1975) and confirmed by many other laboratories (Russell and Johnson, 1975; Chester and Meadow, 1975; Morrison and Leive, 1975; Garfield et al., 1976; Kabi, 1976; Leive, 1977; Jarrell and Kropinski, 1977; DiRienzo and MacLeod, 1978; Van der Zwan et al., 1978; Palva and Makela, 1980; DeDur et al, 1980; Goldman and Leive, 1980). Today it is rather firmly established that all procedures of isolation will extract not one component from the highly complex bacterial cell wall, but several. This was to be expected. It also became clear that the conventional steps of purification are insufficient. Purification must include the most powerful procedures at our disposal, with the best possible resolution, to isolate endotoxins in chemically pure form. We are only at the beginning of such efforts.

This heterogeneity is significant for several reasons. First, it makes results of chemical analyses but particularly chemical structural proposals, tenuous. Second, since we inject or apply mixtures of components, we cannot determine which one of them causes the observed biological reaction. This can be decided only after isolating the components and by repeating the biological assays with pure components. Therefore, the search for the active site in endotoxins should begin with the search for active "contaminants" and only after we eliminate this possibility, can we initiate efforts to identify the active sites of endotoxins.

V. The Chemical Structure of Endotoxin

In the following very brief survey we shall first refer to the major chemical studies of the various moieties of the endotoxic macromolecule that led to our present understanding of its structure. Second, we shall discuss the still-existing ambiguities and missing data, which are very much needed for the complete clarification of the chemical structure.

The major, and probably the best known, part of the LPS is the polysaccharide moiety. The most investigated bacterial strains were the Salmonellae, and the achievements of Otto Lüderitz and Otto Westphal, in collaboration with Fritz Kaufmann, Anne Marie Staub, and several other collegues, revealed the existence of unusual carbohydrates (Fromme et al., 1957; Westphal and Lüderitz, 1960; Lüderitz, 1970) and established the chemical basis for serological specificities (Westphal, 1960; Staub and Westphal, 1964; Lüderitz et al., 1968). While the major emphasis in these studies was on the LPS of various Salmonellae, others who followed their methodologies selected other bacterial strains. Due to these studies, we know today the exact structure of the polysaccharide moieties of a few LPS preparations (Kenne et al., 1977; Kontrohr and Kocsis, 1978; Hoffman et al., 1980).

The search for a toxic component in the endotoxin structure led the early investigators into studies of the lipid moiety (Boivin et al., 1933b). Systematic studies were initiated by Westphal and Lüderitz in 1954. Chemical analyses of the lipid A preparation were carried out and the first findings led to the conclusion that it is a nitrogen-containing phospholipid. Next the long-chain carboxylic acid content of the LPS was analyzed by circular, reversed-phase paper chromatography. A great similarity in the fatty acid composition of eleven gram-negative LPS preparations was established (Nowotny et al., 1958).

On the other hand, the lipid A preparations themselves were found to be heterogeneous first by simple two-dimensional paper chromatography and later by silicic-acid-impregnated paper and SiO_2 column chromatographic methods (Nowotny 1961, 1963; Nowotny et al., 1963). For the first analyses of the lipid A components a fraction was selected, which was soluble both in pyridine and in anhydrous methanol. Later this preparation was further fractionated on a SiO_2 column and one of the major components which showed some pyrogenicity in rabbits was collected from several runs and subjected to chemical analyses. Results of this work, which was started in the Wander Research Institute, Freiburg, Germany, and continued in the City of Hope Medical Center, Duarte, California, identified a new class of lipids where the long-chain carboxylic acids are either O-acyl- or amide-bound to phosphorylated glycosamines (Nowotny, 1959, 1961, 1963, 1971b; Nowotny, et al., 1963; Closse, 1960). Figures 1a and 1b show the first proposals for the major structural features of these lipids. Burton and Carter (1964) confirmed by their detailed studies the existence of such subunits, but they left open the possibility of glycosidic linkages between the glucosamines (Fig. 1c).

Further details of the structure were elaborated soon thereafter. Gmeiner et al. (1969, 1971) studied *Salmonella minnesota* R595 rough mutant endotoxin which is reported to consist of a complete lipid moiety without the polysaccharides (Lüderitz et al., 1966). Gmeiner et al. used hydrazinolysis of the R595 mutant endotoxin. By this method, they found that the molecule consists of 2 glycosamines which are in β 1→6 glycosidic linkage and they also assumed from their findings that the three KDO molecules which complete the structure are in C3 position attached to the nonreducing glycosamine as an oligosaccharide side chain. The position of the phosphate groups on the glucosamines were Cl and C4 (Fig. 1d). The position of the fatty acids was studied by Rietschel et al. (1972) as shown in Fig. 1e. Several other mutant endotoxins as well as wild-type LPS preparations were also studied by Hase and Rietschel (1976). This time alkaline hydrolysis was used to cleave the O-acyl linkages and hydrazinolysis to remove the amide linked carboxylic acids. The authors concluded that the β 1→6 glycosidic linkages of Gmeiner et al. (1969) occur in most, if not all, lipid moieties. Similar conclusions were reached by other laboratories (Adams and Singh, 1969; Rosner et al., 1979a), thus confirming the basic features of this new class of lipids where aminosugars are ester- or amide-linked with long-chain carboxylic acids. The exact position of the phosphoric acid groups, as well as of the carboxylic acids, remains somewhat uncertain. After continued and detailed studies Hase and Rietschel (1977) described more complicated lipid structures where P-diesters are also indicated in the lipid moiety of *Chromobacterium violaceum* (Fig. 1f). It is clear today, that the lipid moieties of gram-negative LPS preparations are not all identical chemically or biologically, although great similarities exist between them (Ng et al. 1974, 1976a; Rietschel et al., 1977;

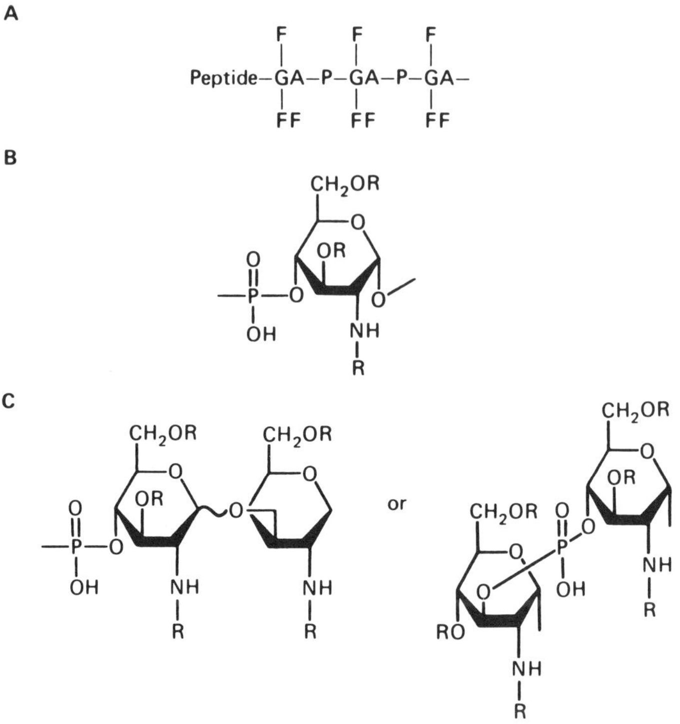

A

$$\text{Peptide}-\overset{\overset{\displaystyle F}{|}}{\underset{\underset{\displaystyle FF}{|}}{GA}}-P-\overset{\overset{\displaystyle F}{|}}{\underset{\underset{\displaystyle FF}{|}}{GA}}-P-\overset{\overset{\displaystyle F}{|}}{\underset{\underset{\displaystyle FF}{|}}{GA}}-$$

B

C

Figure 1. Structural studies of the lipid moiety in LPS. Abbreviations: F and R = fatty acid radicals {-Co(CH$_2$)n-CH$_3$}; Rm = 3-OH C$_{14}$ acid radical; KDO = 2-keto-3-deoxyoctonic acid; PS = polysaccharide. Some of the structures were redrawn from the

Lüderitz *et al.*, 1978, Westphal *et al.*, 1980), as will be discussed later in more detail.

One neglected part of the LPS structure is the covalently bound peptides. So far every LPS or mutant endotoxin preparation we have analyzed contained at least 0.5% amino acids. It is known that the tri-chloroacetic-acid-extracted preparations may contain as much as 10% peptides; our enzyme-extracted and -purified endotoxin from *Serratia marcescens* contained 27% amino acids, and it is curious to note that this was the most potent endotoxin in both toxic and beneficial effects we have ever tested (Lehrer and Nowotny, 1972). A considerable percent-

D

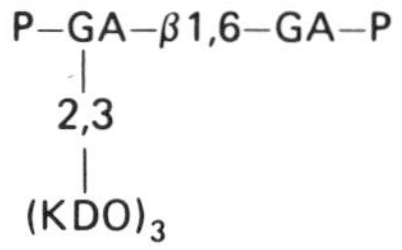

$$P-GA-\beta 1,6-GA-P$$
$$\mid$$
$$2,3$$
$$\mid$$
$$(KDO)_3$$

E

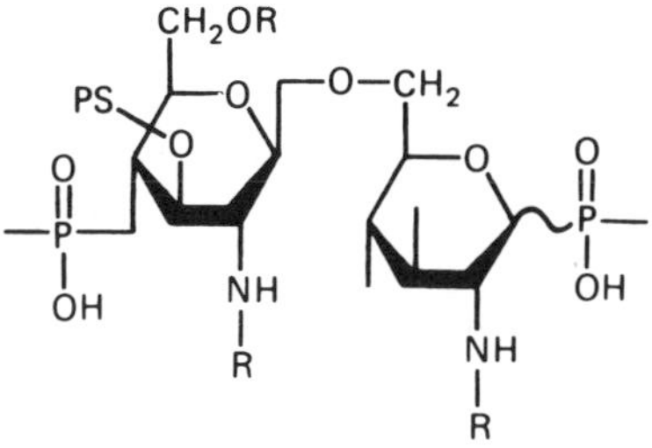

F

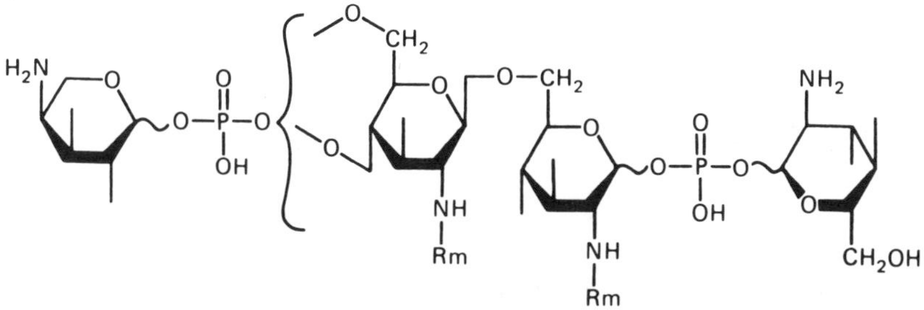

original references in order to facilitate comparisons: (A) Nowotny, 1961; (B) Nowotny *et al.*, 1963; (C) Burton and Carter, 1964; (D) Gmeiner *et al.*, 1969; (E) Hase and Rietschel, 1976; (F) Hase and Rietschel, 1977.

age of the amino acids present in various LPS or endotoxin prepara- tions appears to be covalently bound. Peptide containing phosphorylat- ed D-glucosamine was one of the components that we isolated in pure form from the partial hydrolysate of a lipid A fraction and the amino acids of this compound could be liberated only by extended hydrolysis with 2 N HCl. (Nowotny, 1961, 1963, 1971b). More detailed studies of Wober and Alaupovic (1971) dealt with the protein moiety of LPS. The discovery of the "endotoxin protein," a mitogenic component of some endotoxin preparations, renewed interest in this constituent (Sultzer and Goodman, 1976, Morrison *et al.*, 1976). Creech *et al.* (1964) found that a polysaccharide-peptide complex of *Serratia marcescens* free of lipid A was highly effective in causing complete regression of sarcoma 37 in mice. Ribi and associates (1979) claimed recently that the peptides of endotoxin preparations are essential for their antitumor effect. In spite of these important observations, our present knowledge of the structure as well as of the biological role of the bound peptides is highly insufficient.

These summarize what we know about the major features of the LPS structure. The best known is the composition of the polysaccharide moiety, great progress was made in the clarification of the lipid moiety's structure. Least known are the locations of some amino compounds (amino acids, aliphatic amines) in the macromolecule and the type of linkages which bind them to the complex structure. It is evident that in spite of significant advances, we still do not know the entire structure. We do not know whether we accounted for all components (see "missing percentage," Nowotny, 1969, 1971a,b) and we have just started to realize that side chains or other substituents in specific positions not only provide highly characteristic structural features to some LPS molecules, but may also be the determinants of unique biological activities (Nowotny, 1971a; Ng *et al.*, 1976a; Westphal *et al.*, 1980).

Most importantly we know very little if anything about the three-dimensional arrangement of the moieties. It is entirely possible that only a certain spatial coordination of the functional groups renders the macromolecule toxic while another orientation of the same components will result in a change in the biological potency. We referred to this possibility a long time ago as "toxic conformation" (Tripodi and Nowotny, 1966). Extending the same theoretical possibility for the elicitation of beneficial effects one could talk more generally about "active (or inactive) conformations." Whether these exist or not is unknown. We can test this hypothesis only after we have elucidated the three dimensional structure. Such studies, quite clearly, will require the application of all available physical and physicochemical analytical procedures, but first of all, a full understanding of the chemical composition of the structure.

VI. Relationship of Structure to Function

A. One Site for All Activities

An overwhelming majority of publications assign both toxic and beneficial effects to the lipid moiety (for reviews see Westphal, 1960; Galanos *et al.*, 1977; Westphal *et al.*, 1958, 1977, 1980). Before we discuss the results and their interpretations, we should survey the lipid A preparations used in most of these experiments.

The first lipid A preparation, as described in the classical work of Westphal and Lüderitz (1954), was obtained from smooth *Salmonella* lipopolysaccharides, which were extracted from the bacteria by the very efficient phenol–water procedure. This LPS preparation was purified and hydrolyzed in 1.0 N HCl for 30 min at boiling water bath temperature. The precipitate obtained was called *lipid A*.

The most frequently tested lipid A preparation was obtained from *S. minnesota* R345 (an Rb mutant) endotoxin obtained by phenol–chloroform–petroleum ether extraction. This endotoxin was hydrolyzed with 1% acetic acid for 1 or 2 hr and the precipitate obtained was complexed with BSA. It was reported by several laboratories that the above lipid A–BSA complex is as active as the parent endotoxin. This mutant R345 has been described by Lüderitz *et al.* (1966) and by Schmidt *et al.* (1969) as a "leaky" mutant, wherein the biosynthesis of the O-polysaccharide chain is not completely blocked and therefore every culture contains a significant percentage of cells with a complete polysaccharide moiety.

Another preparation, the Re mutant endotoxin of *S. minnesota* considered to be composed of lipid moiety and KDO only, was frequently taken as a model of lipid A and tested in a number of biological assays. This preparation, particularly after purification by chromatography, bears probably the greatest structural resemblance to the lipid moiety of the intact, smooth endotoxic LPS.

The next question is whether the above three classes of preparations are identical in chemical composition and whether they are true representatives of the native lipid moiety in endotoxins.

It has been properly documented by independent laboratories that acid-hydrolyzed lipid A precipitates (more recently called *free lipid A*) are highly heterogeneous mixtures. TLC easily reveals many components and also provides a fingerprintlike characteristic pattern of the various lipid A precipitates. For example, lipid A from phenol–water-extracted endotoxin differs from TCA-soluble lipid A in this system, although both were obtained from the same bacterial strain and hydrolyzed under identical conditions. The same hydrolysis applied to a rough mutant endotoxin gives again a different TLC pattern. Hydrolysis with 1% acetic acid at 100°C for 2 hr also gave a heterogeneous mixture of various lipid breakdown products (Chang and Nowotny, 1975). These indicate that the various "free lipid A" preparations are different mixtures of various molecular entities.

It also becomes questionable whether the lipid moieties of the various intact endotoxins are identical. If the lipid moieties in all three

types of the above starting materials were identical, the use of the same
hydrolytic conditions should result in lipid A precipitates that are at
least quite similar, if not identical. This was not the case. From this observation alone, one has to assume that the lipid moieties of various
gram-negative endotoxins are not identical. Further support for this assumption came from comparative chemical and biological studies of
five Re mutant endotoxic glycolipids (Ng *et al.*, 1976a), which were analyzed by preparative TLC, and the molar ratios of the various components determined by the method of Chen *et al.* (1975). We found that
the above chromatographically purified lipids of the Re mutants are
both chemically and biologically different. Some of these, such as *E. coli*
F515, were inert in enhancing nonspecific resistance to *S. typhi* infection but were toxic and could elevate resistance to tumor challenge. *S.
minnesota* R595 Re mutant endotoxin was very potent in enhancing nonspecific resistance to *S. typhi* infection but could not protect significantly against the tumor and was a relatively weak inducer of CSF (Ng *et al.*,
1974, 1976a; Nowotny, 1977). *S. minnesota* R345, the leaky mutant endotoxin, appears to be potent in all assays. We interpreted the findings
by assuming that the presence or absence of certain hitherto
unidentified functional groups within or attached to the lipid moiety
may be the real determinants of the beneficial activities (Nowotny,
1969, 1971a,b; Ng *et al.*, 1976a).

   The acceptance of the lipid moiety's diversity met considerable resistance initially, but soon thereafter confirmation came from Rietschel
et al. (1977) and Lüderitz *et al.* (1978), who showed by elegant experiments that considerable chemical as well as biological differences may
exist in the lipid moiety of gram-negative endotoxins.

   The major points to be emphasized here are: (1) As has been
shown by all laboratories who subjected acid-hydrolyzed "free lipid A"
precipitates to chromatography, these preparations are quite heterogeneous. The number and the chemical nature of the components they
contain depend upon the starting material used and the hydrolytic conditions chosen. (2) The lipid moieties of various smooth or rough endotoxins are different, although great structural similarities may exist.
(3) It only follows that the "free lipid A" and the lipid moiety in the intact endotoxin are not identical chemically and therefore they should
differ in their biological activities as well. This is now confirmed. (4)
Accordingly, we cannot deduce generalized conclusions regarding the
role of lipid moieties in biological reactions since the results obtained
may apply only to the preparation tested and only to the activity measured.

   After having said that, the most important question to be discussed
is whether all biological activities can be elicited by the lipid moiety.

It is true that there are over 40 inert components in most "free lipid A" preparations, and they form approximately 90% of it. The remaining 10% contains fractions that are as active as the parent endotoxin. The molar ratio of one of these fractions (RESI #2) is somewhat similar to the molar ratio of Re mutant endotoxin (Chang and Nowotny, 1975). If this fraction is a hydrolytic breakdown product of the lipid moiety, and elicits all the toxic *and* all the beneficial effects, the assumption that this moiety is the sole active site may hold true.

Furthermore, in spite of the fact that chemically detoxified preparations (for details see below) still have almost unchanged beneficial effects, both toxic and beneficial reactions may be elicited by one active site in the intact endotoxin. The fact that it was possible to reduce or eliminate toxic properties while maintaining beneficial effects may be due to differences in structural requirements to elicit toxic and beneficial effects. As will be discussed later, it may very well be that an almost intact structure is required for the elicitation of grave toxic effects, while partially destroyed remnants of it may be still capable of initiating reactions leading to activated host defenses.

It may be clearer now to see what remains to be established before we can conclude that there is only one active site in the endotoxin structure. First, it has to be proven that all *active* fractions in a beneficial preparation were parts of the lipid moiety of an intact endotoxin. Second, the possibility that nonlipid components of the endotoxin preparations are active in these assays must be unambiguously excluded. What is needed are experiments that do not employ mixtures of compounds but rather chemically pure entities for both chemical and biological analyses. The major thrust of our research has been and still is to obtain such pure and well-defined compounds.

B. The Toxic Site

While we question the validity of the "one site for all activities" theory, there is little or no doubt regarding the location of the toxic site.

The toxic effects seem to be induced by the lipid moiety. The origin of this claim can be traced back to the findings of Boivin *et al.* (1933b), who hydrolyzed their TCA-extracted endotoxin with 0.2 N acetic acid at boiling water bath temperature. This partial hydrolysis precipitated a lipid, called *fraction A* and left in the supernatant a polysaccharide, called *fraction B*. The lipid fraction A was toxic when suspended in alkaline solution. The polysaccharide fraction B was neither toxic nor immunogenic, but reacted with O antiserum.

Several other hydrolytic procedures were applied to various endotoxin preparations to cleave it into smaller components. Miles and Pirie (1939) and Binkley *et al.* (1945) used 0.1 N HCl, Morgan and Partridge (1940) and Freeman (1943) used 0.1 N acetic acid, Bendich and Chargaff (1946) used glacial acetic acid, and Westphal and Lüderitz (1954) used 1.0 N HCl to break down the large complex into more manageable small components. Pursuing the fundamental observations of Boivin *et al.* (1933b), some of the above workers attempted to identify the subunit in the endotoxin complex that is the carrier of toxicity. Goebel and associates (Binkley *et al.*, 1945) isolated endotoxin from *Shigella paradysenteriae,* complexed it with protein and subjected it either to acidic or to alkaline hydrolysis. When acidic hydrolysis was applied, they obtained a toxic protein and a nontoxic polysaccharide. In the case of alkaline hydrolysis, they isolated a toxic polysaccharide and a nontoxic protein. They assumed the existence of a toxic T component that is neither protein nor polysaccharide. The most significant achievements in this field are due to the extensive investigations initiated by Westphal, Lüderitz, and their associates. Westphal and Lüderitz (1954) hydrolyzed their phenol–water-extracted endotoxic LPS with 1.0 N HCl into a lipid precipitate, which they called lipid A, and a partially degraded polysaccharide. They reported that lipid A is soluble in chloroform and some other organic solvents. Similarly to Boivin *et al.* (1933b), they dispersed the lipid A fraction and determined its toxicity by pyrogenicity measurement. It was found that this preparation retained approximately 1/10 of the toxicity of the native LPS, as determined on a weight basis. It was (Westphal *et al.*, 1958) concluded that the lipid portion is most probably the toxic component of the LPS.

Disagreements were expressed by several investigators. Ribi and associates elaborated their arguments in numerous papers, the major points being as follows. First, they determined the lipid (fatty acid) content of a large number of preparations by quantitative analytical chemical procedures and found no correlation between percent lipid content and toxicity (Ribi *et al.*, 1961a). They also subjected endotoxin preparations to mild hydrolysis with acetic acid and found that inactivation of the pyrogenicity precedes the cleavage of fatty acids from the complex. The conclusion was that not fatty acids but other acid-sensitive components are responsible for the elicitation of toxic reactions (Ribi *et al.*, 1961b). They also isolated lipid A according to the method of Westphal and Lüderitz (1954), used the chloroform-soluble components of it and measured the toxicity of these in a number of semi-quantitative biological assays. They found that this preparation had only 0.1 to 1% of the starting activity (Ribbi *et al.*, 1961b) in contrast to the 10% or more reported by Westphal and co-workers. We offered an explanation of this discrepancy. We could show that the chloroform-soluble components of lipid A are inactive but the chloroform-insoluble residue of the lipid A mixture contains an acid-resistant and incompletely degraded structural residue of the LPS, which is fully active in a number of typical endotoxicity assays (Chang and Nowotny, 1975).

Our support regarding the important role of the lipid moiety in the toxicity of endotoxins comes from other series of experiments. One of these was immunochemical detoxification using antibodies to endotoxins. We incubated *in vitro* Boivin-type endotoxin of *Serratia marcescens* with high-titer rabbit O antiserum to the same strain. Complete and highly efficient neutralization of three toxicities (pyrogenicity, Shwartzman skin reactivity, and mouse lethality) could be achieved by the formation of endotoxin–antibody complexes. If the O antiserum was absorbed by chemically detoxified endotoxin, which had its lipid moiety partially destroyed, the remaining serum antibodies were still efficient in neutralizing toxicity. If the same chemically detoxified endotoxin was used as an immunogen, the serum obtained could not completely abolish endotoxin toxicity, since it did not contain immunoglobulins to intact and active lipid moieities. Our interpretation was that the rabbit O antiserum contained antibodies specific to the lipid moiety, and these antibodies were efficient enough in covering and thereby neutralizing the toxic site, the lipid moiety (Nowotny *et al.*, 1965; Radvany *et al.*, 1966). One contradictory finding (Kim and Watson, 1966) and several confirming results were published (Tate *el al.*, 1966; Berczi, 1967; Ralovich and Lang, 1971; Rietschel and Galanos, 1977) indicating that antibodies to the lipid moiety can neutralize toxicity. Furthermore, it was reported that such antibodies can protect experimental animals against infections with a variety of virulent gram-negative bacteria (Braude *et al.*, 1977; Young *et al.*, 1975; McCabe, 1972; Chedid *et al.*, 1968), since it was assumed that the lipid moiety of all gram-negative endotoxins is the same or at least very similar. Accordingly, antibody to the lipid portion of one endotoxin should cross-react and cross-neutralize the endotoxins of serotypically different virulent bacteria. It was also assumed that endotoxin is the most significant agent during infection with virulent bacteria. We showed that a limited cross-reacitivity exists, which was not sufficient to give a wide range of cross-protection (Ng *et al.*, 1974, 1976a). Furthermore, we assumed that the protection observed by using cross-reactive anti-lipid-A antibodies must be restricted to those bacteria where the lipid moiety of the endotoxin was sufficiently exposed (Ng *et al.*, 1976b). Similar negative results were published by Mullan *et al.* (1974), Greisman *et al.* (1978, 1979), and Mattsby-Baltzer and Kaijser (1979), indicating that such cross-protection must indeed be restricted to a few microorganisms.

The most convincing proof for the dominant role of the lipid moiety in the toxicity of endotoxins came from the measurements of biological activities of mutant endotoxins. Lüderitz provided us with endotoxins isolated from rough mutants of *S. minnesota* (Lüderitz *et al.*, 1966). One of these, the R595 strain, is an Re mutant, without hexoses and heptoses. This mutant LPS, called *endotoxic glycolipid*, was fully toxic in all assays (except mouse lethality, which was reduced) and also active in a number of beneficial effects (Tripodi and Nowotny, 1965, 1966; Lüderitz *et al.*, 1966; Kasai and Nowotny, 1967). Identical results were obtained by Kim and Watson (1967).

Experiments using chemical detoxification as an approach led us to the same conclusion. We introduced alterations on some known functional groups of the chemical structure, as did Noll and Braude (1961), and determined the consequences of these on the biological activities of the endotoxin (Nowotny, 1963). We saw that several chemical changes can eliminate toxicity while maintaining some of the beneficial effects (Johnson and Nowotny, 1964). All procedures that detoxified the endotoxins had one feature in common: they all cleaved ester linkages in the lipid moiety (Nowotny, 1964).

More detailed studies of the mechanism of alkaline detoxification were carried out (Tripodi and Nowotny, 1966; Cundy and Nowotny, 1968). Kinetic studies of the detoxification revealed that some reactions considered to be manifestations of toxicity could be eliminated rapidly, such as chick embryo lethality or pyrogenicity, but during the same time of hydrolysis the local Shwartzman reactivity was enhanced. We concluded from these observations that the integrity of different structural subunits is required for the elicitation of the above parameters of endotoxic activities. While intact molecules are required for lethality and pyrogenicity, the Shwartzman lesions can still be elicited by partially degraded ones. At any rate, it is clear that the elicitation of the various endotoxic activities requires different chemical structures, different functional groups or group distributions.

C. Beneficial Effects and Their Probable Sites

In searching for a molecular explanation of the beneficial effects, a number of assumptions may serve as working hypotheses. The first is that two or more components in the heterogeneous preparation carry two or more biological activities. One is toxic, another (or several) elicits beneficial effects. If this is so, one should be able to separate them from each other (for a few references where such separations were successfully carried out see below). The other possibility is that one macromolecule carries all the biological activities but has separate active sites for toxic and for beneficial effects, like the O-antigenic site, which is in the polysaccharide, and the toxic site, which seems to be in the lipid moiety. Finally, we have to assume the possibility that all biological reactions are elicited by a single active site of one macromolecule.

In the following discussion of findings, an attempt is made to support or eliminate the above possibilities using experimental data relevant to the above points and treating them critically, including our own results.

a. Heading the list of beneficial effects of endotoxins is immunogenicity. Immunogenicity is beneficial provided it does not lead to hypersensitivity or to other pathological consequences. Humoral immunity to endotoxin was shown to neutralize several toxic effects *in vitro*, as discussed above. Chapter 4 deals with this aspect in detail.

Once again, the most important discoveries in this field are those of Westphal and Lüderitz, who initiated and completed research in this field. Together with their collaborators and students, they clearly established the chemical structure of the O-antigenic determinants of gram-negative bacteria. Their thorough work resulted in the evolvement of "chemotypes," which grouped gram-negatives according to the presence or absence of certain carbohydrates in the polysaccharide moiety of the endotoxic LPS. It was demonstrated that the chemotype classification accurately agreed with the "serotype" grouping of the Kaufman-White scheme (Lüderitz *et al.*, 1968, Lüderitz, 1970; Westphal, 1960). This work was equally important for carbohydrate chemists, since among the accomplishments we find the first descriptions of new hexoses (3,6-bisdeoxyhexoses) and a number of related compounds. This work is and will remain a classical chapter of immunochemistry.

Besides the O-specific polysaccharides, other parts of the endotoxin complexes can also elicit a humoral immune response. I have already referred to our earlier findings, which can be considered as indirect evidence for the existence of anti-lipid moiety antibodies in hyperimmunized rabbit O antiserum (Radvany *et al.*, 1966). Immunization with lipid A precipitates or with heptoseless Re mutant endotoxins was unsuccessful in several laboratories. The first description of antilipid antibodies can be found in the Ph.D. thesis of Rank (1969), which was followed by a publication (Rank *et al.*, 1969). This author used *S. minnesota* R mutants, heated the bacteria for 5 min at 100°C, and injected the suspension i.p. into mice. Another preparation used was extracted mutant endotoxin mixed with aluminum hydroxide adjuvant. Not all animals responded but some showed a significant titer against the Re

mutant 595. Rb mutant R345 was the most immunogenic in these experimental series. Galanos *et al.* (1971) acid-treated *S. minnesota* R595
bacteria and injected this preparation to induce anti-lipid moiety immunoglobulins. Ng *et al.* (1976b) used chloroform–methanol-extracted and
purified R595 mutant endotoxin to coat autologous erythrocytes of
rabbits. Reinjection of these coated erythrocytes into the rabbits induced the production of highly specific antisera to R595.

The beneficial effects of the various antisera have already been discussed as has the neutralization of toxic properties by them. With regard to the observation that antibody-neutralized endotoxins still can
elicit beneficial effects, we observed (Nowotny *et al.*, 1965; Radvany *et
al.*, 1966) that nontoxic precipitates of endotoxin–antibody complex
were fully active in inducing enhanced nonspecific resistance to virulent
bacterial infections. More recently, Behling in our laboratories found
that a similar complex was also active in providing protection to lethally irradiated mice (Nowotny *et al.*, 1982). The interpretation of these
findings is not simple. It is feasible that the O antiserum of rabbits
does not contain blocking antibodies to the beneficial site or beneficial
contaminants of the endotoxin preparation. It appears to be more likely that the presence of immunoglobulins changes the *in vivo* fate of the
endotoxin. It was observed that opsonizing antibodies facilitate the uptake of chemically detoxified endotoxin by the RES (Golub *et al.*, 1968,
1970). It is most likely that similar consequences alter the fate of antibody-neutralized endotoxins. This may mean that either by faster clearance or through the blockade of toxic sites, the endotoxin cannot react
with those primary targets that could trigger toxic manifestations. It
will bypass these, but not those targets that must be reached in order
to elicit beneficial effects. These latter targets may be subcellular components of the RES, where a rapid intracellular dissociation of the endotoxin–antibody complex may follow their uptake, thus allowing the
endotoxin to act and trigger effects considered to be beneficial. A clear
answer to this problem can be expected only from the use of
monoclonal antibodies directed to certain structural subunits of the
complex endotoxin molecule. Such experiments are now being conducted in our laboratories.

b. It would be beyond the scope of this chapter to discuss under
which condition one should consider mitogenicity harmful or beneficial. If we consider lymphocyte proliferation and the generation of certain lymphokines desirable, as they often are, we should include
mitogenicity in the list of beneficial effects.

After it was discovered that LPS can induce mitosis of splenocytes,
several laboratories investigated the active site of the LPS molecule that

is mitogenic in mice. They used rough mutant endotoxins, assuming that the difference between the smooth parent LPS and the various rough mutant preparations lies only in the polysaccharide content. It was also assumed that all these preparations have an identical lipid moiety in common. The first report was published by Andersson *et al.* (1973), which was followed by several other investigators using very similar approaches. The conclusion of these experiments was that it is the lipid moiety that elicits a mitogenic response in normal murine spleen cells. It is known that trace amounts of amino acids are present in most endotoxic glycolipids, such as *S. minnesota* R595 or others. Whether these trace amounts of amino acids are residues of the so-called "endotoxin protein" remains to be investigated.

A mitogenic protein was discovered in certain endotoxins by Sultzer and Goodman (1976) and simultaneously by Morrison *et al.* (1976). Sultzer studied the genetics of responsiveness to LPS using C3H/HeJ nonresponder and the corresponding C3H responder strains of mice (Sultzer, 1968, 1969). He discovered that C3H/HeJ mice responded to TCA endotoxin but not to phenol–water-extracted preparations. Skidmore *et al.* (1975) arrived at the same conclusion. The major difference between the two preparations is the higher amino acid content in the TCA than in the phenol–water extracts. Careful extraction of proteins from TCA endotoxin yielded a preparation that was mitogenic in C3H/HeJ mice as well as in other strains. Leive and Morrison (1972) extracted endotoxin with butanol–water. This endotoxin also contained a peptide that was mitogenic for C3H/HeJ mice. Extensive studies of this phenomenon by several laboratories followed and it was shown, among other things, that the protein preparation does not contain endotoxin. The nature of the linkage between this protein and the endotoxin requires more study. The discovery of mitogenic proteins in some endotoxin preparations was further evidence that active sites do exist outside of the lipid moiety.

The location of this mitogenic site in the endotoxic structure may be very close to the lipid moiety. Peptides covalently bound to phosphorylated glycosamine were reported (Nowotny, 1961, 1971a,b), and studied in more detail (Wober and Alaupovic, 1971). This peptide appears to be linked to the lipid moiety. Whether the mitogenic protein is identical with this is unknown at this time. At any rate, the mitogenic endotoxin protein is a nonendotoxic but biologically active site, not identical with the lipid moiety.

The mitogenic protein does not elicit the characteristic toxic effects of endotoxin, such as lethality, Shwartzman-type lesions, pyrogenicity, hypotension, disseminated intravascular coagulation, or others.

The mitogenic effect of a nontoxic synthetic model of glycolipids, resembling some characteristic structural components of the lipid moiety of endotoxins, could be demonstrated also. Asselineau and Asselineau (1967, unpublished) synthesized N-palmitoyl-D-glucosamine, which was nontoxic in several parameters, but could induce mitogenicity, provided it was properly dispersed (Rosenstreich et al., 1974). The beneficial effects of this and synthetic glycolipid homologs will be discussed in greater depth later.

c. At this point, the discussion of colony formation by endotoxin-induced CSF is relevant. We intended to study the component of the endotoxin that is responsible for CSF generation. We isolated a lipid A preparation by short hydrolysis with 1.0 N HCl, according to Westphal and Lüderitz (1954). The supernatant, which contained partially degraded polysaccharides and other soluble split-products, was dialyzed or ion-exchanged and lyophilized. The preparation was called PS. Chloroform–methanol-extracted Re mutant endotoxin was obtained from S. minnesota R595. Using these as CSF-inducing agents, we found that PS has a reduced but still considerable activity as compared to untreated LPS. The lipid A and the Re mutant endotoxin preparations were less active. This report was the first in which biological activity of a PS preparation was described (Chang et al., 1974). In continuation of this work, we observed that PS also has a reduced but significant protective effect against lethal irradiation (Nowotny et al., 1975).

The results of Pluznik et al. (1976) were quite the opposite. They found that lipid A was as active as the parent smooth LPS and that their PS preparation was inert. In a subsequent paper, the same laboratory reported (Apte et al., 1977) that another PS preparation had approximately 40% of the LPS activity, and the mutant glycolipids and the Galanos-type lipid A–BSA complex were as good or better than LPS.

This discrepancy in the findings has been resolved recently. We subjected endotoxin preparations to carefully controlled hydrolysis with 0.2 N HCl or 0.2 N acetic acid and followed the loss of toxicity and CSF activity in the PS preparation during hydrolysis. It was found that toxicity (chick embryo lethality) decreases more rapidly than CSF activity. In 30 min, less than 1% of the original toxicity remains in the PS preparation, while almost 90% of the CSF-inducing potency is still present. Continued hydrolysis destroyed the CSF activity also. It was reduced to 40% in 60 min and to barely detectable levels in 90 min, indicating that the CSF-inducing structure is also quite acid sensitive. A somewhat prolonged hydrolysis during the preparation of a PS sample will inevitably lead to an inert material (Nowotny et al., 1982). Details of this work, carried out in collaboration with Pluznik and associates, will be published elsewhere.

The conclusions of these studies are twofold. Probably the more important is that comparisons of preparations made in different laboratories are meaningful only if the same starting materials were used and the methods of isolation were strictly identical. This has been experienced in the past by comparing "lipid A" samples that turned out to be quite different (Chang and Nowotny, 1975) and more recently by the use of PS samples that were prepared under dissimiliar conditions.

The other conclusion is not entirely new, namely that toxicity of preparations is not a prerequisite for CSF generation. Several nontoxic preparations are well known to elicit CSF, such as some urine fractions or lung culture filtrates (Burgess *et al.*, 1977) and several others. What makes the above observations important is that PS is a breakdown product of fully toxic endotoxin, thus showing that nontoxic sites with beneficial activities may exist in some endotoxin derivatives.

d. Protection against lethal irradiation is one of the oldest known beneficial effects of endotoxins (Mefferd *et al.*, 1953; Smith *et al.*, 1957, 1959, 1961). As already mentioned, 10 μg of PS obtained from *Serratia marcescens* Boivin-type endotoxin by 30 min hydrolysis with 1.0 N HCl could protect 50% of the lethally irradiated (700 R) mice. Ten micrograms of parent endotoxin protected 70% of the animals (Nowotny *et al.*, 1975). At the same time, *S. minnesota* R595 mutant endotoxin (chloroform–methanol extraction and purification) was hardly effective. 1 N HCl-hydrolyzed lipid A from *Serratia marcescens* had no protective effect at all. That PS preparations can be protective was also shown by Urbaschek (1980), who reported that a Freeman-type polysaccharide preparation, obtained by acetic acid hydrolysis (Freeman, 1943), was fully active both in CSF induction and in protecting mice from the consequences of lethal irradiation.

These results clearly require more thorough study and therefore it is premature to draw too many conclusions. It is rather obvious that the pressence of a fully active toxic site is not a prerequisite for radioprotective effect. Preliminary results also allow the assumption that the CSF-inducing component in the PS mixture is not the same as the component eliciting protection against radiation. Measuring the optimal time for the release of this latter component during hydrolysis with 1.0 N HCl from *Serratia marcescens* endotoxin, we found that the PS preparation obtained after 30 min hydrolysis has considerably less protection than a PS after 60 min hydrolysis. This latter PS was almost void of CSF activity, while 30-min PS was potent in CSF induction. In other words, the optimal hydrolytic conditions to obtain an active PS preparation for radiation protection and for CSF activity are quite different (Behling and Nowotny, 1979, unpublished).

e. The immune adjuvant effect of endotoxin was discovered by Johnson et al. (1956), and a brief review of the history of the adjuvant action of endotoxins is found in Chapter 12 by Johnson. Also, several chapters in this volume deal with the possible mechanisms of this phenomenon. Our work concentrated on the identification of the active site of this action. Chemically detoxified preparations, rough mutant endotoxins, and PS samples were used. The mutant endotoxins were highly active in enhancing the humoral immune response to T-dependent and T-independent immunogens (Chen et al., 1973). Chemically detoxified (CH_3OK deacylated) endotoxoid was less potent than intact endotoxin but still active as adjuvant for human gamma globulin immunogen in mice (Johnson and Nowotny, 1964). PS preparation (from Serratia marcescens Boivin-type endotoxin hydrolyzed with 1.0 N HCl for 30 min) was tested as immune adjuvant for SRBC-injected ICR mice. In the first experiments, PS was given simultaneously with SRBC, and no adjuvant action was measured. Later, we varied the time interval between PS and SRBC injections and found that proper timing of PS injection (1 to 3 days before SRBC) gave a significant enhancement of plaque-forming spleen cells, as measured by the Jerne assay. Continuation of these time sequence studies led to the discovery that a single injection of endotoxin elicits oscillationlike changes of positive and negative effects. Certain time intervals between LPS (or PS) and SRBC injection led to immunosuppression, while other time intervals gave a positive enhancement of the immune response. Souvannavong and Adam (1980) reported similar effects by using MDP adjuvant. Alternating enhancement or suppression was observed also in the endotoxin-induced protection against radiation, or protection against transplantable murine tumors (Behling and Nowotny, 1977, 1980; Nowotny, 1977; Behling, 1982).

In vitro immune adjuvant effect was measured by using SRBC as the immunogen, incubated in tissue culture tubes with murine spleen cells in the presence of absence of endotoxin, or CH_3 OK detoxified endotoxoid (MEX) or PS preparations. These experiments gave results similar to those in the above in vivo measurements, showing that the PS preparation is only slightly less active than the parent endotoxin and that the MEX samples had a more reduced but still measurable adjuvant effect (Frank et al., 1977, 1980).

PS or endotoxin could also compensate for the immunosuppressive effect of a few tumor chemotherapeutic drugs, such as vincristine and dianhydrogalactitol. Gaál in our laboratories studied the effect of various time intervals on the immunosuppressive potential of the above-named and other cytostatic drugs in an anti-SRBC system. It was established that endotoxin or the PS preparation could not only compensate for suppression caused by the drug but that if PS and the drug were given at a certain time schedule they could exert a highly significant immune enhancement (Gaál and Nowotny, 1979). Although further studies are needed to elucidate the underlying mechanisms, it is clear that nontoxic, nonmitogenic, fatty acid-free PS preparations can serve as potent immune adjuvants, provided they are used at certain time intervals before SRBC immunogen.

f. The enhancement of nonspecific resistance to bacterial infections by endotoxin derivatives was also studied. We pretreated mice with detoxified endotoxin (MEX) obtained from *Serratia marcescens* Boivin-type endotoxin. As challenge, 10^7 viable *S. typhi* O901 bacteria were injected into mice i.p., 24 hr after pretreatment with MEX. The mouse LD_{50} of the MEX preparation could not be determined, since no mice died from a 200 mg/kg dose. Chick embryo lethality showed an over 400-fold reduction in toxicity. The MEX pretreatment, given in the same dose ranges, was about half as active as the toxic parent endotoxin in protecting mice from lethal challenge with *S. typhi* O901 (Johnson and Nowotny, 1964; Rote, 1974). PS preparations have not yet been studied in sufficient detail to include the results in this review.

Similar results were obtained by Urbaschek and Urbaschek (1975) using the same type of detoxified preparation made from *E. coli* endotoxin. Many more details of this phenomenon were elaborated by these authors and the results were published in a number of valuable publications. Several other detoxified preparations were used with similar results. McIntire *et al.* (1976) used phthalylated endotoxin. Bertók and co-workers detoxified their endotoxin by γ-irradiation (Bertók *et al.*, 1973; Füst *et al.*, 1977; Bertók, 1980). The use of this attenuated endotoxin (tolerin) provided comparable protection against infection with virulent microorganisms. For more details on this, see Chapter 10.

g. Tolerance to endotoxin could successfully be induced by a single injection of MEX (Urbaschek and Nowotny, 1968). This was shown by reduced lethality as well as by the prevention of microcirculatory changes caused by toxic endotoxin (Urbaschek *et al.*, (1968). In another series of experiments, Abdelnoor *et al.* (1981) used the measurement of hemodynamic changes in rabbits as a parameter of shock induced by toxic endotoxin. It was found that the endotoxin detoxified by the slime mold *Dictyostelium discoideum* enzymes (probably by deacylation) induced a refractoriness to toxic endotoxin. It is important that repeated injection of a *Serratia marcescens* PS preparation was equally effective.

h. Necrosis of tumors is probably the most investigated beneficial effect of endotoxin. Endotoxin is one of the active components in the "mixed bacterial toxins" of Coley (1893, 1898) (for review see Nauts *et al.*, 1953). It was present in "Shear's polysaccharide" (Shear, 1941; Shear *et al.*, 1943; O'Malley *et al.*, 1963) and in the tumor-necrotizing component isolated by Ikawa *et al.* (1952).

Regarding the chemical nature of the component that elicits tumor necrosis, considerable controversy and no solid data exist. Shear and co-workers considered their preparation to be a polysaccharide; Ikawa and co-workers isolated necrosamine, a sphingosine-like component, and assumed this to be the active component in their *E. coli* culture filtrate. Creach *et al.* (1964) and Homma and Abe (1972) isolated a protein-rich preparation, which lost its activity as an antitumor agent when treated with protease (Tanamoto *et al.*, 1979). They concluded that neither the lipid nor the polysaccharide alone can induce antitumor effects. The same authors reported that the lipid A portion of the molecule is responsible for adjuvant action, and that complete deacylation of the O-ester-linked fatty acids in the lipid moiety did not diminish the interferon-inducing capacity of the preparation. Ribi *et al.* (1979) concluded that the presence of polypeptides in the LPS is required for antitumor effect.

We used CH_3OK-detoxified *Serratia marcescens* endotoxin to induce necrosis of sarcoma 37 subcutaneous tumor in mice (Nowotny *et al.*, 1971), using the method of Shear. In the same series we also used rough mutant endotoxin of *S. minnesota* R595. The CH_3OK-treated preparation had profound changes in its lipid moiety, but it was still active, and the *S. minnesota* R595 mutant LPS was the most active. As we know, the lipid moiety comprises a major percentage of this preparation. These results were interpreted to signify that while the lipid moiety is where the tumor-necrotizing structural subunits are located, partial removal of the carboxylic acids only reduces but does not eliminate the antitumor effect. The role of peptides present in these preparations was not investigated.

In the continuation of these studies, we investigated the conditions that lead to the prevention of tumor take (Grohsman and Nowotny, 1972; Yang and Nowotny, 1974). The following tumors were used: TA3-Ha (a nonspecific ascites tumor developed from a spontaneous mammary adenocarcinoma), leukemia 1210, and Lewis lung tumor. We found that the CH_3OK preparation had a long-lasting protective effect against TA3-Ha challenge. A high dose (250 μg/mouse) still protected the mice from a lethal challenge 3 weeks later; a lower dose could protect if it was given 24 hr earlier than the tumor (Yang-Ko *et al.*, 1974). Single injection of a PS preparation had only a slight retarding effect on the take of TA3-Ha tumor. If PS was given repeatedly several days before the tumor challenge, a moderate protection was achieved

(Nowotny and Butler, 1980). Very strong protection (70–90% survival vs. 12% in the control group) was obtained if the animals were infected on day −18 with BCG and treated with PS before tumor inoculation. The most potent treatment of all was BCG infection on day −18, endotoxin or PS injection on day 0, and bleeding of the mice 2 hr after endotoxin or PS. Passive transfer of 0.5 ml of this post-BCG-PS serum i.p. to mice followed by a viable TA3-Ha tumor challenge i.p. protected 95–100% of the animals (Butler et al., 1978; Nowotny and Butler, 1980). Similar sera could also induce hemorrhagic necrosis of sarcoma 37 tumor, as described earlier by O'Malley et al. (1963).

In summary, we have surveyed the preparations that could elicit the following beneficial effects of endotoxins: (a) antigenicity, (b) mitogenicity, (c) leukopoietic effect, (d) protection against lethal irradiation, (e) immune adjuvant effect, (f) enhancement of nonspecific resistance to viable infections, (g) induction of tolerance to endotoxic reactions, and (h) elevation of nonspecific resistance to experimental tumors. The preparations used were complete endotoxins, incompletely synthesized mutants, or breakdown products of chemically altered preparations and synthetic glycolipids. All of the results listed here suggest that beneficial effects can be elicited by nontoxic derivatives. In the next paragraphs we will elaborate further on nontoxic preparations that were active in one or several beneficial assay systems.

D. Beneficial Contaminants

We have already discussed a few nontoxic derivatives of endotoxins, such as the chemically detoxified endotoxoids or the PS preparations. Similar detoxified preparations were obtained by other laboratories and Sultzer reviewed all of these in an excellent article in 1971. The nontoxic lipid-free "endotoxin protein" of Sultzer was also discussed above; this is a nontoxic, lipid-free substance that elicits mitogenicity.

In addition to these, there are a number of gram-negative bacterial products reported in the literature that are nontoxic but retain one or another of the beneficial effects of endotoxins. Since they may occur as contaminants in the conventional endotoxin preparations, they will be discussed here, in the chronological order of their publication.

Ribi and co-workers (Anacker et al., 1964; Rudbach et al., 1967, 1969) reported the isolation of a polysaccharide-rich macromolecular substance from the protoplasm of S. enteritidis and E. coli. Since it reacted with O antiserum to S. enteritidis, the preparation was named native hapten and later protoplasmic hapten. This hapten was nontoxic and nonpyrogenic. We received samples of it through the courtesy of Dr.

Ribi, and tested it for nonspecific enhancement of tumor resistance using TA3-Ha murine adenocarcinoma. A single injection of the hapten
was almost without effect, but five injections on consecutive days before challenge with viable tumor protected 80% of ICR mice from this
allogeneic tumor. Mortality in the saline-treated control group was
92%. This substance was particularly active if its application was preceded 14 days earlier by infection with viable BCG bacteria. At the
same time, no toxic effects of the hapten could be observed, not even
in the BCG-sensitized mice (Butler and Nowotny, 1979).

Marsh and Crutchley (1967) and Crutchley *et al.* (1968) isolated
"free endotoxin" from the culture supernatant of *E. coli* O78 and fractionated it by column chromatography. They observed the presence of
several components; one major constituent was fully endotoxic but another fraction, present in sizable amounts, was nontoxic and nonimmunogenic, gave no Shwartzman skin lesions, but was fully active in enhancing the nonspecific resistance to infection with viable bacteria.
Chemical analyses of this component showed that it consists of amino
acids, amino sugars, ester-bound fatty acids, and no KDO.

Berger *et al.* (1968) disrupted *E. coli* cells and separated the protoplasmic content from the cell walls. Both fractions were subjected to
phenol–water extraction; the phenol phase of the protoplasmic extract
yielded a protein called *protodyne*, that could protect mice to subsequent
challenges with viable *S. typhimurium*, *S. typhosa*, *P. aeruginosa*, *K. pneumoniae*, *Streptococcus mastidis*, and *Diplococcus pneumoniae* bacteria. The
protodyne preparation was negative in pyrogenicity and in Shwartzman
skin assays. It was free of lipids and polysaccharides. Berger assumed
earlier (1967) that the protective action of bacterial extracts could be
best explained by assuming the existence of at least two substances in
the usual endotoxin preparations. One is endotoxin, to which tolerance
develops rapidly if injected repeatedly; the other, which also enhances
nonspecific resistance, does not induce tolerance and is not identical
with endotoxin. It was assumed that protodyne may fulfill the role of
the latter. The authors also showed by chemical analysis that their
preparation is not identical with the "protoplasmic hapten" discussed
above.

A rather new development in this line of research was brought by
our results, when we tested the effectiveness of protodyne in inducing
CSF in mice. Protodyne had an activity (on a weight basis) quite similar
to endotoxins when injected i.p. to ICR mice. The tested protodyne
preparations did not contain detectable amounts of endotoxins as measured by chick embryo lethality or Shwartzman skin test (Nowotny and
Berger, 1982, unpublished).

Also quite recently, we obtained an aliquot of the mitogenic "endotoxin protein" from Sultzer. This substance was also tested for CSF-inducing capacity, and considerable activity was found when 20 μg was injected i.p. to ICR mice (Nowotny and Sultzer, 1982, unpublished). Details of these experiments will be published elsewhere. Neither protodyne nor the mitogenic protein induced the stimulation of bone marrow cell colony formation when they were added directly to the cell culture.

Choay and Sakouhi (1979) solubilized lipid-extracted *S. enteritidis* bacteria with 4% SDS at 100°C. The dialyzed supernatant was precipitated with ethanol, and deproteinized with phenol extraction. When this preparation was digested with RNase and DNase and filtered through an Amicon PM10 membrane, the filtrate contained a nontoxic fraction that could protect mice from experimental infection with a lethal dose of *K. pneumoniae*. The same preparation induced endotoxin tolerance. The substance had a molecular weight between 2000 and 10,000, absorbed UV light from 254 to 270 nm, was heat resistant up to 107°C, and stable between pH 5 and 9.5.

Ribi *et al.* (1979) reported that tumor regression-inducing activity of various endotoxins may be separable from their toxicity. Continuing these investigations, Takayama *et al.* (1981) described the isolation of a nontoxic lipid A from *S. typhimurium* Re mutant endotoxin by very mild hydrolysis at pH 4.5. The preparation was highly heterogeneous, but they succeeded in isolating a nontoxic fraction from it by DEAE column chromatography that cured 78% of the line 10 guinea pig hepatocarcinomas when injected intratumorally together with trehalose dimycolate and nontoxic lipid from Re mutants (Takayama *et al.*, 1981).

A mitogenic but nontoxic, barely pyrogenic and Shwartzman-negative lipid A and lipid A–BSA complex was isolated and studied by Raichvarg *et al.* (1981). The same authors reported earlier on experiments with PS isolated from *H. influenzae* and injected into human volunteers. The nontoxic PS caused enhanced phagocytosis of viable *H. influenzae* bacteria (Raichvarg *et al.*, 1980).

Isolation of lipid A preparations that were not active in some assays was reported recently by Westphal *et al.* (1980). Continued thorough studies of the structure of the lipid moiety of various bacteria species by this group revealed some unusual structures in *Chromobacterium violaceum, Rhodopseudomonas viridis, R. plaustris, Pseudomonas diminenta,* and *P. vesicularis.* Investigating the biological activities of the above and other irregular lipid structures in mitogenicity, pyrogenicity, lethality, and complement-activating assays, it was found that lipid A from *C. violaceum* did not react with complement although it was highly toxic and pyrogenic. In contrast, lipid A from *R. viridis* was neither toxic nor pyrogenic but activated complement (Lüderitz *et al.*, 1978). Measuring the capabilities of various lipid A preparations to induce necrosis of Meth A fibrosarcomas in mice, further differences could be established. The so-called "free lipid A" solubilized by triethylamine was without effect, while complete LPS induced hemorrhagic necrosis (Westphal *et al.*, 1980). These results were interpreted as indicating that additional polysaccharides are required to elicit all of the effects of endotoxins.

The Freeman-type polysaccharide (FPS) was tested by Urbaschek (1980) in colony formation stimulation and in radiation-protection assays, as discussed earlier. The FPS had high activity in both test systems. We prepared FPS by hydrolyzing *S. minnesota* 1114, *S. minnesota* R5, *S. minnesota* R7, and *S. minnesota* R595 whole cells directly with 0.2 N acetic acid for 60 min at 100°C. The supernatants were dialyzed, centrifuged at 10,000*g*, and lyophilized. The polysaccharide-rich preparations were negative in the Shwartzman assay at 40-μg dose, nonpyrogenic at 2 μg dose, and nontoxic to 11-day-old chick embryos at 5 μg dose, given i.v. When testing the CSF-inducing potency of these, large differences were seen. The most active was FPS from R5, less active was 1114 FPS, even less but still active was FPS from the R595 Re mutant. R7 was barely active. A preview of these unpublished findings is given in Table 1.

All the above-listed findings bring us back to the discussion of the heterogeneity of endotoxins. As described earlier, all endotoxin preparations, including the best LPS samples, are mixtures of several bacterial products. Among these are found some that are very toxic and also active in nonspecific resistance enhancement. Others are nontoxic but still elicit beneficial effects; further fractions appear to be inert in all conventional assays of endotoxic reactions (Nowotny *et al.*, 1966). It has been assumed that these components of the endotoxin preparations may contribute either directly, or indirectly like cofactors, to the multiplicity and to the intensity of the biological reactions we measure (Nowotny, 1971a).

Some of these components may be identical with others discussed in this chapter. Protoplasmic hapten, protodyne, or the compounds isolated by other laboratories may very well occur as contaminants of endotoxin preparations. As far as the possible origin of these components is concerned, they may be precursors or decay products of endotoxins, but in several instances they seem to be unrelated to it. Again, only chemical analyses of isolated fractions will be able to shed light on the

Table 1. Toxicity and CSF Activity of Various Bacterial Fractions

	CE LD_{50}[a]	Shwartzman[b]	CSF[c]
Freeman-type polysaccharides			
S. minnesota 1114	4.6 μg	0.19	16
S. minnesota R5	> 10.0 μg	Neg. at 40 μg	73
S. minnesota R7	> 10.0 μg	Neg. at 40 μg	3
S. minnesota R595	1.0 μg	Neg. at 40 μg	5
Se. marcescens O8	0.32 μg	0.30	52
Protodyne [d]	NT[d]	Neg. at 100 μg	24
Endotoxin protein[e]	NT[e]	NT[e]	44
S. minnesota 1114 LPS	0.02	3.2	51

[a] Chick embryo LD_{50}. The preparations were given i.v. to 11-day-old embryos.
[b] Shwartzman skin lesions on rabbits elicited by 40 μg given i.d. Values are cm^2.
[c] CSF release induced by 25 μg given i.p. The number of colonies formed from 10^5 bone marrow cells are indicated.
[d] Protodyne was kindly provided by Dr. F. Berger. The toxicity of this sample was not tested, but previous studies of similar preparation showed no toxicity.
[e] Endotoxin protein is the mitogenic peptide isolated by Dr. B. Sultzer and provided for these experiments. The toxicity of this batch was not tested, but similar preparations were found to be void of endotoxicity.

possible origin of these beneficial contaminants of endotoxin preparations.

E. Synthetic Compounds

Final proof of the correctness of any structural proposal always comes from complete synthesis of the assumed structure and its testing in biological assays. If the proposal was correct, and the organic chemists could synthesize the exact structure, the preparation has to be as active as the natural product. Therefore, we should review the results obtained so far with synthetic model compounds, resembling some or all parts of the known structure of the lipid moiety.

One of the most unusual linkages in the lipid moiety is the N-acyl linkage between hexosamines and long-chain carboxylic acids (Nowotny, 1961; Nowotny et al., 1963; Burton and Carter, 1964; Gmeiner et al., 1971). As a simple model of these, Asselineau and Asselineau synthesized N-palmitoyl-D-glucosamine, according to the procedure of Fieser *et al.* (1956). Fine dispersion of this glycolipid was tested for endotoxicity in chick embryo and mouse lethality, pyrogenicity and Shwartzman assays in rabbits, but no toxicity could be detected. Later, the mitogenic effect of the same preparation was measured and it was found that it was mitogenic, provided it was dispersed in a special way, which was not easily reproducible (Rosenstreich *et al.*, 1974). Further homologs of such glycolipids were synthesized by us and brought into stable liposome form. The adjuvant effect of these was measured using bovine gamma globulin and SRBC as immunogens. The immune response was measured by passive hemagglutination, by rosette formation, and by the Jerne plaque assay. It was found that some of these synthetic glycolipids were potent immune adjuvants (Behling *et al.*, 1976). They were also used alone and in combination with various endotoxin preparations and mycobacterial cell wall products to enhance the nonspecific resistance of mice to TA3-Ha tumor challenge. Certain combinations of these glycolipids could protect a significant percentage of the mice (Butler and Nowotny, 1979). None of these glycolipids showed any toxic properties.

Gorbach *et al.* (1979) synthesized 2-(DL-3-hydroxytetra-decanoyl) amino-2-deoxyglucose-6-phosphate and its nonhydroxylated analog. Only serological tests were reported, which showed that both compounds could inhibit the reaction between lipid A and anti-lipid-A rabbit serum.

In the last few years, several Japanese research teams reported synthesis of models of the lipid A structure.

Kiso and co-workers further extended the synthesis of *N*-acylated D-glucosamine derivatives (Kiso *et al.*, 1980, 1981, 1982), coupled 3-OH-myristic acid to the primary amino group of glucosamine and prepared various derivatives of this compound. One of these included a phosphorylated, 1→6 linked disaccharide of the above glycolipid, thus synthesizing the most characteristic part of the proposed structure. Biological activities were measured, such as *Limulus* lysate clotting, mitogenicity, adjuvant effect, and tumor necrotization. The compound was active in *Limulus* lysate clotting, showed reduced activity as a tumor-necrotizing agent but was not mitogenic, and showed no adjuvant effect and no cytotoxicity (Kiso *et al.*, 1982). No *in vivo* toxic effects were reported.

Shiba and associates (Inage *et al.*, 1980; Shiba, 1982) synthesized larger models of lipid A, but the preparation was nontoxic, and showed only moderate activity as a spleen cell mitogen and as an immune adjuvant.

These results might have a number of explanations, one being that the structural proposal was incomplete. It is well known that even in the most firmly established structures there are often linkages or specific steric arrangements through which several alternate possibilities may exist but only one of these will yield an active product. This may obviously apply to naturally occurring and to synthetic models of endotoxins. This means that the activities require a highly specific three-dimensional steric arrangement of all of the known components.

In any case, synthesis of endotoxin will be successful only if all fine details of the structure of the polysaccharide and the lipid moieties, their possible substituents and appendages become fully known. On the other hand, we cannot exclude the fact that continued organic chemical efforts result in synthetic compounds that elicit not all but only some of the truly characteristic endotoxicity reactions. Such compounds will be invaluable tools in studies aimed at elucidating the molecular events of these reactions.

VII. Summary and Conclusion

Results of excellent studies over several decades obtained by Westphal, Lüderitz, their students, and co-workers have shown beyond a doubt that the major immunodeterminants of the endotoxic LPS macromolecules reside in their polysaccharide moiety. Some functional groups in the lipid moiety are considerably less immunogenic. Every piece of information available to us in the past decade also shows that the lipid moiety of the endotoxin is the part of the molecular complex that is responsible for the toxic effects elicited by these substances, thus supporting the claim of Westphal and Lüderitz, expressed first almost three decades ago, in 1954.

Regarding the existence of separate beneficial site(s), much encouraging data are available but no final conclusion can be offered. It is possible that a structural entity of the LPS molecule, possessing a certain critical chemical composition and physicochemical properties, can elicit these effects and that this entity has no resemblance to the lipid moiety. Several published findings that were discussed in this chapter allow this assumption.

Another possibility is that there is only one active site in the endotoxin and that this is the lipid moiety. If small quantities are given the effect will be beneficial but an overdose may lead to shock and death.

There is no reason to argue against the low vs. high dose effects. Whatever dose is beneficial for mice, is 1/10 or less of an LD_{50}, and the same will apply to other species, including humans. But these apply only to the use of complete LPS or endotoxin. If we compare a parent LPS and a chemically detoxified derivative of it, and do a quantative dose-response determination in a beneficial assay system, we will see that on a weight basis the derivative will be similar to the parent in the beneficial effect, while the toxicity of it will be reduced by several magnitudes.

One could also assume that such detoxified products are beneficial due to the presence of incompletely destroyed endotoxin residues in them. This argument is fully logical, but quantitative data make it unlikely. If the toxicity of a detoxified preparation or a PS goes below 1% of the original and still maintains 40 to 90% of the starting beneficial effects, the latter cannot be attributed to the less than 1% residual toxic endotoxin content. Furthermore, 1% of a beneficial dose of toxic endotoxins usually does not elicit beneficial effects. Needless to say, such quantitative determinations are valid only if close-to-linear dose-response relationship can be established in the applied biological assay.

It is reasonable to expect that if a single structural subunit of the lipid moiety would elicit both the harmful and the beneficial effects, chemical alterations of this subunit should result in changes of both effects. This is not the case. As a matter of fact, chemical changes distinguished them from each other. Accordingly the lipid moiety can be the carrier of both activities only if different structural subunits in it are required to elicit both effects. When the lipid moiety is chemically detoxified as a result of partial destruction, it may still be capable of inducing events leading to beneficial outcome if the structure still contains the necessary elements to trigger such reactions. The same reasoning may apply to explain beneficial differences among similarly toxic mutant endotoxins.

That incomplete structures, partially degraded or attenuated, can have valuable activities is also supported by the existence of naturally occurring nontoxic but beneficially active, endotoxinlike components. These may be precursors isolated before the completion of their biosynthesis, or products of cell decay that lost their full potency during the process. These are expected to be coextracted with the endotoxins and they probably form some of the several fractions present in the conventional endotoxin preparations.

This line of thought leads us to the possible existence of beneficial contaminants in the conventional endotoxin preparation. It is possible that while endotoxin is sensitive to acid or other treatments, the beneficial contamninant is not. The potential role of the several nonendo-toxic components of crude endotoxin in the biological effects of such preparations was discussed earlier (Nowotny, 1971). It was assumed that these compounds, although not eminently active per se, may con-tribute, like cofactors, to the development of endotoxic reactions. One may assume that several of these "contaminants," acting in concert, can lead to effects from which the host will benefit.

These sets of evidence are circumstantial at best and several if not all of them need better substantiation. What undoubtedly emerges from the arguments presented here is that toxic manifestations do not have to accompany beneficial effects. This may have several explana-tions, but considering them from the point of view of a chemist, the most likely of these is that *different structural requirements are needed to elicit the toxic and the beneficial effects.*

Our worthy task is to identify the chemical structures that induce reactions leading to enhanced host resistance, but it is also obvious that the pursuit of this goal faces considerable difficulties. Oversimplifica-tions are condemnable and although they make attractive lectures and often-quoted publications, they inevitably mislead the audience through avoidance of confounding realities. It is evident that this incomplete re-view has brought up many more new problems than it has solved old ones, as every honest look into the complexities of similar biological phenomena should do. What one can still hope for from such expo-sures of admittedly disheartening perplexity is the realization of still open possibilities in the search for better understanding of underlying mechanisms. Only through improved interpretation can we hope for better control of beneficial effects, and only with a broadsighted ap-proach will we have better chances of reaching the point where previ-ous deliberations rather than serendipity will lead to findings with practical significance. We think that endotoxins, although the subject of research for many decades, are still unexploited for therapeutic uses. We believe that only well-defined molecular entities of the complex en-dotoxin structure, or synthetic molecular probes resembling endotoxins, will give us the cleanest, finest molecular tools to study, understand, and command the events leading to some of the beneficial effects discussed in this volume.

VIII. References

Abdelnoor, A. M., Chang, H. L., Pham, P. H., and Nowotny, A., 1980, Lack of relation-ship between toxicity and bone marrow cell colony stimulating activity of endotoxin preparations, *Proc. Soc. Exp. Biol. Med.* **163**:156.

Abdelnoor, A. M., Johnson, A. G., Anderson-Imbert, A., and Nowotny, A., 1981, Immunization against bacteria- and endotoxin-induced hypotension, *Infect. Immun.* **32**:1093.

Adams, G. A., 1967, Extraction of lipopolysaccharides from gram-negative bacteria with dimethyl sulfoxide, *Can. J. Biochem.* **45**:422.

Adams, G. A., and Singh, P. P., 1970, The chemical constitution of lipid A from *Serratia marcescens*, *Can. J. Biochem.* **48**:55.

Addison, P. D., and Berry, L. J., 1981, Passive protection against X-irradiation with serum from zymosan-primed and endotoxin-injected mice, *J. Reticuloendothelial Soc.* **30**: 301.

Adorini, L., Ruco, L., Uccini, S., Soravito De Franceschi, G., Baroni, C. D., and Doria, G., Biological effects of *Escherichia coli* lipopolysaccharide (LPS) *in vivo*. II. Selection in the mouse thymus of PHA- and Con A-responsive cells, *Immunology* **31**:225.

Alexander, P., and Evans, R., 1971, Endotoxin and double-stranded RNA render macrophages cytotoxic, *Nature New Biol.* **232**:76.

Allison, A. C., Davies, P., and Page, R. C., 1973, Effects of endotoxin on macrophages and other lymphoreticular cells, *J. Infect. Dis.* **128**(Suppl.):212.

Anacker, R. L., Finkelstein, R. A., Haskins, W. T., Landy, M., Milner, K. C., Ribi, E., and Stashak, R. W., 1964, Origin and properties of naturally occurring hapten from *Escherichia coli*, *J. Bacteriol.* **88**:1705.

Andersson, J., Möller, G., and Sjöberg, O., 1972, Selective induction of DNA synthesis in T and B lymphocytes, *Cell. Immunol.* **4**:381.

Andersson, J., Melchers, F., Galanos, C., and Lüderitz, O., 1973, The mitogenic effect of lipopolysaccharide on bone marrow-derived mouse lymphocytes: Lipid A as the mitogenic part of the molecule, *J. Exp. Med.* **137**:943.

Apte, R. N., Hertogs, C. F., and Pluznik, D. H., 1977, Regulation of lipopolysaccharide-induced granulopoiesis and macrophage formation by spleen cells. I. Relationship between colony-stimulating factor release and lymphocyte activation *in vitro*, *J. Immunol.* **118**:1435.

Baker, E. E., Goebel, W. F., and Perlman, E., 1949, The specific antigens of variants of *Shigella sonnei*, *J. Exp. Med.* **89**:325.

Banerji, B., and Alving, C. R., 1979, Lipid A from endotoxin: Antigenic activities of purified fractions in liposomes, *J. Immunol.* **123**:2558.

Barsi, I., 1947, A new treatment of rheumatoid arthritis, *Br. Med. J.* **2**:252.

Behling, U. H., and Nowotny, A., 1977, Immune adjuvancy of lipopolysaccharide and a nontoxic hydrolytic product demonstrating oscillating effects with time, *J. Immunol.* **118**:1905.

Behling, U. H., and Nowotny, A., 1978, Long-term adjuvant effect of bacterial endotoxin in prevention and restoration of radiation-caused immunosuppression, *Proc. Soc. Exp. Biol. Med.* **157**:348.

Behling, U. H., and Nowotny, A., 1980, Cyclic changes of positive and negative effects of single endotoxin injections, in: *Bacterial Endotoxins and Host Response* (M. K. Agarwal, ed.), pp. 11–26, Elsevier/North-Holland, Amsterdam.

Behling, U. H., and Nowotny, A., 1982, Bacterial endotoxins as modulators of specific and nonspecific immunity, in: *Regulatory Implications of Oscillatory Dynamics in the Immune System* (C. DeLisi and J. Hiernaux, eds.), CRC Press, Cleveland, Ohio.

Behling, U. H., Campbell, B., Chang, C. M., Rumpf, C., and Nowotny, A., 1976, Synthetic glycolipid adjuvants, *J. Immunol.* **117**:847.

Behling, U. H., Pham, P. H., Madani, F., and Nowotny, A., 1980, Components of lipopolysaccharide which induce colony stimulation, adjuvancy and radioprotection, in: *Microbiology 1980* (D. Schlessinger, ed.), pp. 103–107, American Society for Microbiology, Washington, D.C.

Bendich, A., and Chargaff, E., 1946, The isolation and characterization of two antigenic fractions of *Proteus* OX-19, *J. Biol. Chem.* **166**:283.

Berczi, I., 1967, Endotoxin neutralizing effect of antisera to *Escherichia coli* endotoxin, *Z. Immunitaetsforsch.* **132**:303.

Berger, F. M., 1967, The effect of endotoxin on resistance to infections and disease, *Adv. Pharmacol.* **5**:19.

Berger, F. M., Fukui, G. M., Ludwig, B. J., and Rosselet, J. P., 1968, Increase of non-specific resistance to infection by protodyne, a protein component derived from bacterial protoplasm, *Proc. Soc. Exp. Biol. Med.* **127**:556.

Bernstein, C., and Klotz, S. D., 1947, Fever therapy in herpes zoster, *J. Lab. Clin. Med.* **32**:1544.

Berry, L. J., 1971, Metabolic effects of bacterial endotoxin, in: *Microbial Toxins*, Vol. V (S. Kadis, G. Weinbaum, and S. J. Ajl, eds.), pp. 165–208, Academic Press, New York.

Bertók, L., 1980, Radio-detoxified endotoxin as a potent stimulator of nonspecific resistance, *Perspect. Biol. Med.* **24**:61.

Bertók, L., Kocsár, L., Várterész, V., Bereznai, T., and Antoni, F., 1973, Procedure for preparation and use of radio-detoxified bacterial endotoxin [Hungarian], Patent 162,973, Budapest.

Billroth, T., 1865, Beobachtungsstudien über Wundfieber und (accidentelle) Wundkrankheiten, *Arch. Klin. Chir.* **6**:372.

Binkley, F., Goebel, W. F., and Perlman, E., 1945, Studies on the Flexner group of dysentery bacilli. II. The chemical degradation of the specific antigen of type Z *Shigella paradysenteriae* (Flexner), *J. Exp. Med.* **81**:331.

Bishop, D. G., and Work, E., 1965, An extracellular glycolipid produced by *Escherichia coli* grown under lysine-limiting conditions, *Biochem. J.* **96**:567.

Boivin, A., Mesrobeanu, J., and Mesrobeanu, L., 1933a, Extraction d'un complexe toxique et antigenique á partir du *bacille d'Aertrycke*, *C. R. Soc. Biol.* **114**:307.

Boivin, A., Mesrobeanu, J., and Mesrobeanu, L., 1933b, Technique pour la préparation des polysaccharides microbiens spécifiques, *C. R. Soc. Biol.* **113**:490.

Braude, A. I., Ziegler, E. J., Douglas, H., and McCutchan, J. A., 1977, Antibody to cell wall glycolipid of gram-negative bacteria: Induction of immunity to bacteremia and endotoxemia, *J. Infect. Dis.* **136**(Suppl.):167.

Braun, V., and Rehn, K., 1969, Chemical characterization, spatial distribution and function of a lipoprotein (murein-lipoprotein) of the *E. coli* cell wall: The specific effect of trypsin on the membrane structure, *Eur. J. Biochem.* **10**:426.

Burdon-Sanderson, J., 1876, On the process of fever, *Practitioner* **16**:257.

Burgess, A. W., Camakaris, J., and Metcalf, D., 1977, Purification and properties of colony-stimulating factor from mouse lung-conditioned medium, *J. Biol. Chem.* **252**:1998.

Burton, A., and Carter, H. E., 1964, Purification and characterization of the lipid A component of the lipopolysaccharides from *Escherichia coli*, *Biochemistry* **3**:411.

Butler, R. C., and Nowotny, A., 1976, Colony stimulating factor (CSF) containing serum has anti-tumor effect, *IRCS Med. Sci.* **4**:206.

Butler, R. C., and Nowotny, A., 1979, Combined immunostimulation in the prevention of tumor take in mice using endotoxins, their derivatives and other immune adjuvants, *Cancer Immunol. Immunother.* **6**:255.

Butler, R. C., Nowotny, A., and Friedman, H., 1977, Stimulation of an *in vitro* antibody response by endotoxin-induced soluble factors, *J. Reticuloendothelial Soc.* **22**:28.

44   A. Nowotny

Butler, R. C., Abdelnoor, A. M., and Nowotny, A., 1978, Bone marrow colony-stimulating factor and tumor resistance-enhancing activity of post-endotoxin mouse sera, *Proc. Natl. Acad. Sci. USA* **75:**2893.

Butler, R. C., Nowotny, A. and Friedman, H., 1979, Macrophage factors that enhance the antibody response, *Ann. N.Y. Acad. Sci.* **332:**564.

Butler, R. C., Friedman, H., and Nowotny, A., 1980, Restoration of depressed antibody responses of leukemic splenocytes treated with LPS-induced factors, *Adv. Exp. Med. Biol.* **121A:**315.

Buttke, T. M., and Ingram, L. O., 1975, Comparison of lipopolysaccharides from *Agmenellum quadruplicatum* to *Escherichia coli* and *Salmonella typhimurium* by using thin-layer chromatography, *J. Bacteriol.* **124:**1566.

Carswell, E. A., Old, L. J., Kassel, R. L., Green, S., Fiore, N., and Williamson, B., 1975, An endotoxin-induced serum factor that causes necrosis of tumors, *Proc. Natl. Acad. Sci. USA* **72:**3666.

Centanni, E., 1894, Über infektiöses Fieber: Das Fiebergift der Bakterien, *Dtsch. Med. Wochenschr.* **20:**148.

Chang, C. M., and Nowotny, A., 1975, Relation of structure to function in bacterial O-antigens. VII. The biological activity of "lipid A" preparations, *Immunochemistry* **12:**19.

Chang, H., Thompson, J. J., and Nowotny, A., 1974, Release of colony stimulating factor (CSF) by non-endotoxic breakdown products of bacterial lipopolysaccharides, *Immunol. Commun.* **3:**401.

Chedid, L., Parant, M., Parant, F., and Boyer, P., 1968, A proposed mechanism for natural immunity to enterobacterial pathogens, *J. Immunol.* **100:**292.

Chen, C. H., and Nowotny, A., 1974, Direct determination of molar ratios of various chemical constituents in endotoxic glycolipids in silicic acid scrapings from thin-layer chromatographic plates, *J. Chromatogr.* **97:**39.

Chen, C. H., Johnson, A. G., Kasai, N., Key, B. A., Levin, J., and Nowotny, A., 1973, Heterogeneity and biological activity of endotoxic glycolipid from *Salmonella minnesota* R595, *J. Infect. Dis.* **128**(Suppl.):43.

Chen, C. H., Chang, C. M., Nowotny, A. M., and Nowotny, A., 1975, Rapid biological and chemical analyses of bacterial endotoxins separated by preparative thin-layer chromatography, *Anal. Biochem.* **63:**183.

Chester, I. R., and Meadow, P. M., 1975, Heterogeneity of lipopolysaccharide from *Pseudomonas aeruginosa*, *Eur. J. Biochem.* **58:**273.

Choay, J., and Sakouhi, M., 1979, Fraction capable of inducing in vivo a resistance to bacterial infections, process for obtaining said fraction from bacteria and drugs containing said fraction, U.S. Patent 4,148,877, April 10, 1979.

Closse, A., 1960, Chemische Bausteinanalyse der Lipoid A-Komponente in bakteriellen Endotoxinen. PhD Thesis, Albert Ludwigs Universität, Freiburg i. Br., Germany.

Coley, W. B., 1893, The treatment of malignant tumors by repeated inoculations of erysipelas, with a report of ten original cases, *Am. J. Med. Sci.* **105:**487.

Coley, W. B., 1898, The treatment of inoperable sarcoma with the mixed toxins of erysipelas and *Bacillus prodigiosus*, *J. Am. Med. Assoc.* **31:**389.

Colobert, L., 1958, Étude de la lyse de Salmonelles pathogènes provoginée par le lysozyme, après délipidation partielle de la paroi externe, *Ann. Inst. Pasteur* **95:**156.

Creech, H. J., Breuninger, E. R., and Adams, G. A., 1964, Polysaccharide-peptide complexes from Serratia marcescens cells, *Canadian J. Biochem.* **42:**593.

Crutchley, M. J., Marsh, D. G., and Cameron, J., 1968, Biological studies on free endotoxin and a non-toxic material from culture supernatant fluids of *Escherichia coli* 078K80, *J. Gen. Microbiol.* **50:**413.

Cundy, K. R., and Nowotny, A., 1968, Quantitative comparison of toxicity parameters of bacterial endotoxins, *Proc. Soc. Exp. Biol. Med.* **127**:999.

Currie, G. A., and Basham, C., 1975, Activated macrophages release a factor which lyses malignant cells but not normal cells, *J. Exp. Med.* **142**:1600.

Davies, D. A. L., 1960, Polysaccharides of gram-negative bacteria, *Adv. Carbohydr. Chem.* **15**:271.

DeDur, A., Chaby, R., and Szabo, L., 1980, Isolation of two protein-free and chemically different lipopolysaccharides from *Bordetella pertussis* phenol-extracted endotoxin, *J. Bacteriol.* **143**:78.

DeLaunay, A., and Lasfargues, E., 1945, Sur le pouvoir necrosant de l'exotoxine staphylococcique et de l'endotoxine typhique, observe comperativement *in vivo* et *in vitro*, *C. R. Soc. Biol.* **139**:363.

DiRienzo, J. M., and MacLeod, R. A., 1978, Composition of the fractions separated by polyacrylamide gel electrophoresis of the LPS of a marine bacterium, *J. Bacteriol.* **136**:158.

Doe, W. F., and Hensen, P. M., 1978, Macrophage stimulation by bacterial lipopolysaccharides. I. Cytolytic effect on tumor target cells, *J. Exp. Med.* **148**:544.

Doe, W. F., Yang, S. T., Morrison, D. C., Betz, S. J., and Henson, P. M., 1978, Macrophage stimulation by bacterial lipopolysaccharides, II. Evidence for independent differentiation signals delivered by lipid A and a protein-rich fraction of LPS, *J. Exp. Med.* **148**:557.

Douglas, S. R., and Fleming, A., 1921, On the antigenic properties of acetone-extracted bacteria, *Br. J. Exp. Pathol.* **2**:131.

Elin, R. J., and Wolff, S. M., 1976, Biology of endotoxin, *Annu. Rev. Med.* **27**:127.

Favorite, G. O., and Morgan, H. R., 1946, Therapeutic induction of fever and leukocytosis, using a purified typhoid pyrogen, *J. Lab. Clin. Med.* **31**:672.

Fieser, M., Fieser, L., Toromanoff, E., Hirata, Y., Heymann, H., Teft, M., and Bhattacharya, S., 1956, Synthetic emulsifying agents, *J. Am. Chem. Soc.* **78**:2825.

Frank, S., Specter, S., Nowotny, A., and Friedman, H., 1977, Immunocyte stimulation *in vitro* by nontoxic bacterial lipopolysaccharide derivatives, *J. Immunol.* **119**:855.

Frank, S. J., Specter S. J., Nowotny, A., and Friedman, H., 1980, Effect of immune sera upon enhanced *in vitro* antibody response, *Adv. Exp. Med. Biol.* **121**:B261.

Freeman, G. G., 1943, The components of the antigenic complex of *Salmonella typhimurium*, *Biochem. J.* **37**:601.

Freeman, G. G., and Anderson, T. H., 1941, The hydrolytic degradation of the antigenic complex of *Bact. typhosum* Ty 2, *Biochem. J.* **35**:564.

Friedman, H., Specter, S., Butler, R. C., and Nowotny, A., 1980, Humoral factors in the adjuvant effect of lipopolysaccharides, in: *Microbiology 1980* (D. Schlessinger, ed.), American Society for Microbiology, Washington, D.C.

Fromme, I., Himmelspach, K., Lüderitz, O., and Westphal, O., 1957, Analyse der Zuckerbausteine in bakteriellen Lipopolysacchariden Abequose and Tyvelose als 3.6-Bisdeoxy-aldohexosen, *Ange. Chemie* **20**:643.

Füst, G., Bertók, L., and Juhász-Nagy, S., 1977, Interactions of radio-detoxified *Escherichia coli* endotoxin preparations with the complement system, *Infect. Immun.* **16**:26.

Gaál, D., and Nowotny, A., 1979, Immune enhancement by tumor therapeutic drugs and endotoxin, *Cancer Immunol. Immunother.* **6**:9.

Galanos, C., Lüderitz, O., and Westphal, O., 1969, A new method for the extraction of R lipopolysaccharides, *Eur. J. Biochem.* **9**:245.

Galanos, C., Lüderitz, O., and Westphal, O., 1971, Preparation and properties of antisera aganist the lipid-A component of bacterial lipopolysaccharides, *Eur. J. Biochem.* **24:** 116.

Galanos, C., Lüderitz, O., Rietschel, E. T., and Westphal, O., 1977, Newer aspects of the chemistry and biology of bacterial lipopolysaccharides, with special reference to their lipid A component, in: *Biochemistry of Lipids II* (T. W. Goodwin, ed.), *International Review of Biochemistry*, Vol. 14, pp. 239–335, University Park Press, Baltimore.

Garfield, D. J., Rebers, R. A., and Heddleston, K. L., 1976, Immunogenic and toxic properties of a purified lipopolysaccharide–protein complex from *Pasteurella multocida, Infect. Immun.* **14:**990.

Gately, C. L., Gately, M. K., and Mayer, M. M., 1976, Separation of lymphocyte mitogen from lymphotoxin and experiments on the production of lymphotoxin by lymphoid cells stimulated with partially purified mitogen: A possible amplification mechanism of cellular immunity and allergy, *J. Immunol.* **116:**669.

Gery, I., Kruger, J., and Spiesel, S. Z., 1972, Stimulation of B-lymphocytes by endotoxin: reactions of thymus-deprived mice and karyotypic analysis of dividing cells in mice bearing T_6T_6 thymus grafts, *J. Immunol.* **108:**1088.

Gmeiner, J., Lüderitz, O., and Westphal, O., 1969, Biochemical studies on lipopolysaccharides of Salmonella R mutants. 6. Investigations on the structure of the Lipid A component, *Eur. J. Biochem.* **7:**370.

Gmeiner, J., Simon, M., and Lüderitz, O., 1971, The linkage of phosphate groups and of 2-keto-3-deoxyoctonate to the Lipid A component in a *Salmonella minnesota* lipopolysaccharide, *Eur. J. Biochem.* **21:**355.

Goebel, W. F., Binkley, F., and Perlman, E., 1945, Studies on the Flexner group of dysentery bacilli. I. The specific antigens of *Shigella paradysenteriae* (Flexner), *J. Exp. Med.* **81:**315.

Goldman, R. C., and Leive, L., 1980, Heterogeneity of antigenic side chain length in LPS from *Escherichia coli* Olll and *Salmonella typhimurium* LT2, *Eur. J. Biochem.* **107:**145.

Golub, S., Gröschel, D. H. M., and Nowotny, A., 1968, Factors which affect the reticuloendothelial system: Uptake of bacterial endotoxins, *J. Reticuloendothelial Soc.* **5:**324.

Golub, S., Gröschel, D., and Nowotny, A., 1970, Studies on the opsonization of bacterial endotoxoid, *J. Reticuloendothelial Soc.* **7:**518.

Gorbach, V. I., Krasikova, I. N., Lukyanov, P. A., Razamakhnina, O. Y., Solov'eva, T. F., and Ovodov, Y. S., 1979, Structural studies on the immunodominant group of lipid A from LPS of *Yersinia pseudotuberculosis, Eur. J. Biochem.* **98:**83.

Gray, G. W., and Wilkinson, S. G., 1965, The effect of ethylenediaminetetra-acetic acid on the cell walls of some gram-negative bacteria, *J. Gen. Microbiol.* **39:**385.

Green, S., Dobrjansky, A. E., Carswell, E. A., Kassel, R. L., Old, L. J., Fiore, N. and Schwartz, M. K., 1976, Partial purification of a serum factor that causes necrosis of tumors, *Proc. Natl. Acad. Sci. USA* **73:**381.

Greisman, S. E., DuBuy, J. B., and Woodward, C. L., 1978, Experimental gram-negative bacterial sepsis: Reevaluation of the ability of rough-mutant antisera to protect mice, *Proc. Soc. Exp. Biol. Med.* **158:**482.

Greisman, S., E., DuBuy, J. B., and Woodward, C. L., 1979, Experimental gram-negative bacterial sepsis: Prevention of mortality not preventable by antibiotics alone, *Infect. Immun.* **25:**538.

Grohsman, J., and Nowotny, A., 1972, The immune recognition of TA3 tumors, its facilitation by endotoxin, and abrogation by ascites fluid, *J. Immunol.* **109:**1090.

Hase, S., and Rietschel, E. T., 1976 , Isolation and analysis of the lipid A backbone lipid A structure of lipopolysaccharides from various bacterial groups, *Eur. J. Biochem.* **63:** 101.

Hase, S., and Rietschel, E. T., 1977, The chemical structure of the Lipid A component of lipopolysaccharides from *Chromobacterium violaceum* NCTC 9694, *Eur. J. Biochem.* **75**:23.

Heilman, D. H., and Bernton, H. W., 1961, The effect of endotoxin on tissue cultures of spleen of normal and tuberculin-sensitive animals, *Am. Rev. Respir. Dis.* **84**:862.

Hibbs, J. B., Jr., Taintor, R. R., Chapman, H. A., Jr., and Weinberg, J. B., 1977, Macrophage tumor killing: Influence of the local environment, *Science* **197**:279.

Hoffman, J., Lindberg, B., Glowacka, M., Derylo, M. and Lorkiewicz, 1980, Structural studies of the lipopolysaccharide from Salmonella typhimurium 902 (CoIIb drd2), *Eur. J. Biochem.* **105**:103.

Hoffmann, M. K., Green, S., Old, L. J., and Oettgen, H. F., 1976, Serum containing endotoxin-induced tumour necrosis factor substitutes for helper T-cells, *Nature (London)* **263**:416.

Homma, J. Y., and Abe, C., 1972, Differences in chemical nature of the endotoxins derived from dissociants, types la and sm, *Pseudomonas aeruginosa* strain N10, *Jpn. J. Exp. Med.* **42**:23.

Hort, E. C., and Penfold, W. J., 1911, The dangers of saline injections, *Br. Med. J.* **2**:1589.

Ikawa, M., Koepfli, J. B., Mudd, S. G., and Niemann, C., 1952, An agent from *E. coli* causing hemorrhage and regression of an experimental mouse tumor. I. Isolation and properties. *J. Natl. Cancer Inst.* **13**:157.

Inage, M., Chaki, H., Kusumoto, S., and Shiba, T., 1980, Chemical synthesis of bisdephospho lipid A of *Salmonella* endotoxin, *Chem. Lett.* **1980**:1373.

Jann, B., Reske, K., and Jann, K., 1975, Heterogeneity of lipopolysaccharides: Analysis of polysaccharide chain lengths by sodium dodecylsulfate-polyacryl-amide gel electrophoresis, *Eur. J. Biochem.* **60**:239.

Jarrell, K., and Kropinski, A. M., 1977, The chemical composition of the LPS from *Pseudomonas aeruginosa* strain PAO and a spontaneously derived rough mutant, *Microbios* **19**:103.

Johnson, A. G., and Nowotny, A., 1964, Relationship of structure to function in bacterial O-antigens. III. Biological properties of endotoxoids, *J. Bacteriol.* **87**:809.

Johnson, A. G., Gaines, S., and Landy, M., 1956, Studies on the O-antigen of *Salmonella typhosa*. V. Enhancement of antibody response to protein antigens by the purified lipopolysaccharide, *J. Exp. Med.* **103**:225.

Jona, J. L., 1916, A contribution to the experimental study of fever, *J. Hyg.* **15**:169.

Kabir, S., 1976, Electrophoretic studies of smooth and rough *Salmonella* lipopolysaccharides on polyacrylamide gel, *Microbios Lett.* **1**:79.

Kasai, N., 1966, Chemical studies on the lipid component of endotoxin with special emphasis on its relation to biological activities, *Ann. N.Y. Acad. Sci.* **133**:486.

Kasai, N., and Nowotny, A., 1967, Endotoxic glycolipid from a heptoseless mutant of *Salmonella minnesota*, *J. Bacteriol.* **94**:1824.

Kasai, N., Hayashi, Y., and Komatsu, N., 1970, Studies on the endotoxic glycolipid of a heptoseless *S. minnesota* mutant. 3. A direct extraction method of the glycolipid with chloroform–methanol, *Jpn. J. Bacteriol.* **25**:163.

Kenne, L., Lindberg, B., Petersson, K., and Romanowska, E., 1977, Basic structure of the oligosaccharide repeating-unit of the *Shigella flexneri* O-antigens, *Carboh. Res.* **56**:363.

Kim, Y. B. and Watson, D. W., 1966, Role of antibodies in reactions to gram-negative bacterial endotoxins, *Ann. N.Y. Acad. Sci.* **133**:727.

Kim, Y. B., and Watson, D. W., 1967, Biologically active endotoxins from *Salmonella* mutants deficient in O- and R-polysaccharides and heptose, *J. Bacteriol.* **94**:1320.

Kiso, M., Nishiguchi, H., and Hasegawa, A., 1980, Application of ferric chloride-catalyzed glycosylation to a synthesis of glycolipids, *Carbohydr. Res.* **81**:C13.

Kiso, M., Nishiguchi, H., Murase, S., and Hasegawa, A., 1981, A convenient synthesis of 2-deoxy-2-(D- and -(L-3-hydroxytetradecanoylamino)-D-glucose: Diastereoisomers of the monomeric, lipid A component of the bacterial LPS, *Carbohyd. Res.* **88:**C5.

Kiso, M., Hasegawa, A., Okumura, H., and Azuma, I., 1982, Synthetic and biological studies on the lipid A component of the bacterial lipopolysaccharides, in: *Immunomodulation by Microbial Products and Related Synthetic Compounds* (Y. Yamamura and S. Kotani, eds.), Elsevier, Amsterdam.

Konno, S., 1974, Studies on the lipid components of endotoxin. I. Thin-layer chromatographic analyses and purification of lipids in acid hydrolyzed endotoxin of *Salmonella typhosa*, *Kitasato Arch. Exp. Med.* **47:**115.

Konno, S., 1978, Studies on the lipid components of endotoxin. II. Chemical analyses and biological properties of neutral, polar-I and polar-II subfractions, *Microbiol. Immunol.* **22:**287.

Kontrohr, T., and Kocsis, B., 1978, Structure of the hexose region of Shigella sonnei Phase II lipoplysaccharide with 3-Deoxy-D-manno-octulosonic acid as possible immunodeterminant and its relation to Escherichia coli R₁ core, *Eur. J. Biochem.* **88:** 267.

LaGrange, P. H., and Mackaness, G. B., 1975, Effects of bacterial lipopolysaccharide on the induction and expression of cell-mediated immunity. II. Stimulation of the efferent arc, *J. Immunol.* **114:**447.

Lehrer, S., 1971, Enzymatic effects on bacterial endotoxins. Ph.D. thesis, Temple University, Philadelphia.

Lehrer, S., and Nowotny, A., 1972, Isolation and purification of endotoxin by hydrolytic enzymes, *Infect. Immun.* **6:**928.

Leive, L., 1965, Release of lipopolysaccharide by EDTA treatment of *E. coli*, *Biochem. Biophys. Res. Commun.* **21:**290.

Leive, L., 1977, Domains involving nonrandom distribution of lipopolysaccharide in outer membrane of *Escherichia coli*, *Proc. Natl. Acad. Sci. USA* **74:** 5065.

Leive, L., and Morrison, D. C., 1972, Isolation of lipopolysaccharides from bacteria, *Methods Enzymol.* **28:**254.

Lüderitz, O., 1970, Recent results on the biochemistry of the cell wall lipopolysaccharides of *Salmonella* bacteria, *Angew. Chem.* **9:**649.

Lüderitz, O., Galanos, C., Risse, H. J., Ruschmann, E., Schlecht, S., Schmidt, G., Schulte-Holthausen, H., Wheat, R., and Westphal, O., 1966, Structural relationships of *Salmonella* O and R antigens, *Ann. N.Y. Acad. Sci.* **133:**349.

Lüderitz, O., Jann, K., and Wheat, R., 1968, Somatic and capsular antigens of gram-negative bacteria, *Compr. Biochem.* **26A:**105.

Lüderitz, O., Galanos, C., Lehmann, V., Mayer, H., Rietschel, E. T., and Weckesser, J., 1978, Chemical structure and biological activities of lipid A's from various bacterial families, *Naturwissenschaften* **65:**578.

McCabe, W. R., 1972, Immunization with R mutants of *S. minnesota*. I. Protection against challenge with heterologous gram-negative bacilli, *J. Immunol.* **108:**601.

McIntire, F. C., Hargie, M. P., Schenck, J. R., Finley, R. A., Sievert, H. W., Rietschel, E. T., and Rosenstreich, D. L., 1976, Biologic properties of nontoxic derivatives of a lipopolysaccharide from *Escherichia coli* K235, *J. Immunol.* **117:**674.

McNeill, T. A., 1970, Antigenic stimulation of bone marrow colony forming cells. III. Effect *in vivo*, *Immunology* **18:**61.

Maddy, A. H., 1964, The solubilization of the protein of the ox-erythrocyte ghost, *Biochim. Biophys. Acta* **88:**448.

Männel, D. N., Moore, R. N., and Mergenhagen, S. E., 1980, Macrophages as a source of tumoricidal activity (tumor-necrotizing factor), *Infect. Immun.* **30:**523.

Marsh, D. G., and Crutchley, M. J., 1967, Purification and physico-chemical analysis of fractions from the culture supernatant of *Escherichia coli* O78K80: Free endotoxin and a non-toxic fraction, *J. Gen. Microbiol.* **47**:405.

Martin, A. R., 1934, The toxicity for mice of certain fractions isolated from *Bact. aertrycke*, *Br. J. Exp. Pathol.* **15**:137.

Mattsby-Baltzer, I., and Kaijser, B., 1979, Lipid A and anti-lipid A, *Infect. Immun.* **23**:758.

McGhee, J. R., Farrar, J. J., Michalek, S. M., Mergenhagen, S. E., and Rosenstreich, D. L., 1979, Cellular requirements of lipopolysaccharide adjuvanticity: a role for both T lymphocytes and macrophages for in vitro responses to particulate antigen, *J. Exp. Med.* **149**:793.

Mefferd, R. B., Henkel, D. T., and Loeffer, J. B., 1953, Effect of piromen on survival of irradiated mice, *Proc. Soc. Exp. Biol. Med.* **83**:54.

Metcalf, D., 1971, Acute antigen-induced elevation of serum colony stimulating factor (CSF) levels, *Immunology* **21**:427.

Miles, A. A., and Pirie, N. W., 1939, The properties of antigenic preparations from *Brucella melitensis*. I. Chemical and physical properties of bacterial fractions, *Br. J. Exp. Pathol.* **20**:83.

Möller, G., Andersson, J., and Sjöberg, O., 1972, Lipopolysaccharides can convert heterologous red cells into thymus-independent antigens, *Cell. Immunol.* **4**:416.

Morgan, W. T. J., 1937, Studies in immunochemistry. II. The isolation and properties of a specific substance from *B. dysenteriae* (Shiga), *Biochem. J.* **31**:2003.

Morgan, W. T. J., and Partridge, S. M., 1940, Studies in immunochemistry. 4. The fractionation and nature of antigenic material isolated from *Bact. dysenteriae* (Shiga), *Biochem. J.* **34**:169.

Morgan, W. T. J., and Partridge, S. M., 1941, Studies in immunochemistry. VI. The use of phenol and of alkali in the disintegration of antigenic material from *Bact. dysenteriae* (Shiga), *Biochem. J.* **35**:1140.

Morrison, D. C., and Leive, L., 1975, Fractions of lipopolysaccharide from *E. coli* Olll-B4 prepared by two extraction procedures, *J. Biol. Chem.* **250**:2911.

Morrison, D. C., and Ryan, J. L., 1979, Bacterial endotoxins and host immune response, *Adv. Immunol.* **28**:293.

Morrison, D. C., and Ulevitch, R. J., 1978, Effects of bacterial endotoxins on host mediation systems, *Am. J. Pathol.* **93**:526.

Morrison, D. C., Betz, S. J., and Jacobs, D. M., 1976, Isolation of a lipid A bound polypeptide responsible for "LPS-initiated" mitogenesis of C3H/HeJ spleen cells, *J. Exp. Med.* **144**:840.

Mullan, W. A., Newsome, P. M., Cunnington, P. G., Palmer, G. H., and Wilson, M. E., 1974, Protection against gram-negative infections with anti-serum to lipid A from *Salmonella minnesota* R595, *Infect. Immun.* **10**:1195.

Müller, P. T., 1911, Über den Bakteriengehalt des in Apotheken erhältlichen destilliesten Wassers, *Dtsch. Med. Wochenschr.* **58**:2739.

Nauts, H. C., Fowler, G. A., and Bogatko, F. H., 1953, A review of the influence of bacterial infection and of bacterial products (Coley's toxins) on malignant tumors in man, *Acta Med. Scand.* **145**(Suppl. 276):103.

Ng, A. K., Chang, C.-M., Chen, C.-H., and Nowotny, A., 1974, Comparison of the chemical structure and biological activities of the glycolipids of *Salmonella minnesota* R595 and *Salmonella typhimurium* SL1102, *Infect. Immun.* **10**:938.

Ng, A. K., Butler, R. C., Chen, C.-H., and Nowotny, A., 1976a, Relationship of structure to function in bacterial endotoxins. IX. Differences in the lipid moiety of endotoxic glycolipids, *J. Bacteriol.* **126**:511.

Ng, A. K., Chen, C.-H., Chang, C.-M., and Nowotny, A., 1976b, Relationship of structure to function in bacterial endotoxins: Serologically cross-reactive components and their effect on protection of mice against some gram-negative infections, *J. Gen. Microbiol.* **94:**107.

Noll, H., and Braude, A. I., 1961, Preparation and biological properties of a chemically modified *Escherichia coli* endotoxin of high immunogenic potency and low toxicity, *J. Clin. Invest.* **40:**1935.

Nowotny, A., 1959, On the chemistry of the lipid components (lipid A) of endotoxins. Lecture, International Conference on Biological Effects of Endotoxins in Relation to Immunity, Freiburg i. Br., Germany.

Nowotny, A., 1961, Molecular structure of a phosphomucolipid and its occurrence in some strains of *Salmonella, J. Am. Chem. Soc.* **83:**501.

Nowotny, A., 1963, Relation of structure to function in bacterial O-antigens. II. Fractionation of lipids present in Boivin-type endotoxin of *Serratia marcescens, J. Bacteriol.* **85:**427.

Nowotny, A., 1963, Endotoxoid preparations, *Nature (London)* **197:**721.

Nowotny, A., 1964, Chemical detoxification of bacterial endotoxins, in: *Bacterial Endotoxins* (M. Landy and W. Braun, eds.), p. 29, Rutgers University Press, New Brunswick, N.J.

Nowotny, A., 1966, Heterogeneity of endotoxic bacterial lipopolysaccharides revealed by ion exchange column chromatography, *Nature (London)* **210:**278.

Nowotny, A., 1969, Molecular aspects of endotoxic reactions, *Bacteriol. Rev.* **33:**72.

Nowotny, A., 1971a, Chemical and biological heterogeneity of endotoxins, in: *Microbial Toxins* Vol. IV (G. Weinbaum, S. Kadis, and S. J. Ajl, eds.), p. 309, Academic Press, New York.

Nowotny, A., 1971b, Relationship of structure and biological activity of bacterial endotoxins, *Naturwissenschaften* **58:**397.

Nowotny, A., 1977, Relation of structure to function in bacterial endotoxins, in: *Microbiology 1977* (D. Schlessinger, ed.), American Society for Microbiology, Washington, D. C.

Nowotny, A., 1983, Shedding bacteria, in: *Biomembranes* Vol. 11 (A. Nowotny ed.), p. 1, Plenum Press, New York.

Nowotny, A. and Butler, R. C., 1980, Studies on the endotoxin induced tumor resistance, *Adv. Exp. Med. Biol.* **121B:**455.

Nowotny, A., Lüderitz, O., und Westphal, O., 1958, Rundfilter-Chromatographie langkettiger Fettsäuren bei der Analyse bakterieller Lipopolysaccharide, *Biochem. Z.* **330:**47

Nowotny, A., Thomas, S., and Duron, O. S., 1963, Chemistry of firmly-bound cell-wall lipids in gram-negative bacteria, in: *Biochemical Problems of Lipids, BBA Library* **1:**422.

Nowotny, A., Radvany, R., and Neale, N., 1965, Neutralization of toxic bacterial O-antigens with O-antibodies while maintaining their stimulus on nonspecific resistance, *Life Sci.* **4:**1107.

Nowotny, A., Cundy, K., Neale, N., Nowotny, A. M., Radvany, R., Thomas, S., and Tripodi, D., 1966, Relation of structure to function in bacterial O-antigens. IV. Fractionation of the components, *Ann. N.Y. Acad. Sci.* **133:**586.

Nowotny, A., Kasai, N., and Tripodi, D., 1967, The use of mutants for the study of the relationship between structure and function in endotoxins, in: *La structure et les effets biologiques des produits bacteriens provenants de germes Gram-négatifs* (L. Chedid ed.), p. 79, Colloq. Int. sur les Endotoxines, CNRS No. 174, Paris.

Nowotny, A., Golub, S., and Key, B., 1971, Fate and effect of endotoxin derivatives in tumor-bearing mice, *Proc. Soc. Exp. Biol. Med.* **136:**66.

Nowotny, A., Behling, U. H. and Chang, H. L., 1975, Relation of structure to function in bacterial endotoxins. VIII. Biological activities in a polysaccharide-rich fraction, *J. Immunol.* **115:**199.

Nowotny, A., Behling, U. H., and Nowotny, A. M., 1982, Molecular aspects of endotoxin induced beneficial effects, in: *Immunomodulation by Microbial Products and Related Synthetic Compounds* Y. Yamamura and S. Kotani, eds.), Elsevier, Amsterdam.

Nowotny, A. M., Thomas, S., Duron, O. S., and Nowotny, A., 1963, Relations of structure to function in bacterial O-antigens. I. Isolation methods. *J. Bacteriol.* **85:**418.

O'Malley, W. E., Achinstein, B., and Shear, M. J., 1963, Action of bacterial polysaccharide on tumors. III. Repeated response of sarcoma 37 in tolerant mice, to *Serratia marcescens* endotoxins, *Cancer Res.* **23:**890.

Ormsbee, R. A., Bell, E. J., and Lackman, D. B., 1962, Antigens of *Coxiella burnetii*. I. Extraction of antigens with nonaqueous organic solvents, *J. Immunol.* **88:**741.

Ozato, K., Adler, W. H., and Ebert, J. D., 1975, Synergism of bacterial lipopolysaccharides and concanavalin A in the activation of thymic lymphocytes, *Cell. Immunol.* **17:** 532.

Palmer, J. W., and Gerlough, T. D., 1940, A simple method for preparing antigenic substances from the typhoid bacillus, *Science* **92:**155.

Palva, E. T., and Mäkelä, P. H., 1980, LPS heterogeneity in *Salmonella typhimurium* analyzed by sodium dodecyl sulfate polyacrylamide gel electrophoresis, *Eur. J. Biochem.* **107:**137.

Parant, M. A., Parant, F. J., and Chedid, L. A., 1980, Enhancement of resistance to infections by endotoxin-induced serum factor from *Mycobacterium bovis* BCG-infected mice, *Infect. Immun.* **28:**654.

Peavy, D. L., Adler, W. H., and Smith, R. T., 1970, The mitogenic effects of endotoxin and staphylococcal enterotoxin B on mouse spleen cells and human peripheral lymphocytes, *J. Immunol.* **105:**1453.

Peavy, D. L., Adler, W. H., Shands, J. W., and Smith, R. T., 1974, Selective effects of mitogens on subpopulations of mouse lymphoid cells, *Cell. Immunol.* **11:**86.

Pfeiffer, R., 1892, Untersuchungen über das Cholera Gift, *Z. Hyg. Infektionskr.* **11:**393.

Pluznik, D. H., Apte, R. N., and Galanos, C., 1976, Lipid A, the active part of bacterial endotoxins in inducing serum colony stimulating activity and proliferation of splenic granulocyte/macrophage progenitor cells, *J. Cell. Physiol.* **87:**71.

Quesenberry, P., Morley, A., Stohlman, F., Jr., Rickard, K., Howard, D., and Smith, M., 1972, Effect of endotoxin on granulopoiesis and colony-stimulating factor, *N. Engl. J. Med.* **286:**227.

Radvany, R., Neale, N., and Nowotny, A., 1966, Relation of structure to function in bacterial O-antigens. VI. Neutralization of endotoxic O-antigens by homologous O-antibody, *Ann. N.Y. Acad. Sci.* **133:**763.

Raichvarg, D., Guenounou, M., Brossard, C., Alouf, J. E., and Agernay, J., 1980, Endotoxin-like substances in bacterial cell-outer membrane: Correlation between structure and biological activities of *Haemophilus influenzae* type A endotoxin, in: *Bacterial Endotoxins and Host Response* (M. K. Agarwal, ed.), Elsevier/North-Holland, Amsterdam.

Raichvarg, D., Guenounou, M., Brossard, C., and Agernay, J., 1981, Characteristics of a lipid preparation (lipid A) from *Haemophilus influenzae* type A lipopolysaccharide, *Infect Immun.* **33:**49.

Raistrick, H., and Topley, W. W. C., 1934, Immunizing fractions isolated from *Bact. aertrycke, Br. J. Exp. Pathol.* **15:**113.

Ralovich, B., and Lang, C., 1971, Antipyrogenic effect of immune sera, *Z. Immunitaetsforsch. Exp.* **141:**251.

Rank, W. R., 1969, Untersuchungen zur Antikörper-Biosynthese, Ph.D. Thesis, Max-Planck-Institut für Biochemie, München, Germany.

Rank, W. R., Flugge, U., and DiPauli, R., 1969, Inheritance of the lipid-A induced 19S plaque-forming cell response in mice: Evidence for three antigen recognition mechanisms, *Behringwerk Mitt.* **49**(Suppl):222.

Repaske, R., 1958, Lysis of gram-negative organisms and the role of versene, *Biochim. Biophys. Acta* **30**:225.

Ribi, E., Milner, K. C., and Perrine, T. D., 1959, Endotoxic and antigenic fractions from the cell wall of *Salmonella enteritidis:*Methods for separation and some biological activities, *J. Immunol.* **82**:75.

Ribi, E., Haskins, W. T., Landy, M., and Milner, K. C., 1961a, Preparation and host-reactive properties of endotoxin with low content of nitrogen and lipid, *J. Exp. Med.* **114**: 647.

Ribi, E., Haskins, W. T., Landy, M., and Milner, K. C., 1961b, Symposium on bacterial endotoxins. I. Relationship of chemical composition to biological activity, *Bacteriol. Rev.* **25**:427.

Ribi, E., Parker, R., Strain, S. M., Mizuno, Y., Nowotny, A., Von Eschen, K., Cantrell, J. L., McLaughlin, C. A., Hwang, K. M., and Goren, M. B., 1979, Peptides as requirement for immunotherapy of the guinea pig line-10 tumor with endotoxins, *Cancer Immunother.* **7**:43.

Rietschel, E. T., and Galanos, C., 1977, Lipid A antiserum-mediated protection against lipopolysaccharide- and lipid A-induced fever and skin necrosis, *Infect. Immun.* **15**:34.

Rietschel, E. T., Gottert, H., Lüderitz, O., and Westphal, O., 1972, Nature and linkage of the fatty acids present in the lipid A component of *Salmonella* lipopolysaccharides, *Eur. J. Biochem.* **28**:166.

Rietschel, E. T., Hase, S., King, M.-T., Redmond, J., and Lehmann, V., 1977, Chemical structure of lipid A, in: *Microbiology* 1977 (D. Schlessinger, ed.), American Society for Microbiology, Washington, D.C.

Roberts, R. S., 1949, The endotoxin of *Bact. coli*, *J. Comp. Pathol. Ther.* **59**:284.

Roberts, R. S., 1966, Preparation of endotoxin, *Nature (London)* **209**:80.

Rosenstreich, D. L., Asselineau, J., Mergenhagen, S. E., and Nowotny, A., 1974, A synthetic glycolipid with B-cell mitogenic activity, *J. Exp. Med.* **140**:1404.

Rosner, M. R., and Khorana, H. G., 1979, The structure of lipopolysaccharide from a heptose-less mutant of *Escherichia coli* K-12, *J. Biol. Chem.* **254**:5918.

Rosner, M. R., Tang, J.-Y., Barzilay, I., and Khorana, H. G., 1979a, Structure of the lipopolysaccharide from an *Escherichia coli* heptose-less mutant. I. Chemical degradations and identification of products, *J. Biol. Chem.* **254**:5906.

Rosner, M. R., Verret, R. C., and Khorana, H. G., 1979b, The structure of lipopolysaccharide from an *Escherichia coli* heptose-less mutant, *J. Biol. Chem.* **254**:5926.

Rote, N. S., 1974, The role of long-chain carboxylic acids endotoxin, Ph.D. thesis, Temple University, Philadelphia.

Roussy, G., 1889, Recherches expérimentales: Substances calorigènes et frigorigènes d'origine microbienne; pyretogènine et frigorigènine, *Gaz. Hop. (Paris)* **62**:171.

Rudbach, J. A., Anacker, R. L., Haskins, W. T., Milner, K. C., and Ribi, E., 1967, Physical structure of a native protoplasmic polysaccharide from *Escherichia coli*, *J. Immunol.* **98**:1.

Rudbach, J. A., Milner, K. C., and Ribi, E., 1969, Occult endotoxin in bacterial protoplasm, *J. Bacteriol.* **99**:51.

Russell, R. R. B., 1976, Free endotoxin—A review, *Microbios Lett.* **2**:125.

Russell, R. R. B., and Johnson, K. G., 1975, SDS-polyacrylamide gel electrophoresis of lipopolysaccharides, *Can. J. Microbiol.* **21**:2013.

Salton, M. R. J., 1971, Bacterial membranes, *CRC Crit. Rev. Microbiol.* **1**:161.

Samelson, S., 1912, Über das sogenannte Kochsalzfieber, *Monatsschr. Kinderheilkd.* **11**:125.

Scheid, M. P., Hoffmann, M. K., Kumuro, K., Hämmerling, U., Abbott, J., Boyse, E. A., Cohen, G. H., Hooper, J. A., Schulof, R. S., and Goldstein, A. L., 1973, Differentiation of T cells induced by preparations from thymus and by nonthymic agents: The determined state of the precursor cell, *J. Exp. Med.* **138**:1027.

Schmidt, G., Schlecht, S., Lüderitz, O., and Westphal, O., 1969, Untersuchungen zur Typisierung von Salmonella-R-Formen. I. Mitteilung, Mikrobiologische und serologische Untersuchungen an Salmonella minnesota-Mutanten, *Zentralbl. Bakteriol. Parasitenkd. Infektionskr. Hyg.* **209**:483.

Schmidtke, J. R., and Najarian, J. S., 1975, Synergistic effects on DNA synthesis of phytohemagglutinin or concanavalin A and lipopolysaccharide in human peripheral blood lymphocytes, *J. Immunol.* **114**:742.

Scibienski, R. J., 1980, Immunologic properties of protein-lipopolysaccharide complexes-III. Role of carbohydrate in the LPS adjuvant effect, *Molecular Immunol.* **17**:21.

Seibert, F. B., 1925, The cause of many febrile reactions following intravenous infections, *Am. J. Physiol.* **71**:621.

Shear, M. J., 1941, Effect of a concentrate from *B. prodigiosus* filtrate on subcutaneous primary induced mouse tumors, *Cancer Res.* **1**:731.

Shear, M. J., and Turner, F. C., 1943, Chemical treatment of tumors. V. Isolation of the hemorrhage-producing fraction from *Serratia marcescens (Bacillus prodigiosus)* culture filtrate, *J. Natl. Cancer Inst.* **4**:81.

Shear, M. J., Perrault, A., and Adams, J. R., Jr., 1943, Chemical treatment of tumors. VI. Method employed in determining the potency of hemorrhage-producing bacterial preparations, *J. Natl. Cancer Inst.* **4**:99.

Shiba, T., 1982, Synthetic approach to elucidation for chemical structure and biological activity of lipid A, in: *Immunomodulation by Microbial Products and Related Synthetic Compounds* (Y. Yamamura and S. Kotani, ed.), Elsevier, Amsterdam.

Shwartzman, G., 1928, Studies on *Bacillus typhosus* toxic substance. I. The phenomenon of local reactivity to *B. typhosus* culture filtrate, *J. Exp. Med.* **48**:247.

Skidmore, B. J., Morrison, D. C., Chiller, J. M., and Weigle, W. O., 1975, Immunologic properties of bacterial lipopolysaccharide (LPS). II. The unresponsiveness of C3H/HeJ mouse spleen cells to LPS-induced mitogenesis is dependent on the method used to extract LPS, *J. Exp. Med.* **142**:1488.

Smith, W. W., Alderman, I. M., and Gillespie, R. E., 1957, Increased survival in irradiated animals treated with bacterial endotoxins, *Am. J. Physiol.* **191**:124.

Smith, W. W., Marston, R. A., and Cornfield, J., 1959, Patterns of hemopoietic recovery in irradiated mice, *Blood* **14**:737.

Smith, W. W., Alderman, I. M., and Cornfield, J., 1961, Granulocyte release by endotoxin in normal and irradiated mice, *Am. J. Physiol.* **201**:396.

Smith, W. W., Brecher, G., Fred, S., and Budd, R. A., 1966, Effect of endotoxin on the kinetics of hemopoietic colony-forming cells in irradiated mice, *Radiat. Res.* **27**:710.

Souvannavong, V., and Adam, A., 1980, Opposite effects of the synthetic adjuvant *N*-acetyl-muramyl-L-alanyl-D-isoglutamine on the immune response in mice depending on experimental conditions, *Eur. J. Immunol.* **10**:654.

Spitznagel, J. K., and Allison, A. C., 1970, Mode of action of adjuvants: Effects on antibody responses to macrophage-associated bovine serum albumin, *J. Immunol.* **104**:128.

Staub, A. M., and Westphal, O., 1964, Etude chimique et biochimique de la specificite immunologique des polyosides bacteriens, *Bulletin de la Societe de Chimie biologique*, **XLVI**, **12**:1647.

Sultzer, B. M., 1968, Genetic control of leucocyte responses to endotoxin, *Nature (London)* **219**:1253.

Sultzer, B. M., 1969, Genetic factors in leucocyte responses to endotoxin: Further studies in mice, *J. Immunol.* **103**:32.

Sultzer, B. M., 1971, Chemical modification of endotoxin and inactivation of its biological properties, in: *Microbial Toxins*, Vol. V (S. Kadis, G. Weinbaum, and S. J. Ajl, eds.), pp. 91–126, Academic Press, New York.

Sultzer, B. M., and Goodman, G. W., 1976, Endotoxin protein: A B-cell mitogen and polyclonal activator of C3H/HeJ lymphocytes, *J. Exp. Med.* **144:**821.

Sutherland, A. M., and Wolf, H. G., 1940, Experimental studies on headache: Further analysis of mechanism of headache in migraine, hypertension and fever, *Arch. Neurol. Psychiatry* **44:**929.

Takano, T., Yoshioka, O., Mizuno, D., Watanabe, K., Ohtani, T., and Kageyama, K., 1968, Dynamic state of the spleen cells of mice after administration of the endotoxin of *Proteus vulgaris*. II. Cellular differentiation into RE-cells 48 hours after administration of the endotoxin, *Jpn. J. Exp. Med.* **38:**241.

Takayama, K., Ribi, E., and Cantrell, J. L., 1981, Isolation of a nontoxic lipid A fraction containing tumor regression activity, *Cancer Res.* **41:**2654.

Tanamoto, K.-I., Abe, C., Homma, J. Y., and Kojima, Y., 1979, Regions of the LPS of *Pseudomonas aeruginosa* essential for antitumor and interferon-inducing activities, *Eur. J. Biochem.* **97:**623.

Tate, W. J., III, Douglas, H., Braude, A. I., and Wells, W. W., 1966, Protection against lethality of *E. coli* endotoxin with "O" antiserum, *Ann. N.Y. Acad. Sci.* **133:**746.

Terry, G. G., 1939, *Fever and Psychoses*, Harper & Row (Hoeber), New York.

Tripodi, D., and Nowotny, A., 1965, Relation of structure to function in bacterial O-antigens. V. Nature of active sites in endotoxic lipopolysaccharides of *Serratia marcescens*, Information Exchange Group #5, Memo No. 67.

Tripodi, D., and Nowotny, A., 1966, Relation of structure to function in bacterial O-antigens. V. Nature of active sites in endotoxic lipopolysaccharides of *Serratia marcescens*, *Ann. N.Y. Acad. Sci.* **133:**604.

Tucker, J., 1946, Typhoid shock therapy, *Cleveland Clin. Q.* **13:**67.

Urbaschek, B., and Nowotny, A., 1968, Endotoxin tolerance induced by detoxified endotoxins (endotoxoid), *Proc. Soc. Exp. Biol. Med.* **127:**650.

Urbaschek, B., and Urbaschek, R., 1975, Some aspects of the effects of endotoxin, endotoxoid and biogenic amines, in: *Gram-negative Bacterial Infections* (B. Urbaschek, R. Urbaschek, and E. Neter, eds.), p. 323, Springer-Verlag, Berlin.

Urbaschek, B., Branemark, P.-I., and Nowotny, A., 1968, Lack of endotoxin effect on the microcirculation after pretreatment with detoxified endotoxin (endotoxoid), *Experientia* **24:**170.

Urbaschek, R., 1980, Effects of bacterial products on granulopoiesin, *Adv. Exp. Med. Biol.* **121B:**51.

Van der Zwan, J. C., Dankert, J., DeVries, K., Orie, N. G. M., and Kauffman, H. F., 1978, Purification of specific precipitinogen and extraction of endotoxin from *Haemophilus influenzae*, *J. Clin. Pathol.* **31:**370.

Webster, M., Sagin, J. F., Landy, M., and Johnson, A. G., 1955, Studies on the O-antigen of *Salmonella typhosa*. I. Purification of the antigen, *J. Immunol.* **74:**455

Westphal, O., 1960, Récentes rescherches sur la chimie et la biologie des endotoxines des bactéries a gram négatif, *Ann. Inst. Pasteur* **98:**789.

Westphal, O., and Lüderitz, O., 1954, Chemische Erforschung von Lipopolysacchariden gram negativer Bakterien, *Angew. Chem.* **66:**407.

Westphal, O., Lüderitz, O., and Bister, F., 1952, Über die Extraktion von Bakterien mit Phenol/Wasser, *Z. Naturforsch.* **7b:**148.

Westphal, O., Nowotny, A., Lüderitz, O., Hurni, H., Eichenberger, E., and Schönholtzer, G., 1958, Die Bedeutung der Lipoid-Komponente (Lipoid A) für die biologischen Wirkungen bakterieller Endotoxine. *Pharm. Acta Helv.* **33:** 401.

Westphal, O., Westphal, U., and Sommer, T., 1977, The history of pyrogen research, in: *Microbiology 1977* (D. Schlessinger, ed.), pp. 221–235, American Society for Microbiology, Washington, D.C.

Westphal, O., Lüderitz, O., Rietschel, E. T., and Galanos, C., 1980, Bacterial lipopolysaccharide and its lipid A component: Some historical and some current aspects, *Biochem. Soc. Trans.* **9:**191.

Wober, W., and Alaupovic, P., 1971, Studies on the protein moiety of endotoxin from gram-negative bacteria, *Eur. J. Biochem.* **19:**340.

Work, E., Knox, K. W., and Vesk, M., 1966, The chemistry and electron microscopy of an extracellular lipopolysaccharide from *Escherichia coli, Ann. N.Y. Acad. Sci.* **133:**438.

Wu, M.-C., and Health, E. C., 1973, Isolation and characterization of lipopolysaccharide protein from *Escherichia coli, Proc. Natl. Acad. Sci. USA* **70:**2572.

Yang, C., and Nowotny, A., 1974, Effect of endotoxin on tumor resistance in mice, *Infect. Immun.* **9:**95.

Yang-Ko, C., Grohsman, J., Rote, N., Jr., and Nowotny, A., 1974, Non-specific resistance induced in allogeneic mice to the transplantable TA3-Ha tumor by endotoxin and its derivatives, 8th International Congress on Chemotherapy, Athens, pp. 193–197.

Young, L. S., Stevens, P., and Ingram, J., 1975, Functional role of antibody against "core" glycolipid of Enterobacteriaceae, *J. Clin. Invest.* **56:**850.

2

Effect of LPS on Carbohydrate and Lipid Metabolism

Judy A. Spitzer and John J. Spitzer

I. Introduction

Administration of endotoxin, a lipopolysaccharide (LPS) extracted from cell walls of gram-negative bacteria, elicits alterations in various metabolic parameters in many mammalian species, including man. Studies of the mechanism of metabolic alterations induced by LPS are important because release of endotoxin may contribute to the effects of sepsis, and clinical or experimental infections may alter the course of other types of trauma, shock, or injury. The overall metabolic and endocrine effects of endotoxicosis are predominantly catabolic. In general, the endocrine changes appear to be directed toward water and salt conservation, maintenance of blood pressure, and the mobilization of fuel primarily from carbohydrate stores. Eventually the breakdown of muscle protein follows, providing additional substrate for gluconeogenesis and satisfying the energy requirements of peripheral tissues and the brain.

An understanding of the metabolic changes following LPS administration requires a consideration of the effects of starvation. Endotoxin-treated animals usually do not consume food and water. Glucose is derived from glycogenolysis of liver glycogen and from gluconeogenesis using primarily lactate, amino acids, and glycerol as glucose precursors. The long-term, chronic postendotoxin phase differs from the adaptive changes in starvation in not being able to mount the physiologic response of excessive breakdown of adipose stores with subsequent fatty

Judy A. Spitzer and John J. Spitzer • Department of Physiology, Louisiana State University Medical Center, New Orleans, Louisiana 70112.

acid and ketone utilization, but rather catabolizing muscle proteins and thereby providing branched-chain amino acids for satisfying fuel needs *in situ* and amino acid precursors for gluconeogenesis (Kinney and Felig, 1979).

The nature of the metabolic response to endotoxicosis depends on the time elapsed after LPS administration. For example, initial hyperglycemia (Menten and Manning, 1924) followed by a marked fall in blood glucose (Cameron *et al.*, 1940) has been known for a long time and documented repeatedly (Berry, 1971). Different phases of endotoxin shock also reflect different rates of gluconeogenesis; and actual elevation in gluconeogenesis from lactate and alanine is observed a few hours after introduction of LPS into the system (Kuttner and Spitzer, 1978; Wolfe *et al.*, 1977a), while preterminally or with lethal doses of LPS a significant decline in gluconeogenesis is recorded (Filkins and Cornell, 1974; Wolfe *et al.*, 1977b).

Although alterations of protein metabolism are equally important, the aim of this review is limited to highlighting endotoxin-induced adaptive changes in various pathways of carbohydrate and lipid metabolism. The role of hormonal modulators of these processes at several levels of organization from whole body to individual cells and subcellular organelles also will be discussed. It was our intention to primarily consider metabolic alterations elicited by relatively small doses of LPS, thus to deal with conditions that cause longer-lasting and milder injury followed by eventual recovery in the majority of individuals. Only in some instances will we make reference to preterminal events.

II. Hemodynamic Alterations following LPS

Although this review deals with the metabolic effects of LPS, inclusion of a brief reference to some of the hemodynamic alterations observed following LPS administration is justified and perhaps even necessary, since the distribution of cardiac output, and thus organ and tissue blood flow changes exert a marked effect on the overall metabolism of all major substrates. In alluding to hemodynamic changes, it should be kept in mind that a considerable variability exists among different species in their hemodynamic responses to LPS (Ferguson *et al.*, 1978). Also, in most instances blood flow changes are LPS dose dependent, and thus a certain amount of oversimplification had to be introduced in the following brief discussion.

Blood pressure decreases precipitously following LPS administration. The marked decrease is followed by a variable rebound towards control values (Raymond *et al.*, 1981; Romanosky *et al.*, 1980a; Scott *et al.*, 1973; Wolfe *et al.*, 1977a). Cardiac output also falls shortly after LPS as seen in Fig. 1 and either remains low (Ferguson *et al.*, 1978) or returns to near control levels (Romanosky *et al.*, 1980a; Wolfe *et al.*, 1977a) depending on the severity of LPS insult. There is a general tendency on the part of the body to protect myocardial, cerebral, and hepatic arterial blood flow (Ferguson *et al.*, 1978). Thus, coronary sinus blood flow in the dog following LPS administration may not change at all (Spitzer *et al.*, 1974), or may be moderately decreased (Scott *et al.*, 1973), depending on the dose of LPS given. Hepatic arterial blood flow

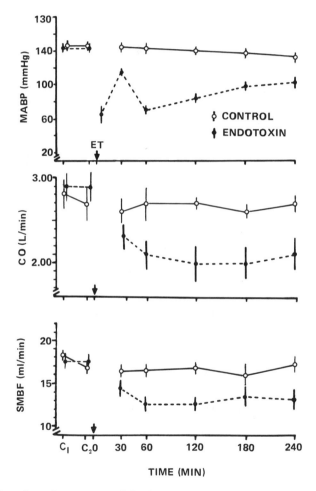

Figure 1. Hemodynamic responses following endotoxin administration (data from Romanosky *et al.*, 1980a).

Table. 1. Changes of Cardiac Output and Blood Flow in Selected Organs after
Endotoxin Administration[a]

	Control	4 hr postendotoxin
Cardiac output	1.65 ± 0.13	1.21 ± 0.32
(liter/min)		(-27%)
Mesenteric adipose	14.1 ± 1.8	3.0 ± 1.2
tissue (ml/min/100 g)		(-79%)
Pancreas	49.6 ± 6.8	10.0 ± 3.1
(ml/min/200 g)		(-80%)
Hepatic (arterial)	18.4 ± 6.9	32.9 ± 9.9
(ml/min/100 g)		$(+79\%)$

[a] Data from Spitzer *et al.* (1980).

is increased in the face of a fairly severe fall in cardiac output (Ferguson *et al.*, 1978), thus protecting O_2 supply to the liver. Blood flow to
skeletal muscle is usually diminished in proportion to changes in cardiac output (Wolfe *et al.*, 1977a). Endotoxin affects adipose tissue and
pancreatic blood flow even more markedly than the cardiac output
(Ferguson *et al.*, 1978; Spitzer *et al.*, 1973b, 1980; Wolfe *et al.*, 1977a),
and thus blood flow may be severely compromised in these organs (Table 1).

III. Alterations in Carbohydrate Metabolism following LPS

Glucose transport. The influence of endotoxin on glucose transport
and uptake had been investigated at the cellular level and in isolated
muscle preparations both *in vivo* and *in vitro*. Endotoxin-induced changes in glucose transport were studied in isolated adipocytes (Leach and
Spitzer, 1981). Six hours after *E. coli* endotoxin injection, the basal uptake of 3-*O*-methylglucose was significantly increased, with no alteration
in insulin responsiveness as seen in Fig. 2. Using the glucose analog
2-deoxyglucose, basal uptake was not significantly altered; however, an
inappropriately low response to a submaximal dose of insulin was observed. Spermine-stimulated uptake was similarly depressed. Endotoxin
seemingly did not effect the hexokinase-mediated phosphorylating system. The cytochalasin B-insensitive uptake—i.e., the diffusion component of total transport—of both 2-deoxyglucose and 3-*O*-methylglucose
increased significantly in endotoxic cells. These results are consistent
with the hypothesis that (1) a site of endotoxin-induced insulin resistance is at the cell membrane level; (2) endotoxin does not compro-
mise the hexokinase system; (3) the cell membrane-localized effect of

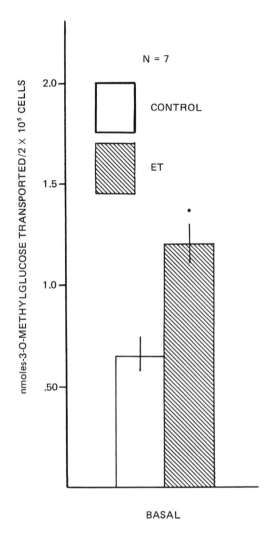

Figure 2. Basal 3-O-methylglucose uptake in adipocytes isolated from rats 6 hr after the injection of 0.9% saline (control) or an LD_{50} dose of LPS (endotoxin, ET). Substrate concentration was 2 mM, and the assay was performed at 24°C (Leach and Spitzer, 1981).

endotoxin on hexose transport is not necessarily mediated by the insulin receptor; and (4) the entry of 2-deoxyglucose and 3-O-methylglucose may involve two separate transport systems, differentially modulated by endotoxin.

Elevated uptake of glucose by skeletal muscle was observed following *E. coli* endotoxin administration (Raymond *et al.*, 1981; Romanosky *et al.*, 1980a) as illustrated in Fig. 3. Under some circumstances, LPS-caused hypoxia may be responsible for increased glucose uptake, depending on prevailing conditions. Hemidiaphragms obtained from

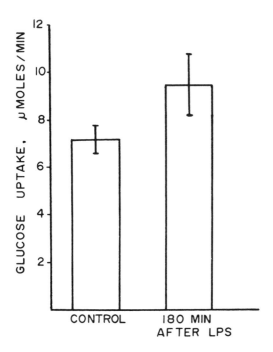

Figure 3. Change in skeletal muscle glucose uptake following endotoxin administration (data from Romanosky *et al.*, 1980a).

endotoxin-injected rats also demonstrated higher glucose uptake than control tissue (Filkins and Buchanan, 1977). Endotoxin-induced increase in glucose uptake and lactate release was also shown in white blood cells (Hinshaw *et al.*, 1977).

Glycolysis. The stimulatory effect of endotoxin—similar to the action of insulin—on aerobic and anaerobic glycolysis in several types of tumor cells was noted by Woods *et al.* (1961, 1964). Insulin often masked glycolytic stimulation by endotoxin, suggesting a commonly shared site of action; conversely, glycolysis of insulin-insensitive Krebs-2 carcinoma was not stimulated by entotoxin. The ability of endotoxin preparations to enhance cellular glycolysis was found to parallel their biological activity, and enzymatic (Landy *et al.*, 1962) or acid hydrolysis (Ribi *et al.*, 1961) reduced or abolished toxicity and the glycolysis-stimulating effect to a similar extent (Woods *et al.*, 1961). The glycolysis-stimulating effect of endotoxin is not limited to malignant cells and tissues; it was also observed in normal polymorphonuclear leukocytes (Cohn and Morse, 1960; Martin *et al.*, 1955), bone marrow, and spleen (Woods *et al.*, 1964).

Glycogenolysis. Endotoxin-induced hypoglycemia and rapid liver glycogen depletion have been repeatedly described (Berry, 1971; Hinshaw, 1976). The mechanism of accelerated glycogenolysis in endotoxin-poisoned animals has not yet been unequivocally elucidated; it has been attributed by various investigators to activation of catechol-amine-responsive adenylate cyclase (Bitensky *et al.*, 1971; Gimpel *et al.*, 1974), to an increase in liver phosphorylase activity (Hamosh and Shapiro, 1960), unspecified glycogenolytic enzymes (Zwadyk and Snyder, 1973a), or pyruvate kinase (Snyder *et al.*, 1971). However, results obtained in different laboratories are not always in agreement with these findings (Gartner, 1975; Zwadyk and Snyder, 1973b).

Hepatocytes isolated from fed rats 1 hr after *S. enteritidis* endotoxin administration demonstrated significantly increased rates of glycogenol-ysis (Filkins and Buchanan, 1977). *In vitro* addition of the same endotoxin to isolated hepatocytes was without effect. These results tend to support a mediated mechanism in bringing about endotoxin-induced changes in glycogenolysis.

Glucose kinetics and oxidation. The hypoglycemia that occurs in endotoxin shock (Hinshaw, 1976) could be a consequence of decreased hepatic output of glucose and/or increased extraction by the tissues of glucose from plasma. All evidence gathered so far, whether in the whole animal (glucose kinetics) or in isolated cells or tissue from endotoxin-treated animals (glucose oxidation), indicates not only a lack of impairment, but actually an increase of peripheral extraction and utilization of glucose. The effect of *E. coli* endotoxin on glucose kinetics was studied in dogs (Wolfe *et al.*, 1977a), guinea pigs (Merrill and Spitzer, 1978), and rats (Kelleher *et al.*, 1982) by means of the primed constant infusion of [6 − ^{3}H]glucose. All of these studies demonstrated significantly increased rates of disappearance (R_d) after endotoxin administration. At the isolated tissue level, basal glucose oxidation by fat pads of endotoxin-treated rats was reported to be enhanced (Filkins and Buchanan, 1977). The insulin-like action of endotoxin on glucose oxidation by isolated adipocytes was demonstrated both as a result of endotoxin injection several hours prior to sacrifice and cell isolation, and of *in vitro* endotoxin exposure of cells derived from saline-treated rats (Holley and Spitzer, 1980; Spitzer and Holley, 1979). The *in vitro* experiments, later confirmed by others (Witek-Janusek and Filkins, 1981), buttress the argument for a direct action of endotoxin in its insulinomimetic effect on glucose oxidation. LPS increases oxidation by isolated myocytes as well. Since this action requires relatively higher concentrations of LPS, it is not clear whether the same control mechanisms are affected when LPS is given *in vitro* and *in vivo* (Liu *et al.*, 1981). *In vitro* exposure of leukocytes to endotoxin has also been reported to result in increased glucose oxidation (Cline *et al.*, 1968).

Gluconeogenesis. Aside from alterations in the hormonal milieu, the most important factor influencing gluconeogenesis is the availability of the biosynthetic precursors. The lack of uniformity in the data in the shock literature is likely due to several factors, e.g., phases or intensities of the shock condition in which the measurements were made, differences in kinetic assumptions, and differences in the researchers' vantage point, whether one considers activities of key gluconeogenic enzymes, or the production of glucose under conditions prevailing in shock.

Several studies have implicated impaired gluconeogenesis as the primary factor leading to hypoglycemia after endotoxin administration (Filkins and Cornell, 1974; LaNoue et al., 1968; Shands et al., 1969), without actually demonstrating reduced hepatic glucose output in vivo. Investigation of glucose and lactate kinetics in dogs revealed significant elevation in the rate of appearance (R_a) and the rate of disappearance (R_d) of glucose for 150 min after endotoxin administration, after which both returned to control levels, but neither fell below the preinfusion value. The increased recycling of lactate (Wolfe et al., 1977a) was accompanied by increased incorporation of alanine to glucose (Kuttner and Spitzer, 1978). With studies using lethal doses of LPS, or preterminally, a significant decline in gluconeogenesis is apparent (Filkins and Cornell, 1974; Wolfe et al., 1977b). The eventual failure of hepatic gluconeogenesis in the face of marked hypoglycemia may be explained in part at the enzymatic level. Berry et al. (1968) have shown that endotoxin prevents the induction by cortisone of phosphoenolpyruvate carboxykinase (PEPCK) activity. The depletion of carbohydrate in endotoxin-poisoned animals may also be facilitated by increased pyruvate kinase activity (Snyder et al., 1971), which accelerates the loss of phosphoenolpyruvate, that can continue along the energy-yielding glycolytic and tricarboxylic acid pathways, or can be converted via the gluconeogenic pathway into glucose.

IV. Alterations in Lipid Metabolism following LPS

Although lipid metabolism accounts for a major portion of the energy produced in the body, LPS-induced changes in lipid metabolism have not been studied extensively. A short review was published on this subject a few years ago (Spitzer, 1979). The current discussion builds and expands on that review.

The major lipid metabolites found in plasma are free fatty acids (FFA), triglyceride fatty acids (TGFA), and ketone bodies. FFA are water insoluble and are transported in plasma in albumin-bound form. TGFA are solubilized in lipoprotein complexes, while ketone bodies are water soluble. The source of FFA is adipose tissue, where release is regulated by hormone-sensitive lipase and modulated by the action of various hormones (primarily catecholamines and insulin). TGFA are of hepatic origin and their production is under the influence of a very intricate set of regulatory controls. Peripheral removal of TGFA takes place through enzymatic action of tissue lipoprotein lipase (LPL). The hepatic production of ketone bodies is also regulated by a set of hormonally controlled regulatory factors. Arterial concentration appears to be the major determinant of the peripheral utilization of ketone bodies.

Lipolysis of triglyceride in adipose tissue results in the production of FFA and glycerol. Lipolytic activity in the adipose tissue is best assessed by the release of glycerol as the freed FFA may be reesterified in the tissue, while glycerol is not reutilized (due to unavailability of glycerol kinase). Following LPS administration, lipolysis in adipose tissue decreases as indicated by decreased glycerol release (Spitzer et al., 1973b). It has been demonstrated that decreased blood flow in adipose tissue also markedly inhibited lipolysis (Kovach et al., 1976). Thus, it is assumed that the severe blood flow restriction imposed on adipose tissue that accompanies an LPS insult is at least partially responsible for the decreased lipolysis observed in vivo. It is of interest to note, however, that isolated adipocytes from LPS-injected animals or normal adipocytes treated with LPS in vitro exhibit increased response to the lipolytic stimuli of catecholamines (Spitzer et al., 1973b). Fat cells treated with endotoxin in vitro also have a significantly higher cAMP level, which can be elevated severalfold by subsequent norepinephrine stimulation (Spitzer, 1974). Thus, the capability for an increased response to lipolytic stimuli is present in the cells and can be fully expressed in the proper environment. These findings accentuate the physiologic significance of maintaining adequate blood flow for the ultimate manifestation of response to lipolytic stimuli.

In addition to decreased lipolytic activity in adipose tissue in vivo, there is some reason to believe that an increased reesterification of newly lipolyzed fatty acids may also take place. Although proof of this phenomenon is presently not available, the increased availability of α-glycerophosphate along with the lowered pH and elevated lactate concentration in adipose tissue would favor such reesterification (Fredholm, 1971; Issekutz et al., 1965). Furthermore, there is some basis for assuming that lipolysis in various fat pads may be more affected by LPS than adipose tissue depots interspersed in skeletal muscle (Romanosky et al., 1980b). However, actual experimental data in support of this concept are as yet lacking. LPS administration markedly diminishes the capacity of adipose tissue to remove fatty acids from plasma TGFA as indicated by a decreased LPL activity in this tissue following LPS (Scholl et al., 1982).

Although the liver plays a key role in fat metabolism, the experimental data that are available concerning possible changes of FFA oxidation and TGFA synthesis in the liver following LPS administration are rather sparse. Arterial concentration of FFA is an important regulator of the flux of these metabolites. Changes in arterial FFA following LPS are not completely predictable; while this parameter often decreases (Romanosky *et al.*, 1980b; Spitzer *et al.*, 1972), other times it remains unchanged following LPS (Scott *et al.*, 1973; Spitzer *et al.*, 1974). As in the case under control conditions, arterial FFA concentration appears to be the major determinant of FFA turnover following LPS administration (Romanosky *et al.*, 1980b). Thus, it has been demonstrated that for a given arterial concentration, FFA turnover remains unchanged in LPS-treated animals.

Myocardium and skeletal muscle are among the major consumers of lipid metabolites and are the two peripheral tissues whose metabolism has been studied most extensively following LPS administration. Even if myocardial blood flow remains unchanged, FFA uptake and oxidation decrease markedly following LPS (Spitzer *et al.*, 1974). At the same time, the utilization of lactate by the myocardium is clearly elevated. Thus, a greater fraction of the metabolic CO_2 is derived from lactate, and a lesser fraction from FFA under these conditions (Table 2). Although direct effects of LPS on myocardial metabolism remain a possibility, as indicated by the increased lactate oxidation rate of myocardi-

Table 2. Changes of FFA Flux, Myocardial Blood Flow, and Substrate Utilization after Endotoxin Administration ($N = 6$)[a]

				Change after endotoxin	
	Control			100–140 min	170–240 min
MABP (mm Hg)	122	±	6	−47%	−29%
FFA flux (μmoles/min)	683	±	216	−15%[b]	−27%
Coronary sinus blood flow (ml/min/100 g)	106	±	11	−14%[b]	−15%[b]
FFA oxidation (μmoles/min/100 g)	19.3	±	2.5	−69%	−40%
Lactate uptake (μmoles/min/100 g)	48.5	±	10.5	+81%	+78%
Arterial FFA concn. (μmoles/ml)	0.659	±	0.093	0[b]	0[b]
Arterial lactate concn. (μmoles/ml)	1.31	±	0.24	+264%	+279%

[a] Data from Spitzer *et al.* (1974).
[b] Change not significant.

al homogenates following LPS administration (Liu and Spitzer, 1977), it seems that the increased arterial lactate concentration accompanying LPS administration is responsible for the altered substrate utilization by the myocardium under these conditions (Spitzer *et al.*, 1974). One could speculate about the "usefulness" of the myocardial utilization of a water-soluble, amply available substrate during a condition that may herald impending hypoxia (J. J. Spitzer, 1981). However, the actual mechanism of such "switch over" of substrates remains to be elucidated. TGFA serve as readily used myocardial substrates, albeit minor ones (Little *et al.*, 1970). The capacity of the myocardium to remove TGFA decreases following LPS administration, as indicated by diminished myocardial LPL activity in this tissue (Bagby and Spitzer, 1980).

Recent investigations have shown that trauma and injury decrease ketogenesis in the liver (Neufeld *et al.*, 1976). Investigation of peripheral removal of ketone bodies following LPS administration in dogs (Romanosky *et al.*, 1982) has shown that at any given arterial ketone concentration, the removal of ketone bodies was identical in saline-treated and LPS-treated animals.

V. Alterations in Endocrine Balance following LPS

Hormonal changes. The rates of appearance and utilization of metabolic fuels are significantly modulated by several hormones. The synthesis of macromolecules is promoted by insulin, while the so-called counter regulatory hormones (glucagon, catecholamines, somatotropin, glucocorticoids) stimulate the breakdown of glycogen (provision of glucose and lactate), triglycerides (source of FFA and ketone bodies as fuel substrates, glycerol as a gluconeogenic precursor), and protein (for gluconeogenic precursors). Circulating plasma levels of insulin, glucagon, catecholamines, somatotropin, ACTH, and glucocorticoids have been measured in endotoxin poisoning.

One of the outstanding endocrine features of serious bacterial infections is the fall in the molar ratio of plasma insulin to glucagon (Spitzer *et al.*, 1980), due primarily to the significant rise in plasma glucagon as seen in Fig. 4. Plasma insulin levels of endotoxin-injected animals do not deviate significantly from normal (Holley and Spitzer, 1980), except for a transient rise induced by the early hyperglycemia (Zenser *et al.*, 1974) and the hyperinsulinemia observed after glucose loading (Adelye *et al.*, 1981; Blackard *et al.*, 1976). The plasma levels of corticosteroids, ACTH, catecholamines, somatotropin, and pancreatic polypeptide are elevated during endotoxemia (Kelleher *et al.*, 1981; Motsay *et al.*, 1971; Rayfield *et al.*, 1977), while somatostatin-like activity remains unchanged (Spitzer *et al.*, 1980).

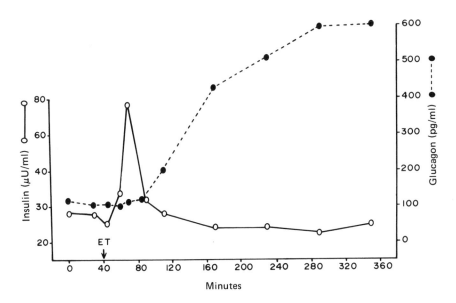

Figure 4. Plasma insulin and glucagon concentration changes following endotoxin (ET) administration (data from Spitzer *et al.*, 1980).

There is a lack of agreement concerning the effect of endotoxin on insulin concentrations and mediation of glucose dyshomeostasis by elevated insulin levels. Buchanan and Filkins (1976) assign a key role to the alterations in plasma insulin concentration in precipitating the metabolic derangements of endotoxin shock, while work by other investigators does not support this view (Zenser *et al.*, 1974).

Endotoxin-induced altered sensitivities of tissues to insulin. The issue of endotoxin possibly altering tissue sensitivity to insulin not solely by changing circulating hormone concentration, but by exerting its influence on insulin's membrane-bound receptor, was addressed by Holley and Spitzer (1980). These investigators found that adipocytes isolated from rats 6 hr after endotoxin injection had higher affinities for insulin in the low hormone dose range, implying that the insulin receptor may be a site for modulation of target cell sensitivity in endotoxemia as seen in Fig. 5.

Numerous studies indicate altered insulin function within the pathophysiologic framework of endotoxicosis. Abnormal glucose tolerance has been reported following piromen administration in normal human controls and in juvenile diabetics (Rayfield *et al.*, 1977). Insulin responsiveness and sensitivity in endotoxemia also have been studied at the tissue and isolated cellular level. Epididymal fat pads, taken from

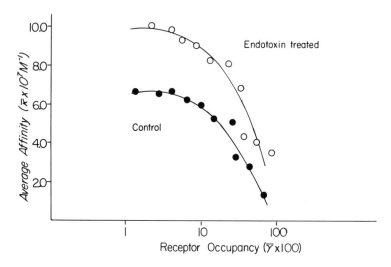

Figure 5. Average affinity profile based on displacement plots derived from four experiments in which rats were injected with LPS, 2 mg/100 g body wt, 6 hr prior to the isolation of adipocytes. This plot relates mean receptor affinity $\bar{\kappa}$ to different levels of receptor occupancy (Holley and Spitzer, 1980).

rats 3 hr after administration of a low dose of endotoxin, oxidized glucose at an elevated basal rate and also had an increased insulin response (Filkins and Figlewicz, 1979). Adipocytes isolated from endotoxic rats 6 hr after injection of an LD_{60} dose of endotoxin demonstrated increased basal glucose oxidation, i.e., an insulin-like effect, but a depressed level of insulin-stimulated glucose oxidation. Insulin sensitivity of these toxic cells was also reduced (i.e., the increment due to insulin stimulation above the basal level) as compared to cells of normal saline-treated animals (Holley and Spitzer, 1980). Thus, by definition these cells were insulin resistant in terms of glucose oxidation. Similar results were obtained by exposing adipocytes of saline-injected rats to endotoxin *in vitro* (Spitzer and Holley, 1979). However, when another physiologic function of insulin, namely its lipolysis-opposing, or antilipolytic effect, was examined in fat cells of similarly treated rats, a response greater than that of normal adipocytes was elicited, indicating elevated insulin sensitivity (J. A. Spitzer, 1981). Sensitivity to the antilipolytic action of insulin could be perturbed also by *in vitro* exposure of adipocytes to endotoxin. The experiments demonstrate that endotoxin-induced alterations in insulin sensitivity are not necessarily uniform in terms of the various physiologic activities of the hormone, even at the same target cell level.

Various lines of evidence implicate Ca^{2+} translocation in the cellular mechanism of action of endotoxin (Connor *et al.*, 1973; Hikawyj-Yevich and Spitzer, 1977; Nicolas *et al.*, 1972; Person, 1977; Soulsby *et al.*, 1978). In recent years, work by several investigators also linked alterations in cellular calcium fluxes to the action of catecholamines, insulin, glucagon, and vasopressin. (Blackmore *et al.*, 1979; Chen *et al.*, 1978; McDonald *et al.*, 1978; Schimmel, 1975; Vydelingum *et al.*, 1978). Perturbation by LPS of cellular Ca^{2+} homeostasis may contribute to the concurrent alterations in hormone actions. Exploration of this possibility appears to be promising for gaining further insight into the molecular mechanism of action of LPS in cellular metabolism.

VI. Summary of LPS-Induced Alterations in Carbohydrate and Lipid Metabolic Homeostasis

Carbohydrate metabolism. (1) Initial hyperglycemia—due primarily to increased glycogenolysis, mediated by elevated sympathetic discharge. (2) Increased hepatic output of glucose fueled mainly by lactate and alanine as gluconeogenic precursors, coexisting with increased rate of disappearance, i.e., peripheral utilization of glucose. Elevated plasma lactate concentration is supplied by muscle glycogenolysis and enhanced rate of metabolism of glucose. The sources of plasma alanine are muscle protein breakdown and amination of muscle pyruvate. (3) Preterminally, or with lethal doses of LPS, the rate of gluconeogenesis fails to keep up with prevailing demands and severe hypoglycemia ensues.

Lipid metabolism. The supply of the major lipid fuel substrate, FFA, is heavily dependent on blood flow. Strong lipolytic hormonal stimuli are unquestionably present; however, curtailment of blood flow in the adipose regions interferes with the delivery of FFA for oxidation by peripheral tissues. Glycerol resulting from lipolysis of triglycerides also enters the gluconeogenic pathway. Peripheral utilization of FFA or ketone bodies does not appear to be altered directly by LPS.

Decreased tissue LPL activities are likely contributing factors to the hypertriglyceridemia observed in endotoxicosis, since LPL is instrumental in the removal of lipids from the circulation.

In conclusion, the overall impairments of carbohydrate and lipid metabolism following LPS administration may be characterized as follows: the supply of fuels is compromised by depletion of glycogen stores, by inappropriate matching of gluconeogenesis to the increased utilization of glucose, and by a decreased net release of FFA from adipose tissue (primarily due to blood flow restriction in this organ). At the same time, the ability of peripheral organs to utilize glucose, lactate, FFA, and ketone bodies appears to be unaltered and in some cases even enhanced.

ACKNOWLEDGMENTS

Some of the work reported in this review was supported by ONR Contracts N0014-76-C-0132 and N0014-80-C-0466, and NIH Grants HL07098 and GM07029.

VII. References

Adelye, A. G., Al-Fibouri, L. M., Furman, B. L., and Parratt, J. R., 1981, Endotoxin-induced metabolic changes in the conscious, unrestrained rat: Hypoglycemia and elevated blood lactate concentrations without hyperinsulinemia, *Circ. Shock* **8**:543–550.
Bagby, G. J., and Spitzer, J. A., 1980, Lipoprotein lipase activity in rat heart and adipose tissue during endotoxic shock, *Am. J. Physiol.* **238**:H325–H330.
Berry, L. J., 1971, Metabolic effects of bacterial endotoxin, in: *Microbial Toxins*, Vol. V (S. Kadis, G. Weinbaum, and S. J. Ajl, eds.), pp. 165–208, Academic Press, New York.
Berry, L. J., Smythe, D. S., and Colwell, L. S., 1968, Inhibition of hepatic enzyme induction as a sensitive assay for endotoxin, *J. Bacteriol.* **96**:91–99.
Bitensky, M. W., Gorman, R. E., and Thomas, L., 1971, Selective stimulation of epinephrine-response adenyl-cyclase in mice by endotoxin, *Proc. Soc. Exp. Biol. Med.* **138**:773.
Blackard, W. G., Anderson, J. H., and Spitzer, J. J., 1976, Hyperinsulinism in endotoxin shock dogs, *Metabolism* **25**:675–684.
Blackmore, P. F., Assimacopoulos-Jeannet, F., Chan, T. M., and Exton, J. H., 1979, Studies on α-adrenergic activation of hepatic glucose oxidation, *J. Biol. Chem.* **254**:2828–2834.
Buchanan, B. J., and Filkins, J. P., 1976, Insulin secretion and the carbohydrate metabolic alterations of endotoxemia, *Circ. Shock* **3**:267–280.
Cameron, G. R., Relefield, M. E., and Wilson, J., 1940, Pathological changes produced in rats and mice by a toxic fraction derived from *Bacterium typhimurium*, *J. Pathol. Bacteriol.* **51**:223–233.
Chen, J. J., Babcock, D. F., and Lardy, H. A., 1978, Norepinephrine, vasopressin, glucagon and A23187 induce efflux of calcium ion from an exchangeable pool in isolated rat hepatocytes, *Proc. Natl. Acad. Sci. USA* **75**:2234–2238.
Cline, M. J., Melmon, K. L., Davis, W. C., and Williams H. E., 1968, Mechanism of endotoxin interaction with human leukocytes, *Br. J. Haematol.* **15**:539–547.
Cohn, Z. A., and Morse, S. I., 1960, Functional and metabolic properties of polymorphonuclear leucocytes. II. The influence of lipopolysaccharide endotoxin, *J. Exp. Med.* **3**:689–704.
Connor, J., Fine, J., Kusano, K., McCrea, M. J., Parnas, I., and Prosser, C. L., 1973, Potentiation by endotoxin of responses associated with increases in calcium conductance, *Proc. Natl. Acad. Sci. USA* **70**:3301–3304.
Ferguson, J. L., Spitzer, J. J., and Miller, H. I., 1978, Effects of endotoxin on regional blood flow in the unanesthetized guinea pig, *J. Surg. Res.* **23**:236–243.

Filkins, J. P., and Buchanan, B. J., 1977, *In vivo* vs *in vitro* effects of endotoxin on glyco-genolysis, gluconeogenesis and glucose utilization, *Proc. Soc. Exp. Biol. Med.* **155:**216–218.

Filkins, J. P., and Cornell, R. P., 1974, Depression of hepatic gluconeogenesis and the hypoglycemia of endotoxic shock, *Am. J. Physiol.* **227:**778–782.

Filkins, J. P., and Figlewicz, D. P., 1979, Increased insulin responsiveness in endotoxico-sis, *Circ. Shock* **6:**1–6.

Fredholm, B. B., 1971, The effect of lactate in canine subcutaneous adipose tissue *in situ,* *Acta Physiol. Scand.* **81:**110–123.

Gartner, S. L., 1975, Hepatic levels of cyclic AMP in normal and lead-sensitized rats after treatment with bacterial endotoxin, *Experientia* **31:**566.

Gimpel, L., Hodgins, D. S., and Jacobson, E. D., 1974, Effects of endotoxin on hepatic adenylate cyclase activity, *Circ. Shock* **1:**21.

Hamosh, M., and Shapiro, B., 1960, The mechanism of glycogenolytic action of endotox-in, *Br. J. Exp. Pathol.* **41:**372.

Hikawyj-Yevich, I., and Spitzer, J. A., 1977, The role of adrenergic receptors and Ca^{2+} in the action of endotoxin on human fat cells, *J. Surg. Res.* **23:**233–238.

Hinshaw, L. B., Archer, L. T., Beller, B. K., White, G. L., Schroeder, T. M., and Holmes, D. D., 1977, Glucose utilization and role of blood in endotoxin shock, *Am. J. Physiol.* **233:**E71–E79.

Holley, D. C., and Spitzer, J. A., 1980, Insulin action and binding in adipocytes exposed to endotoxin *in vitro* and *in vivo, Circ. Shock* **7:**3–13.

Issekutz, B., Jr., Miller, H. I., Paul, P., and Rodahl, K., 1965, Effect of lactic acid on free fatty acids and glucose oxidation in dogs, *Am. J. Physiol.* **209:**1137–1144.

Kelleher, D. L., Gagby, G. J., and Spitzer, J. J., 1981, Metabolic and endocrine alterations following endotoxin administration in normal and diabetic rats, in: *Advances in Physio-logical Sciences,* Vol. 26 (Z. Biro, A. G. B. Kovach, J. J. Spitzer, and H. B. Stoner, eds.), pp. 181–189, Pergamon Press/Akademiai Kiado, Budapest.

Kelleher, D. L., Bagby, G. J., Fong, B. C., and Spitzer, J. J., 1982, Glucose turnover 5 hours following endotoxin administration to normal and diabetic rats, *Metabolism* **31:**252–257.

Kinney, J. M., and Felig, P., 1979, The metabolic response to injury and infection, in: *En-docrinology* (L. J. DeGroot, ed.), pp. 1963–1985, Grune & Stratton, New York.

Kovach, A. G. B., Kovach, E., Sandor, P., Spitzer, J. A., and Spitzer, J. J., 1976, Metabolic responses to localized ischemia in adipose tissue, *J. Surg. Res.* **20:**37–44.

Kuttner, R., and Spitzer, J. J., 1978, Gluconeogenesis from alanine in endotoxin-treated dogs, *J. Surg. Res.* **25:**166.

Landy, M., Whitby, J. L., Michale, J. G., Woods, M. W., and Newton, W. L., 1962, Effect of bacterial endotoxin in germ free mice, *Proc. Soc. Exp. Biol. Med.* **109:**352–356.

LaNoue, K., Mason, A. D., Jr., and Daniels, J., 1968, The impairment of gluconeogenesis by gram negative infection, *Metabolism* **17:**606–611.

Leach, G. J., and Spitzer, J. A., 1981, Endotoxin induced alterations in glucose transport in isolated adipocytes, *Biochim. Biophys. Acta* **648:**71–79.

Little, J. R., Goto, M., and Spitzer, J. J., 1970, Effect of ketones on metabolism on FFA by dog myocardium and skeletal muscle, *Am. J. Physiol.* **219:**1458–1463.

Liu, M.-S., and Spitzer, J. J., 1977, Myocardial fatty acid and lactate kinetics after *E. coli* endotoxin administration, *Circ. Shock* **4:**191–200.

Liu, M.-S., Long, W. M., and Spitzer, J. J., 1981, Influence of *E. coli* endotoxin on palmi-tate, glucose and lactate utilization by isolated dog heart myocytes, in: *The Pathology of Endotoxin at the Cellular Level* (J. A. Majda and R. J. Person, eds.), pp. 115–121, Liss, New York.

McDonald, J. M., Burns, D. E., and Jarett, L., 1978, Ability of insulin to increase calcium uptake by adipocyte endoplasmic reticulum, *J. Biol. Chem.* **253:**3504–3508.

Martin, S. P., McKinney, G. R., and Green, R., 1955, The metabolism of human polymorphonuclear leukocytes, *Ann. N.Y. Acad. Sci.* **59:**996–1002.

Menten, M. L., and Manning, H. M., 1924, Blood sugar studies on rabbits injected with organisms of the enteritidis-parathyroid B group, *J. Med. Res.* **44:**676–681.

Merrill, G. F., and Spitzer, J. J., 1978, Glucose and lactate kinetics in guinea pigs following *Escherichia coli* endotoxin administration, *Circ. Shock* **5:**11–21.

Motsay, G. J., Alho, A. V., Dietzmann, R. H., and Lillehei, R. C., 1971, Pathophysiology and therapy of endotoxin (septic) shock, in: *Emergency Medical Management* (S. Spitzer, W. W. Oaks, and J. H. Moyer, eds.), pp. 247–267, Grune & Stratton, New York.

Neufeld, H. A., Pace, J. A., and White, F. E., 1976, The effect of bacterial infections on ketone concentrations in rat liver and blood and on free fatty acid concentrations in rat blood, *Metabolism* **25:**877–884.

Nicolas, G. G., Mela, L. M., and Miller, L. D., 1972, Shock-induced alterations of mitochondrial membrane transport, *Ann. Surg.* **176:**579–584.

Person, R. J., 1977, Endotoxin alters spontaneous transmitter release at frog neuromuscular junction, *J. Neurosci. Res.* **3:**63–72.

Rayfield, E. J., Curnow, R. T., Reinhard, D., and Kochiceril, N. M., 1977, Effects of acute endotoxemia on glucoregulation in normal and diabetic subjects, *J. Clin. Endocrinol. Metab.* **45:**513–521.

Raymond, R. M., Harkema, J. M., and Emerson, T. E., Jr., 1981, Mechanism of increased glucose uptake by skeletal muscle during *E. coli* endotoxin shock in the dog, *Circ. Shock* **8:**77–93.

Ribi, E., Haskins, W. T., Landry, M., and Milner, K. C., 1961, Symposium on bacterial endotoxin. I. Relationship of chemical composition to biological activity, *Bacteriol. Rev.* **25:**427–436.

Romanosky, A. J., Bagby, G. J., Bockman, E. L., and Spitzer, J. J., 1980a, Increased skeletal muscle glucose uptake and lactate release following *E. coli* endotoxin administration, *Am. J. Physiol.* **239:**E311–E316.

Romanosky, A. J., Bagby, G. J., Bockman, E. L., and Spitzer, J. J., 1980b, Free fatty acid utilization by skeletal muscle after endotoxin administration, *Am. J. Physiol.* **239:**E391–E395.

Romanosky, A. J., McGuinness, O., and Spitzer, J. J., 1982, Ketone body (KB) clearance following endotoxin (ET) administration in dogs, *Fed. Proc.* **41:**1133.

Schimmel, R. J., 1975, The role of calcium ion in epinephrine activation of lipolysis, *Horm. Metab. Res.* **8:**195–201.

Scholl, R. A., Lang, C. H., Spitzer, J. J., and Bagby, G. J., 1982, Tissue lipoprotein lipase (LPL) activities in rats challenged with endotoxin, live *E. coli,* or peritonitis, *Fed. Proc.* **41:**1134.

Scott, J. C., Weng, J. T., and Spitzer, J. J., 1973, Myocardial metabolism during endotoxic shock, in: *Neurohumoral and Metabolic Aspects of Injury* (A. G. B. Kovach, B. Stoner, and J. J. Spitzer, eds.), pp. 375–386, Plenum Press, New York.

Shands, P. W., Jr., Miller, W., Martin, H., and Senterfilt, V., 1969, Hypoglycemic activity of endotoxin. II. Mechanism of the phenomenon in BCG-infected mice, *J. Bacteriol.* **98:**495–501.

Snyder, J. S., Deters, M., and Ingle, J., 1971, Effect of endotoxin on pyruvate kinase activity in mouse liver, *Infect. Immun.* **4:**138–142.

Soulsby, M. E., Bruin, F. D., Looney, T. J., and Hess, M. L., 1978, Influence of endotoxin on myocardial calcium transport and the effect of augmented venous return, *Circ. Shock* **5:**23–34.

Spitzer, J. A., 1974, Endotoxin-induced alterations in isolated fat cells: Effect on norepinephrine-stimulated lipolysis and cyclic 3'5'-adenosine monophosphate accumulation, *Proc. Soc. Exp. Biol. Med.* **145:**186–191.

Spitzer, J. A., 1981, Altered insulin sensitivity in endotoxin and septic shock, in: *Advances in Physiological Sciences,* Vol. 26 (Z. Biro, A. G. B. Kovach, J. J. Spitzer, and H. B. Stoner, eds.), pp. 143–150, Pergamon Press/Akademiai Kiado, Budapest.

Spitzer, J. A., and Holley, D. C., 1979, Alterations in insulin action by endotoxin *in vitro,* in: *Advances in Shock Research,* Vol. 2 (W. Schumer, J. J. Spitzer, and B. E. Marshall, eds.), pp. 129–136, Liss, New York.

Spitzer, J. A., Archer, L., Greenfield, L. J., Hinshaw, L. B., and Spitzer, J. J., 1972, Metabolism of the nonhepatic splanchnic area in baboons and the effects of endotoxin, *Proc. Soc. Exp. Biol. Med.* **141:**21–25.

Spitzer, J. A., Kovach, A. G. B., Sandor, P., Spitzer, J. J., and Storck, R., 1973a, Adipose tissue and endotoxin shock, *Acta Physiol. Acad. Sci. Hung.* **44:**183–194.

Spitzer, J. A., Kovach, A. G. B., Rosell, S., Sandor, P., Spitzer, J. J., and Storck, R., 1973b, Influence of endotoxin on adipose tissue metabolism, in: *Advances in Experimental Medicine and Biology,* Vol. 33 (A. G. B. Kovach, H. B. Stoner, and J. J. Spitzer, eds.), pp. 337–344, Plenum Press, New York.

Spitzer, J. J., 1979, Lipid metabolism in endotoxic shock, *Circ. Shock* **1**(Suppl.):69–79.

Spitzer, J. J., 1981, Lactate–fatty acid competition in the heart during shock, in: *Advances in Physiological Sciences,* Vol. 8 (A. G. B. Kovach, E. Monos, and G. Rubanyi, eds.), pp. 175–178, Pergamon Press/Akademiai Kiado, Budapest.

Spitzer, J. J., Bechtel, A. A., Archer, L. T., Black, M. R., and Hinshaw, L. B., 1974, Myocardial substrate utilization in dogs following endotoxin administration, *Am. J. Physiol.* **227:**132–136.

Spitzer, J. J., Ferguson, J. L., Hirsch, H. J., Loo, S., and Gabbay, K. W., 1980, Effects of *E. coli* endotoxin on pancreatic hormones and blood flow, *Circ. Shock* **7:**353–360.

Vydelingum, N., Kissebah, A. H., and Wijn, V., 1978, Importance of cyclic guanosine 3'5'-monophosphate and calcium ions in insulin stimulation of lipoprotein lipase activity and protein synthesis in adipose tissue, *Horm. Metab. Res.* **10:**38–46.

Witek-Janusek, L., and Filkins, J. P., 1981, Insulin-like action of endotoxin: Antagonism by steroidal and non-steroidal anti-inflammatory agents, *Circ. Shock* **8:**573–583.

Wolfe, R. R., Elahi, D., and Spitzer, J. J., 1977a, Glucose and lactate kinetics after endotoxin administration in dogs, *Am. J. Physiol.* **232:**E180–E185.

Wolfe, R. R., Elahi, D., and Spitzer, J. J., 1977b, Glucose kinetics in dogs following a lethal dose of endotoxin, *Metabolism* **26:**847–850.

Woods, M. W., Burke, D., Howard, T., and Landy, M., 1961, Insulin-like action of endotoxins in normal and leukemic leukocytes and other tissues, *Proc. Am. Assoc. Cancer Res.* **3:**279.

Woods, M. W., Landy, M., Burke, D., and Howard, T., 1964, Effect of endotoxin on cellular metabolism, in: *Bacterial Endotoxins* (M. Landy and W. Braun, eds.), pp. 160–181. Rutgers University Press, New Brunswick, N.J.

Zenser, T. V., DeRobertis, F. R., George, D. T., and Rayfield, E. J., 1974, Infection-induced hyperglucagonemia and altered hepatic response to glucagon in the rat, *Am. J. Physiol.* **227:**1299–1305.

Zwadyk, P., Jr., and Snyder, L. S., 1973a, Effects of endotoxin on hepatic glycogen metabolism *in vitro,* Proc. Soc. Exp. Biol. Med. **142:**299.

Zwadyk, P., Jr., and Snyder, L. S., 1973b, Effects of endotoxin on glycogenolytic enzymes of mouse liver, *Proc. Soc. Exp. Biol. Med.* **143:**864–868.

3

Effect of LPS On the Clotting System

Alexander M. Abdelnoor

I. Introduction

The activation of the clotting system is one of the body's natural defense mechanisms. In addition to the formation of a fibrin clot that seals an injured vessel, other systems are triggered such as fibrinolytic, kinin, and complement. As a consequence, several mediators are produced that are involved in nonspecific resistance to infections and tumors.

Platelet aggregation and the interplay of the intrinsic and extrinsic pathways lead to the formation of fibrin (Fig. 1). Following injury to a blood vessel, platelets adhere to the exposed collagen and aggregate (primary aggregation). The aggregated platelets then release biologically active substances, notably ADP, thromboxanes, serotonin, and prostaglandin E_2 and $F_{2\alpha}$. In addition, platelet factor 3 (PF3), factor V, and possibly other factors that participate in the activation of the intrinsic pathway are made available. The released ADP causes further platelet aggregation (secondary aggregation). The platelet plug that is formed at the site of vessel injury is strengthened by fibrin, which is produced when the intrinsic and/or extrinsic pathways are activated.

The intrinsic pathway is activated in a specific sequence of events and is initiated following the activation of factor XII (Hageman factor). Activated factor XII (XII_a) activates factor XI. Activated factor XI (XI_a) activates factor IX. Activated factor IX (IX_a) in the presence of factor VIII activates factor X. Activated factor X (X_a) in the presence of Ca^{2+}, factor V, and PF3 converts prothrombin to thrombin. Thrombin acti-

Alexander M. Abdelnoor ● Department of Microbiology, Faculty of Medicine, American University of Beirut, Beirut, Lebanon.

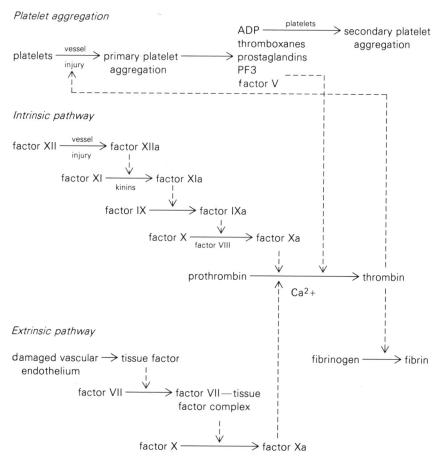

Figure 1. The clotting system. Vessel injury leads to platelet aggregation, activation of factor XII, and the availability of tissue factor. The interplay of intrinsic and extrinsic pathways and platelet aggregation results in the formation of fibrin.

vates factor XIII and cleaves fibrinogen into fibrin monomers and fibrinopeptides. Thrombin also acts on platelets causing their aggregation and release of platelet factors. Activated factor XIII (XIII_a) acts on fibrin monomers. As a result of this interaction, cross-linked peptide bonds between fibrin monomers are produced. Factor XII_a also activates the kinin, fibrinolytic, and complement systems.

The extrinsic pathway is activated by tissue factor (thromboplastin), which is made available following tissue injury. Tissue factor forms a complex with factor VII. This complex activates factor X. From this step on, the events are similar to the intrinsic pathway.

Bacterial lipopolysaccharide (LPS) is capable of activating the clotting system. There are at least three ways in which LPS may achieve this, namely: (1) it acts on platelets causing them to aggregate and/or release platelet factors; (2) it acts on leukocytes causing the release from these cells of tissue factor; and (3) it activates factor XII (Fig. 2). These mechanisms will be reviewed in this chapter.

II. LPS–Platelet Interactions

Although Muller-Berghans (1978) proposed that the effect of LPS on platelets does not play a primary role in activating the clotting system, it has been shown without doubt that LPS interacts directly or indirectly with platelets both *in vivo* and *in vitro* causing them to aggregate and release factors, some of which are needed for fibrin formation. Most studies have dealt with LPS–rabbit or LPS–human platelet interactions. It is well established that LPS causes rabbit and human platelet aggregation *in vivo*. *In vitro* studies indicate that LPS interacts with rabbit platelets, but variable results have been reported where human platelets were used.

A. Effect of LPS on Rabbit and Human Platelets

Davis *et al.* (1960) showed that the injection of LPS to animals was followed by thrombocytopenia and the appearance of platelet aggregates in the blood vessels of several organs. They (Davis *et al.*, 1961) later showed that these reactions were accompanied by the appearance of serotonin and histamine in the plasma. In man, the situation appears

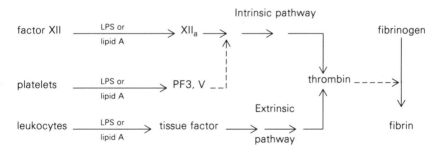

Figure 2. LPS–clotting system interactions. LPS or lipid A can act on factor XII, platelets, and leukocytes. Any one or all of these interactions lead to thrombin formation. Thrombin converts fibrinogen to fibrin, and similar to LPS, acts on platelets causing them to aggregate and release factors, some of which participate in the intrinsic pathway.

to be similar since intravenous injection of LPS to volunteers resulted in an acute reduction of circulating platelets. Furthermore, patients with gram-negative septicemia had low platelet counts (Horder and Manrique, 1965; Cohen and Gardner, 1966). Des Prez *et al.* (1961) showed that LPS added to rabbit platelet-rich plasma (PRP) resulted in platelet aggregation and release of serotonin, histamine, and PF3. This finding was confirmed by others (Nagayama *et al.*, 1971; Brown and Lachmann, 1973). However, variable results were reported on studies investigating LPS–human platelet interactions *in vitro*. Ream *et al.* (1965) reported that washed human platelets underwent aggregation in the presence of LPS and factor V. Nagayama *et al.* (1971) showed that there was a selective release of serotonin by human platelets when LPS was added to PRP. However, they were unable to demonstrate aggregation or enhancement of PF3. Bjornson and Hill (1973) showed that some of their LPS preparations when added to human PRP caused aggregation. Binding of LPS to human platelets was suggested by Das *et al.* (1973). They showed that platelets isolated from patients with gram-negative septicemia were reactive in the *Limulus* lysate assay. Hawiger *et al.* (1977) separated human platelets of plasma proteins by Sepharose 2B filtration. Using this plasma protein-free platelet preparation, they were able to show that LPS interacted with human platelets resulting in the release of serotonin and adenine nucleotides. Semeraro *et al.* (1978b) examined the effect of 16 LPS preparations on human PRP and washed human platelet suspensions prepared by albumin density gradient centrifugation. All of the preparations failed to induce aggregation or serotonin release. However, as will be described later, they suggested that LPS may interact with washed human platelets to promote their factor X-activating activity.

The variability of results obtained in studies using human platelets as well as the differences between LPS–human and LPS–rabbit platelet interactions may be explained in part when one looks at the mechanisms by which LPS interacts with platelets. Studies have revealed that there are at least three ways in which LPS and platelets can interact. These are complement-dependent, complement-independent, and antibody-dependent interactions.

B. Complement-Dependent Interactions

Spielvogel (1967) performed electron microscopic studies on rabbit platelets in plasma treated with LPS. He showed that LPS adhered to platelets. This adherence resulted in platelet degranulation followed by death. However, no LPS adherence took place when the platelets were suspended in heat-inactivated plasma or plasma prepared by using EDTA. Likewise, no LPS adherence occurred when the plasma was pretreated with either zymosan or antigen–antibody complexes. In this study, Spielvogel concluded that complement is one of the factors involved in LPS–platelet interactions. Fong and Good (1971) showed that the Sanarelli–Shwartzman reaction was prevented in rabbits decomplemented with cobra venom factor (CVF). In another study, they showed that CVF inhibited LPS–platelet interactions both *in vivo* and *in vitro* and the inhibition was dose and time dependent. Furthermore, they demonstrated that CVF treatment had no effect on collagen or ADP interactions with platelets, indicating that these substances were not involved in LPS–platelet interactions (Fong *et al.*, 1974). Similar results were obtained using dogs (Garner *et al.*, 1974) and guinea pigs (Kane *et al.*, 1973). Brown and Lachmann (1973) studied the *in vivo* behavior of chromium-labeled platelets in normal, C6-deficient and C3–9-depleted rabbits following a lethal dose of LPS. In normal rabbits, most of the labeled platelets were destroyed and the rabbits died soon after. In the C3–9-depleted rabbits, the platelets were not destroyed and the rabbits survived. The degree of platelet destruction in the C6-deficient rabbits varied, and when platelet recovery was greater than 65% the rabbit survived. They concluded that complement depletion correlated with normal platelet survival rate, absence of mortality, and failure to activate PF3.

It therefore seems apparent from the above-mentioned studies that complement is involved in certain LPS–platelet interactions. Morrison and Kline (1977) have demonstrated that LPS is capable of activating both the classical and the alternate pathway. In a study to identify which of the two pathways was involved in LPS–platelet interactions, Semeraro *et al.* (1978a) showed that LPS failed to induce platelet aggregation when added to PRP containing EGTA. Since EGTA blocks the classical and has no effect on the alternate pathway, they concluded that the classical and not the alternate pathway is involved in complement-dependent LPS–platelet interactions.

Activation of the complement system results in cleavage of its components. It has been proposed that C3b, a split product of C3, binds to receptors on rabbit platelets, resulting in aggregation (Muller-Berghans, 1978). The fact that human platelets do not possess receptors for C3b (Henson, 1969; Des Prez and Marney, 1971) explains in part why some investigators failed to observe LPS–human platelet interactions in their experimental systems.

C. Complement-Independent Interactions

It has already been mentioned that LPS interacts with protein-free, washed human platelets (Hawiger *et al.*, 1977), and that Das *et al.* (1973) were able to demonstrate the presence of LPS on human platelets using the *Limulus* lysate assay. Ulevitch *et al.* (1975) studied the effect of LPS on the disappearance from the circulation of ^{51}Cr-labeled platelets in normal and CVF-treated rabbits. In normal rabbits, there was a biphasic disappearance of circulating platelets, whereas only one platelet disappearance phase was observed in CVF-treated rabbits. They suggested that one of the phases was independent of complement. The results of a similar study using rhesus monkeys instead of rabbits indicated that complement was not involved in the thrombocytopenia that occurred following the administration of LPS since CVF-treated monkeys reacted in a manner similar to that of normal monkeys (Ulevitch *et al.*, 1978). Semeraro *et al.* (1978b) studied the effect *in vitro* of several bacterial LPS on washed human platelets and human PRP. Although the LPS preparations studied failed to induce aggregation or release of serotonin, an LPS–platelet interaction probably occurred since washed platelets obtained from PRP preincubated with LPS had a higher platelet coagulant activity than platelets obtained from PRP preincubated with buffer. Since PF3 and factor V activity were not influenced by LPS treatment, they suggested that LPS enhanced the factor X activator activity of human platelets. It can be noted here that complement was not involved in the interaction observed since they were working with washed platelets.

These results indicate that in the rabbit, LPS interacts with platelets by two mechanisms, one being a direct interaction and the other being mediated by complement. In man and the subhuman primate, only the former mechanism has been observed.

D. Antibody-Dependent Interactions

In addition to the above-described interactions, a third mechanism involving antibodies has been observed by some investigators. Humphrey and Jaques (1955) demonstrated that antigen–antibody complexes may induce platelet aggregation and release of platelet constituents. Des Prez and Bryant (1966) compared LPS–platelet and immune complex–platelet interactions. They immunized rabbits with ovalbumin and studied the effect of ovalbumin on citrated immune PRP. In a parallel experiment, they studied the effect of LPS on citrated nonimmune rabbit PRP. Whereas LPS interacted with platelets in citrated PRP, ovalbumin did not interact in the immune PRP. However, ovalbumin interacted in immune PRP when divalent ions were made available. From their results they suggested that antigen–antibody complexes formed interacted with complement in the presence of divalent ions; which in turn interacted with platelets. Complement, they believe, was not involved in LPS–platelet interactions. They concluded that the end result of immune complex and LPS interactions with platelets may be similar; that is, platelets phagocytize LPS and immune complexes and as a consequence aggregation and release reaction occurs. Spielvogel (1967) showed that LPS–platelet interactions in PRP did not occur when the plasma was preabsorbed with LPS. He suggested that antibodies may be needed for the interaction and were removed by absorption. Abdelnoor *et al.* (1980) showed that preformed LPS–antibody complexes were capable of inducing rabbit and human platelet aggregation when added to PRP. This type of interaction was different from LPS–rabbit platelet interactions also observed in this study since the former and not the latter was inhibited by guinea pig serum. From the results of this study, it appears that at least two mechanisms are involved in LPS–rabbit platelet interactions, one dependent and the other independent of antibodies. When human PRP was used, only the antibody-requiring mechanism was operational.

The role of complement in antibody-dependent interactions is questionable. Both Des Prez and Bryant (1966) and Spielvogel (1967) suggest that complement is involved. However, Lüscher and Pfueller (1978) demonstrated that immune complexes can interact with platelets independently of complement. Although the subclasses of IgG in their aggregated form differ with respect to their ability to activate complement, it was shown that all of the subclasses interacted with platelets. In addition, these investigators showed that loss of complement-activating ability of IgG by removal of carbohydrate did not affect its ability to interact with platelets.

The blood of most mammals contains antibodies to LPS since they are in constant contact with gram-negative bacteria (Michael *et al.*, 1962; Landy and Weidanz, 1964). Anti-LPS antibody levels may vary from one individual to another. This is another factor to consider in explaining the variable results obtained by several investigators when they studied LPS–human platelet interactions.

III. LPS–Leukocyte Interactions

A. Leukocytes

It has been demonstrated that LPS interacts with leukocytes causing a release from these cells of tissue factor (thromboplastin). Rickles *et al.* (1977) showed that mononuclear cells in culture generated tissue factor following stimulation with LPS. As little as 0.0001 μg/ml of *E.*

coli LPS had a stimulatory effect, and inhibition of DNA synthesis in cell culture by cytosine arabinoside or nonlethal irradiation failed to impair the generation of tissue factor. Hiller *et al.* (1977) determined the subpopulation of leukocytes responsible for the generation of tissue factor. They separated blood leukocytes by Ficoll density centrifugation into granulocytes, lymphocytes, and monocytes. Each of these subpopulations was suspended in culture medium to which 10 μg/ml of *S. enteritidis* LPS was added. The monocytes developed a high tissue factor activity, whereas negligible activity was demonstrated in the neutrophil suspension. Tissue factor activity demonstrable in the lymphocyte suspension was believed by the authors to be due to contamination with monocytes. That the monocyte is the primary producer of tissue factor has also been demonstrated by Tonooka (1978) and Edwards *et al.* (1979). On the other hand, Lerner *et al.* (1977) have demonstrated that neutrophils were primarily responsible for the production of tissue factor. They showed that LPS added to citrated whole blood led to the production of tissue factor. Tissue factor was not produced if the blood was depleted of neutrophils prior to the addition of LPS. Furthermore, tissue factor generation by neutrophils was claimed to be dependent on protein synthesis since cycloheximide inhibited its production.

B. Mechanisms

There are few reports that deal with mechanisms involved in LPS–leukocyte interactions. Leukocytes possess LPS receptors on their surfaces (Springer and Adye, 1975; Hawiger *et al.*, 1977). It has been suggested by Muller-Berghans (1978) that LPS binds to these receptors resulting in the release of tissue factor. Prydz *et al.* (1977) reported that the complement fragment C3b, produced upon activation of the complement system, triggers a marked increase of tissue factor in cultured human monocytes. They proposed that the monocytes are activated when C3b binds to receptors on these cells. Rothberger *et al.* (1977) demonstrated that the leukocyte procoagulant activity of human leukocytes was increased after they were incubated with immune complexes. Thus, it is probable that LPS–antibody complexes may also be involved in the generation of tissue factor by leukocytes.

C6-deficient rabbits show an impaired availability of PF3 and yet they develop disseminated intravascular coagulation (DIC) following injection of LPS. This led Muller-Berghans (1978) to investigate clotting mechanisms other than LPS–platelet interactions. Since leukopenia as well as thrombocytopenia are induced following injection of LPS, Muller-Berghans focused on the role of the leukocyte in triggering the clotting system. He showed that LPS induced DIC in rabbits with thrombocyto-penia and normal leukocyte counts, but did not induce DIC in rabbits with leukopenia and normal platelet counts. He suggested that LPS interacted with leukocytes causing the release of tissue factor, which triggers the extrinsic pathway. The role of platelets, he believes, is to accelerate an already activated clotting system.

IV. LPS–Factor XII Interactions

Several studies provided indirect evidence that LPS interacts with factor XII resulting in its activation (Rodriguez-Erdmann, 1964; Mason et al., 1970; Kimball et al., 1972; Muller-Berghans and Schneberger, 1971). Rodriguez-Erdmann (1964) showed that factor XII activity decreased following the administration of LPS. He proposed that factor XII was directly activated by LPS to initiate clotting. Muller-Berghans and Schneberger (1971) showed that factor XII activity in rabbits decreased following the development of the generalized Shwartzman reaction. This decrease in factor XII activity was prevented by pretreating the rabbits with coumarin. Coumarin inhibits the synthesis of vitamin K-dependent coagulation factors and should not affect factor XII. Furthermore, they showed that inhibition of factor XII activation by lysozyme infusion did not prevent the generalized Shwartzman reaction. They concluded from these results that although factor XII activation induced by LPS in the generalized Shwartzman reaction may occur, other mechanisms leading to intravascular coagulation are involved.

That factor XII is activated by a direct interaction with LPS has been demonstrated by Morrison and Cochrane (1974). They prepared a mixture of LPS and purified precursor factor XII, and showed that this mixture converted prekallikrein to its active form and reduced the clotting time of factor XII-deficient plasma. Using sucrose gradient ultracentrifugation and SDS-polyacrylamide gel electrophoresis, they demonstrated that LPS forms a complex with factor XII and that no cleavage to lower-molecular-weight substances occurred. They concluded that this LPS–factor XII interaction results in initiation of the intrinsic pathway, the kinin and perhaps the fibrinolytic systems.

An interesting finding is that of Walker et al. (1979), who showed that chromium chloride-treated LPS was less toxic to mice than the parent molecule and yet had an enhanced ability to activate factor XII. This suggests that toxicity may not be required for this process, a significant point to bear in mind when one thinks of the beneficial uses of LPS in stimulating the clotting system.

V. Relation of Clotting, Complement, Kinin, and Fibrinolytic Systems and Their Role in Natural Defense

In addition to the clotting system, complement, kinin, and fibrinolytic systems are activated by LPS. It has already been mentioned that LPS can activate both the classical and the alternate pathway (Morrison and Kline, 1977). Also, patients suffering from infection with LPS-containing organisms had decreased serum hemolytic complement activity, indicating that complement was activated possibly by LPS (Galloway *et al.*, 1977). We have seen earlier that complement degradation products, in particular C3b, interact with platelets and monocytes resulting in the activation of the clotting system.

Aasen *et al.* (1978a) and Gallimore *et al.* (1978) reported that plasma prekallikrein levels were considerably reduced in dogs during the later stages of endotoxin-induced shock. They believe that activation of the plasma kinin system is associated with circulating collapse of endotoxin shock in dogs. Furthermore, as pointed out earlier, Morrison and Cochrane (1974) have shown that LPS-activated factor XII activated prekallikrein.

Plasminogen levels, plasmin activity, and antiplasmin activity were determined during endotoxin shock in dogs (Aasen *et al.*, 1978b). It was demonstrated that antiplasmin activity and plasminogen levels decreased and plasmin activity appeared.

From these studies it appears that the complement, kinin, and fibrinolytic systems are activated by LPS and the link between these systems is LPS-activated factor XII (Fig. 3).

As a consequence of activating these systems, several mediators

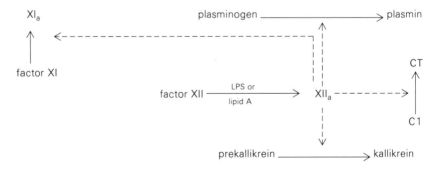

Figure 3. Relationship of clotting, kinin, fibrinolytic, and complement systems. Factor XII, activated by LPS or lipid A, converts factor XI to XI$_a$ (intrinsic pathway), prekallikrein to kallikrein (kinin), plasminogen to plasmin (fibrinolytic), and C1 to CT (complement).

that play prominent roles in enhancing the body's natural defense against infections and tumors are produced. Some of these mediators and their functions will be mentioned here. Fibrinopeptides are chemotactic for leukocytes and enhance the action of bradykinin. Fibrin enhances the movement of phagocytes and their ingestion of particles. Bradykinin increases vascular permeability and causes contraction of smooth muscle. Prostaglandin F causes smooth muscle contraction and Prostaglandin E_2 produces dilation of bronchial smooth muscle. Kallikrein is chemotactic for neutrophils, monocytes, and basophils. C3b and C4b are opsonins. C3a and C5a are chemotactic and they can also bind to basophils and mast cells causing the release from these cells of vasoactive amines.

VI. Structure–Function Relationship

LPS is composed mainly of three regions, namely; the O-specific polysaccharide (responsible for serologic specificity), the basal core oligosaccharide, and lipid A (Lüderitz et al., 1966). Cleavage of LPS into a polysaccharide-rich fraction (PS) and lipid A can be attained by mild acid hydrolysis (Nowotny et al., 1975). The PS can be purified from the supernatant of the hydrolysate and lipid A remains in the sediment. Lipid A can also be extracted from mutants such as the R595 S. minnesota mutant. These mutants synthesize incomplete polysaccharide because they lack certain transferases and/or synthetases. Lipid A extracted from R595 consists of ethanolamine, ketodeoxyoctonate (KDO), fatty acids, and phosphorus. They lack the basal core and O-specific side chain. The relation of the structure of LPS to its ability to interact with the clotting system has been investigated using the above-mentioned preparations and most reports indicate that this function is related to the lipid moiety. The PS preparation did not cause human or rabbit platelet aggregation when added to PRP. Lipid A extracted from R595 induced platelet aggregation in rabbit PRP (Abdelnoor et al., 1980). It has been demonstrated that lipid A activates complement by the classical pathway and that PS is responsible for the activation of complement by the alternate pathway (Morrison and Kline, 1977). Futhermore, Semeraro et al. (1978a) have demonstrated that the classical and not the alternate pathway was involved in complement-dependent LPS–platelet interactions. Hence, it appears that lipid A is responsible for this function. Likewise, it has been demonstrated that lipid A is involved in complement-independent interactions since both LPS and lipid A were active in the experimental systems studied (Ulevitch et al., 1975; Semeraro et al., 1978a). The size of LPS was another factor considered by Hawiger (1978) in studying structure–function relationships. He demonstrated an increase in LPS–platelet interactions following mild alkaline hydrolysis of LPS. Such treatment reduced the size of LPS and facilitated its contact with receptors on the platelet surface.

Like LPS, lipid A was capable of interacting with factor XII to produce an active product and interact with leukocytes resulting in the generation of tissue factor (Morrison and Cochrane, 1974; Muller-Berghans, 1978).

Another factor to consider in the study of structure–function relationship is the heterogeneity of the LPS and lipid A preparations used in the various studies. Using thin-layer chromatography and ion-exchange chromatography, LPS and lipid A could be separated into several components. The fractions obtained varied in their biological properties (Nowotny *et al.*, 1966; Nowotny, 1969, 1971) as well as in their chemical properties (Nowotny, 1977). Semeraro *et al.* (1978a,b) found that different LPS preparations varied in their ability to interact with platelets. They attributed these variations to differences in the anticomplementary effect of the preparations. These differences may be due to the heterogeneity of the LPS preparations used in their study.

VII. Beneficial Role of LPS

In considering the beneficial uses of LPS through its action on the clotting system, one must think of at least two important factors: the toxicity of LPS, and the fact that it can prevent as well as promote bleeding. It has been well established that the lipid moiety is responsible for its toxic properties. The nontoxic PS preparation has been shown to have beneficial properties. It protects against lethal radiation and causes the induction of bone marrow cell colony-stimulating factor. It also protects against the hypotensive effect of LPS (Chang *et al.*, 1974; Nowotny *et al.*, 1975; Abdelnoor *et al.*, 1981). In one study, it was demonstrated that PS does not interact with platelets (Abdelnoor *et al.*, 1980). There are no reports on the interaction of PS with factor XII or leukocytes.

Eichenberger (1955) and Von Kaulla (1958) have described the fibrinolytic properties of LPS and its possible use as a therapeutic agent in man. If LPS is to be used as an anticoagulant or as a coagulant, one of these properties should be abolished without affecting the other. Furthermore, its toxic properties should be reduced. The chromium chloride-treated LPS described by Walker *et al.* (1979) seems to be promising. Since lipid A and LPS are heterogeneous, it is probable that certain fractions possess one property or the other. This as well as dosage and route of administration may be important factors to consider.

VIII. Conclusions

LPS activates the clotting system by several ways. These include LPS interaction with platelets, leukocytes, and factor XII. Using rabbit platelets, complement-dependent, complement-independent, and antibody-dependent interactions have been observed. Complement-independent and antibody-dependent interactions have been observed with human platelets. The results of these interactions are platelet aggregation and/or release of platelet factors.

LPS interacts with leukocytes, mainly monocytes and possibly nontrophils, either by binding to cell receptors or by activating the complement system. C3b produced binds to receptors on monocytes. Immune complexes may be involved as well. Release of tissue factor is a result of these interactions.

LPS forms a complex with factor XII. This complex can initiate the extrinsic pathway, kinin, complement, and possibly the fribrinolytic systems.

The lipid moiety (lipid A) is responsible for all these interactions. Since lipid A is toxic and causes fibrin formation as well as fibrinolysis, these activities should be taken into consideration if LPS is to be used to modulate coagulation.

IX. References

Aasen, A. O., Frolich, W., Sangstad, O. D., and Amundsen, E., 1978a, Plasma kallikrein activity and prekallikrein levels during endotoxin shock in dogs, *Eur. Surg. Res.* **10**:50.

Aasen, A. O., Ohlsson, K., Larsbraaten, M., and Amundsen, E., 1978b, Changes in plasminogen levels, plasmin activity and activity of antiplasmins during endotoxin shock in dogs, *Eur. Surg. Res.* **10**:63.

Abdelnoor, A. M., Kassem, H., Bikhazi, A. B., and Nowotny, A., 1980, Effect of gram-negative bacterial lipopolysaccharide derived polysaccharides, glycolipids, and lipopolysaccharides on rabbit and human platelets *in vitro*, *Immunobiology* **157**:145.

Abdelnoor, A. M., Johnson, A. G., Anderson-Imbert, A., and Nowotny, A., 1981, Immunization against bacteria- and endotoxin-induced hypotension, *Infect. Immun.* **32**:1093.

Bjornson, H. S., and Hill, E. O., 1973, Bacteriodaceae in thromboembolic disease: Effect of cell wall components on blood coagulation *in vivo* and *in vitro*, *Infect. Immun.* **8**:911.

Brown, D. L., and Lachmann, P. J., 1973, The behaviour of complement and platelets in lethal endotoxin shock in rabbits, *Int. Arch. Allergy Appl. Immunol.* **45**:193.

Chang, H., Thompson, J. J., and Nowotny, A., 1974, Release of colony stimulating factor (CSF) by non-endotoxic breakdown product of bacterial lipopolysaccharides, *Immunol. Commun.* **3**:401.

Cohen, P., and Gardner, F. M., 1966, Thrombocytopenia as a laboratory complication of gram negative bacteremic infection, *Arch. Internal Med.* **117**:113.

Das, J., Schwartz, A. A., and Folkwon, J., 1973, Clearance of endotoxin by platelets: Role in increasing the accuracy of the *Limulus* gelation test and in combating experimental endotoxemia, *Surgery* **74**:235.

Davis, R. B., Mecker, W. R., and McQuarrie, D. G., 1960, Immediate effects of intravenous endotoxin on serotonin concentrations and blood platelets, *Circ. Res.* 8:234.

Davis, R. B., Mecker, W. R., and Bailey, W. L., 1961, Serotonin release after injection of *E. coli* endotoxins in the rabbit, *Fed. Proc.* 20:261.

Des Prez, R. M., and Bryant, R. E., 1966, Effect of bacterial endotoxin on rabbit platelets. IV. The divalent ion requirement of endotoxin-induced and immunologically induced platelet injury, *J. Exp. Med.* 124:971.

Des Prez, R. M., and Marney, S. R., Jr., 1971, in: *The Circulating Platelets* (S. A. Johnson, ed.), p. 415, Academic Press, New York.

Des Prez, R. M., Horowitz, H. I., and Hook, E. W., 1961, Effect of bacterial endotoxin on rabbit platelets. I. Platelet aggregation and release of platelet factors *in vitro, J. Exp. Med.* 114:857.

Edwards, R. L., Rickles, F. R., and Babrove, A. M., 1979, Mononuclear cell tissue factor: Cell of origin and requirements for activation, *Blood* 54:359.

Eichenberger, E., 1955, Fibrinolyse nach intravenöser Injektion bakterieller Pyrogene, *Acta Neuroveg.* 11:201.

Fong, J. S. C., and Good, R. A., 1971, Prevention of the localized and generalized Shwartzman reaction by an anticomplementary agent: Cobra venom factors, *J. Exp. Med.* 134:642.

Fong, J. S. C., White, J. G., and Good, R. A., 1974, Inhibition of endotoxic lipopolysaccharide-mediated platelet aggregation by cobra venom anticomplementary factor, *Blood* 44:399.

Gallimore, M. J., Aasen, A. O., and Amundsen, E., 1978, Changes in plasma levels of prekallikrein, kallikrein, high molecular weight kininogen and kallikrein inhibitors during lethal endotoxin shock in dogs, *Haemostasis* 7:79.

Galloway, R. E., Levin, J., Butler, T., *et al.*, 1977, Activation of protein mediators of inflammation and evidence for endotoxemia in *Borrelia recurrentis* infection, *Am. J. Med.* 63:933.

Garner, R., Chater, B. V., and Brown, D. L., 1974, The role of complement in endotoxic shock and disseminated intravascular coagulation: Experimental observations in the dog, *Br. J. Haematol.* 28:393.

Hawiger, J., 1978, Platelets in Sanarelli–Shwartzman reaction and in human sepsis, in: *Platelets: A Multidisciplinary Approach* (G. de Gaetano and S. Garattini, eds.), p. 321, Raven Press, New York.

Hawiger, J., Hawiger, A., Steckley, S., Timmons, S., and Cheng, C., 1977, Membrane changes in human platelets induced by lipopolysaccharide endotoxin, *Br. J. Haematol.* 35:285.

Henson, P. M., 1969, The adherence of leucocytes and platelets induced by fixed IgG antibody or complement, *Immunology* 16:107.

Hiller, E., Saal, J. G., and Riethmueller, G., 1977, Procoagulant activity of activated monocytes, *Haemostasis* 6:347.

Horder, M. H., and Manrique, R., 1965, Aktivierung der Blutgerinnung durch ein bakterielles Endotoxin beim Menschen unter Kumarintherapy, *Klin. Wochenschr.* 43:397.

Humphrey, J. H., and Jacques, R., 1955, The release of histamine and 5-hydroxytryptamine (serotonin) from platelets by antigen–antibody reactions *(in vitro), J. Physiol. (London)* 128:9.

Kane, M. A., May, J. E., and Frank, M. M., 1973, Interaction of the classical and alternate complementary pathway with endotoxin lipopolysaccharide: Effect on platelets and blood coagulation, *J. Clin. Invest.* 52:370.

Kimball, H. R., Melmon, K. L., and Wolfe, S. M., 1972, Endotoxin-induced kinin production in man, *Proc. Soc. Exp. Biol. Med.* **139**:1078.

Landy, M., and Weidanz, W. P., 1964, Natural antibodies against gram negative bacteria, in: *Bacterial Endotoxins* (M. Landy and W. Braun, eds.), p. 275, Rutgers University Press, New Brunswick, N.J.

Lerner, R. G., Goldstein, R., and Cummings, G., 1977, Endotoxin induced disseminated intravascular clotting: Evidence that it is mediated by neutrophil production of tissue factor, *Thromb. Res.* **11**:253.

Lüderitz, O., Staub, A. M., and Westphal, O., 1966, Immunochemistry of O and R antigens of *Salmonella* and related Enterobacteriaceae, *Bacteriol. Rev.* **30**:192.

Lüscher, E. F., and Pfueller, S. L., 1978, Platelets as targets of immunological reactions, in: *Platelets: A Multidisciplinary Approach* (G. de Gaetano and S. Garattini, eds.), p. 261, Raven Press, New York.

Mason, J. W., Kleeberg, U., Dolan, P., and Colman, R. W., 1970, Plasma kallikrein and Hageman factor in gram negative bacteremia, *Ann. Intern. Med.* **73**:545.

Michael, J. G., Whitby, J. L., and Landy, M., 1962, Studies on natural antibodies to gram negative bacteria, *J. Exp. Med.* **115**:3.

Morrison, D. C., and Cochrane, C. G., 1974, Direct evidence for Hageman factor (factor XII) activation by bacterial lipopolysaccharides (endotoxins), *J. Exp. Med.* **140**:797.

Morrison, D. C., and Kline, L. F., 1977, Activation of the classical and properdin pathways of complement by bacterial lipopolysaccharides (LPS), *J. Immunol.* **118**:362.

Muller-Berghans, G., 1978, The role of platelets, leucocytes and complement in the activation of intravascular coagulation by endotoxins, in: *Platelets: A Multidisciplinary Approach* (G. de Gaetano and S. Garattini, eds.), p. 303, Raven Press, New York.

Muller-Berghans, G., and Schneberger, R., 1971, Hageman factor activation in the generalized Shwartzman reaction induced by endotoxins, *Br. J. Haematol,* **21**:513.

Nagayama, M., Zucker, M. B., and Beller, F. K., 1971, Effect of a variety of endotoxins on human and rabbit platelet function, *Thromb. Diath. Haemorrh.* **26**:467.

Nowotny, A., 1969, Molecular aspects of endotoxic reactions, *Bacteriol. Rev.* **33**:72.

Nowotny, A., 1971, Relation of structure and biological activity of bacterial endotoxins, *Naturwissenschaften* **58**:397.

Nowotny, A., 1977, Relation of structure to function in bacterial endotoxins, in: *Microbiology 1977* (D. Schlessinger, ed.), American Society for Microbiology, Washington, D.C.

Nowotny, A., Cundy, K., Neale, N., Nowotny, A. M., Radvany, R., Thomas, S., and Tripodi, D., 1966, Relation of structure to function in bacterial O-antigens. IV. Fractionation of the components, *Ann. N.Y. Acad. Sci.* **133**:586.

Nowotny, A., Behling, U. H., and Chang, H. L., 1975, Relation of structure to function in bacterial endotoxins. VIII. Biological activities in a polysaccharide-rich fraction, *J. Immunol.* **115**:199.

Prydz, H., Allison, A. C., and Schorlemmer, H. U., 1977, Further link between complement activation and blood coagulation, *Nature (London)* **270**:173.

Ream, V. J., Deykin, D., Gurewich, V., and Wessler, S., 1965, The aggregation of human platelets by bacterial endotoxins, *J. Lab. Clin. Med.* **66**:245.

Rickles, F. R., Levine, J., Hardin, J. A., *et al.*, 1977, Tissue factor generation by human mononuclear cells: Effects of endotoxin and dissociation of tissue factor generation from mitogenic response, *J. Lab. Clin. Med.* **89**:792.

Rodriguez-Erdmann, F., 1964, Studies on pathogenesis of the generalized Shwartzman reaction. III. Trigger mechanism for the activation of the prothrombin molecule, *Thromb. Diath. Haemorrh.* **12**:471.

Rothberger, H., Zimmerman, T. S., Spiegelberg, H. L., and Vaughan, J. H., 1977, Leuco-

cyte procoagulant activity: Enhancement of production *in vitro* by IgG and antigen–
antibody complexes, *J. Clin. Invest.* **59:**549.

Semeraro, N., Colucci, M., Fumarola, D., and Vermylen, J., 1978a, Platelets and endotox-
ins: Complement-dependent and complement-independent interactions, in: *Platelets:
A Multidisciplinary Approach* (G. de Gaetano and S. Garattini, eds.), Raven Press, New
York.

Semeraro, N., Fumarola, D., Merters, F., and Vermylen, J., 1978b, Evidence that endotoxins enhance the factor X activation activity of washed human platelets, *Br. J.
Haematol.* **38:**243.

Spielvogel, A. R., 1967, An ultrastructural study of the mechanism of platelet–endotoxin
interaction, *J. Exp. Med.* **126:**235.

Springer, G. F., and Adye, J. C., 1975, Endotoxin-binding substances from human leucocytes and platelets, *Infect. Immun.* **12:**978.

Tonooka, T., 1978, Development of tissue factor activity in mononuclear cells cultured *in
vitro, Tohoku J. Exp. Med.* **125:**213.

Ulevitch, R. J., Cochrane, C. G., Henson, P. M., Morrison, D. C., and Doc, W. F., 1975,
Mediation systems in bacterial lipopolysaccharide-induced hypotension and disseminated intravascular coagulation, *J. Exp. Med.* **142:**1570.

Ulevitch, R. J., Cochrane, C. G., Bangs, K., *et al.*, 1978, The effect of complement depletion on bacterial lipopolysaccharide induced hemodynamic and hematologic changes
in the rhesus monkey, *Am. J. Pathol.* **92:**227.

Von Kaulla, K. N., 1958, Intravenous protein-free pyrogen: A powerful fibrinolytic agent
in man, *Circulation* **17:**187.

Walker, R. I., Porvaznik, M., Egan, J. E., and Miller, A. M., 1979, Hageman factor activation and tight junction disruption in mice challenged with attenuated endotoxin,
Experientia **35:**759.

4

Immunogenicity of Endotoxin

Erwin Neter

I. Introduction

Endotoxin or lipopolysaccharide (LPS) is one of the most extraordinary bacterial products because of the variety of biologic effect it engenders in susceptible animals. Some of these reactions are undesirable, since they may be associated with pathologic changes and even with death. Others are beneficial when the result is enhanced resistance or immunity to toxemia and/or infection. It must be emphasized, however, that, depending upon the degree of reaction and the particular circumstances, one and the same effect may be either beneficial or harmful. Thus, high fever may have undesirable results or, as utilized some time ago, may be curative when used in patients to cure certain diseases or to enhance resistance. Pyrogenicity has been studied for more than two centuries, as reviewed in depth by Westphal *et al.* (1977). Similarly, an immune response to a given antigenic determinant may have either beneficial, harmful, or indifferent effects.

This chapter deals with the immunogenicity of endotoxin and its components. Of particular import are the LPS macromolecules of smooth strains; the core lipopolysaccharides of rough strains shared by unrelated serotypes, serogroups, or species; the lipid A moiety; and the lipid A-associated protein. It must be emphasized that antigenicity does not necessarily indicate immunogenicity of a given molecule and it is immunogenicity that is the subject of this chapter.

Erwin Neter ● Departments of Microbiology and Pediatrics, State University of New York at Buffalo, and Laboratory of Bacteriology, Children's Hospital, Buffalo, New York 14222. Dedicated to Professor Dr. Bernhard Urbaschek, respected colleague and valued friend, on the occasion of his sixtieth birthday.

In addition to serving as antigenic determinants for a specific immune response, lipopolysaccharides or certain of its components also affect the immune system by (1) stimulating DNA synthesis in B cells, (2) producing a polyclonal response resulting in maturation of antibody-forming cells, (3) serving as an adjuvant, enhancing the immune response, and (4) exerting immunosuppressive effects. A review of endotoxins and the immune response, summarizing earlier data, was published by Neter (1969).

Analysis of such dual effects of LPS, namely, specific immunogenicity and nonimmunogenic effects on the immune system, is not always easily accomplished. For example, if LPS given to an animal were to suppress an immune response to an unrelated antigen, the phenomenon is logically described under the heading of "immunosuppression." The possibility must be considered, however, that under special circumstances LPS serves as an adjuvant by enhancing the production of suppressor T cells. Thus, adjuvanticity may result in immunosuppression. An unusual type of immunosuppression was encountered during the analysis of the common enterobacterial antigen existing in an immunogenic and a nonimmunogenic form. The lack of immunogenicity of the latter is due to the presence of LPS and its removal renders the antigen immunogenic (Neter, 1969).

Far less is known about lipid A and the lipid A-associated protein than about LPS itself. Westphal, during his early studies, notably with Lüderitz, visualized the existence within the large LPS macromolecule of a moiety that accounts for the seemingly nonspecific toxicity (Westphal et al., 1958; Westphal, 1975). The existence of an immunogenic common moiety independent of the serologic specificity was postulated and documented many years ago by Watson (1964). Information on the immunologic properties of the lipid A-associated protein has been obtained primarily by Morrison and his co-workers (Morrison and Rudbach, 1981; Morrison and Ryan, 1979) and by Sultzer and Goodman (1976).

It is well known that serum from various animal species, even in the absence of known past or present infection, contains antibodies against numerous LPS antigens of various bacterial species, serogroups, and serotypes. The question whether the presence of these "normal" antibodies is the result of exposure of the host to the antigen present in a given microorganism cannot be a subject for discussion in this chapter. It must be emphasized, however, that the level of these antibodies increases with age, that in man IgG but not IgM antibodies are transmitted through the placenta, and that, particularly with regard to the antibodies against the polysaccharide component of LPS, the production of IgM antibodies prevails for a considerable length of time but is followed by the production of IgG antibodies upon continued immunization or infection. The possible protective role of secretory antibodies can be mentioned only in passing.

A variety of serologic procedures has been used to identify and quantitate antibodies against the various LPS components, particularly against the polysaccharide determinants. Such studies are of particular importance, since the polysaccharides characterize numerous species, serogroups, and/or serotypes of LPS-producing bacteria. In addition to classical agglutination and precipitation tests, the passive or indirect hemagglutination procedure has been widely employed. The hemagglutination test can be modified as a Coombs test by the use of antiglobulins directed against the immunoclass of the LPS antibodies. More recently, the enzyme-linked immunosorbent assay (ELISA), also referred to as enzyme immunoassay or EIA, was developed for the titration of immunoglobulins of various classes (IgG, IgM, IgA) in both serum and secretions (e.g., Vos et al., 1979; Mutchnick and Keren, 1981; Granfors et al., 1981). The radioimmunoassay (RIA) procedure was compared to conventional methods by Horsfall and Rowley (1979a) and proved to be more sensitive, particularly for the detection of mucosal antibodies. Other procedures available include coagglutination, protein A-dependent radioimmunoprecipitation, and complement-mediated bactericidal tests.

Far more information is available on the humoral immune response to LPS and its components than about cellular immunity (delayed-type hypersensitivity).

II. Immunogenicity of Endotoxin in Natural Infections

It has been known for decades that many pathogenic members of LPS-producing microorganisms induce a specific antibody response against the haptenic determinant, identified as O antigen. The Widal test represents an early example of the diagnostic value of the documentation of an LPS-specific antibody response in typhoid fever. This test was introduced after it was shown by Gruber that experimentally produced antiserum is useful for the identification of *S. typhi*. Shiga used this principle when he provided evidence that a hitherto unknown microorganism proved to be a cause of dysentery. Since that time an enormous amount of information has become available for the characterization of the numerous O antigens by the corresponding antibodies. An increase in LPS-specific antibody titers during and following acute infection has been used to provide supporting evidence for the role of suspected pathogens, to differentiate between relapse and reinfection, for the documentation of subclinical infections, and for epidemiologic purposes (Neter, 1971). Under appropriate conditions, LPS-producing microorganisms also engender a local antibody response with the production of IgA, IgM, and/or IgG antibodies. It is the IgA antibodies that play a significant role (e.g., Sohl Åkerlund et al., 1979). In this connection it is of interest to point out that locally produced antibodies in patients with pyelonephritis may become attached to bacterial cells and can be demonstrated by the FA technique. Observations on patients with urinary tract infection also revealed the incisive effect of the location of the infection and the local antibody response: In patients with pyelonephritis, such antibodies are produced more frequently and in higher titers than in subjects with cystitis or bacteriuria (cf. Neter, 1975). In the study of the immune response of patients, it is important to keep in mind the effects of numerous factors, such as the age of the subjects. Although infants may respond less actively to antigenic stimulation by LPS of infecting microorganisms, even these young subjects may mount a specific antibody response, as was recently shown by Giacoia et al. (1981). Among the numerous factors affecting the antibody response to LPS are malignancies and immunosuppressive drugs; however, even under these conditions a specific antibody response to LPS can be documented in some of these patients (Han et al., 1967; Griffiths et al., 1977a; Baliah and Neter, 1980; Gannon et al., 1980).

It is not the purpose to discuss here the protective effects of O antibodies in natural infections of man and animals.

III. Immunogenicity of Endotoxin in Experimental Infections and Immunization

It is well known that serum of healthy individuals contains antibodies against numerous O antigens of LPS-producing bacteria, and it is assumed that the production of these antibodies is the result of natural exposure to the respective antigens. It is important to keep in mind that cross-reactions among polysaccharide determinants are seen quite frequently. Evidence for the induction of antibody formation by oral exposure to LPS-producing microorganisms was provided experimentally both in adults and in infants. Jodal et al. (1977) studied the immune response of adults exposed to *E. coli* O83 and observed that an increase in O-specific antibody titers could be documented by means of a modified Farr ammonium sulfate precipitation test but not by indirect hemagglutination and ELISA tests. The differences in these results may be due to differences in antigens used as indicators rather than in the sensitivity of the procedures themselves. It is conceivable that antigens in the intestinal tract of infants may be more immunogenic than those in adults. Lodinová *et al.* (1973) reported that antibody production, both systemic and mucosal, by infants may be effected by experimental colonization with a nonenteropathogenic strain of *E. coli.* Some of the related observations will be briefly mentioned. As shown by Hohmann *et al.* (1979), one of the major determinants of immunogenicity of LPS-producing bacteria is the ability of the strains to survive and grow in the Peyer's patches of the small intestine of mice. An interesting study on the effects of an attenuated strain of *S. typhimurium* given orally to mice was reported by Moser *et al.* (1980). The authors used a UDP-galactose-4-epimerase-deficient strain. This immunization resulted in the production of LPS-specific antibodies in both serum and intestinal fluid and in significant protection against challenge with a fully virulent strain. The antibodies in the intestinal fluid were largely of the IgA immunoclass, as expected. A detailed study on the antibody response of the O determinant of LPS of mice infected orally with *S. typhimurium* was carried out by Sziegoleit (1976) and revealed that intestinal antibodies were largely of the IgA and IgG and serum antibodies were largely of the IgM and IgG immunoclasses. Biozzi *et al.* (1979) published an excellent review of the genetics of immunoresponsiveness to natural antigens in the mouse and included the results of their own studies on the response of selectively bred mice to formalin-killed *Salmonella* cells. Schlecht and Westphal (1979) carried out a careful study on the antibody-inducing and protective capacity of enterobacterial S and R forms in mice. They observed that the protective capacity against *S. typhimurium* infection, in order of decreasing potency, is *S. typhimurium* S form, *Salmonella* R mutants, heterologous *Salmonella* S forms, and *E. coli* R mutants and that no direct relationship could be documented between immunizing capacity and antibody-inducing potency. Porter *et al.* (1974) studied the immune response in young pigs and Eddie *et al.* (1971) in rabbits. Striking differences in the response of eight mouse strains to *Salmonella* vaccine were documented by DiPauli (1977), and differences were even noted between individuals of the identical strain. It is of considerable interest to note that, as shown by Holmgren and Ahlstedt (1974) and Ahlstedt and Holmgren (1975), experimental immunization with a given O group of *E. coli* engenders more IgG antibodies in mice and rabbits previously exposed to unrelated O antigens.

The subject of the beneficial effects of experimental infection or immunziation with rough or attenuated strains is beyond the scope of this chapter.

IV. Immunogenicity of LPS

Although endotoxin or LPS connotes bacterial products with common features and biologic activities, the immunogenicity of these preparations may differ from product to product, even when LPS is isolated from the same O group and has identical antigenicity. LPS has been extracted from various microorganisms by the phenol–water, trichloroacetic acid (TCA), diethylene glycol, SDS, hypertonic NaCl–sodium citrate solution, aqueous ether, butanol–water, and phenol–chloroform–petroleum ether methods. That preparations obtained by the identical procedure from the same microorganism may differ in immunogenicity may be surprising. Horsfall and Rowley (1979b) observed striking differences in immunogenicity in mice of Boivin-type antigens obtained by the trichloroacetic acid extraction method from *Vibrio cholerae*.

It has been recognized for many years that LPS is a T-cell-independent immunogen, as evident, for example, from the observation of Manning *et al.* (1972), who showed that nude mice respond to immunization with phenol–water-extracted LPS as well as control animals. Under suitable conditions, LPS is immunogenic even when minute amounts are injected into suitable hosts. In the study of Benner and van Oudenaren (1976), it was shown that *E. coli* LPS in a dose as small as 0.001 μg given intravenously to mice resulted in the appearance of specific plaque-forming cells in the spleen. Vos *et al.* (1979) provided evidence of the high immunogenicity of commercially available phenol-extracted LPS from *E. coli* O127 in rats. The antigen was immunogenic when injected intravenously into young adult animals in doses from 10 pg to 1 mg, the optimal dose being 0.1 to 1000 μg. Intravenous immunization with commercially obtained LPS from *E. coli* O55 revealed that bone marrow of the mouse is capable of an immune response as evident from the appearance of plaque-forming cells (Benner and van Oudenaren, 1977). In the mouse, as shown by Barasoain *et al.* (1978), 25–30 μg per mouse given intraperitoneally yielded a good antibody response. It is not always appreciated that purified LPS primes animals for a secondary response to the polysaccharide part of the molecule. Thus, Rudbach (1971) found that in mice, picogram amounts of *E. coli* LPS sensitize the animals effectively, the magnitude of the secondary response depending upon the amount used for the booster injection.

So far as the duration of the antibody response to LPS is concerned, Citron and Michael (1981) made the significant observation that adherent cells, in part at least, contribute to the cyclic features of antibody production in selected strains of mice, presumably acting as immunosuppressants for limited periods of time. LPS, instead of enhancing the immune response to a given antigen, may act as an immunosuppressant agent. For example, as shown by Winchurch *et al.* (1981), LPS may activate suppressor cells in newborn mice.

Even with a single preparation, the immunogenicity can be altered experimentally. Heating or treatment with alkali of LPS may significantly reduce the immunogenicity while maintaining the antigenicity (Whang *et al.*, 1971; Neter *et al.*, 1973). Interestingly, this loss of immunogenicity is reversible to a significant degree by repeated freezing and thawing, suggesting that immunogenicity, but not antigenicity, is altered by deaggregation and reaggregation rather than by a chemical process. Rudbach (1971) observed that dissociation of purified *E. coli* LPS with sodium deoxycholate results in a product that is nonimmunogenic for rabbits but stimulates the production of antibodies in mice. As early as 1968, Nowotny showed that detoxified LPS, compared to the toxic parent, remains a good immunogen in rabbits.

Numerous factors affect the antibody response to LPS, including species of animal, the strain of mouse, such as BALB/c, SJL/L, and C3H, and age of the animals (Smith, 1976). As with other polysaccharide antigens, the immune response to LPS depends, in part, upon the animal species; for example, antibodies of different specificities may be induced by certain *Salmonella* serogroups in rabbits and horses. Rudbach and Reed (1977) characterized the immune response of C3H/HeJ mice to LPS and made the significant observation that, in this particular strain, only a primary response was elicited by phenol–water-extracted LPS from *E. coli* O113. Detailed analysis of the mode of action of endotoxin utilized the endotoxin-refractory C3H/HeJ strain by means of cell transfer after irradiation. Thus, chimeras are produced and the effects of endotoxin can be analyzed. Reference is made to the study of Michalek *et al.* (1980) and to other chapters of this volume. Very recently, Pier *et al.* (1981) observed that the reactivity of C3H/HeJ mice to LPS differs with preparations obtained from Enterobacteriaceae and from *Pseudomonas aeruginosa*. These observations suggest reinvestigation of the fundamental problems associated with biologic activities of LPS. McGhee *et al.* (1979) have clearly shown that phenol–water-extracted LPS engenders polysaccharide-specific antibodies in responder strains (C3H/HeN or C57BL/6J) but not in the nonresponder strain C3H/HeJ. These authors also documented the fact that F_1 progeny of responder and nonresponder parents showed an intermediate response. Of special interest is the observation that these differences were not noted when butanol–water-extracted LPS was used. The presence of lipid A-associated protein or, possibly, of other peptides may affect the immunogenicity of various preparations in various animals.

It is not the purpose to discuss the fundamental aspects of the immunogenicity of LPS. It should be pointed out that the analysis is rendered more complex by the fact that lipid A and, when applicable, the lipid A-associated protein also have nonimmunologic effects on the immune system. The latter were critically and imaginatively discussed recently by Morrison and Rudbach (1981). It is of interest to note that artificial antigens have been prepared by combining the haptenic saccharides with a carrier, such as bovine serum albumin (Svenson *et al.*, 1979; Svenson and Lindberg, 1981; Jorbeck *et al.*, 1981). This approach is particularly useful when studying the possible protective effects of polysaccharide-specific antibodies to infection, since such preparations do not contain lipid A or lipid A-associated protein, which may significantly affect the immune response. A similar approach is based on the development of nonpathogenic strains (e.g., *E. coli*) following conjugation with a pathogenic strain (e.g., *S. flexneri*) resulting in the acquisition of the genes by the former organism responsible for the production of surface antigen (LPS) of the latter. Theoretically, such a viable vaccine should induce immunity without causing serious illness (Levine *et al.*, 1977). Izui *et al.* (1980) showed that lipid A-associated protein did not markedly affect the primary antibody response of LPS and, thus, showed that, depending on the mouse strains used, *in vitro* and *in vivo* effects do not necessarily parallel each other. To indicate the great complexity, Skelly *et al.* (1980a,b) provided evidence that in certain strains of mice, LPS coupled to a haptenic determinant stimulated LPS antibody to a lesser degree than the unsubstituted LPS preparation. Differences in immunogenicity of *S. enteritidis* LPS were shown by Hepper *et al.* (1979) in experiments in which BALB/c mice were immunized with phenol–water- or TCA-extracted preparations, and the antibody response measured by a plaque-forming assay. These differences, to a significant extent, are explained on the basis of the presence of protein in the latter preparations. These authors showed that digestion with trypsin and Pronase of the TCA–LPS preparation markedly reduced its immunogenicity.

In view of the immunogenicity of various LPS preparations, it is not surprising that, in addition to circulating IgM and IgG antibodies, secretory antibodies of the IgA variety may also be engendered. For example, Reynolds *et al.* (1974), utilizing LPS preparations from the major serotypes of *P. aeruginosa*, observed that intranasal immunization induced a local immune response in rabbits, notably the production of secretory IgA. Ebersole and Molinari (1978) performed a detailed study on the comparative immunogenicity of LPS and nonviable organisms for topical immunization, regarding both the primary and the secondary immune responses and the immunoclasses of antibodies appearing in blood and saliva of mice. It took longer for the appearance of antibodies in saliva and the peak titers obtained were lower following immunization with LPS compared to the particulate antigen. Danneman and Michael (1976) showed that LPS as antigen did not induce antibodies of IgE specificity.

In view of striking differences in immunogenicity of various LPS preparations, it is important that in fundamental studies of endotoxin careful attention be paid to the particular preparations used.

V. Immunogenicity of R Mutants and Their Lipopolysaccharides

It has been known for many decades that the mutations occur rather frequently in enteric bacteria and that the R mutants are less virulent or nonvirulent when compared to the S parent strains. In spite of loss of virulence, endotoxicity remains. Much has been learned regarding the chemical composition of LPS from S and R strains. That the LPS preparations from R mutants induce antibodies has been clearly established and, as discussed elsewhere in this volume, investigations have been directed toward utilization of these nonvirulent strains to induce protection against infection by virulent microorganisms.

Sensitive methods for the quantitation of these antibodies include RIA and EIA (Bruins *et al.*, 1978; Zeckel *et al.*, 1980).

There is no question that immunization with the highly rough mutant, Re 595 of *S. minnesota*, induces a specific antibody response in the rabbit. Interestingly, these antibodies do not react with lipid A. It must be kept in mind that antibodies to unrelated O antigens may be stimulated by this immunization, possible due to the mitogenic effect of LPS. As discussed elsewhere in this volume, animals immunized with the Re mutant develop resistance to and antisera protect against experimental infection, including disease caused in rabbits made granulocytopenic (Bruins *et al.*, 1977; Ziegler *et al.*, 1973).

Braude *et al.* (1977a,b) induced core glycolipid antibodies in rabbits by immunization with killed cells of the UDP-galactose-deficient mutant (J5) of *E. coli* O111. It is of considerable interest that, using the same approach, Davis *et al.* (1978) showed that such an antiserum also neutralizes the effects of meningococcal endotoxin. This observation is of particular interest since human fulminating meningococcal disease resembles endotoxic changes more closely than many infections caused by LPS-producing microorganisms. Similarly, Diena *et al.* (1978) observed that immunization with R-type LPS from *Neisseria gonorrhoeae* induced significant protection in mice and chicken. In the latter model, embryonated eggs obtained from immunized hens were protected against challenge in contrast to the controls. Detailed studies on the protective effects of antibodies to the core LPS were carried out by McCabe *et al.* (1977) and could be ascribed to the 2-keto-3-deoxyoctonate (KDO) rather than to lipid A. Young and Stevens (1977) showed that immunization with the core glycolipid antigen obtained from the Re mutant *S. minnesota* R595 provided less protection than type-specific immunization in dogs and mice. More than a decade ago, Chedid *et al.* (1968) suggested the role of core antibody. Galactose-deficient rough mutants probably have clinical relevance, since Ryan *et al.* (1973) isolated galactose-deficient, endotoxin-producing strains from patients with urinary tract infection.

VI. Immunogenicity of Lipid A

Lipid A has been recognized as the major component of the LPS molecule responsible for toxicity and certain other biological activities. It is present in both S and R forms of LPS-producing microorganisms. An excellent critical review of our knowledge on the chemistry, biology, and immunogenicity of lipid A is part of the review published by Galanos, Lüderitz, Rietschel, and Westphal (1977). Much of the information now available has been due to the pioneering efforts of this group of researchers.

That lipid A is not identical in LPS produced by all bacterial genera and species has been clearly established, both by chemical and by immunological analysis. Such studies have extended even to photosynthetic prokaryotes (Weckesser *et al.*, 1979). Lipid A is immunogenic in the rabbit, particularly when R mutants are injected. Antibodies have been produced in rabbits, dogs, mice, and goats, among other animal species. It is important to point out that special immunization schedules may be required, particularly a lengthy interval between primary and secondary immunization. Interestingly, according to Westenfelder *et al.* (1975), lipid A is nonimmunogenic in puppies, although it is immunogenic in adult dogs. Immunization has been established with a variety of preparations. For example, Johns *et al.* (1977) engendered antibodies by the injection of lipid A from certain enterobacterial and nonenterobacterial strains after the antigen was solubilized with triethylamine or complexed with rabbit serum albumin. The immune response differed from that elicited by the Re mutant of *S. typhimurium*. Interestingly, lipid A antibodies could be engendered also by liposomes containing the antigen.

A variety of tests have been used for the quantitation of lipid A antibodies, including hemagglutination, precipitation, diffusion, passive hemolysis, and ELISA tests (Mattsby-Baltzer and Kaijser, 1979; Fink and Galanos, 1981). Attention should be called to the fact that immunization of rabbits with lipid A may also stimulate the production of antibodies to unrelated antigens, possibly because of the mitogenicity of the antigen (Bruins *et al.*, 1977). In addition to these various animal species, lipid A antibodies are also encountered in sera of patients with a variety of gram-negative infections, including those of the urinary tract (Simon *et al.*, 1974).

The early expectation that lipid A may be used in man for active immunization against most or all bacteria sharing this moiety has not been fulfilled. Nor has the evidence provided hope for the expectation that the corresponding antisera are curative in such infections. In fact, preliminary evidence suggests that such antibodies may be pathogenetic (Westenfelder *et al.*, 1977). These authors also observed that the immune response plays a role in the development of lipid A-induced nephritis. More recently, Hemstreet *et al.* (1981) showed that lipid A stimulates the production in rats of cytotoxic antibodies, which are responsible for injury to kidney tubular cells. As an immunogen, glycolipid from *S. minnesota* Re 595 was used. The mechanism by which cell injury is induced, either because the antibody cross-reacts with kidney antigen, serves as adjuvant to break tolerance, or becomes attached to cells and thus acts as neoantigen, has not been unequivocally determined.

VII. Immunogenicity of Lipid A-Associated Protein (Endotoxin Protein)

Although it has been known for decades that LPS as part of the outer membrane of microorganisms is associated with peptides or proteins, immunologic studies of LPS-associated protein have been carried out only recently, particularly in the laboratories of Morrison *et al.* (1980) and Sultzer *et al.* (1980). It has been established that this protein is associated with lipid A (lipid A-associated protein) and can reassociate with LPS or lipid A. It is an active B-cell stimulant; proliferation and differentiation of B cells is the result. In view of the fact that LPS is a mitogen, it was important to exclude that LPS contamination is responsible for B-cell mitogenicity of the lipid A-associated protein. Betz and Morrison (1977) provided this evidence by showing that the protein is active even in the lipid A-unresponsive C3H/HeJ mouse. This protein also affects on a nonimmunologic basis the immune response to the polysaccharide component of LPS. Thus, it was shown by Hepper *et al.* (1979) that the secondary response to the polysaccharide component differs depending on whether Boivin-type or protein-free

LPS preparations were used for antigenic stimulation. In contrast, as shown by Izui *et al.* (1980), the primary polysaccharide antibody response in mice was influenced only to an insignificant degree by lipid A-associated protein or by lipid A itself.

This lipid A-associated protein differs from the lipoprotein described by Braun and Wolff (1970; Braun, 1973); the antibody response to this particular bacterial protein of patients with a variety of enterobacterial infections has been described by Griffiths *et al.* (1977b). It is not the purpose of this contribution to discuss all protein components present in the cell wall of the numerous LPS-producing microorganisms. Suffice it to mention a few of these contributions: Hansen *et al.* (1981) and Taplits and Michael (1979). Homma *et al.* (1976) and Yamamoto *et al.* (1979) studied in considerable detail the protein moiety, referred to as OEP, of endotoxin of *P. aeruginosa* and made the interesting observations that this common antigen is immunogenic, that the corresponding antibodies are protective, and that the antigen is also produced by *V. cholerae*. It is well known that infection or immunization with LPS-containing bacteria may result in enhanced antibody response to a serologically unrelated LPS. As reported by Ahlstedt and Lindholm (1977), this effect may be due to the lipid A-associated protein.

At the present time, definitive information on the immunogenicity of this protein under various conditions, such as natural and experimental infections and immunization, and of the corresponding immune response (humoral and/or cellular) as it affects susceptibility and resistance to infection by LPS-producing microorganisms, remains to be elucidated.

VIII. The Immune Response to Endotoxin and Its Diagnostic Significance

Obviously, infections caused by LPS-producing microorganisms have to be classified as undesirable events, irrespective of the role of LPS. From a clinical point of view, the immune response to LPS is being utilized to the advantage of infected individuals and animals, since its documentation may be used effectively for specific diagnosis and, therefore, for appropriate therapy. That the O antigen, the LPS of *S. typhi*, is a good immunogen has been known since the end of the last century, when Widal introduced documentation of the antibody response in the diagnostic armamentarium. Since that time the same approach has been utilized for the diagnosis of other infections caused by LPS-producing microorganisms, including brucellosis, tularemia, salmonellosis, shigellosis, and others. Indeed, evidence for a humoral anti-body response to LPS has been extensively explored for the differentiation between single and mixed infections, when the suspected pathogens are present on mucous membranes or in the intestinal tract; for the differentiation between relapse and reinfection; for the diagnosis of subclinical infection; and for clarification of epidemiologic aspects of diseases (Neter, 1971, 1979).

IX. The Immune Response to Endotoxin and Its Therapeutic Implications

It is not the purpose of this chapter to review the numerous studies on active and passive immunization against infection by LPS-producing microorganisms. It is well established that typhoid vaccine protects significantly, albeit not absolutely, against this infection, although the immune response to the O antigen certainly alone is not responsible for the prophylactic effects. Only limited application has been made of immunization with isolated LPS. Suffice it to mention that LPS obtained from *P. aeruginosa* has been utilized for active immunization by, among others, Pennington and Kuchmy (1980). To what extent oral immunization with attenuated or R mutants of enteric bacteria protects against enteric infection due to a specific immune response to LPS rather than due to microbial antagonism has not been completely elucidated. Extracts with O specificity were studied extensively for immunoprophylaxis by Ocklitz and Mochmann (Mochmann *et al.*, 1974, 1980).

It must be emphasized that antibodies directed against the polysaccharide component of LPS do not regularly elicit a favorable response. Production of such antibodies, even in high titers, by patients with typhoid fever is frequently encountered and yet the disease continues and may even be fatal. Similar observations have been made in subjects with urinary tract infections. In fact, bacteria present in the bladder and originating from the kidney may be coated with O-specific antibodies of the major immunoclasses IgG, IgM, and IgA without termination of the infection.

X. Outlook

It remains for future studies to analyze, in depth, cell-mediated immunity and delayed-type hypersensitivity to the various components of LPS, including the polysaccharide determinants characterizing smooth strains, the core polysaccharide, lipid A, and lipid A-associated protein.

Even with a single pathogen, such as *S. typhimurium*, the role of delayed hypersensitivity, with particular reference to LPS determinants and genetic host factors, has not been definitively characterized (Hormaeche *et al.*, 1981). It remains to be determined to what extent this immune response affects susceptibility and resistance to the numerous infections caused by LPS-producing bacteria. In addition, the cellular events, including the stimulation of T cells and the subsets of these immunocytes, need clarification. With the enormous progress made in the field of cellular immunology and with the utilization of new techniques, such as the hybridoma method, further progress can be expected in the near future.

XI. References

Ahlstedt, S., and Holmgren, J., 1975, Alteration of the antibody response to *Escherichia coli* O antigen in mice by prior exposure to various somatic antigens, *Immunology* **29:** 487.

Ahlstedt, S., and Lindholm, L., 1977, Antibody response to *Escherichia coli* O antigens and influence of the protein moiety of the endotoxin, *Immunology* **33:**629.

Baliah, T., and Neter, E., 1980, The immune response to typhoid vaccine of patients with idiopathic minimal lesion nephrosis, *Int. J. Pediatr. Nephrol.* **1:**212.

Barasoain, I., Rojo, J. M., and Portolés, A., 1978, Transfer of antibacterial immunity by "immune" RNA from animals treated with *Pseudomonas* lipopolysaccharide, *Z. Immunitaetsforsch.* **154:**34.

Benner, R., and van Oudenaren, A., 1976, Antibody formation in mouse bone marrow. V. The response to the thymus-independent antigen *Escherichia coli* lipopolysaccharide, *Immunology* **30:**49.

Benner, R., and van Oudenaren, A., 1977, Antibody formation in mouse bone marrow. VI. The regulating influence of the spleen on the bone marrow plaque-forming cell response to *Escherichia coli* lipopolysaccharide, *Immunology* **32:**513.

Betz, S. J., and Morrison, D. C., 1977, Chemical and biologic properties of a protein-rich fraction of bacterial lipopolysaccharides. I. The *in vitro* murine lymphocyte response, *J. Immunol.* **119:**1475.

Biozzi, G., Mouton, D., Sant'Anna, O. A., Passos, H. C., Gennari, M., Reis, M. H., Ferreira, V. C. A., Heumann, A. M., Bouthillier, Y., Ibanez, O. M., Stiffel, C., and Siqueira, M., 1979, Genetics of immunoresponsiveness to natural antigens in the mouse, *Curr. Top. Microbiol. Immunol.* **85:**31.

Braude, A. I., Ziegler, E. J., Douglas, H., and McCutchan, J. A., 1977a, Antibody to cell wall glycolipid of gram-negative bacteria: Induction of immunity to bacteremia and endotoxemia, *J. Infect. Dis.* **136** (Suppl.):167.

Braude, A. I., Ziegler, E. J., Douglas, H., and McCutchan, J. A., 1977b, Protective properties of antisera to R core, in: *Microbiology 1977* (D. Schlessinger, ed.), pp. 253–256, American Society for Microbiology, Washington, D.C.

Braun, V., 1973, Molecular organization of the rigid layer and the cell wall of *Escherichia coli, J. Infect. Dis.* **128** (Suppl.):9.

Braun, V., and Wolff, H., 1970, The murein-lipoprotein linkage in the cell wall of *Escherichia coli, Eur. J. Biochem.* **14:**387.

Bruins, S. C., Stumacher, R., Johns, M. A., and McCabe, W. R., 1977, Immunization with R mutants of *Salmonella minnesota*. III. Comparison of the protective effect of immunization with lipid A and the Re mutant, *Infect. Immun.* **17:**16.

Bruins, S. C., Ingwer, I., Zeckel, M. L., and White, A. C., 1978, Parameters affecting the enzyme-linked immunosorbent assay of immunoglobulin G antibody to a rough mutant of *Salmonella minnesota*, *Infect. Immun.* **21:**721.

Chedid, L., Parant, M., Parant, F., and Boyer, F., 1968, A proposed mechanism for natural immunity to enterobacterial pathogens, *J. Immunol.* **100:**292.

Citron, M. O., and Michael, J. G., 1981, Regulation of the immune response to bacterial lipopolysaccharide by adherent cells, *Infect. Immun.* **33:**519.

Danneman, P. J., and Michael, J. G., 1976, Adjuvant and immunogenic properties of bacterial lipopolysaccharide in IgE and IgG antibody formation in mice, *Cell. Immunol.* **22:**128.

Davis, C. E., Ziegler, E. J., and Arnold, K. F., 1978, Neutralization of meningococcal endotoxin by antibody to core glycolipid, *J. Exp. Med.* **147:**1007.

Diena, B. B., Ashton, F. E., Ryan, A., and Wallace, R., 1978, The lipopolysaccharide (R type) as a common antigen of *Neisseria gonorrhoeae*. I. Immunizing properties, *Can. J. Microbiol.* **24:**117.

DiPauli, R., 1977, Natural history of the immune response to *Salmonella* polysaccharides in inbred strains of mice, in: *Microbiology 1977* (D. Schlessinger, ed.), pp. 280–285, American Society for Microbiology, Washington, D.C.

Ebersole, J. L., and Molinari, J. A., 1978, The induction of salivary antibodies by topical sensitization with particulate and soluble bacterial immunogens, *Immunology* **34:**969.

Eddie, D. S., Schulkind, M. L., and Robbins, J. B., 1971, The isolation and biologic activities of purified secretory IgA and IgG anti-*Salmonella typhimurium* "O" antibodies from rabbit intestinal fluid and colostrum, *J. Immunol.* **106:**181.

Fink, P. C., and Galanos, C., 1981, Determination of anti-lipid A and lipid A by enzyme immunoassay, *Immunobiology* **158:**380.

Galanos, C., Lüderitz, O., Rietschel, E. T., and Westphal, O., 1977, Newer aspects of the chemistry and biology of bacterial lipopolysaccharides, with special reference to their lipid A component, in: *Biochemistry of Lipids II* (T. W. Goodwin, ed.), *International Review of Biochemistry*, Vol. 14, pp. 239–335, University Park Press, Baltimore.

Gannon, P. J., Surgalla, M. J., Fitzpatrick, J. E., and Neter, E., 1980, Immunoglobulin G and immunoglobulin M antibody responses of patients with malignancies to the O antigens of bacteria causing bacteremia, *J. Clin. Microbiol.* **12:**60.

Giacoia, G. P., Neter, E., and Ogra, P., 1981, Respiratory infections in infants on mechanical ventilation: The immune response as a diagnostic aid, *J. Pediatr.* **98:**691.

Granfors, K., Viljanen, M. K., and Toivanen, A., 1981, Measurement of immunoglobulin M, immunoglobulin G, and immunoglobulin A antibodies against *Yersinia enterocolitica* by enzyme-linked immunosorbent assay: Comparison of lipopolysaccharide and whole bacterium as antigen, *J. Clin. Microbiol.* **14:**6.

Griffiths, E. K., Surgalla, M. J., Fitzpatrick, J. E., and Neter, E., 1977a, Antibody response to *Serratia marcescens* isolated from patients with malignant diseases, *J. Clin. Microbiol.* **6:**499.

Griffiths, E. K., Yoonessi, S., and Neter, E., 1977b, Antibody response to enterobacterial lipoprotein of patients with varied infections due to Enterobacteriaceae, *Proc. Soc. Exp. Biol. Med.* **154:**246.

Han, T., Sokal, J. E., and Neter, E., 1967, Salmonellosis in disseminated malignant diseases, *N. Engl. J. Med.* **276:**1045.

Hansen, E. J., Frisch, D. F., McDade, R. L., Jr., and Johnston, K. H., 1981, Identification of immunogenic outer membrane proteins of *Haemophilus influenzae* type b in the infant rat model system, *Infect. Immun.* **32:**1084.

Hemstreet, G. P., Enoch, P. G., Fine, P. R., and Wheat, R., 1981, Lipid A induction of cytotoxic antibody to cultured syngeneic rat kidney tubular cells, *Kidney Int.* **19**:275.

Hepper, K. P., Garman, R. D., Lyons, M. F., and Teresa, G. W., 1979, Plaque-forming cell response in BALB/c mice to two preparations of LPS extracted from *Salmonella enteritidis*, *J. Immunol.* **122**:1290.

Hohmann, A., Schmidt, G., and Rowley, D., 1979, Intestinal and serum antibody responses in mice after oral immunization with *Salmonella*, *Escherichia coli*, and *Salmonella–Escherichia coli* hybrid strains, *Infect. Immun.* **25**:27.

Holmgren, J., and Ahlstedt, S., 1974, Enhancement of the IgG antibody production to *Escherichia coli* O antigen by prior exposure to serologically different *E. coli* bacteria, *Immunology* **26**:67.

Homma, J. Y., Abe, C., Okada, K., Tonamoto, K., and Hirao, Y., 1976, The biological properties of the protein moiety of the endotoxin of *Pseudomonas aeruginosa*, in: *Animal, Plant and Microbial Toxins*, Vol. 1 (A. Ohsaka and K. Hayoshi, eds.), pp. 449–508, Plenum Press, New York.

Hormaeche, C. E., Fahrenkrog, M. C., Pettifor, R. A., and Brock, J., 1981, Acquired immunity to *Salmonella typhimurium* and delayed (footpad) hypersensitivity in BALB/c mice, *Immunology* **43**:547.

Horsfall, D. J., and Rowley, D., 1979a, A comparison of methods for measuring intestinal antibodies against *Vibrio cholerae* in mice, *Aust. J. Exp. Biol. Med. Sci.* **57**:61.

Horsfall, D. J., and Rowley, D., 1979b, Intestinal antibody to *Vibrio cholerae* in immunised mice, *Aust. J. Exp. Biol. Med. Sci.* **57**:75.

Izui, S., Morrison, D. C., Curry, B., and Dixon, F. J., 1980, Effect of lipid A-associated protein and lipid A on the expression of lipopolysaccharide activity. I. Immunological activity, *Immunology* **40**:473.

Jodal, U., Ahlstedt, S., Hanson, L. Å., Lidin-Janson, G., and Sohl Åkerlund, A., 1977, Intestinal stimulation of the serum antibody response against *Escherichia coli* 083 in healthy adults, *Int. Arch. Allergy Appl. Immunol.* **53**:481.

Johns, M. A., Bruins, S. C., and McCabe, W. R., 1977, Immunization with R mutants of *Salmonella minnesota*. II. Serological response to lipid A and the lipopolysaccharide of Re mutants, *Infect. Immun.* **17**:9.

Jörbeck, H. J. A., Svenson, S. B., and Lindberg, A. A., 1981, Artificial *Salmonella* vaccines: *Salmonella typhimurium* O-antigen-specific oligosaccharide–protein conjugates elicit opsonizing antibodies that enhance phagocytosis, *Infect. Immun.* **32**:497.

Levine, M. M., Woodward, W. E., Formal, S. B., Gemski, P., Jr., DuPont, H. L., Hornick, R. B., and Snyder, M. J., 1977, Studies with a new generation of oral attenuated shigella vaccine: *Escherichia coli* bearing surface antigens of *Shigella flexneri*, *J. Infect. Dis.* **136**:577.

Lodinová, R., Jouja, V., and Wagner, V., 1973, Serum immunoglobulins and coproantibody formation in infants after artificial intestinal colonization with *Escherichia coli* 083 and oral lysozyme administration, *Pediatr. Res.* **7**:659.

McCabe, W. R., Bruins, S. C., Craven, D. E., and Johns, M., 1977, Cross-reactive antigens: Their potential for immunization-induced immunity to gram-negative bacteria, *J. Infect. Dis.*, **136**:S161.

McGhee, J. R., Michalek, S. M., Moore, R. N., Mergenhagen, S. E., and Rosenstreich, D. L., 1979, Genetic control of *in vivo* sensitivity to lipopolysaccharide: Evidence for condominant inheritance, *J. Immunol.* **122**:2052.

Manning, J. K., Reed, N. D., and Jutila, J. W., 1972, Antibody response to *Escherichia coli* lipopolysaccharide and type III pneumococcal polysaccharide by congenitally thymusless (nude) mice, *J. Immunol.* **108**:1470.

Mattsby-Baltzer, I., and Kaijser, B., 1979, Lipid A and anti-lipid A, *Infect. Immun.* **23**:758.

Michalek, S. M., Moore, R. N., McGhee, J. R., Rosenstreich, D. L., and Mergenhagen, S. E., 1980, The primary role of lymphoreticular cells in the mediation of host responses to bacterial endotoxin, *J. Infect. Dis.* **141**:55.

Mochmann, H., Ocklitz, H. W., Weh, L., and Heinrich, H., 1974, Oral immunization with an extract of *Escherichia coli* enteritidis, *Acta Microbiol. Acad. Sci. Hung.* **21**:193.

Mochmann, H., Zwierzchowski, J., Hering, L., Molenda, J., Ocklitz, H. W., Bocianowski, M., Austenat, L., Wałachowski, W., and Janas, Z., 1980, The protective effect of an EDTA-sodium-extract-vaccine obtained from swine-pathogenic *E. coli* in a field trial. 1. Communication: Direct and indirect immunization of suckling piglets, *Zentralbl. Bakteriol. Parasitenkd. Infektionskr. Hyg. Abt. 1 Orig. Reihe A* **247**:192.

Morrison, D. C., and Rudbach, J. A., 1981, Endotoxin–cell membrane interactions leading to transmembrane signaling, in: *Contemporary Topics in Molecular Immunology*, Vol. 8 (F. P. Inman and W. J. Mandy, eds.), pp. 187–218, Plenum Press, New York.

Morrison, D. C., and Ryan, J. L., 1979, A review–bacterial endotoxins and host immune function, *Adv. Immunol.* **28**:293.

Morrison, D. C., Wilson, M. E., Raziuddin, S., Betz, S. J., Curry, B. J., Oades, Z., and Munkenbeck, P., 1980, Influence of lipid A-associated protein on endotoxin stimulation of nonlymphiod cells, in: *Microbiology 1980* (D. Schlessinger, ed.), pp. 30–35, American Society for Microbiology, Washington, D.C.

Moser, I., Hohmann, A., Schmidt, G., and Rowley, D., 1980, Salmonellosis in mice: Studies on oral immunization with live avirulent vaccines, *Med. Microbiol. Immunol.* **168**:119.

Mutchnick, M. G., and Keren, D. F., 1981, *In vitro* synthesis of antibody to specific bacterial lipopolysaccharide by peripheral blood mononuclear cells from patients with alcoholic cirrhosis, *Immunology* **43**:177.

Neter, E., 1969, Endotoxins and the immune response, *Curr. Top. Microbiol. Immunol.* **47**:82.

Neter, E., 1971, The immune response of the host: An aid to etiology, pathogenesis, diagnosis, and epidemiology of bacterial infections, *Yale J. Biol. Med.* **44**:241.

Neter, E., 1975, Estimation of *Escherichia coli* antibodies in urinary tract infection: A review and perspective, *Kidney Int.* **8**(Suppl.):23.

Neter, E., 1979, Differential diagnosis of urinary tract infection by immunological methods, *Contr. Nephrol.* **16**:22.

Neter, E., Whang, H. Y., and Mayer, H., 1973, Immunogenicity and antigenicity of endotoxic lipopolysaccharides: Reversible effects of temperature on immunogenicity, *J. Infect. Dis.* **128**(Suppl.):56.

Nowotny, A., 1968, Immunogenicity of toxic and detoxified endotoxin preparations, *Proc. Soc. Exp. Biol. Med.* **127**:745.

Pennington, J. E., and Kuchmy, D., 1980, Mechanism for pulmonary protection by lipopolysaccharide *Pseudomonas* vaccine, *J. Infect. Dis.* **142**:191.

Pier, G. B., Markham, R. B., and Eardley, D., 1981, Correlation of the biologic responses of C3H/HeJ mice to endotoxin with the chemical and structural properties of the lipopolysaccharides from *Pseudomonas aeruginosa* and *Escherichia coli*, *J. Immunol.* **127**:184.

Porter, P., Kenworthy, R., Noakes, D. E., and Allen, W. D., 1974, Intestinal antibody secretion in the young pig in response to oral immunization with *Escherichia coli*, *Immunology* **27**:841.

Reynolds, H. Y., Thompson, R. E., and Devlin, H. B., 1974, Development of cellular and humoral immunity in the respiratory tract of rabbits to *Pseudomonas* lipopolysaccharide, *J. Clin. Invest.* **53**:1351.

Rudbach, J. A., 1971, Molecular immunogenicity of bacterial lipopolysaccharide antigens: Establishing a quantitative system, *J. Immunol.* **106**:993.

Rudbach, J. A., and Reed, N. D., 1977, Immunological responses of mice to lipopolysaccharide: Lack of secondary responsiveness by C3H/HeJ mice, *Infect. Immun.* **16**:513.

Ryan, J. L., Braude, A. I., and Turck, M., 1973, Galactose-deficient endotoxin from urinary *Escherichia coli, Infect. Immun.* **7**:476.

Schlecht, S., and Westphal, O., 1979, Protective role of *Salmonella* R mutants in salmonella infection in mice, *Zentralbl. Bakteriol. Parasitenkd. Infektionskr. Hyg. Abt. 1 Orig. Reihe A* **245**:71.

Simon, G., Reindke, B., and Marget, W., 1974, Lipoid-A-Antikörpertiter bei Pyelonephritis und anderen Infektionen mit gram-negativen Bakterien, *Infection* **2**:178.

Skelly, R. R., Munkenbeck, P., and Morrison, D. C., 1980a, Immune responses to hapten–lipopolysaccharide conjugates. 1. Modulation of *in vivo* antibody responses to the polysaccharide, *Cell. Immunol.* **49**:99.

Skelly, R. R., Munkenbeck, P., and Morrison, D. C., 1980b, Immune responses to hapten–lipopolysaccharide conjugates in mice. II. Characterization of the molecular requirements for the induction of antibody synthesis. *J. Immunol.* **124**:468.

Smith, A. M., 1976, The effects of age on the immune response to type III pneumococcal polysaccharide (SIII) and bacterial lipopolysaccharide (LPS) in BALB/c, SJL/J, and C3H mice, *J. Immunol.* **116**:469.

Sohl Åkerlund, A., Ahlstedt, S., Hanson, L. Å., and Jodal, U., 1979, Antibody responses in urine and serum against *Escherichia coli* antigen in childhood urinary tract infection, *Acta Pathol. Microbiol. Scand. Sect. C* **87**:29.

Sultzer, B. M., and Goodman, G. W., 1976, Endotoxin protein: A B-cell mitogen and polyclonal activator of C3H/HeJ lymphocytes, *J. Exp. Med.* **144**:821.

Sultzer, B. M., Goodman, G. W., and Eisenstein, T. K., 1980, Endotoxin protein as an immunostimulant, in: *Microbiology 1980* (D. Schlessinger, ed.), pp. 61–65, American Society for Microbiology, Washington, D.C.

Svenson, S. B., and Lindberg, A. A., 1981, Artificial *Salmonella* vaccines: *Salmonella typhimurium* O-antigen-specific oligosaccharide–protein conjugates elicit protective antibodies in rabbits and mice, *Infect. Immun.* **32**:490.

Svenson, S. B., Nurminen, M., and Lindberg, A. A., 1979, Artificial *Salmonella* vaccines: O antigenic oligosaccharide–protein conjugates induce protection against infection with *Salmonella typhimurium, Infect. Immun.* **25**:863.

Sziegoleit, A., 1976, Secretory immunoglobulins and serum antibodies after oral infection of the white mouse with *Salmonella typhimurium, Z. Immunitaetsforsch.* **152**:244.

Taplits, M., and Michael, J. G., 1979, Immune response to *Escherichia coli* B surface antigens, *Infect. Immun.* **25**:943.

Vos, J. G., Buys, J., Hanstede, J. G., and Hagenaars, A. M., 1979, Comparison of enzyme-linked immunosorbent assay and passive hemagglutination method for quantification of antibodies to lipopolysaccharides and tetanus toxoid in rats, *Infect. Immun.* **24**:798.

Watson, D. W., and Kim, Y. B., 1964, Immunological aspects of pyrogenic tolerance, in: *Bacterial Endotoxins* (M. Landy and W. Braun, eds.), pp. 522–536, Rutgers University Press, New Brunswick, New Jersey.

Weckesser, J., Drews, G., and Mayer, H., 1979, Lipopolysaccharides of photosynthetic prokaryotes, *Annu. Rev. Microbiol.* **33**:215.

Westenfelder, M., Galanos, C., and Madsen, P. O., 1975, Experimental lipid A-induced nephritis in the dog, *Invest. Urol.* **12**:337.

Westenfelder, M., Galanos, C., Madsen, P. O., and Marget, W., 1977, Pathological activities of lipid A: Experimental studies in relation to chronic pyelonephritis, in: *Microbiology 1977* (D. Schlessinger, ed.), pp. 277–279, American Society for Microbiology, Washington, D.C.

Westphal, O., 1975, Bacterial endotoxins, *Int. Arch. Allergy Appl. Immunol.* **49**:1.

Westphal, O., Nowotny, A., Lüderitz, O., Hurni, H., Eichenberger, E., and Schönholzer, G., 1958, Die Bedeutung der Lipoid-Komponente (Lipoid A) für die biologischen Wirkungen bakterieller Endotoxine (Lipopolysaccharide), *Pharm. Acta Helv.* **33**:401.

Westphal, O., Westphal, U., and Sommer, T., 1977, The history of pyrogen research in: *Microbiology 1977* (D. Schlessinger, ed.), pp. 221–235, American Society for Microbiology, Washington, D.C.

Whang, H. Y., Mayer, H., and Neter, E., 1971, Differential effects on immunogenicity and antigenicity of heat, freezing and alkali treatment of bacterial antigens, *J. Immunol.* **106**:1552.

Winchurch, R. A., Hilberg, C., Birmingham, W., and Munster, A. M., 1981, Bacterial lipopolysaccharides activate immune suppressor cells in newborn mice, *Cell. Immunol.* **58**:458.

Yamamoto, A., Homma, J. Y., Ghoda, A., Ishihara, T., and Takeuchi, S., 1979, Common protective antigen between *Pseudomonas aeruginosa* and *Vibrio cholerae*, *Jpn. J. Exp. Med.* **49**:383.

Young, L. S., and Stevens, P., 1977, Cross-protective immunity to gram-negative bacilli: Studies with core glycolipid of *Salmonella minnesota* and antigens of *Streptococcus pneumoniae*, *J. Infect. Dis.* **136**:S174.

Zeckel, M., Bruins, S. C., Kohler, R. B., Wheat, L. J., and White, A., 1980, Radioimmunoassay for detection of IgG antibodies against enterobacterial antigens, *Am. J. Med. Sci.* **280**:151.

Ziegler, E. J., Douglas, H., Sherman, J. E., Davis, C. E., and Braude, A. I., 1973, Treatment of *E. coli* and *Klebsiella* bacteremia in agranulocytic animals with antiserum to a UDP-Gal-epimerase-deficient mutant, *J. Immunol.* **111**:433.

5

Protection against Gram-Negative Bacteremia with Antiserum to Endotoxins

Abraham I. Braude and Elizabeth J. Ziegler

Although the mechanisms of the toxic and immunologic properties of endotoxins have been extensively examined by many investigators, only a few have worked on the possible value of an antitoxin against endotoxins. There have been at least two deterrents to such research that might explain the lag in development of an antitoxin to these lipopolysaccharide (LPS) poisons of gram-negative bacteria. The first is the question of whether or not LPS causes, or contributes to, the toxic manifestations of gram-negative bacterial infections. Until the role of LPS in human infection appeared to be a reasonable possibility, there would be little justification for undertaking the task of developing an antitoxin against LPS for human use. The second deterrent is related to the apparent antigenic heterogeneity of the endotoxins in the many species and serotypes of pathogenic gram-negative bacteria. It would seem impractical, on first consideration, to prepare antisera against each of the different serotypes of O antigens presented by all these gram-negative bacteria.

In this review we attempt to dispose of these deterrents by giving our reasons for believing that endotoxin contributes to the manifestations by gram-negative bacteremias, and by presenting a solution to the problem of antigenic diversity in lipopolysaccharides. We also show that antibody against LPS can function as an antitoxin and that such an antitoxin can diminish the mortality from gram-negative bacteremias.

Abraham I. Braude and Elizabeth J. Ziegler • Department of Medicine and Pathology, University of California, San Diego, California 92103.

I. A Consideration of the Toxic Effects of Endotoxin in Human Infection

Countless observations have shown that intravenous injection of minute amounts of LPS into animals or man causes changes in blood pressure, clotting, body temperature, circulating leukocytes, and metabolism. Since all gram-negative bacteria carry endotoxin on their surface, it follows that during bacteremia with these organisms, LPS is always in contact with the circulating blood and that endotoxemia is inevitable. As expected, injection of living or dead gram-negative bacteria intravenously reproduces all of the manifestations of intravenous LPS, and spontaneously occurring gram-negative bacteremia does the same. It is true that bacteremias produced by gram-positive bacteria have many clinical similarities to those produced by gram-negatives, even though gram-positives never have endotoxins. The striking difference between the two forms of bacteremia, however, is the incidence of shock. Shock is rare in bacteremia caused by staphylococci, pneumococci, or other streptococci, but occurs frequently in gram-negative bacteremia and has a high mortality rate. Since the clinical picture of shock from gram-negative bacterial septicemia is identical to that produced experimentally with intravenous endotoxin, its occurrence emphasizes the importance of LPS in gram-negative bacteremia.

The hypotensive effects of endotoxin in gram-negative bacteremia were probably first suspected in 1949 in our clinical studies with chlortetracycline in the treatment of *Brucella melitensis* infections (Spink *et al.*, 1948). In contrast to most patients, who recovered quickly after treatment, some of those given the drug suddenly went into shock with high fever. This alarming reaction occurred as the blood became sterile, and was followed by recovery from both shock and brucellosis. Since this paradoxical response to treatment took place when the *Brucella* organisms were being killed by the drug, we proposed the idea that an endotoxin was suddenly released by the action of chlortetracycline on the bacteria. Later studies showed that a minute intravenous dose of purified *Brucella* LPS would produce fever and hypotension in brucellosis.

Further clinical evidence that endotoxin caused shock in patients with gram-negative bacteremia took the form of violent transfusion reactions from contaminated blood (Braude, 1958; Borden and Hall, 1951). The bacteria causing these reactions were cryophilic gram-negative bacilli that grew well in refrigerated blood but died at the temperature of the human body (Stevens *et al.*, 1953). Since the bacteria causing lethal shock were unable to survive after the onset of bacteremia, it was clear that their toxins, rather than infection, killed the patient. These saprophytic bacteria were found to possess toxic lipopolysaccharides that reproduced experimentally all of the signs of endotoxemia seen in the victims of the transfusion accidents from contaminated blood. In contrast to these reactions produced by gram-negative bacteria, those transfusion accidents from blood contaminated with gram-positive organisms were never accompanied by shock and were never serious (Braude, 1958).

An opportunity to observe in patients the effects of purified LPS came from the studies of Brues and Shear (1944), who gave intravenous *Serratia marcescens* endotoxin for the treatment of cancer in an attempt to reproduce in man the damage to tumors seen in certain animal experiments. The intravenous injection of LPS into these patients did not arrest their cancers but did reproduce the characteristic effects of endotoxemia seen in experimental animals and in patients with gram-negative bacteremia occurring secondary to urinary infections, pneumonia, peritonitis, diverticulitis, endometritis, burns, and infected wounds.

These clinical and experimental observations make a strong case for the importance of endotoxin as the key factor in the pathogenesis of shock, fever, clotting disturbances, and other phenomena seen in patients with gram-negative bacteremia. The fact that all gram-negative bacteria carry endotoxin on their outer membrane accounts for the uniform clinical picture of bacteremia regardless of the etiologic species.

II. The Problem of Antigenic Diversity in Lipopolysaccharides and Its Solution

In theoretical terms, the solution to the problem of antigenic diversity in the endotoxins of gram-negative bacteria is simple: eliminate the O antigens and expose the core sugars.

The O antigen is the problem because it constitutes the outermost region of the lipopolysaccharides, and contains long chains of repeating tetrasaccharides or pentasaccharides that have amazing immunologic specificity so that very little cross-reactions are seen between the countless O antigens of gram-negative bacteria. These differences, which represent tremendous evolutionary diversification, depend not only on the presence of rare sugars, such as 6-dideoxyhexoses and 3,6-dideoxyhexoses, but also on differences in the sequential arrangement of identical sugars in the oligosaccharide unit. The importance of sugar sequences is seen from the fact that *Salmonella typhimurium* LPS and *S. newport* LPS have identical sugars in their repeating O units but show no significant immunologic cross-reaction. Both contain abequose, mannose, rhamnose, galactose, and glucose but *S. typhimurium* belongs to serogroup B and *S. newport* to C_2 (Lüderitz *et al.*, 1966). This versatility among gram-negative bacteria in achieving many immunologically distinct surface antigens makes them nearly invulnerable, for practical purposes, to treatment with O antisera because it would be useless to attempt production of polyvalent antibodies that would cover the enormous spectrum of antigenic possibilities.

The sugars in the core region of LPS are another matter. In contrast to the extreme variation in structure of O side chains, the structure of the core oligosaccharide is remarkably similar in different gram-negative bacteria. The core polysaccharide contains not only KDO, the link to lipid A, but also heptose, phosphate, ethanolamine, and three hexoses. The hexoses, which consist of galactose, glucose and *N*-acetyl-glucosamine, comprise the outer core, and the remainder are designated the inner core. There is almost no variation in structure of the core oligosaccharides in *Salmonella* organisms and those of *E. coli* and *Shigella* differ only slightly (Nikaido, 1973). Moreover, even when there are structural differences, the core LPS of different species contain identical terminal sugars. Thus, the terminal sugar in the LPS of the Ra mutant of either *Shigella* or *Salmonella* is glucosamine, in the Rc mutant glucose, and in the Rd mutant heptose.

Even less variation in chemical structure is found among preparations of lipid A from different gram-negative bacteria. Among the Enterobacteriaceae, which are the leading causes of lethal human bacteremia, there is no evidence for gross variations in the lipid A structure (Lüderitz, 1977). In addition to its chemical homogeneity among different bacteria, it is considered to be the primary toxic moiety of LPS. Despite these two potential advantages as a target for a protective antitoxin, lipid A suffers two disadvantages that discouraged us from undertaking immunization studies. One is that it is a poor antigen because it is insoluble in water. The other is that it resides so deeply in the LPS molecule that its antigenic determinants would be harder to reach by antibody than less cryptic core sugars. Some immunogenicity of lipid A has been demonstrated after whole bacterial cells have been treated with mild acid, which cleaves the linkage of KDO to the glucosamine backbone of lipid A. This maneuver releases the polysaccharide, which is removed by washing, and leaves the lipid A on the cell surface. Vaccination with such cells has produced antisera that react with lipid A-coated red cells, but the protective properties of such antibodies against LPS or bacteremia have not been easy to demonstrate, and occasionally caused the reverse effect, i.e., enhancement of lipid A toxicity (Galanos *et al.*, 1977; Hodgin and Dreivs, 1976; Bruins *et al.*, 1977).

Taking into account the various properties of these regions in LPS, i.e., O sugars, core oligosaccharide, and lipid A, we decided that core oligosaccharide offered the most realistic potential as an antigen for preparing antisera against LPS. Granting that lipid A is the toxic moiety, we felt that immunoglobulins reaching the core sugars might neutralize the toxic neighboring lipids and also promote clearance from the circulation of the whole LPS molecule. We also anticipated that antibodies to the core sugars could be produced with vaccines composed of rough cells of gram-negative bacteria, and that these antibodies would reach the core antigens in smooth bacteria through three possible mechanisms: (1) exposure of core sugars when the cells divide; (2) exposure of core oligosaccharide antigen after cleavage of O sugars by enzymes in body fluids as proposed by Chedid *et al.* (1968); (3) exposure of core sugars through inconstant attachment of O side chains, which leaves unsubstituted core stubs.

In order to prepare antibody to the core, we immunized with the M-mutant of *E. coli* O111 B4, known as J5. M mutants are those that are defective in UDP-galactose synthesis and synthesize a core structure up to the point where galactose would be incorporated by the smooth organism (Lüderitz *et al.*, 1966). The core structure proximal to the point of defective synthesis in M-mutants is that of the Rc LPS, containing

$$\text{glucose} 1 \rightarrow 3 \text{heptose} 1 \rightarrow 3 \text{heptose} 1 \rightarrow 8 \text{KDO}$$

We have found that the core polysaccharide in the J5 mutant of *E. coli* is an excellent antigen and can generate antisera with considerable protective power against LPS toxicity. We have produced these protective antisera by immunization either with vaccines composed of boiled cells of *E. coli* J5 or with its purified core LPS.

III. Neutralization of Endotoxin by Antitoxins

A. Assay Systems

A realistic antitoxin against endotoxin should be able to prevent those manifestations of LPS intoxication that are prominent in patients with gram-negative bacteremia, namely, fever, clotting disturbances, and lethal shock. Each of these can be reproduced in certain animals and therefore subjected to controlled assays under vigorous experimental conditions. The mouse is best for examining lethal endotoxin shock because it is the only laboratory animal that displays the linear dose–response curve to endotoxin that is required for accurate statistical analysis of its lethal action in the presence or absence of antibody. Fever and clotting disturbances, on the other hand, are best studied in the rabbit. The dose response to the pyrogenic effect of LPS in rabbits is better standardized than that to any of its other biologic effects. After i.v. injection of 0.1 μg LPS, for example, the temperature begins to rise in about 10–20 min and reaches a peak in 70 min. After the first peak, the temperature declines slightly, but starts to rise again at 2 hr and reaches a second peak at about 3 hr. The second peak is dose dependent, and can be used quantitatively to assay neutralization of the pyrogenic effect of LPS by antibody (Milner, 1973).

The action of antitoxin against the clotting effects of LPS can be measured in two experimental models: the dermal Shwartzman reaction and the Sanarelli–Shwartzman reaction. Both are produced by two injections of LPS, given 18–24 hr apart. In the dermal reaction, the first is given i.d. and the second i.v. Soon after the i.v. injection, the site of the i.d. injection undergoes hemorrhagic necrosis secondary to local thrombosis and infarction. Injection of LPS into the skin causes an Arthus-like inflammation about the vessels, and injection i.v. initiates coagulation so that the inflamed vessels undergo thrombosis. This lesion has a sharp end point for measuring protection by antitoxin against LPS.

In the Sanarelli–Shwartzman phenomenon, both injections are given i.v. and invariably produce severe disseminated intravascular coagulation (DIC) that culminates in bilateral necrosis of the renal cortex. This dramatic form of kidney damage results from glomerular deposits of fibrin and provides another sharp end point for testing the protective ability of antitoxin to LPS (Lee and Stetson, 1965).

Both the dermal Shwartzman and the Sanarelli–Shwartzman reactions are realistic models of the clotting disturbances that characterize gram-negative bacteremia in patients. The dermal reaction appears to be the exact counterpart of meningococcal purpura, one of the most serious and common manifestations of meningococcal septicemia (Davis and Arnold, 1974). Similarly, the DIC in the Sanarelli–Shwartzman reaction is identical in its features to that seen in all forms of severe gram-negative bacteremias. LPS initiates clotting through both intrinsic and extrinsic pathways (Morrison and Ulevitch, 1978). Clotting via the intrinsic pathway begins when the negatively charged phosphate residues of lipid A activate Hageman factor (XIII). Activated Hageman factor initiates the reaction that leads to conversion of prothrombin to thrombin. The importance of the extrinsic pathway is evident from the fact that blood leukocytes are required for both the dermal Shwartzman and the Sanarelli–Shwartzman reactions. Among the leukocytes, monocytes are considered to be the important source of tissue factor (TF), which initiates clotting via the extrinsic pathway. TF is a lipoprotein in the plasma membrane of various cells that complexes with factor VII and calcium ions to activate factor X. Activated factor X then converts prothrombin to thrombin.

During the evolution of the Sanarelli–Shwartzman reaction, DIC depletes fibrinogen, platelets, and prothrombin, and fibrinolysis is activated so that fibrin degradation fragments (or fibrin split-products) accumulate in the blood. These fragments became anticoagulants since they inhibit both proteolysis of fibrinogen by fibrin and the polymerization of fibrin monomer to form a clot. The fibrin degradation products also inhibit platelet aggregation. This dual disturbance by fibrin split-products on clotting and platelets causes severe bleeding in patients who lose their clotting factors during the DIC that occurs in meningococcemia, *E. coli* bacteremia of pregnancy, and other severe infections caused by gram-negative bacteria. Since fibrinogen depletion, platelet depletion, split-product formation, and consumption of other clotting factors can all be prevented by antitoxin against LPS, they are valuable markers for measuring antiserum effectiveness.

B. Protection against Endotoxin with Antibody to LPS (Antitoxin)

Either active or passive immunization will diminish the toxicity of LPS in these assay systems.

Active Immunization. Active immunization with LPS causes two-phase increase in resistance to endotoxin: the first phase occurs within 12 hr after one dose of LPS when no antibody is detectable; the second at about seven days when high titers of antibody are found. The mechanism of early resistance is unknown and because antibody is not involved it has been designated *tolerance*, a term that implies a nonimmune or pharmacologic mechanism for enhanced detoxification. The early stage of tolerance to the pyrogenic action of LPS, for example, has been explained by the theory that the preceding dose depleted the endogenous pyrogen in leukocytes (Atkins and Bodel, 1972). Whatever the mechanism of early tolerance, it has been well established that the resistance to LPS occurring after conventional periods of immunization is mediated immunologically and can be transferred with serum containing antibody (Freedman, 1960).

Passive Immunization: Fever. After immunization with smooth LPS, rabbits develop less fever than nonimmune rabbits when challenged with the same dose of LPS, and this resistance to fever can be transferred with serum containing antibody to the homologous LPS (Milner, 1973; Freedman, 1960). The diminished febrile response in these immunized rabbits is characterized by a lower second peak in the fever curve, or by its disappearance. Antiserum against smooth LPS can reduce somewhat the pyrogenicity of LPS from heterologous species of bacteria, but is less antipyretic than antiserum against the core region of LPS (Greisman and Hornick, 1973). The greater antipyretic action of such core-antisera against heterologous LPS is evidence that the O side chains conceal the core antigens and interfere with the production of antibody to the core region which is needed for heterologous protection.

Passive Immunization: Lethality. Intraperitoneal injection of human or rabbit antisera prevents death from i.v. endotoxin in mice (Davis *et al.*, 1969; Freedman, 1959; Ziegler *et al.*, 1973a). The degree of protection is a function of antiserum dosage and follows a first-order reaction when the dose of antiserum is plotted against the dose of endotoxin. Rabbit antisera against the homologous smooth LPS reduce the lethal potency of *E. coli, Serratia marcescens,* and *Salmonella abortus equi* endotoxin by 50–80%. Equal protection occurs against heterologous LPS with antisera generated against smooth or rough LPS, providing further proof that O antibody is not necessary for preventing death from endotoxin.

Passive Immunization: The Dermal (Local) Shwartzman Reaction. Passive immunization with rabbit antiserum to endotoxin prevented the dermal Shwartzman reaction in animals given either homologous or heterologous endotoxins (Braude and Douglas, 1972; Davis *et al.*, 1978). In these LPS experiments, the antiserum was given 2 hr before the provocative i.v. dose of LPS. The protection was spectacular as illustrated by the drop in frequency of dermal reactions from 98% to 9.6% with antiserum to smooth *E. coli* O111 in large-scale experiments employing nearly 100 rabbits. Similarly, antiserum to *S. typhimurium* LPS reduced the frequency of Shwartzman reactions from 100% to 30.3% in animals challenged with homologous *S. typhimurium* endotoxin. In experiments with heterologous endotoxin, the *S. typhimurium* antiserum lowered the rate of local Shwartzman reactions from 98% to 50%.

Excellent protection against the local Shwartzman reaction was also seen when antiserum against the core LPS of the J5 mutant was used to neutralize the toxin *in vivo*. The J5 antibodies not only prevented such reactions in rabbits inoculated with *E. coli* O111 LPS, but were also impressive in their ability to prevent hemorrhagic necrosis induced in the Shwartzman reaction by LPS from meningococci belonging to the three major groups, A, B, and C (Davis *et al.*, 1978). As noted earlier in this review, the neutralization of meningococcal toxin by core antibody has important clinical implications because the dermal Shwartzman reaction is the experimental equivalent of meningococcal purpura (Davis and Arnold, 1974). Thus, it is especially noteworthy that specific antisera prepared by vaccination with each of these three groups of meningococcus were less potent and showed less cross-protection than core antiserum. In fact, in some experiments there was no cross-protection between meningococcal antisera of different groups, a finding that implied concealment of the meningococcal core LPS and emphasized the potential value of core vaccines for providing broader protection than the current A and C vaccines.

Passive Immunization: Prevention of Intravascular Coagulation (the Generalized Shwartzman Reaction). Protection against the generalized Shwartzman reaction with antiserum to endotoxin is as dramatic as that described for the dermal Schwartzman reaction (Braude *et al.*, 1973). As was the case in the dermal reaction, antiserum was given 2 hr before the second dose of LPS in rabbits primed for the generalized reaction. With such protection, the number of animals developing renal cortical necrosis fell from 90% to 18%, and the precipitous drop in fibrinogen and platelets was prevented. The DIC and renal cortical necrosis were prevented by antisera against homologous and heterologous LPS by J5 antiserum and by both 19 S and 7 S immunoglobulins. Passive protection lasted for at least 3 weeks. Concentrated gamma globulin from the antiserum retained full protective power against LPS after complement-binding sites were treated with 2-mercaptoethanol in the presence of iodoacetate (to prevent reaggregation of 7 S dimers). Thus, it may be safe to administer the gamma globulin fraction i.v. after such chemical manipulation because the danger of i.v. gamma globulin resides in its ability to fix complement *in vivo.* Intravenous gamma globulin is an important consideration for patients who might benefit from antitoxin against LPS after developing gram-negative bacteremia.

Antiserum probably protects against DIC and renal cortical necrosis at two different steps in the pathogenesis of the Shwartzman reaction: initiation of clotting and blockade of the reticuloendothelial system (RES).

Both steps appear to be necessary for the generalized reaction to materialize (Lee and Stetson, 1965). After the first dose of LPS triggers clotting by activation of Hageman factor, the fibrin polymers are removed fast enough by the RES to prevent their deposition in glomerular capillaries. The second dose, however, not only blocks clearance of fibrin by the RES but creates a new supply of fibrin that remains in the circulation until it is filtered out by the glomeruli and occludes their capillaries. The ability of antiserum to neutralize the clotting action of LPS is brought out by the sharp reduction in fibrin split-products when antiserum is given before the first dose of endotoxin (Braude *et al.*, 1973). Antiserum against LPS must also prevent RES blockade, and maintain renal blood flow, i.e., it protects the kidneys from the sluggish glomerular flow of shock that would promote deposition of fibrin.

C. Protection against Experimental Bacteremia with Antibody to LPS

After demonstrating that anti-core LPS, or core antitoxin, consistently neutralized the various manifestations of LPS toxicity that would be encountered in bacteremia, we set out to determine if core antitoxin would also protect against bacteremia. In formulating these experiments, we took into account two separate needs for such an antitoxin in clinical medicine: (1) the need to treat patients in whom gram-negative bacteremia had already occurred, and (2) the need to prevent bacteremia in subjects who were at high risk of developing gram-negative bacteremia.

One of the problems in experimental work on gram-negative bacteremia is to devise a laboratory model that would be realistic from the standpoint of bacteremia in patients. An important criterion for such a realistic model is that bacteremia develop spontaneously after immunosuppression, rather than artificially after injection of massive numbers of bacteria into healthy animals. We found that overwhelming lethal gram-negative bacteremia developed spontaneously in neutropenic rabbits if they were fed gram-negative bacteria in their drinking water (Braude *et al.*, 1969). Since rabbits normally have *no* enteric gram-negative bacilli in their bowel, they can be colonized with any gram-negative organism. The bacteria penetrate the gut and produce high fever with overwhelming bacteremia and fatal shock. *Pseudomonas* will do the same if dropped into the conjunctival sac of neutropenic rabbits and will produce the classic picture of *Pseudomonas* vasculitis, ecthyma gangrenosum, and fatal bacteremia (Ziegler and Douglas, 1979). The neutropenic rabbit model thus mirrors closely the spontaneous development of endogenous bacteremia and its clinical manifestations in immunosuppressed patients, and avoids the unrealistic procedure of injecting the huge inocula of gram-negative bacteria that are needed to kill healthy mice. This requirement for huge inocula in order to kill mice with gram-negative bacteremia results in unacceptably steep dose–response curves, often with less than one log difference between sublethal and 100% lethal doses. In addition, most studies have been done with rabbit sera that are toxic for mice and would nullify some protection by antiserum, so that a variety of conflicting results have been reported in mouse studies. The neutropenic rabbit model avoids the problem of heterologous species serum because potent rabbit antiserum is produced conveniently and in large volumes.

By using the neutropenic rabbit model, various investigators have found that vaccination with core endotoxin or passive administration of core antisera has given consistent protection against the wide range of heterologous gram-negative bacteria encountered in lethal human bacteremia and shock (Ziegler *et al.*, 1973b, 1975; Marks *et al.*, 1978; Bruins *et al.*, 1977). Antiserum to core LPS can prevent death from bacteremia in neutropenic rabbits when given therapeutically after the onset of bacteremia due to *P. aeruginosa*, *E. coli*, and *K. pneumoniae*. In experiments with survival rates below 10% in controls given nonimmune rabbit serum, survival rose to 40–70% among rabbits treated with one injection of J5 core antiserum after the onset of bacteremia, and *no* antibiotics (Ziegler *et al.*, 1973b, 1975). Gamma globulin from J5 antiserum was also effective therapeutically. Active or passive protection remains undiminished for at least 1 month. The survival rate from *Pseudomonas* bacteremia was increased from 27% to 92% (p < 0.0005) after vaccination with the J5 core antigen even though no significant protection against *Pseudomonas* bacteremia occurred after immunization with a vaccine composed of *E. coli* O111, the parent organism from which the J5 mutant was derived (Ziegler *et al.*, 1975). This failure of *E. coli* O111 vaccine supports the concept that the core LPS is concealed by O111 oligosaccharide antigens so that core antibodies are not generated by vaccination.

These favorable results in rabbits have also been noted in dogs after vaccination with core glycolipid from the Re mutant of *S. minnesota*. The vaccinated animals enjoyed excellent protection against fatal shock from *E. coli* or *S. marcescens* bacteremia (Young *et al.*, 1978).

In mice, the results have been variable. Vaccination with the *S. minnesota* Re antigen protected mice against *Salmonella* organisms, *E. coli*, *K. pneumoniae*, and *Pasteurella multocida* (McCabe, 1972; Hodgin and Dreivs, 1976). Lipid A was less effective than the Re antigen in mice and failed altogether to stimulate active immunity in rabbits (Bruins *et al.*, 1977). Moreover, antisera with high titers of lipid A antibody gave no protection to mice challenged with *E. coli* O4. The difference in effectiveness between Re and lipid A antibody suggests that lipid A is too cryptic in *E. coli* O4 for specific antibody reactions but KDO is not. This explanation would fit with the fact that lipid A has no KDO, which is situated externally to lipid A in the LPS of smooth bacteria like *E. coli* O4.

The failure of core antisera to protect mice against gram-negative bacteremia in other studies can probably be explained by differences in experimental design (Greisman *et al.*, 1978). Thus, negative-protection experiments have been conducted with single, small doses of heterologous (rabbit) antisera given to mice that were challenged at a single lethal dose level in a very steep dose-response that allowed no opportunity for demonstrating protection against a more realistic challenge. In addition, many of the experiments have given antibody i.v. in an attempt to protect against experimental peritonitis. Since the protective antibody to core is IgM, and since this immunoglobulin remains within the vascular system, experimental peritonitis with enteric bacteria is an unreasonable challenge for examining protection against bacteremia. Yet, properly designed studies with adequate doses of core antisera against bacteria that give an acceptable dose–response curve, can show protection against bacteremia in mice challenged i.p. (Marks *et al.*, 1978).

D. Treatment and Prevention of Gram-Negative Bacteremia by Giving J5 Antiserum to Patients

In view of the striking protection against gram-negative bacteremia and endotoxemia obtained with core antiserum in experimental animals, we carried out a double-blind clinical trial of J5 antiserum in patients with gram-negative bacteremia. We were encouraged in this undertaking by the work of Zinner and McCabe (1976), who obtained evidence that high titers of antibody to core LPS (Re) at the onset of human bacteremia were associated with significantly less frequent shock and death. We prepared human antiserum by vaccinating healthy men s.c. with heat-killed cells of the J5 mutant and collecting blood for antiserum preparation at the height of the antibody response against J5 LPS as measured by passive hemagglutination (Ziegler *et al.*, 1982). The antiserum was given i.v. along with antibiotics and other standard therapy for gram-negative bacteremia. Of 304 patients who received J5 antiserum or preimmune control serum near the onset of illness, a diagnosis of gram-negative bacteremia was made in 212 (70%). The death rate from gram-negative bacteremia was lowered from 42/101 (39%) in controls to 23/103 (22%) in those given J5 antiserum, a highly significant difference ($p = 0.011$). In other words, mortality from bacteremia was cut virtually in half. These results are impressive because antiserum or control serum was given only to seriously ill patients and not to those who developed the mild bacteremia of urinary tract infections. Many of the patients were in profound septic shock and when these cases were analyzed we found that 22/39 (56%) of the J5 group recovered from shock in contrast to 11/38 (29%) of the controls ($p = 0.015$). The control and antiserum groups were almost identical with respect to age, sex, race, incidence of neutropenia, severity of underlying disease, antibiotic treatment, granulocyte transfusions, and use of high-dose steroids.

In addition to this successful trial on the therapeutic value of J5 antiserum, we have obtained evidence that J5 antiserum has prophylactic value as well (Wolf *et al.*, 1979). The prophylactic studies were set up in neutropenic patients with leukemia or lymphoma, and the first phase of the study in 16 neutropenic patients has been completed with favorable results. Patients received at random either J5 antiserum or preimmune control serum intravenously in a dose of 3 ml/kg every 21 days during neutropenia ($<$ 500 PMN/mm^3). The J5 antiserum sharply reduced the occurrence of fever so that the incidence of febrile days (over 38.0°C) per days of neutropenia fell from 70/180 (44%) in controls to 35/194 (18%) in those given antiserum ($p < 0.0005$). Since most of these patients were not bacteremic, these results suggest that attacks of fever in neutropenic patients result from endotoxemia during flareups of gram-negative bacterial infections, and that antibody to core glycolipid prevents fever by neutralizing LPS.

IV. Conclusions

These observations in animals and patients have established that antibody against core LPS has at least three special therapeutic properties:

1. It has a broad spectrum of protection against different species of gram-negative bacteria and their endotoxins.
2. It can be used successfully for treatment of established bacteremia and shock.
3. It can reverse profound shock from gram-negative bacteremia.

When combined with the best available antibiotic, pressor, and surgical therapy, J5 antiserum markedly improves the outlook, so that over 80% can expect to recover, despite their grave underlying illness and impaired immunity. Since antibody persists in the circulation for at least 4 weeks, administration of J5 antiserum prophylactically to patients at high risk of developing gram-negative bacteremia might be expected to lower the death rate even further because it could act before shock develops.

V. References

Atkins, E., and Bodel, P., 1972, Fever, *N. Engl. J. Med.* **286**:27.

Borden, C., and Hall, W., 1951, Fatal transfusion reactions from massive bacterial contamination of blood, *N. Engl. J. Med.* **245**:760.

Braude, A., 1958, Transfusion reactions from contaminated blood, *N. Engl. J. Med.* **258**:1289.

Braude, A., and Douglas, H., 1972, Passive immunization against the local Shwartzman reaction, *J. Immunol.* **108**:505.

Braude, A., Douglas, H., and Jones, J., 1969, Experimental production of lethal *Escherichia coli* bacteremia of pelvic origin, *J. Bacteriol.* **98**:979.

Braude, A., Douglas, H., and Davis, C., 1973, Treatment and prevention of intravascular coagulation with antiserum to endotoxin, *J. Infect. Dis.* **128**(Suppl.):157.

Brues, A., and Shear, M., 1944, Reactions of four patients with advanced malignant tumors to injections of a polysaccharide from *Serratia marcescens* culture filtrate, *J. Natl. Cancer Inst.* **5**:95.

Bruins, S. C., Stumacher, R., Johns, M. A., and McCabe, W. R., 1977, Immunization with R mutants of *Salmonella minnesota*. III. Comparison of the protective effect of immunization with lipid A and the Re mutant, *Infect. Immun.* **17**:16.

Chedid, L., Parant, M., Parant, F., and Boyer, F., 1968, A proposed mechanism for natural immunity to enterobacterial pathogens, *J. Immunol.* **100**:292.

Davis, C., and Arnold, K., 1974, Role of endotoxin in meningococcal purpura, *J. Exp. Med.* **140**:159.

Davis, C., Brown, K., Douglas, H., Tate, W., and Braude, A., 1969, Prevention of death from endotoxin with antisera. I. The risk of fatal anaphylaxis to endotoxin, *J. Immunol.* **102**:563.

Davis, C., Ziegler, E., and Arnold, K., 1978, Neutralization of meningococcal endotoxin by antibody to core glycolipid, *J. Exp. Med.* **147**:1007.

Freedman, H., 1959, Passive transfer of protection against lethality of homologous and heterologous endotoxins, *Proc. Soc. Exp. Biol. Med.* **102**:504.

Freedman, H., 1960, Further studies on passive transfer of tolerance to pyrogenicity of bacterial endotoxin: The febrile and leucopenic responses, *J. Exp. Med.* **112**:619.

Galanos, C., Freudenberg, M., Hase, S., Jay, F., and Ruschmann, E., 1977, Biological activities and immunological properties of lipid A, in: *Microbiology 1977* (D. Schlessinger, ed.), American Society for Microbiology, Washington, D.C.

Greisman, S., and Hornick, R., 1973, Mechanisms of endotoxin tolerance with special references to man, *J. Infect. Dis.* **128**(Suppl.):257.

Greisman, S. E., DuBuy, J., and Woodward, C., 1978, Experimental gram-negative sepsis: Reevaluation of the ability of rough mutant antisera to protect mice, *Proc. Soc. Exp. Biol. Med.* **158**:482.

Hodgin, L., and Dreivs, J., 1976, Effect of active and passive immunization with lipid A and *Salmonella minnesota* Re 595 on gram-negative infections in mice, *Infection* **4**:5.

Lee, L., and Stetson, C., 1965, The local and generalized Shwartzman phenomenon, in: *The Inflammatory Process* (B. W. Zweifoch, L. Grant, and R. T. McCluskey, eds.), p. 791, Academic Press, New York.

Lüderitz, O., 1977, Endotoxins and other cell wall components of gram-negative bacteria and their biological properties, in: *Microbiology 1977* (D. Schlessinger, ed.), American Society for Microbiology, Washington, D.C.

Lüderitz, O., Staub, A., and Westphal, O., 1966, Immunochemistry of O and R antigens of *Salmonella* and related Enterobacteriaceae, *Bacteriol. Rev.* **30**:192.

McCabe, W., 1972, Immunization with R mutants of *S. minnesota*, *J. Immunol.* **108**:601.

Marks, M., Ziegler, E., Douglas, H., Corbeil, L., and Braude, A., 1982, Induction of immunity against *Haemophilus influenzae* type B infection by *Escherichia coli* core lipopolysaccharide, *J. Clin. Invest.* **69**:742.

Milner, K., 1973, Patterns of tolerance to endotoxin, *J. Infect. Dis.* **128**(Suppl.):229.

Morrison, D., and Ulevitch, R., 1978, The effects of bacterial endotoxins on host mediation systems: A review, *Am. J. Pathol.* **93**:527.

Nikaido, H., 1973, Biosynthesis and assembly of lipopolysaccharide and the outer membrane layer of gram-negative cell wall, in:*Bacterial Membranes and Walls* (Leive L., ed.), Dekker, New York.

Spink, W., Braude, A., Castaneda, M. R., and Goytia, R., 1948, Aureomycin therapy in human brucellosis due to *Brucella melitensis*, *J. Am. Med. Assoc.* **138**:1145.

Stevens, A., Legg, J., Henry, B., Bille, J., Kirby, W., and Finch, C., 1953, Fatal transfusion reactions from contamination of stored blood by cold growing bacteria, *Ann. Intern. Med.* **39**:1228.

Wolf, J., McCutchan, J., Ziegler, E., and Braude, A., 1979, Prophylactic antibody to core lipopolysaccharide in neutropenic patients, *Proc. 11th Int. Congr. Chemother. 19th Intersci. Conf. Antimimicrob. Agents Chemother.* **19**:65.

Young, L., Stevens, P., and Ingram, J., 1978, Functional role of antibody against "core" glycolipid of Enterobacteriaceae, *J. Clin. Invest.* **56**:850.

Ziegler, E., and Douglas, H., 1979, *Pseudomonas* vasculitis and bacteremia following conjunctivitis: A simple model of fatal *Pseudomonas* infection in neutropenia, *J. Infect. Dis.* **139**:228.

Ziegler, E., Douglas, H., and Braude, A., 1973a, Human antiserum for prevention of the local Shwartzman reaction and death from bacterial lipopolysaccharides, *J. Clin. Invest.* **52**:3236.

Ziegler, E. J., Douglas, H., Sherman, J. E., Davis, C. E., and Braude, A. I., 1973b, Treatment of *E. coli* and *Klebsiella* bacteremia in agranulocytic animals with antiserum to a UDP-Gal-epimerase-deficient mutant, *J. Immunol.* **111**:433.

Ziegler, E. J., McCutchan, J. A., Douglas, H., and Braude, A., 1975, Prevention of lethal *Pseudomonas* bacteremia with epimerase-deficient *E. coli* antiserum, *Trans. Assoc. Am. Physicians* **88**:101.

Ziegler, E. J., McCutchan, J. A., and Braude, A. I., 1978, Clinical trial of core glycolipid antibody in gram-negative bacteremia, *Trans. Assoc. Am. Physicians* **91**:253.

Ziegler, E., McCutchan, J. A., Fierer, J., Glauser, M., Sadoff, J. C., Douglas, H., and Braude, A. I., 1982, Treatment of gram-negative bacteremia and shock with human antiserum to a mutant *Escherichia coli*, *N. Engl. J. Med.* **307**:1225.

Zinner, S., and McCabe, W., 1976, Effects of IgM and IgG antibody in patients with bacteremia due to gram-negative bacilli, *J. Infect. Dis.* **133**:32.

6

The Radioprotective Effect of Bacterial Endotoxin

Ulrich H. Behling

I. Radiation and Immunity

The exposure of a mammal to a single whole-body dose of rapidly delivered ionizing radiation of approximately 100 R or more results in the development of a complex set of clinical signs and symptoms that are collectively termed the *acute radiation syndrome* (Cronkite, 1964; Wald *et al.*, 1962). The time of onset, nature, and severity of this syndrome are a function of the total exposure dose, quality of radiation, and many variables related to the irradiated host. In most of the experimental work described here, X- or γ-radiation was delivered uniformly to the entire body in a single exposure lasting on the order of minutes. The dosage of radiation in these experiments fell within the range of threshold lethal, for this represents the minimum amount of radiation causing death within 30 days of exposure. Peak incidence of mortality for the midlethal dose ($LD_{50/30}$) generally occurs between 10 and 14 days following irradiation. Whole-body irradiation in this dose range produces a complex series of physiological disturbances and morphological changes that are the cumulative results of radiation-induced damage or death of individual cells.

In the acute radiation syndrome, the highly radiosensitive hemopoietic system is most likely to manifest injury. The basic pathological lesion of the hemopoietic tissue is a progressive atrophy of lymph nodes, spleen, and bone marrow. Stem cell division is inhibited, and

Ulrich H. Behling • Oyster Creek Nuclear Generating Station, Forked River, New Jersey 08731, and School of Dental Medicine, University of Pennsylvania, Philadelphia, Pennsylvania 19104.

therefore, these tissues become depleted of cells since mature differentiated cells are progressively eliminated and cannot be replaced by dividing stem cells. The various manifestations of radiation injury resulting from reduced stem cell proliferation do not occur simultaneously but follow a characteristic sequence of events over a period of several weeks. The onset, duration, and severity of changes in cellular composition are largely determined by the respective radiosensitivity and biological life span of the various cell populations found in different hemopoietic tissues. The mitotically active stem cell and the differentiated resting small lymphocyte are the most radiosensitive leukocytes (Ernstrom, 1972; Cronkite, 1973). Their radiosensitivity is defined by a D_{50} value (i.e., the absorbed dose of radiation resulting in the survival of 50% of the irradiated cell population) of about 70 to 90 rads. In serial examinations, leukopenia is observed in the bone marrow earlier than in the peripheral blood due to differences in cell composition. The peripheral blood is composed of more differentiated and hence less radiosensitive cells than the bone marrow. The earliest changes in the circulating blood are characterized by the disappearance of the small mature lymphocytes followed by a slower decline of the remaining cell population. Studies have shown that the disappearance of lymphocytes from the blood and lymphoid tissues is bimodal with respect to time (Hulse, 1959). The initial fall is attributed to the direct destruction of small mature lymphocytes while the second phase is produced by the combined effects of attrition of the remaining cells and of reduced replacement by their mitotically inactive precursors.

II. Radiation-Induced Susceptibility to Infection

The single most important consequence of hemopoietic and lymphoid tissue destruction following radiation exposure is that antimicrobial immunity is severely impaired not only to challenges with pathogenic organisms but also with some of the normal bacterial flora. A detailed description of radiation-enhanced susceptibility to infection is beyond the scope of this review; hence, it seems prudent to reference a few representative review articles and provide a brief summary (Benacerraf, 1960; Bond, 1957; Miller, 1956; Petrov, 1958; Smith, 1963; Talmage, 1955). All of the reviews indicate that following irradiation, every experimental animal, including man, shows enhanced susceptibility to a broad spectrum of microbial agents such as bacteria, viruses, protozoans, rickettsia, and fungi. Experimental approaches to characterize immune suppression usually assess the course of infection after challenge of the irradiated host with either (1) known pathogenic organisms, (2) conditional pathogens (ordinarily one of the normal flora), or (3) the determination of spontaneous indigenous infection(s) without any pathogenic challenge. The last approach supplies the most convincing evidence of reduced antimicrobial immunity in radiation sickness in which infections of conditional pathogenic organisms are often observed. In most instances, the source of infection leading to bacteremia and/or death is of gram-negative or fungal origin. This observation has been repeatedly demonstrated with colon and para-colon bacilli, *Pseudomonas aeruginosa*, type III pneumococci, and a variety of other nonpathogenic bacteria.

III. Radiation-Induced Immune Suppression

In lieu of the invasive nature of enteric organisms in the irradiated host, it is noteworthy that although immune suppression is the primary factor, it is definitely not the only cause for enhanced susceptibility to infection. The increased permeability of the radiosensitive epithelial and blood tissue barriers, which permits the invasion of conditional pathogens from the respiratory and digestive tracts, is also relevant.

As early as 1908, Benjamin and Sluka reported several classical observations of previously irradiated rabbits immunized with beef serum: (1) severely reduced antibody response, (2) reduced antigen elimination in the immunized host, and (3) a critical time-dependence between radiation exposure and antigenic stimulation. Subsequently, Taylor *et al.* (1919) studied the radiosensitivity of lymphocytes and clearly showed that these cells were an exception to the law of Bergonic and Tribondeau because the mature nondividing small lymphocyte was more radiosensitive than its mitotically active and undifferentiated stem cell. In measuring the radiosensitivity of a primary humoral immune response, earlier studies had established that serum antibody levels decrease in a straight-line exponential fashion as a function of radiation dose (Makinodan *et al.*, 1962). Yet, quantitation of serum antibody levels did not prove that the reduced antibody levels were directly proportional to the number of surviving antibody-producing cells. The all-or-none effect of radiation on the individual antibody-secreting cell was first demonstrated with the hemolytic plaque-forming cell assay (PFC) (Kennedy *et al.*, 1965). When the PFC response is measured as a function of radiation dose, a first-order exponential decrease in the number of PFCs is observed with a 50% cell lethality value between 70 and 90 rads. This low D_{50} value suggests that the high radiosensitivity of the PFC response depends on clonal expansion since no cellular process other than cell division could account for a sensitivity of this magnitude. Interestingly, a whole-body exposure dose of 500 R, generally sublethal for mice, would represent approximately six times the D_{50} value for the animals' lymphocyte population. This indicates that the immune system could withstand a 99% reduction of the proliferative capacity of antigen-stimulated lymphocytes without inducing death. The importance of cell proliferation or clonal expansion in humoral immunity was further demonstrated by the classical studies of Taliaferro *et al.* (1964) in which the timing of irradiation relative to antigenic stimulation was found to be critical. When antigen was given prior to irradiation, an increase in antibody production was observed. X-Ray exposures of 500 R given from 2 hr to 48 hr after antigen stimulation resulted in antibody levels equal to or greater than nonirradiated controls. In contrast, the response to antigen was drastically reduced if radiation preceded antigen injection, with maximal suppression at -24 hr. Many other researchers have confirmed the time-dependent relationship between irradiation and immunization since the original observation of Taliaferro *et al.* The experimental data imply that normally radiosensitive small lymphocytes become significantly more radioresistant following antigen stimulation. Irradiation after the induction of an immune response must therefore differentially destroy antigen-unstimulated cells and leave antigen-stimulated radioresistant cells in lymphoid-depleted tissues. This sequential combination of antigen stimulation and irradiation causes intense hemopoietic repair, which favors the clonal expansion of antigen-primed lymphocytes.

IV. Radiation Effects on Cellular Immunity

In general, the effects of irradiation on cell-mediated immunity parallel those effects observed for antibody-mediated immunity. In 1918, Murphy and Taylor discovered that transplanted tumors in irradiated animals showed prolonged growth or, more precisely, reduced host rejection. Since then, graft rejection and delayed hypersensitivity have been recognized as immunological events mediated by thymus-derived lymphoid cells bearing the unique Θ surface antigen. Many authors have reviewed this subject, e.g., Gell and Benacerraf (1961), Lawrence (1959), and Micklem and Loutit (1966). In summary, experimental results have demonstrated that initiation of cell-mediated immunity, like antibody-mediated immunity, is more sensitive when radiation is given prior to antigen challenge. Cellular immunity, however, can be distinguished from humoral immunity because of its relative radioresistance. Radiation exposure doses capable of nearly eliminating a humoral response will only have a marginal and transitory effect on a cell-mediated response. Salvin and Smith (1959) observed no significant reduction in delayed hypersensitivity in guinea pigs exposed to 300 R and injected with toxoid from 72 hr before to 18 hr postirradiation. Visakorpi (1972) measured the skin reaction of rats immunized with BSA 10 days before receiving 800 rads. Skin reactivity to the test antigen was almost entirely suppressed within 24 hr after irradiation but recovery was nearly completed 9 days after irradiation. Blood monocytes were severely depleted at 48 hr and, furthermore, only the transfer of bone marrow cells restored the response. These data suggest that a radiosensitive nonlymphoid cell was affected instead of specifically sensitized effector lymphocytes. These conclusions are consistent with the idea that delayed hypersensitivity skin reactions require the participation of radiosensitive nonlymphoid cells of the bone marrow. The precursor to the functional mature macrophage has been proposed as the radiosensitive element in cell-mediated immunity.

Although whole-body irradiation potentially damages all living cells, it differentially affects hemopoietic and lymphoid tissues and thereby destroys the function of the immune system. Immunologically uncensored microbial infections are therefore the primary cause of death in animals exposed to midlethal doses of radiation.

V. Radioprotective Properties of LPS

The capacity of bacterial products to nonspecifically enhance host resistance to infectious microbes has been recognized for many years. Using isolated *E. coli* cell walls, Rowley (1955) demonstrated significant resistance to a variety of gram-negative pathogens. His observations soon prompted others to study the role of somatic antigens of gram-negative organisms in nonspecific resistance. Subsequent studies determined that lipopolysaccharides from gram-negative bacteria gave rise to nonspecific transitory resistance to experimental infections with various pathogens (Landy, 1956; Landy and Pillemer, 1956; Dubos and Schaedler, 1956; Kiser *et al.*, 1956).

The ability of endotoxin to protect against the deleterious effects of radiation and to decrease susceptibility to postirradiation infections was a logical expectation. This was first reported by Mefferd *et al.* (1953) and later examined in greater detail by Ainsworth and Chase (1959), Perkins *et al.* (1958), and Smith *et al.* (1957). Their studies revealed that a single low dose of endotoxin (10 μg) provided significant protection and that the time of hemopoietic recovery was the critical parameter in determining the survival within the midlethal dose range. The patterns of hemopoietic recovery, as measured by the advance in bone marrow cellularity, granulocytes, lymphocytes, platelets, and hemoglobin in endotoxin-treated animals (Savage, 1964; Kinosita *et al.*, 1963; Smith *et al.*, 1958a), closely resemble the hemopoietic recovery in animals receiving bone marrow transfer (Cole and Ellis, 1953; Lorenz *et al.*, 1952). During the first 3 days after irradiation, the hemopoietic tissues in the bone marrow of endotoxin-treated animals show no significant histological differences from controls. Between the first and the second week, continued degenerative processes eventually lead to the death of irradiated controls. Regenerative foci of developing erythrocytes and leukocytes appear throughout the bone marrow of endotoxin-treated animals as early as the 5th day; by the 10th day, hemopoietic cells are visible and increase in number until the 15th day when the hemopoietic elements appear normal. This may be followed by a temporary enlargement or hypercellularity due to an excess of hemocytoblasts and dividing hemopoietic cells. Maintenance of near-normal levels of hemoglobin in irradiated animals treated with endotoxin is also believed to result from an early increase in the number of stem cells differentiating into erythrocytes in response to endogenous erythropoietin. Although irradiation-induced anemia elevates the plasma concentration of erythropoietin, endotoxin treatment increases the response of irradiated mice to erythropoietin (Linman and Bethall, 1957; Gurney, 1963).

The onset of hemopoietic regeneration in the spleen occurs later than in the bone marrow. After an interval of approximately 1 week postirradiation, Savage (1964) reported the presence of hemocytoblasts in the spleens of irradiated animals previously treated with endotoxin. The subsequent development of nodular swellings on the splenic surface represented hemopoietic centers within the red pulp. Whereas these nodular centers of regeneration varied in size and cellular composition, the majority of nodules consisted of numerous densely packed and mitotically active hemocytoblasts. Rounded reticular cells were consistently present but erythropoietic elements were only occasionally seen in the center of the nodule. These nodules or colony-forming units increased in size and number, and by the 17th day postirradiation the spleen was heavily infiltrated with hemopoietic cells. Before the end of the second week postirradiation, production of neutrophils, eosinophils, and megakaryocytes began; however, these cells did not appear in large numbers until the third week. Lymphocytes were the last cells to be regenerated and were seldom seen before the third week but many appeared thereafter.

In summary, hemopoietic regeneration is more pronounced in the spleens of endotoxin-treated mice from the first to the fourth week after irradiation. Hemopoietic recovery during this time interval is characterized initially by hemocytoblast production followed by gradual replacement by erythrocytes and finally leukocyte formation. In contrast, a lethal dose of ionizing radiation causes severe leukopenia that persists until death in control animals. Thus, the accelerated restoration of functional immunocytes to normal levels in endotoxin-treated mice is believed to be a major factor in the survival of lethally irradiated mice.

In our study of the radioprotective properties of LPS, it was necessary to first establish the lethality dose–response curves for sex, age, and strain of the experimental animal. Female ICR mice, 12–14 weeks old, were placed in groups of 10 per dose ranging from 400 R to 850 R in increments of 50 R (Fig. 1). The source of radiation was a General Electric Maxitron 300 X-ray machine operated at 20 mA, 300 KVP, HVL of 1.10 mm Cu, 60-cm target distance, and 230 R/min rate of exposure. The survival for each exposure dose represents the percentage of mice surviving 30 days after irradiation. The dose–response curve reveals three important values: (1) 450 R is the maximum exposure dose for 100% survival, (2) 565 R is the projected $LD_{50/30}$ value, and (3) 700 R is the minimum dose for 100% lethality. Subsequent studies of radioprotection used 700 R as the 100% lethal dose. Other investigators had previously noted that the time of LPS treatment was a critical variable with respect to irradiation (Smith *et al.*, 1957; Ainsworth and Chase, 1959); therefore, a thorough time-relation study was conducted. Endotoxins for our studies were obtained from *Serratia mar-*

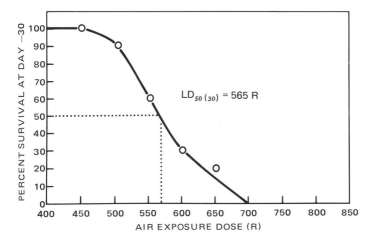

Figure 1. Radiation dose–response of ICR female mice.

cescens by the method described by Nowotny (1979). A single i.p. injection of 10 μg LPS was administered to groups of 10 mice at 24 hr intervals, starting 20 days before and ending 5 days after irradiation (Fig. 2).

Little protection was evident when LPS treatments occurred immediately before or after irradiation; however, a two-cycle biphasic protection response appears when such treatments precede irradiation. The first peak is narrow and corresponds to endotoxin treatments 2–3 days prior to radiation exposure. Protection abruptly declines at −5 days followed by a second broader period of protection with maximal effects between −8 and −14 days.

Endotoxic derivatives obtained by mild acid hydrolysis (Westphal and Lüderitz, 1954) were also examined for their ability to protect from irradiation. Lipid A was obtained from *Salmonella minnesota* 1114 LPS by hydrolysis with 1 N HCl on a boiling water bath for 35 min. The precipitated lipid A was removed by centrifugation, washed twice with distilled water, and lyophilized. For injection, this preparation was dispersed by sonication in saline. The degraded polysaccharide moiety of LPS was isolated by neutralizing the supernatant of the hydrolysate with Dowex 1 anion exchanger in OH⁻form. This preparation was lyophilized and designated PS. In addition, the endotoxic glycolipid (GL) was also extracted from the Re rough mutant of *S. minnesota* R595 using the chloroform—methanol extraction described by Chen *et al.* (1973). When the endotoxicity of *S. minnesota* LPS, its acid hydrolysis products (lipid A and PS), and the GL of its R595 rough mutant were

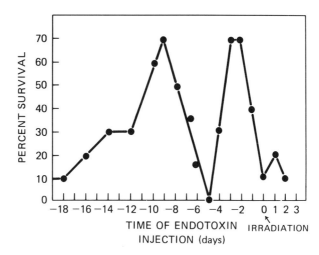

Figure 2. The time-dependent effect of endotoxin on the survival of lethally irradiated (700 R) mice.

Table 1. Parameters of Endotoxicity of LPS and Its Components

| | Preparations[a] | | | | |
| | S. marcescens | | S. minnesota | | |
Endotoxicity assays	LPS	PS	LPS	PS	GL
Mouse LD_{50}[b]	350–450	> 4000	450–500	> 4000	1100
Local Shwartzman[c]	130	0	80	0	110
Chick embryo LD_{50}[d]	0.003	> 10.0	0.015	> 10.0	0.018
B-cell mitogenesis[e]	18.4	0.9	25.1	1.8	34.2
Limulus lysate clotting[f]	10^{-9}	> 10^{-5}	10^{-9}	> 10^{-5}	10^{-8}

[a] Abbreviations: LPS, lipopolysaccharide; PS, water-soluble, carbohydrate-rich hydrolytic product of smooth LPS; GL, endotoxic glycolipid from S. minnesota R595 rough Re mutant.
[b] LD_{50} expressed in μg/mouse.
[c] Size of skin hemorrhage produced by 10-μg i.d. injection. The size is the product of two diameters given in millimeters.
[d] LD_{50} expressed in μg/embryo.
[e] Stimulation index measured by incorporation of [^{3}H] thymidine.
[f] Lowest concentration of preparation in g/ml giving a positive clotting response.

tested in classical endotoxicity assays (mouse and chick embryo lethality, Shwartzman skin reactivity, B-cell mitogenesis, and Limulus lysate clotting), all of the preparations were fully active except PS, the toxicity of which was either absent or reduced by several orders of magnitude (Table 1).

When the endotoxic preparations were tested for their radioprotective properties, a single dose of 10 or 100 μg was injected i.p. (Table 2). The results clearly indicate that the nontoxic PS had nearly the same radioprotective capacity as the parent LPS. Neither the lipid A nor the GL showed significant protection at the 10-μg dose but the GL did show an intermediate level of protection at the 100-μg dose.

The surprising observation of radioprotection by the nontoxic water-soluble PS moiety required further investigations. If the PS moiety of the intact LPS molecule were responsible for radioprotection, then one might reasonably expect PS to express a similar time-dependent radioprotective profile. Using conditions identical to the previous experiments, mice in groups of 10 were treated at various 24-hr intervals with PS before exposure to 700 R (Fig. 3). A time-dependent radioprotective profile for the nontoxic PS preparation was similar to although somewhat reduced compared to the parent LPS. The observation that the characteristic time-dependent radioprotective property of LPS could be reproduced by its nontoxic PS component provided additional evidence that endotoxicity was not essential for all of the biological effects attributed to LPS. This implies that lipid A, which is generally ac-

Table 2. Protective Effect of 10- and 100-μg Test
Preparations Given 48 hr before 700-R Irradiation on
the Survival of ICR Mice

μg preparation	Percent survivors on day			
	5	10	15	20
Endotoxin				
10	100	80	70	70
100	100	80	80	80
Glycolipid				
10	100	70	50	20
100	100	70	50	50
PS fraction				
10	100	70	50	50
100	100	80	80	70
Lipid A				
10	100	20	0	
100	100	30	0	
Control				
10	100	40	5	0
100	90	40	20	0

cepted as the toxic component of LPS, may not necessarily be involved in the nontoxic biological effects of LPS.

Later experiments produced further support for the active role of the PS moiety when LPS was neutralized with homologous antiserum and complete radioprotection was observed (Table 3). It was established earlier that selective neutralization of the lipid moiety of LPS rendered the LPS immune complex nontoxic (Radvany *et al.*, 1966). This was achieved by first absorbing rabbit hyperimmune O antiserum with chemically detoxified endotoxin, thereby selectively removing antibodies reacting with the PS O-specific antigens. When the residual antiserum was reacted with LPS, the endotoxicity of untreated endotoxin was neutralized. It must also be mentioned that the resultant immune complex of PS and O-specific antiserum was active in radioprotection. As a control, rabbit serum–anti-rabbit serum immune complex was tested and found to lack radioprotection. Thus, the possibility of a general nonspecific involvement of immune stimulation by an insoluble immune complex could be dismissed.

We had reported in an earlier publication that both LPS and, more importantly, its nontoxic PS component are active inducers of bone marrow colony stimulation (Nowotny *et al.*, 1975). This *in vitro* method of bone marrow cell proliferation can be induced by a serum factor (CSF) of LPS- or PS-treated animals. Hence, certain endotoxic effects

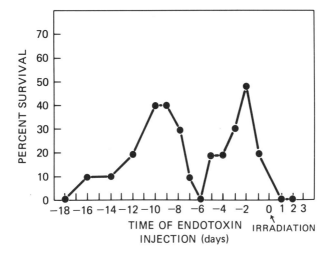

Figure 3. The time-dependent effect of PS on the survival of lethally irradiated (700 R) mice.

must be mediated by humoral factors which can be adoptively transferred. We subsequently tested the sera of mice 3 hr following 5 μg i.v. injection of either LPS or PS for their ability to provide radioprotection. Mice were injected i.p. with 0.5 ml of pooled sera 24 hr before lethal irradiation. Their survival rate for both postendotoxin and post-PS sera were again comparable to those animals treated directly with LPS or PS (Table 4). It must be pointed out that we repeatedly observed moderate to intermediate radioprotection by control sera. This protection may be due to the nonspecific immunoprotective action of serum γ-globulin.

Table 3. The Radioprotective Action of Endotoxin Immune Complexes in Lethally Irradiated Mice

	Percent survivors on day					
Treatment[a]	5	10	15	20	25	30
Control	100	90	0			
LPS	100	100	80	80	80	80
PS	100	100	70	70	70	70
LPS-Ab	100	100	90	90	90	90
PS-Ab	100	100	80	80	80	80
NRS-Ab	100	100	10	0		

[a] Animals in groups of 10 were treated with 10 μg of LPS, PS in free or immune complex form 24 hr before irradiation. An equivalent amount of normal rabbit serum complexed with goat anti-rabbit serum was used as a control.

Table 4. Radioprotective Effect of Postendotoxin and Post-Polysaccharide-Rich Fraction Serum in Mice Receiving 700-R Whole-Body Irradiation

Treatment	Percent survivors on day						
	6	10	14	18	22	26	30
Normal mouse serum[a]	100	100	60	40	40	40	40
Postendotoxin serum	100	100	90	80	80	80	80
Post-PS serum	100	90	70	70	70	70	70
Control	100	80	30	10	0		

[a] Animals were injected with 0.5 ml of serum 24 hr before irradiation.

The ability of endotoxin to elicit the *in vivo* release of a humoral factor(s) into the circulating blood prompted investigation into the origin of such a factor. Our experimental approach was based on the assumption that certain tissues when stimulated *in vivo* by endotoxins would afterwards release this factor *in vitro*. One hour after the injection of LPS or PS in mice, their lungs, liver, kidneys, and spleens were removed. The organs were separately minced and under sterile conditions incubated in serum-free medium (Dulbecco modified Eagle's medium) for 48 hr. Following centrifugation of the culture medium (10,000 g for 30 min), the cell-free supernatant was inactivated at 56°C for 30 min, dialyzed against distilled water for 3 days, and finally lyophilized. From the endotoxin and PS organ culture supernatants tested for radioprotection, only the lung culture of previously LPS- or PS (Table 5). As with normal serum, moderate radioprotection was afforded by normal lung culture supernatant.

Table 5. Radioprotective Effect of Postendotoxin and Post-Polysaccharide-Rich Fraction and Lung Culture Filtrate (LCF) in Mice Receiving Whole-Body Irradiation

Treatment	Percent survivors on day						
	6	10	14	18	22	26	30
Normal mouse LCF[a]	100	80	70	40	30	30	30
Postendotoxin LCF	100	100	90	80	80	80	80
Post-PS LCF	100	90	90	70	60	60	60
Control	100	80	30	10	0		

[a] Animals were injected with 80 μg dissolved in 0.2 ml of saline 24 hr before irradiation.

VI. Cellular and Humoral Endotoxic Effects That Influence the Irradiated Host

Death involving the acute radiation syndrome, as previously stated, is the result of collective damage imposed on the hemopoietic tissues and functional immunocytes. More simply, it is the increased susceptibility to microbial infections that is the primary cause of death in animals exposed to a midlethal dose of ionizing radiation. One of the first major biological effects attributed to endotoxin was the ability of LPS to protect animals against lethal doses of ionizing radiation (Mefferd *et al.*, 1953). At the time, this observation may have been viewed with a certain amount of curiosity with little supportive evidence to provide an acceptable explanation. Based on our current understanding of the mammalian immune system, chemical and biological properties of bacterial endotoxins, and radiation biology, the radioprotective properties of bacterial endotoxins must be anticipated. Most of the literature indicates that endotoxin directly or indirectly affects the immunoregulatory function of nearly all of the presently defined cell populations and associated humoral factors in the mammalian immune system. The indiscriminate destructive action of ionizing radiation on all living cells therefore implies a complex interrelationship within the irradiated LPS-treated host. Thus, certain immunological responses elicited by endotoxin in the irradiated host cannot be attributed to a single effect, but more likely represent the net sum of the various endotoxin radiation-sensitive components of the host.

The altered immune status of endotoxin-treated animals to resist an infectious challenge (nonspecific resistance), a transplantable tumor challenge, or a lethal exposure to radiation merely provides a crude measurement of the collective immunostimulatory effects of endotoxin. The effects that probably contribute most to radioprotection include several humoral and cellular events that tend to counteract the generalized immunosuppressive action of radiation. Kinosita *et al.* (1963) reported granulocytopenia and subsequent leukocytosis; this biphasic response in cellularity of circulating blood was paralleled by similar changes in bone marrow, spleen, thymus, and lymph nodes. Within 24 to 48 hr following endotoxin treatment, cell-rich foci appeared and gave rise to hyperplasia several days later. Kinetic studies demonstrated the presence of two morphologically and kinetically distinct populations of lymphocytes (Takano and Mizuno, 1968). The more proliferative cell line consisted of lymphoblasts with a division time of 20 hr while the other cell line contained lymphocytes and reticular cells with a division time of 72 hr. In general, cell proliferation is an essential feature of the early immune response, and in the absence of a second antigen, endo-toxin leads to B-lymphocyte proliferation and differentiation. Electron microscopic studies by Shohat *et al.* (1973) showed the stimulation to lymphoblast formation, followed by enhanced development of the en-doplasmic reticulum and eventually the maturation of plasma cells.

Although evidence suggests that LPS does not initiate T-cell prolif-eration, it may alter the functional capacity of T lymphocytes. Studies have demonstrated that LPS readily binds to the membranes of thymo-cytes (Möller *et al.*, 1973) and amplifies the proliferative response to selective T-cell activators such as Con A or PHA up to 10-fold (Schmidtke and Najarian, 1975; Ozato *et al.*, 1975). Armerding and Katz (1974) described the *in vitro* action of LPS and how it could be di-rected at B lymphocytes or T lymphocytes depending on culture condi-tions. In the absence of a second antigen, LPS acted predominantly on B lymphocytes; however, in the presence of a second antigen, LPS mainly influenced T-cell function. Some evidence indicates that the po-tent *in vivo* adjuvant effect of LPS results primarily from the stimulation of a T-helper cell line (Newberger *et al.*, 1974; Ness *et al.*, 1976). The discovery of specific murine T-cell surface markers (Lyt, 1, 2, and 3) by Boyse *et al.* (1968) and Kisielov *et al.* (1975) has allowed the identifica-tion of specific T-cell effector cells traditionally referred to as T-helper and T-suppressor cell lines. Using these surface markers, Uchiyama and Jacobs (1978) have shown that LPS induces both T-helper and T-sup-pressor cells when given to mice prior to antigen challenge.

In addition to the various lymphocyte populations affected by LPS, it is well known that LPS also affects phagocytic cells. Several studies have reported that endotoxin alters clearance of foreign substances by the reticuloendothelial system which correlates with changes in resis-tance to infection (Schmidtke and Najarian, 1975; Ozato *et al.*, 1975). The LPS-activated reticuloendothelial system is not only characterized by enhanced phagocytic activity and cytotoxicity but also by a substan-tial increase in the number of phagocytic cells (Spitznagel and Allison, 1970; Arrendondo and Kampschmidt, 1963; Doe and Henson, 1978).

VII. Humoral Changes and Possible Endotoxic Mechanisms

A variety of humoral changes induced by endotoxin are associated with the above-described cellular changes. The discovery of genetically LPS-unresponsive C3H/HeJ mice and their histocompatible LPS-re-sponsive counterpart (C3HBfej or C3HeN) has supplied a unique tool for studying cellular mechanisms and the regulatory role of immune mediators (Sultzer and Nilsson, 1972). This animal model made it possible to show that LPS exerts some of its effects by direct interaction with B cells (i.e., B-cell mitogenicity and generation of B-suppressor cells) and by the generation and/or release of soluble factors (Hoffmann et al., 1979a,b; McGhee et al., 1979). Since B cells from LPS-unresponsive mice exhibited full adjuvancy in the presence of LPS-responsive T cells and macrophages, B cells are not the primary target of endotoxin for immune potentiation. This suggests that macrophages and T-cell mediators must have a critical role. In 1972, Gery and Waksman reported that LPS-stimulated macrophages produced and released a lymphocyte-activating factor (now termed interleukin 1) that induced thymocyte proliferation. Interleukin 1 has also been shown to enhance the antibody response to T-cell-dependent antigens and to replace the requirment for macrophages during the induction of alloantigen-specific T-cell cytotoxicity (Koopman et al., 1978; Wood, 1979). In 1973, Chen and co-workers identified a T-cell-derived mediator (interleukin 2) that was mitogenic for thymocytes, enhanced antibody formation, increased T-cell cytotoxicity, and replaced T-helper cell function (Farrar el al., 1978; Watson et al., 1979).

Other mediators that strongly influence immune status and are greatly affected by endotoxin include interferon (Youngner and Stinebring, 1966), colony-stimulating factor (Metcalf, 1971), tumor-necrotizing factor (Carswell et al., 1975), properdin (Landy and Weidans, 1964), opsonin (Benacerraf, 1960), enzymes (Wahl et al., 1974), and prostaglandins (Rietschel et al., 1980). With the exception of prostaglandins, these mediators can be assumed to collectively contribute to the increased resistance to infection and consequently to the survival of lethally irradiated animals. Although most of these mediators and their relationships to LPS have been extensively described in the literature, prostaglandins have only recently been investigated as possible mediators of LPS action. Prostaglandins represent a class of intercellular mediators that may alter T-cell reactivity to antigens or mitogens, affect generation of lymphokines, influence T-helper and T-suppressor cell function, and interfere with normal clonal expansion of macrophage stem cells (Goodwin and Webb, 1980; Pelus and Strausser, 1977; Ziemecki and Webb, 1976; McConnell et al., 1980).

VIII. Mechanisms of Endotoxin-Induced Oscillations

Although most of the cellular and humoral changes in the endotoxin-treated host are immunostimulatory in nature, some of them are obviously not. Furthermore, the kinetics (time of onset, rate of appearance, magnitude, and disappearance) of these cellular and humoral effector mechanisms are characterized by vastly different time and dose curves. This implies that the altered immune status of the endotoxin-treated host is in a constant state of flux, possibly oscillating from a maximally enhanced to an immune-suppressed state. In addition to the time-dependent radioprotection described here, we have reported that time-dependent immune oscillatory alterations also exist when the endotoxin-treated animals are tested for (1) immune adjuvancy, (2) resistance to transplantation tumor challenge, and (3) SRBC rosetting response (Behling, 1978, 1980, 1982). A comparison of the periodicity of these four examples of time-dependent changes in host immunity shows they are not identical and cannot be superimposed. This lack of temporal synchronization suggests that these host immune responses each represent the integral of a defined set of immunoregulatory mechanisms that may in part(s) be shared. Thus, endotoxin shifts in regulating T-helper/T-suppressor cell functions and their humoral regulators would be primarily suspected in oscillatory changes of immune adjuvancy. No direct link exists for T-helper or T-suppressor roles in radioprotection. Equally, the hemopoietic and mitogenic stimulation induced by endotoxin has no obvious influence on the anti-SRBC rosette-forming cell response; however, they probably represent the primary mechanisms for radioprotection.

Furthermore, the mode of protection or immune enhancement of a given endotoxin-induced alteration may be determined by the nature of trauma imposed on the immune system. Although the resultant endotoxin leukocytosis may be a contributing factor in nonspecific resistance, immune adjuvancy, and radioprotection, the mechanism governing its mode of action may be unique to the specific trauma (antigen stimulation, pathogenic bacterial challenge, radiation exposure, etc.). Athens et al. (1961) studied the kinetic changes of leukocytes after endotoxin treatment and found that early transient leukopenia represents a shift of granulocytes from the circulating granulocyte pool to the marginal granulocyte pool. The latter is attributable to adherence of leukocytes to the vascular epithelium in sites where blood flow is relatively slow (lungs, extremities, etc.). This leukopenia is rapidly followed by leukocytosis. The early stage of leukocytosis is merely a rapid mobilization of cells from the bone marrow into the circulation, followed by an absolute increase in cell number due to stem cell proliferation (Smith et al., 1961; Athens et al., 1961).

The above-described endotoxin-induced mobilization of granulocytes probably represents one of several radioprotective mechanisms. For lower energy photons of the X-ray region, absorption of the damaging X-ray occurs mainly through the photoelectric effect and is quantitatively defined by the atomic number (Z) of the target material to the third power. The effective atomic number of mineralized bone is nearly double that of soft tissue; therefore, the radiation dose to soft tissue immediately adjacent to or encapsulated by bone is greatly enhanced by the large number of secondary electrons emitted from mineral bone after photoelectric absorption (Charlton and Cormack, 1962; Lentsch and Finston, 1967; Spiers, 1949). This dose enhancement in soft tissues produced by secondary electrons from compact bone extends 100 μm from the bone tissue interface and can reach as high as four times the dose to soft tissues away from the bone. Hence, those cells released from the marrow cavities as a result of LPS treatment before irradiation are exposed to substantially lower doses of radiation. This reduction in radiation exposure would significantly increase the number of surviving cells and thereby allow the accelerated repopulation of postirradiation aplastic hemopoietic tissues. Gidali *et al.* (1969) and Smith *et al.* (1966) reported accelerated regeneration of colony-forming units in the spleens of endotoxin-treated mice at various time intervals after irradiation.

A change in radiosensitivity of cells responding to antigenic and mitogenic stimulation by endotoxin may also be a major contributing factor in establishing radioresistance. The resting mature lymphocyte is generally regarded as the most radiosensitive cell of the hemopoietic system. However, upon stimulation it transforms into a mature, nondividing plasma cell that is radioresistant. By virtue of its antigenic and mitogenic properties, endotoxin may establish synchronous cell populations that undergo a series of mitotic divisions characterized by cytological, functional, and radioresistant changes. The role of cell synchronization and its association with radioresistance is supported by two independent observations: (1) daily or repeated LPS treatments do not protect against radiation (Smith *et al.*, 1957; Perkins *et al.*, 1958; Ainsworth and Chase, 1959) and (2) when antigenic stimulation precedes radiation by as little as a few hours, the immune response is extremely radioresistant (Taliaferro *et al.*, 1964).

A final and most obvious factor contributing to radioresistance in endotoxin-treated hosts may simply be the presence of an elevated number of leukocytes at the time of irradiation. Animals exposed to 50% lethal dose retain only 1–2% of the normal anti-SRBC plaque-forming cell response (Kennedy *et al.*, 1965). Therefore, a doubling of the cellular components of hemopoietic tissues would shift the radiation-survival dose–response curve by a value equivalent to the dose required for a 50% kill of individual leukocytes. This value is approximately 100 R and corresponds to the shift in the LD_{50} value for LPS-treated mice.

It is important to remember that the radioprotective mechanisms described above and a variety of associated humoral factors do not occur simultaneously but follow a sequential pattern. This includes leukocyte mobilization, changes in cellular radiosensitivity, and increased numbers of cells over a period extending beyond 2 weeks. Thus, the mode of radio protection is determined by the time interval separating LPS treatment and radiation exposure. A sequential pattern of radio-protective mechanisms must be suspected in order to explain the oscil-latory time-dependent radioprotective property of bacterial endotoxin.

IX. References

Ainsworth, E. J., and Chase, H. B., 1959, Effect of microbial antigens on irradiation mor-tality in mice, *Proc. Soc. Exp. Biol. Med.* **102**:483.

Armerding, D., and Katz, D. H., 1974, Activation of T and B lymphocytes *in vitro*: Regu-latory influence of bacterial lipopolysaccharide (LPS) on specific T helper cell function, *J. Exp. Med.* **139**:24.

Arrendondo, M. T., and Kampschmidt, R. F., 1963, Effect of endotoxins on phagocytic activity of the reticuloendothelial system of the rat, *J. Exp. Biol. Med.* **112**:78.

Athens, J. W., Raab, S. O., Haab, O. P., Mauer, A. M., and Ashenbrucker, H., 1961a, Leukokinetic studies. III. The distribution of granulocytes in the blood of normal subjects, *J. Clin. Invest.* **40**:159.

Athens, J. W., Haab, O. P., Raab, S. O., Mauer, A. M., Ashenbrucker, H., Cartwright, G. E., and Wintrobe, M. M., 1961b, Leukokinetic studies. IV. The total blood, circulat-ing and marginal granulocyte pools and the granulocyte turnover rate in normal subjects, *J. Clin. Invest.* **40**:989.

Behling, U. H., 1982, Bacterial endotoxins as modulators of specific and nonspecific im-munity, in: *Regulatory Implications of Oscillatory Dynamics in the Immune System* (C. DeLisi and A. Hiernaux, eds.), CRC Press, Boca Raton, in press.

Behling, U. H., and Nowotny, A., 1978, Long-term adjuvant effect of bacterial endotoxin in prevention and restoration of radiation-caused immunosuppression, *Proc. Soc. Exp. Biol. Med.* **157**:348.

Behling, U. H., and Nowotny, A., 1980, Cyclic changes of positive and negative effects of single endotoxin injections, in: *Bacterial Endotoxins and Host Response* (M. K. Agarwal, ed.), pp. 11–26, Elsevier/North-Holland, Amsterdam.

Benacerraf, B., 1960, Influence of irradiation on resistance to infection, *Bacteriol. Rev.* **24**: 35.

Benjamin, E., and Sluka, E., 1908, Antikorperbildung Nach Experimenteller Schadigung des Haematopoetischen Systems durch Roentgenstrahlen, *Wien. Klin. Wochenschr.* **21**: 311.

Bond, V. P., 1957, The role of infection in illness following exposure to acute total-body irradiation, *Bull. N.Y. Acad. Med.* **33**:369.

Boyse, E. A., Meyazawa, M., Aoki, T., and Old, L. T., 1968, Ly-A and Ly-B: Two systems of lymphocyte isoantigens in the mouse, *Proc. R. Soc. London Ser. B* **170**:175.

Carswell, E. A., Old, L. J., Kassel, R. L., Green, S., Fiore, N., and Williamson, B., 1975, An endotoxin-induced serum factor that causes necrosis of tumors, *Proc. Natl. Acad. Sci. USA* **72**:3666.

Charlton, D. E., and Cormack, D. V., 1962, Energy dissipation in finite cavities, *Radiat. Res.* **17**:34.

Chen, C. H., Johnson, A. G., Kasai, N., Key, B. A., Levin, J. and Nowotny, A., 1973, Het-erogeneity and biological activity of endotoxic glycolipid from *Salmonella minnesota* R595, *J. Infect. Dis.* **128**(Suppl.):43.

Cole, L. J., and Ellis, M. E., 1953, Age, strain and species factors in post-irradiation protection by spleen homogenates, *Am. J. Physiol.* **173**:487.

Cronkite, E. P., 1964, in: *Atomic Medicine*, pp. 170–188, 192–196, Williams & Wilkins, Baltimore.

Cronkite, E. P., 1973, Radiosensitivity of lymphocytes, *Strahlenschutz Forsch. Prax* **13**:13.

Cronkite, E. P., and Bond, V. P., 1956, Effect of radiation on mammals, *Annu. Rev. Physiol.* **18**:483.

Doe, W. F., and Henson, P. M., 1978, Macrophage stimulation by bacterial lipopolysaccharides. I. Cytolytic effect on tumor target cells, *J. Exp. Med.* **148**:544.

Dubos, R. J., and Schaedler, R. W., 1956, Reversible changes in the susceptibility of mice to bacterial infections, *J. Exp. Med.* **104**:53.

Ernstrom, U., 1972, Effect of irradiation on the release of lymphocytes from the thymus, *Acta Radiol.* **11**:257.

Farrar, J. J., Simon, P. L., Koopman, W. J., and Fuller-Bonar, J., 1978, Biochemical relationship of thymocyte mitogenic factor and factors enhancing humoral and cell-mediated immune responses, *J. Immunol.* **121**:1358.

Gell, P. G. H., and Benacerraf, B., 1961, Delayed hypersensitivity to simple protein antigens, *Adv. Immunol.* **1**:319.

Gery, I., and Waksman, B. H., 1972, Potentiation of the T-lymphocyte response. II. The cellular source of potentiating mediator(s), *J. Exp. Med.* **136**:143.

Gidali, J., Feher, I., and Varteresz, V., 1969, The effect of endotoxin on the growth and differentiation of haemopoietic stem cells, *Atomkernenergie* **14**:235.

Goodwin, J. S., and Webb, D. R., 1980, Regulation of the immune response by prostaglandins, *Clin. Immunol. Immunopathol.* **15**:106.

Gurney, C. W., 1963, Effect of radiation on the mouse stem cell compartment *in vivo*, *Perspect. Biol. Med.* **6**:233.

Hoffmann, M. K., Koenig, S., Mittler, R. S., Oettgen, H. F., Ralph, P., Galanos, C., and Hämmerling, U., 1979a, Macrophage factor controlling differentiation of B-cells, *J. Immunol.* **122**:497.

Hoffmann, M. K., Galanos, C., Koenig, S., and Oettgen, H. F., 1979b, B-Cell activation by lipopolysaccharide: Distinct pathways for induction of mitosis and antibody production, *J. Exp. Med.* **146**:1640.

Hulse, E. Y., 1959, Lymphocyte depletion of the blood and bone marrow of the irradiated rat: A quantitative study, *Br. J. Haematol.* **5**:278.

Kennedy, J. C., Till, J. E., and Siminovitch, L., 1965, Radiosensitivity of the immune response to sheep red cells in the mouse as measured by the haemolytic plaque method, *J. Immunol.* **94**:715.

Kinosita, R., Nowotny, A., and Shikato, T., 1963, Conference on bone marrow transplantation and chemical protection in large animals and man, *Blood* **21**:779.

Kiser, J. S., Lindh, H., and diMello, G. C., 1956, The effect of various substances on resistance to experimental infections, *Ann. N.Y. Acad. Sci.* **66**:312.

Kisielov, P., Hirst, J., Shiku, H., Beverly, P. C. L., Hoffmann, M. K., Boyse, E. A., and Oettgen, H. F., 1975, Ly antigens as markers for functionally distinct subpopulations of thymus derived lymphocytes of the mouse, *Nature (London)* **253**:219.

Koopman, W. J., Farrar, J. J., and Fuller-Bonar, J., 1978, Evidence for the identification of lymphocyte activating factor as the adherent cell-derived mediator responsible for enhanced antibody synthesis by nude mouse spleen cells, *Cell. Immunol.* **35**:92.

Landy, M., 1956, Increased resistance to infection developed rapidly after administration of bacterial lipopolysaccharides, *Fed. Proc.* **15**:598.

Landy, M., and Pillemer, L., 1956, Increased resistance to infection and accompanying alteration in properdin levels following administration of bacterial lipopolysaccharides, *J. Exp. Med.* **104**:383.

Landy, M., and Weidans, W. P., 1964, in: *Bacterial Endotoxins* (M. Landy and W. Braun, eds.), Rutgers University Press, New Brunswick, N.J.

Lawrence, H. S. (ed.), 1959, *Cellular and Humoral Aspects of the Hypersensitive states*, Harper & Row (Hoeber), New York.

Lentsch, J. W., and Finston, R. A., 1967, Increased x-ray dose adjacent to plane bone interfaces as measured by polyethylene, *Phys. Med. Biol.* **12**(4):543.

Linman, J. W., and Bethall, F. H., 1957, The effect of irradiation on the plasma erythropoietic stimulating factor, *Blood* **12**:123.

Lorenz, E., Congdon, C., and Uphoff, D., 1952, Modification of acute irradiation injury in mice and guinea pigs by bone marrow injections, *Radiology* **58**:863.

McConnell, I., Hopkins, J., and Lachman, P., 1980, Lymphocyte traffic through lymph nodes during cell shut-down, *Ciba Found. Symp.* **71**:167.

McGhee, J. R., Farrar, J. J., Michalek, S. M., Mergenhagen, S. E., and Rosenstreich, D. L., 1979, Cellular requirements for lipopolysaccharide adjuvanticity: A role for both T lymphocytes and macrophages for *in vitro* responses to particulate antigens, *J. Exp. Med.* **149**:793.

Makinodan, T. M., Kastenbaum, A., and Peterson, W. I., 1962, Radiosensitivity of spleen cells from normal and pre-immunized mice and its significance to intact animals, *J. Immunol.* **88**:31.

Mefferd, R. B., Henkel, D. T., and Loeffer, J. B., 1953, Effect of piromen on survival of irradiated mice, *Proc. Soc. Exp. Biol. Med.* **83**:54.

Metcalf, D., 1971, Acute antigen-induced elevation of serum colony stimulating factor (CSF) levels, *Immunology* **21**:427.

Micklem, H. S., and Loutit, J. F., 1966, *Tissue Grafting and Radiation*, Academic Press, New York.

Miller, C. P., 1956, The effect of irradiation on natural resistance to infection, *Ann. N.Y. Acad. Sci.* **66**:280.

Möller, G., Andersson, J., Pohlit, H. and Sjöberg, O., 1973, Quantitation of the number of mitogen molecules activating DNA synthesis in T and B lymphocytes, *Clin. Exp. Immunol.* **13**:89.

Murphy, J. B., and Taylor, H. D., 1918, The lymphocyte in natural and induced resistance to transplants. III. The effect of x-rays on artificially induced immunity, *J. Exp. Med.* **28**:1.

Ness, D. B., Smith, S., Talcott, J. A., and Grummet, F. C., 1976, T-Cell requirement for the expression of the lipopolysaccharide adjuvant effect *in vivo*: Evidence for a T-cell dependent and T-cell independent mode of action, *Eur. J. Immunol.* **6**:650.

Newberger, P. E., Hamoka, T., and Katz, D. H., 1974, Potentiation of helper T cell function in IgE antibody responses by bacterial lipopolysaccharide (LPS), *J. Immunol.* **113**: 824.

Nowotny, A., 1979, *Basic Exercises in Immunochemistory*, 2nd ed., pp. 69–72, Springer-Verlag, Berlin.

Nowotny, A., Behling, U. H., and Chang, H., 1975, Relation of structure to function in bacterial endotoxins. VIII. Biological activities in a polysaccharide-rich fraction, *J. Immunol.* **115**:199.

Ozato, K., Adler, W. H., and Ebert, J. D., 1975, Synergism of bacterial lipopolysaccharides and concanavalin A in the activation of thymic lymphocytes, *Cell. Immunol.* **17**: 532.

Pelus, L. M., and Strausser, H. R., 1977, Prostaglandins and the immune response, *Life Sci.* **20**:903.

Perkins, E. H., Marcus, S., Gyi, K. K., and Miya, F., 1958, Effect of pyrogen on phagocytic digestion and survival of x-irradiated mice, *Radiat. Res.* **8**:502.

Petrov, R. V., 1958, Exogenous infections in radiation sickness, *Adv. Mod. Biol.* **46**:48.

Radvany, R., Neale, N., and Nowotny, A., 1966, Relation of structure to function in bacterial O-antigens. VI. Neutralization of endotoxic O-antigens by homologous O-antibody, *Ann. N.Y. Acad. Sci.* **133**:763.

Rietschel, E. T., Schade, U., Lüderitz, O., Fisher, H., and Peskar, B. A., 1980, Prostaglandins in endotoxicosis, in: *Microbiology 1980* (D. Schlessinger, ed.), American Society for Microbiology, Washington, D.C.

Rowley, D., 1955, Stimulation of natural immunity to *Escherichia coli* infections: Observations on mice, *Lancet* **1**:232.

Salvin, S. B., and Smith, R. F., 1959, Delayed hypersensitivity in the development of circulating antibody: The effect of x-irradiation, *J. Exp. Med.* **109**:325.

Savage, A. M., 1964, Hematopoietic recovery in endotoxin-treated lethally x-irradiated Bub mice, *Radiat. Res.* **23**:180.

Schmidtke, J. R., and Najarian, J. S., 1975, Synergistic effects on DNA synthesis of phytohemagglutinin or concanavalin A and lipopolysaccharide in human peripheral blood lymphocytes, *J. Immunol.* **114**:742.

Shohat, M., Janossy, G., and Dourmashkin, R. R., 1973, Development of rough endoplasmic reticulum in mouse splenic lymphocyte stimulated by mitogen, *Eur. J. Immunol.* **3**:680.

Smith, J. C., 1963, Radiation pneumonitis: A review, *Annu. Rev. Respir. Dis.* **87**:647.

Smith, W. W., Alderman, I. M., and Gillespie, R. F., 1957, Increased survival in irradiated animals treated with bacterial endotoxins, *Am. J. Physiol.* **191**:124.

Smith, W. W., Alderman, I. M., and Gillespie, R. E., 1958a, Hemopoietic recovery induced by bacterial endotoxin in irradiated mice, *Am. J. Physiol.* **192**:549.

Smith, W. W., Alderman, I. M., and Gillespie, R. F., 1958b, Resistance to experimental infections and mobilization of granulocytes in irradiated mice treated with bacterial endotoxins, *Am. J. Physiol.* **192**:263.

Smith, W. W., Alderman, I. M., Cornfield, J., 1961, Granulocyte release by endotoxin in normal and irradiated mice, *Am. J. Physiol.* **201**:396.

Smith, W. W., Brecher, G., Fred, S. and Budd, R. A., 1966, Effect of endotoxin on the kinetics of hemopoietic colony-forming cells in irradiated mice, *Radiat. Res.* **27**:710.

Spiers, F. W., 1949, The influence of energy absorption and electron range on dosage in irradiated bone, *Br. J. Radiol.* **12**:521.

Spitznagel, J. K., and Allison, A. C., 1970, Mode of action of adjuvants: Effects on antibody responses to macrophage-associated bovine serum albumin, *J. Immunol.* **104**:128.

Sultzer, B. M., and Nilsson, B. S., 1972, PPD-tuberculin—A B-cell mitogen, *Nature New Biol.* **240**:198.

Takano, T., and Mizuno, D., 1968, Dynamic state of the spleen cells of mice after administration of the endotoxin of *Proteus vulgaris*. I. Cellular proliferation after administration of the endotoxin, *Jpn. J. Exp. Med.* **38**:171.

Taliaferro, W. H., Taliaferro, L. G., and Jaroslow, B. N., 1964, *Radiation and Immune Mechanisms*, Academic Press, New York.

Talmage, D. W., 1955, Effect of ionizing radiation on resistance and infection, *Annu. Rev. Microbiol.* **9**:335.

Taylor, H. D., Witherbee, W. D., and Murphy, J. B., 1919, The effect of radiation on the functional capacity of lymphocytes, *J. Exp. Med.* **29**:53.

Uchiyama, T., and Jacobs, D. M., 1978, Modulation of immune responses by bacterial lipopolysaccharide (LPS): Cellular basis of stimulatory and inhibitory effects of LPS in the *in vitro* IgM antibody response to a T-dependent antigen, *J. Immunol.* **121**:2347.

Visakorpi, R., 1972, Effect of irradiation on established delayed hypersensitivity: Suppression of skin reactions, recovery and the effect of cell transfer, *Acta Pathol. Microbiol. Scand.* **80**:132.

Wahl, L. M., Wahl, S. M., Mergenhagen, S. E., and Martin, G. R., 1974, Collagenase production by endotoxin-activated macrophages, *Proc. Natl. Acad. Sci. USA* **71**:3598.

Wald, N., Thoma, G. E., and Broun, G., 1962, Hematologic manifestations of radiation
exposure in man, *Prog. Hematol.* **3**:1.

Watson, J., Aarden, L., Shaw, J., and Paetkan, V., 1979, Molecular and quantitative analysis of helper T-cell replacing factors on the induction of antigen sensitive B and T
lymphocytes, *J. Immunol.* **122**:1633.

Westphal, O., and Lüderitz, O., 1954, Chemische Erforschung von Lipololysacchariden
gram negativer Bakterien, *Angew. Chem.* **66**:407.

Wood, D. D., 1979, Mechanism of action of human B-cell activating factor. I. Comparison of the plaque-stimulating activity with thymocyte stimulating activity, *J. Immunol.*
123:2400.

Youngner, J. S., and Stinebring, W. R., 1966, Comparison of interferon production in
mice by bacterial endotoxin and statolon, *Virology* **20**:310.

Ziemecki, M., and Webb, D. R., 1976, The regulation of the immune response to T-independent antigens by prostaglandins and B-cells, *J. Immunol.* **117**:2158.

7

Induction of Endotoxin Tolerance

Sheldon Edward Greisman

Resistance to gram-negative bacterial endotoxins occurs in two general forms. One appears genetically determined, the affected species (or strain) exhibiting minimal responses to an initial intravenous injection of relatively massive quantities of endotoxin. Examples are the lethal and pyrogenic unresponsiveness seen in baboons, vervets, and C3H/HeJ mice (Sultzer, 1968; Westphal, 1975). In contrast to such natural resistance, other species respond to an initial intravenous injection of endotoxin with an array of striking physiologic and biochemical alterations, man being one of the most highly responsive (Greisman and Hornick, 1973). Most of these responses decrease progressively when the endotoxin is readministered at appropriate intervals. Such acquired resistance, or tolerance, does not necessarily develop in the same temporal sequence for each response; indeed, particularly during the initial several days, hyperreactivity may occur to one while tolerance appears to another (Greer and Rietschel, 1978). This review will consider some of the mechanisms underlying the induction of endotoxin tolerance to two responses that have been most intensively studied—fever and lethality.

I. Pyrogenic Tolerance

A. Historical Background

Endotoxin tolerance was encountered almost 100 years ago by physicians who employed bacterial vaccines for fever therapy. It was widely recognized by the early 1900s that whenever fever was evoked

Sheldon Edward Greisman • Department of Medicine and Physiology, University of Maryland School of Medicine, Baltimore, Maryland 21201.

with killed gram-negative bacilli, increasing quantities had to be infused intravenously to maintain the necessary elevated core temperatures to produce therapeutic effects. Some patients eventually required as much as 250 ml of typhoid vaccine in a single day (Heyman, 1945). Various methods were used to circumvent this resistance. Nelson (1931) found that resistance could be partly overcome by giving two doses of typhoid vaccine with each treatment; the second dose, given 2 hr later, was described as having the effect of "exploding the charge" supplied by the first. Heyman (1945) recommended continuous intravenous infusions of typhoid vaccine, the infusion rate to be increased as the fever began to decline.

Favorite and Morgan (1942) demonstrated that resistance to the pyrogenic fraction prepared by alcohol precipitation from *Salmonella typhosa* developed in man as it did to the whole typhoid vaccine. Moreover, because circulating antibody titers to this pyrogen did not correlate with resistance, it was suggested that nonimmunologic mechanisms were responsible and the term *tolerance* was employed. Beeson (1946, 1947a,b) performed a series of classic studies on pyrogenic tolerance. Utilizing rabbits, immunologic mechanisms were excluded on the basis of apparent lack of either O specificity or serum transferability, as well as the rapid waning of tolerance after discontinuing the endotoxin injections despite the presence of specific agglutinins. In addition, since tolerance was found to be accompanied by enhanced blood clearance of endotoxin and since "blockade" of the reticuloendothelial system (RES) with thorotrast retarded this enhanced clearance and "abolished" tolerance, it was proposed that tolerance was based upon enhanced RES uptake of circulating endotoxin, thereby protecting more susceptible tissues from injury. This hypothesis was supported by Morgan (1948), who also concluded that pyrogenic tolerance in rabbits was independent of circulating antibody because it extended to heterologous endotoxins, "almost" disappeared while circulating antibody to the toxin still remained elevated, and could be evoked with poorly antigenic, i.e., phenol-extracted, endotoxins. Despite this conclusion, the first indication appears in Morgan's report that O-specific antibody may play some role in pyrogenic tolerance since in one set of experiments, tolerant animals tested with heterologous endotoxins consistently developed more fever than would have been expected if no O specificity of tolerance existed.

The hypothesis that endotoxin tolerance was nonimmunologic and mediated by nonspecific enhancement of RES uptake of circulating endotoxin was extended by Freedman (1959, 1960a,b), who demonstrated that tolerance to the pyrogenic and lethal activities of endotoxin could be passively transferred with serum from endotoxin-tolerant rabbits, and that such passive protection appeared relatable to enhanced RES phagocytic activity as assessed by colloidal carbon clearance.

Developing in parallel with the above concepts was the hypothesis that endotoxin did not act directly to cause fever but rather acted by releasing endogenous pyrogens (EP) from host cells; the EP entered the circulation and then acted upon the thermoregulatory centers to evoke fever (Atkins, 1960). For almost two decades after Beeson (1948) reported the extraction of EP from rabbit polymorphonuclear leuko- cytes, these cells were regarded as the major source of EP in response to endotoxin. Indeed, Herion *et al.* (1961), using nitrogen mustard in rabbits to produce agranulocytosis, concluded that polymorphonuclear leukocytes were the sole source of EP after intravenous endotoxin ad- ministration. Meanwhile, Atkins and Wood (1955) reported that endo- toxin-tolerant rabbits failed to develop circulating EP when given endotoxin intravenously, but did so after pyrogenic tolerance was "re- versed" with thorotrast; moreover, repeated intravenous administration of EP per se did not lead to tolerance (Atkins, 1960). Based upon all these observations, by the early 1960s the hypothesis appeared secure that pyrogenic tolerance was mediated by nonspecific enhancement of RES phagocytic activity, which diverted the circulating endotoxin away from those cells (i.e., polymorphonuclear leukocytes) responsible for EP production (Atkins, 1960).

B. Concept of Hepatic Kupffer Cell as a Major Target for EP Production

A number of observations, however, began to appear that did not fit the above hypothesis. The important studies of Page and Good (1957) with respect to pyrogenic tolerance were generally ignored. These investigators observed that some patients with agranulocytosis, in the absence of circulating neutrophils, developed a febrile response to typhoid vaccine "at least as high as that produced when normal numbers of neutrophils were present in the blood." It was concluded that "injury to neutrophils with liberation of endogenous pyrogen plays no important role in development of fever following the intravenous injection of pyrogens in man." Moreover, when repeated intravenous injections of typhoid vaccine were given, it was concluded that "refrac- toriness to endotoxin develops in the complete absence of neutrophils in the circulating blood or in the blood forming tissues and conse- quently demonstrates that these cells are not essential to this defense reaction." This dilemma seemed resolved when Atkins *et al.* (1967) and Hahn *et al.* (1967) reported that rabbit peritoneal macrophages could produce EP, while Bodel and Atkins (1967) demonstrated that human blood monocytes also synthesized EP. These findings, in conjunction with those of Page and Good (1957) suggest that fever after intravenous endotoxin may be based primarily upon EP release from monocyte/macrophage populations, and that pyrogenic tolerance may therefore result from resistance of these cells to EP synthesis and/or release. Since the RES, in particular the Kupffer cell, is the primary site of localization of intravenously injected endotoxin (Braude *et al.*, 1955; Cremer and Watson, 1957), macrophages of the RES, especially the Kupffer cells themselves, could comprise the major source of EP. If this were true, then pyrogenic tolerance would not be based upon enhanced uptake of endotoxin by the RES, but rather upon reduced responsiveness of the RES per se to the synthesis and/or release of EP in response to endotoxin. Several studies suggest that this indeed is the case.

Fukuda and Murata (1964, 1965) examined the pyrogenic activity of liver extracts from dogs given intravenous endotoxin and concluded "Thus, it is certain that the liver forms an easily extractable pyrogenic factor after taking up endotoxin in its reticuloendothelial cells. . . . The tolerance in fever response might result from the absence of the development of the hepatic pyrogen." Reasons for excluding endotoxin per se as the liver pyrogen appear valid since recipients were endotoxin-tolerant dogs and furthermore, the pyrogen failed to evoke alterations in circulating leukocyte counts characteristic of endotoxemia. However, since no histologic studies were performed, the conclusions could be critized on the basis that contaminating polymorphonuclear leukocytes were not excluded. This objection was overcome by the elegant studies of Dinarello *et al.* (1968) demonstrating that isolated rabbit hepatic Kupffer cells could be stimulated by endotoxin *in vitro* to generate EP. Moreover, isolated Kupffer cells (but not blood leukocytes or lung macrophages) from rabbits made tolerant to intravenously injected endotoxin were found refractory to EP release by endotoxin. It was suggested that hepatic Kupffer cells may comprise the major source of EP after intravenous endotoxin administration and that pyrogenic tolerance may be mediated by increased uptake of endotoxin by Kupffer cells which have become refractory to further EP release.

In vivo studies were performed by Greisman and Woodward (1970) to test the hepatic hypothesis of endotoxin tolerance. By means of chronic indwelling portal vein cannulae, endotoxin was perfused directly into the hepatic circulation of rabbits and pyrogenic responses compared with those after ear vein injection of the toxin. If pyrogenic tolerance were based upon increased RES uptake of endotoxin, thereby protecting other cells, than less fever would be expected after direct perfusion of the liver with the toxin. This did not occur; indeed, great-

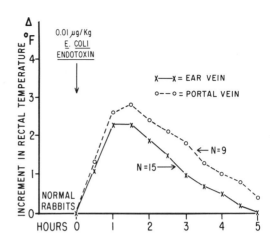

Figure 1. Comparative mean febrile response to 0.01 μg/kg *E. coli* endotoxin administered as an initial infusion via ear vein or via chronically implanted hepatic portal venous cannulae in healthy nontolerant rabbits. (From Greisman and Woodward. 1970.)

er fevers were elicited after intrahepatic infusion of endotoxin (Fig. 1). In contrast, after tolerance was induced by multiple daily ear vein injections of endotoxin, direct hepatic perfusion with the toxin resulted in marked reductions in fever. Proper placement of the portal vein cannulae was verified at the conclusion of each study by perfusion with trypan blue. The findings directly supported the concept that the hepatic Kupffer cell is the major target for endotoxin-induced EP production, and that pyrogenic tolerance is based upon refractoriness of this cell population to such mediator production. These conclusions have recently been reinforced by the studies of Haeseler *et al.* (1977) confirming the ability of isolated rabbit Kupffer cells to produce EP, and by Hanson *et al.* (1980) indicating that macrophages, not polymorphonuclear leukocytes, are the source of EP.

C. Mechanisms Underlying Reduced EP Production during Tolerance

1. Early Pyrogenic Refractory State

Since pyrogenic tolerance appears primarily dependent upon resistance of macrophages, particularly hepatic Kupffer cells, to EP production and/or release in response to intravenously injected endotoxins, the mechanisms underlying such resistance require consideration. If endotoxin is administered to rabbits as a single intravenous injection and retesting performed at varying intervals, pyrogenic tolerance can be observed to develop in a biphasic manner. Tolerance appears within 24 hr, wanes by 48 hr, then reappears after 72 hr (Mulholland *et al.*, 1965;

Greisman *et al.*, 1969b; Milner, 1973). These two distinct phases have
been termed *early* and *late phase pyrogenic tolerance* (Greisman *et al.*,
1969b).

The early phase of tolerance was first noted by Braude *et al.*
(1958), who observed that rabbits exhibited pyrogenic tolerance within
several hours after an initial intravenous endotoxin injection. The
mechanisms underlying this type of tolerance have been intensively
studied using continous intravenous infusions of endotoxin (Greisman
and Woodward, 1965; Greisman *et al.*, 1966). When the toxin is given
in this manner to rabbits or to human volunteers, the initial fever (and
in man the accompanying subjective discomfort, i.e., headache, chills,
myalgia, nausea) begin to wane after several hours despite the continuing infusion (Fig. 2). Volunteers characteristically express disbelief that
the endotoxin is still being infused. This early phase tolerance can be
completely overlooked if endotoxin is given to humans in the usual
fashion as a single intravenous bolus at daily intervals (Fig. 3). The early pyrogenic tolerance state evoked by continuous intravenous infusions

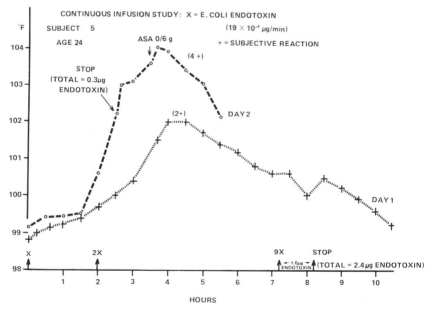

Figure 2. Typical pyrogenic response to an initial continuous intravenous infusion of endotoxin in a healthy volunteer. Note the loss of the early tolerant state when the infusion
is resumed on day 2. (From Greisman and Woodward, 1965, *J. Exp. Med.* **121**:923.)

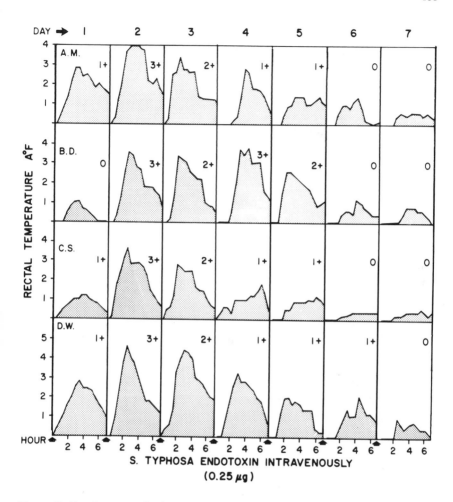

Figure 3. Development of tolerance to daily intravenous injections of endotoxin in healthy volunteers. Subjective toxic responses (headache, myalgia, chills, nausea) are graded from 0 to 4+. Note that tolerance does not appear to develop for several days by this method of administration and that responses are actually enhanced before tolerance develops.

of endotoxin appears based upon acquired refractoriness to endotoxin-induced stimulation of EP synthesis and/or release as evidenced by the following characteristics:

1. It develops within hours.
2. It is transient and requires continuing endotoxin infusions (or closely spaced injections) for maximum maintenance. In man, if the initial infusion is discontinued after tolerance first ap-

pears and is then resumed the following day, volunteers will have regained full responsiveness. Indeed, they usually hyper-react at this time (Fig. 2).

3. It is specific for endotoxins as a class. No tolerance occurs to other pyrogens such as staphylococcal enterotoxin, influenza virus, or tuberculin in specifically sensitized animals (provided massive doses of endotoxin are not given). Thus, EP precursor stores are not exhausted.

4. No specificity exists between endotoxins from diverse bacterial species, i.e., no O specificity is evident.

5. It is not associated with detectable increments in antiendotoxin antibodies and can be readily induced in hosts with impaired antibody synthesis (e.g., splenectomized or 6-mercaptopurine-treated rabbits).

6. Within the sensitive dose–response range, the level of tolerance is directly proportional to the magnitude of the initial pyrogenic response.

7. It is relative, and can be overcome by increasing the rate of endotoxin infusion.

8. It cannot be transferred with plasma.

9. It cannot be overcome with normal plasma or whole blood.

10. It is not associated with circulating granulocytopenia; in fact, granulocytosis occurs by the time tolerance appears.

11. It develops despite detectable circulating levels of endotoxin.

12. Circulating EP become undetectable.

13. Responsiveness persists to preformed EP.

14. It is not associated with augmented ability of liver homogenates or plasma to inactivate endotoxin.

The pyrogenic tolerance that characteristically occurs in rabbits 24 hr after a single intravenous injection of endotoxin exhibits many of the characteristics of the continuous intravenous infusion-induced tolerance described above (Greisman et al., 1969b; Milner, 1973).

2. Enhanced RES Detoxification of Endotoxin

In contrast to the early pyrogenic refractory state described above, if the intravenous endotoxin infusions are continued for several days in rabbits, or given repetitively as single daily intravenous injections, the RES acquires enhanced ability to detoxify endotoxin. Rutenburg et al. (1965) initially demonstrated that detoxification of endotoxin (measured by testing the activity of rabbit splenic homogenates) increased appreciably after repetitive injections of endotoxin, waned within 10 days after the toxin was discontinued despite elevated antiendotoxin antibody titers, and increased again within several days upon resumption of endotoxin administration. These investigators considered such enhanced detoxification by the RES to be the major mechanism underlying endotoxin tolerance and suggested that it represented an "adaptive enzyme" response. These findings were confirmed and extended by Trejo and DiLuzio (1972) using homogenates of mouse livers and spleens. Thus, at least two nonimmunologic mechanisms appear to mediate pyrogenic tolerance, both presumably acting to render RES macrophages less responsive to the EP-generating action of endotoxin. One occurs very rapidly and does not appear to entail enhanced RES detoxification of endotoxin; the other is more delayed and is associated with such enhanced inactivation. It is possible that enhanced detoxification by the RES underlies both phases but that insensitivity of the assay precludes its early detection. In any case, these cellular mechanisms could account for the capability of immunosuppressed hosts to develop pyrogenic tolerance, i.e., humans with agammaglobulinemia (Good and Varco, 1955; Good and Zak, 1956), rabbits treated with 6-mercaptopurine (Wolff *et al.*, 1965b), and splenectomized rabbits and humans (Greisman *et al.*, 1975), although some tolerance-damping effects due to the impaired antibody synthesis can nevertheless be demonstrated upon careful serial quantitative study (Greisman *et al.*, 1975).

The importance of enhanced detoxification of endotoxin in pyrogenic tolerance is supported by the effects of interfering with this mechanism. By use of sublethal doses of carbon tetrachloride, Farrar *et al.* (1968) demonstrated that normal rabbits became about 40 times more susceptible to the pyrogenic activity of endotoxin when tested 2 days afterwards. Five days after the carbon tetrachloride, febrile responses returned to baseline. Normal liver homogenates were capable of detoxifying endotoxin *in vitro*; those from carbon tetrachloride-treated rabbits were not. Other than carbon tetrachloride, hepatotoxins that, like carbon tetrachloride, depressed RES phagocytic function but that did not interfere with fatty acid oxidation did not significantly enhance susceptibility of animals to endotoxin nor destroy the ability of liver homogenates to inactivate endotoxin. Thorotrast also impairs RES capacity to detoxify endotoxin (Wiznitzer *et al.*, 1960). This effect may, in part at least, account for its ability to augment EP production in both normal and tolerant animals as will be subsequently discussed.

3. Antiendotoxin Antibodies

In addition to the above mechanisms of pyrogenic tolerance, which appear to be cellular and which require closely spaced or continuous endotoxin infusions for maximum maintenance, humoral mechanisms also participate. To demonstrate their effects in *actively* immunized animals, the cellular mechanisms have been deliberately permitted to wane, i.e., testing has been delayed for 1 or more weeks after a single intravenous injection of endotoxin or series of such injections; tolerance obtained in this manner has been referred to as *late phase tolerance* (Greisman *et al.*, 1969b). Based upon studies in rabbits and healthy human volunteers, the following characteristics of late phase pyrogenic tolerance have been delineated (Greisman and Young, 1969; Greisman *et al.*, 1964b, 1969b, 1973, 1975):

1. The onset is delayed requiring several days to appear after an initial single intravenous injection of endotoxin.
2. It is enduring, persisting for weeks.
3. It is not directly related to the magnitude of the initial pyrogenic response. Tolerance in healthy volunteers on day 7 was found to be comparable regardless of whether they were intravenously immunized with 0.01 or 0.001 μg/kg *E. coli* endotoxin; the former dose elicited marked febrile and subjective toxic responses, whereas the latter elicited no reactions. The factor common to both doses of endotoxin was their ability to evoke rises in O antibody by the day of tolerance testing.
4. It is relatable to the immunogenicity of the endotoxin preparation. Noll and Braude (1961) demonstrated late phase pyrogenic tolerance in rabbits using a single intravenous injection of a chemically (lithium aluminum hydride) modified *E. coli* endotoxin that was nonpyrogenic but retained immunogenicity. This contrasts with Westphal's (1963) inability to induce late phase pyrogenic tolerance with a preparation of *E. coli* endotoxin treated with hydrogen peroxide to destroy its sugar components but not its pyrogenicity. "Late" tolerance also is not readily produced with poorly antigenic phenol-extracted endotoxin preparations.
5. If endotoxins from smooth gram-negative bacteria are given, it is highly, although not entirely, O specific.
6. It is associated with increments in antiendotoxin antibodies, and is impaired in hosts with impaired antibody-synthesizing capabilites.
7. It is transferable with serum and with both IgG and IgM fractions.
8. "Anamnestic" responses can be transferred with spleen cells from endotoxin-immunized donors. Such "anamnestic" tolerance extends to heterologous endotoxins, but is most marked against the O-specific endotoxin.
9. In man, it can be overcome in part by dividing the dose of endotoxin in two and administering the second portion 2 hr after

the first. This has been attributed to a Danysz reaction, wherein the initial dose of toxin binds a high proportion of protective antibody, allowing the second half to act with less antibody inhibition (Greisman *et al.*, 1964b). The Danysz phenomenon has recently also been demonstrated by Ogata and Kanamori (1978) during *in vitro* neutralization of endotoxin by its O-specific antiserum.

The above observations indicate that immunologic mechanisms, particularly O-specific antibodies, mediate late phase pyrogenic tolerance. These conclusions have been fully confirmed by Milner (1973), who performed extensive studies on the early and late phases of pyrogenic tolerance following a single intravenous injection of endotoxin in rabbits. Confirmation has also been provided by Ralovich *et al.* (1974), who observed that pyrogenic tolerance to a variety of Boivin endotoxin preparations 1 week after discontinuing a series of intravenous injections of gram-negative bacterial vaccines in rabbits was highly O specific.

The most definitive evidence for the capability of "O" specific immunologic mechanisms to confer late phase pyrogenic tolerance has recently been obtained in the laboratories of Lindberg, Svenson, and Greisman (1982) by immunizing rabbits with a pyrogen-free "O" antigenic dodecasaccharide prepared by cleaving the "O" specific polysaccharide side chains from *S. typhimurium* endotoxin by means of a bacteriophage glycanase and then coupling this haptene to human serum albumin. Despite the absence of lipid A and core polysaccharides in this immunogen, and the absence of any pyrogenic or limulus amoebocyte lysate gelation activity, immunized animals developed impressive pyrogenic tolerance to *S. typhimurium* endotoxin. That such tolerance was indeed based upon the immune response elicited to the "O" haptenic polysaccharide was proven by the "O" specificity of the response, i.e., the tolerance did not extend to endotoxin prepared from *S. thompson*.

Most studies of the late phase mechanisms of pyrogenic tolerance have not been performed by waiting for the cellular mechanisms to wane or by immunizing with endotoxin-derived haptenes lacking lipid A as indicated above, but by passive transfer of tolerant phase serum to normal animals. It is critically important that such passive transfer studies be conducted with serum or serum fractions that are pyrogen-free. If not, tolerance will result that is attributed to serum factors but that in fact is actually based upon the early phase tolerance-mediated mechanisms. (Since early phase tolerance is not O specific, endotoxin-contaminated serum or serum fractions will characteristically appear to transfer broad-spectrum tolerance.) This point is illustrated by one classic study in which serum from endotoxin-tolerant rabbits immunized with heat-killed *E. coli* O8 was fractionated on DEAE and

Sephadex G-200 columns and the fractions employed for passive transfer studies of endotoxin tolerance. No precautions to free the columns
of contaminating endotoxin were indicated, nor was the pyrogen content of the serum fractions specified. The findings were interpreted as
evidence for a key role of antibody to common toxophore antigens,
and the unimportance of O-specific antibody in the mediation of pyrogenic tolerance to endotoxin (Kim and Watson, 1965). However, if pyrogen-free precautions are employed, then as expected from the
characteristics of late phase tolerance, sera or serum fractions from rabbits immunized with endotoxins from smooth gram-negative bacteria
transfer pyrogenic tolerance that is consistently most marked against
the homologous O endotoxin (Greisman *et al.*, 1973; Milner, 1973).
Nevertheless, even with pyrogen-free precautions, some passive protection against the pyrogenic activity of heterologous endotoxins is also
evident, just as occurs during the actively induced late phase tolerance.
This was initially documented using rabbits given plasma from *E. coli*
endotoxin-immunized donors and tested with *S. typhosa* endotoxin
(Greisman *et al.*, 1963), as well as human volunteers given plasma from
S. typhosa endotoxin-immunized donors and tested with *Pseudomonas* endotoxin (Greisman *et al.*, 1964b). In quantitative studies, however, heterologous protection against endotoxin pyrogenicity could be demonstrated only by using considerably greater quantities of plasma than
required to provide homologous protection (Greisman *et al.*, 1973).
Passive transfer studies by Muller-Ruchholtz and Sprockhoff (1969)
support these findings. Moreover, neutralization of endotoxin pyrogenicity after *in vitro* preincubation with its homologous O antiserum, initially demonstrated by Nowotny *et al.* (1965), also exhibits high levels
of O specificity (Ralovich *et al.*, 1974; Ogata *et al.*, 1975).

The relatively low levels of pyrogenic tolerance to heterologous
endotoxins transferable with rabbit antiserum were found to be aug-
mentable to a small but significant degree if endotoxins from certain
rough mutants [Rd mutant of *S. typhimurium* and Rc mutant (J5) *E. coli*
O111] rather than those from the parent smooth strains, were used for
immunization (Greisman *et al.*, 1973). This principle was originally described by Braude and Douglas (1972) with relation to passive protection against the Shwartzman reaction, suggesting that in the absence of
O-specific terminal antigenic side chains, common core antigens are
unmasked and that antibodies to these common core antigens can provide broad-spectrum endotoxin protection. However, the level of heterologous protection against endotoxin pyrogenicity transferable with the
rough mutant antisera, although augmented above that provided by
antisera to smooth endotoxins, remained markedly less than that transferable by O-specific antisera to its homologous endotoxin (Greisman

et al., 1973). Thus, it is apparent from late phase active tolerance and the passive transfer studies that both O-specific and common core antibodies can contribute to pyrogenic tolerance and that their relative importance depends upon the antigenic structure of the endotoxin used for immunization as well as that used for testing for tolerance.

How antibodies directed against O-specific side chains of endotoxin and/or common core polysaccharides can reduce the pyrogenic activity of endotoxin and thus contribute to tolerance when that portion of the toxin that stimulates EP production resides in the lipid A moiety (Westphal, 1975) remains unclear. It does not appear based upon enhanced opsonization of the endotoxin so that it is more rapidly removed from the circulation by the RES, thereby protecting other more susceptible cells. As discussed above, RES macrophages, particularly hepatic Kupffer cells, appear primarily responsible for EP production in response to endotoxin. Therefore, antiendotoxin antibodies would be expected to reduce EP production by rendering endotoxin a less effective stimulus to these macrophages. This is consistent with the observations of Dinarello *et al.* (1968) that serum from endotoxin-tolerant, but not from normal rabbits, suppresses EP generation from isolated rabbit Kupffer cells exposed to endotoxin.

Recent studies by Rietschel and Galanos (1977) with antisera against purified lipid A are of great interest. High-titered antisera to purified lipid A did not confer pyrogenic tolerance upon normal rabbits. Moreover, even preincubation of purified lipid A with antiserum to lipid A minimally reduced its pyrogenicity. This contrasts with the striking neutralizing effects of O-specific antisera against its homologous endotoxin as cited earlier. Yet in rabbits recently injected intravenously with either native endotoxins or lipid A, administration of even small quantities (0.25 ml) of the lipid A antiserum led to marked pyrogenic tolerance that extended to all other endotoxins tested. Rietschel and Galanos suggested that this "combined" effect may mediate both the early pyrogenic refractory state and the cross-tolerance after repetitive injections of most endotoxins. Some problems with attributing such a major role to anti-lipid A antibodies are: (1) the immunogenicity of lipid A as it exists in the LPS molecule in its native form is very weak. This seems to be due to the presence of the polysaccharide moiety, which masks the lipid A and suppresses its immunogenicity (Galanos *et al.*, 1977); (2) mice differ from many animal species in that they do not contain naturally occurring anti-lipid A antibodies and do not respond readily to lipid A immunization, even using lipid A-coated bacterial antigens (Galanos *et al.*, 1977). Yet endotoxin tolerance to hypothermia (mice develop hypothermia rather than fever after endotoxin) and lethality can be readily induced in mice (Greer and Rietschel,

1978); (3) pyrogenic tolerance to endotoxin can be induced by daily or by continuous intravenous injections of endotoxin in immunosuppressed animals and humans almost as readily as in normal antibody producers as cited previously.

D. Mechanisms of Thorotrast Effect on Pyrogenic Tolerance

If the hypothesis is valid that pyrogenic tolerance to endotoxin depends primarily upon resistance of RES macrophages to EP release consequent to both cellular effects and to anti-endotoxin antibodies, one important phenomenon must be accounted for: Why does RES "blockade" with thorotrast "abolish" tolerance? Although thorotrast appeared to abolish pyrogenic tolerance (Beeson, 1946, 1947b), subsequent studies demonstrated that this does not actually occur. Rather, the reactivity of both normal and tolerant animals is markedly increased while the striking differences in pyrogenic responses between the normal and tolerant states persist (Greisman et al., 1963; Wolff et al., 1965a). The mechanism whereby thorotrast enhances the reactivity of both normal and tolerant animals, while leaving tolerance intact is not yet clear. Two major possibilities exist: Thorotrast may "blockade" the RES so that endotoxin is diverted to other EP-producing macrophages that have not become resistant. This concept was initially proposed by Dinarello et al. (1968) when it was observed that isolated lung macrophages and blood leukocytes from tolerant rabbits, in contrast to Kupffer and spleen cells, had not acquired resistance to endotoxin-induced EP generation. However, since normal rabbits also exhibit marked enhancement of pyrogenic responses after thorotrast, the concept of diversion of endotoxin from the RES to other EP-producing cells requires the additional assumption that in the nontolerant animal these other cells are intrinsically more responsive than those of the RES macrophages. Available data, both in vitro (Dinarello et al., 1968) and in vivo (Greisman and Woodward, 1970), do not support this required presumption. An alternative possibility is that thorotrast acts directly upon RES macrophages to enhance EP generation in response to endotoxin in both tolerant and normal animals without abolishing the underlying cellular and humoral mechanisms responsible for tolerance. Evidence is available to support this possibility. Although Dinarello et al. (1968) initially reported that in preliminary trials, isolated liver cells from endotoxin-tolerant rabbits given thorotrast did not regain their capacity to release EP in response to endotoxin, subsequent more extensive studies by Dinarello (1969) demonstrated that this actually did occur. As noted earlier, this effect may in part be related to thorotrast's ability to impair RES detoxification of endotoxin. Additional studies of EP production by isolated liver cells from nontolerant animals given thorotrast were inconclusive, since the upper dose–response range of EP production in response to endotoxin was attained prior to thorotrast (Dinarello, 1969). Further studies will therefore be required to determine if thorotrast can augment endotoxin-induced EP production by nontolerant Kupffer cells and thus account for the enhanced pyrogenic responses to circulating endotoxin it provokes in normal animals.

E. Pyrogenic Tolerance during Gram-Negative Bacterial Infections

It is not possible to determine whether the pyrogenic tolerance mechanisms remain functional during acute gram-negative bacterial infections in man. During more chronic febrile gram-negative bacterial infections, with relatively stable fever levels (e.g., typhoid fever and tularemia), tolerance studies have been carried out in informed volunteers (Greisman et al., 1964a, 1969a). During the latter portion of the incubation period of these illnesses, progressively increasing pyrogenic reactivity developed to intravenous endotoxin administration. Such hyperreactivity extended to endotoxins from heterologous gram-negative bacilli. The hyperreactivity persisted throughout illness and then declined progressively during the early phase of convalescence. These changes showed no correlation with the nonspecific phagocytic activity of the RES as measured by blood clearance of radioactively labeled aggregated human serum albumin (Fig. 4). The mechanisms of pyrogenic tolerance were found to remain functional within this hyperreactive framework. Thus, the early pyrogenic refractory state was readily inducible during illness; when continuous intravenous infusions of endotoxin were given, hyperreactive febrile and subjective toxic responses occurred, which after several hours began to subside towards the baseline febrile level despite the continuing endotoxin infusion. When daily single intravenous injections of endotoxin were given during illness, the increments in fever and subjective toxicity diminished progressively as in healthy subjects. Finally, when tolerance was induced prior to illness by repetitive daily intravenous injections of endotoxin, the tolerant volunteers remained significantly less responsive to the endotoxin during illness than did control nontolerant volunteers, although both groups became comparably ill and exhibited hyperreactivity to the endotoxin. These findings simulate those in rabbits given thorotrast, in that the pyrogenic tolerance mechanisms remain functional within a hyperreactive framework as discussed earlier. In typhoid fever, however, the evidence is even more compelling that the hyperreactivity is based upon enhanced RES responsiveness to EP generation, rather than RES "blockade" with diversion of endotoxin to more susceptible tissues, since the blood clearance rates of radiolabeled endotoxin are not impaired (Greisman et al., 1969a).

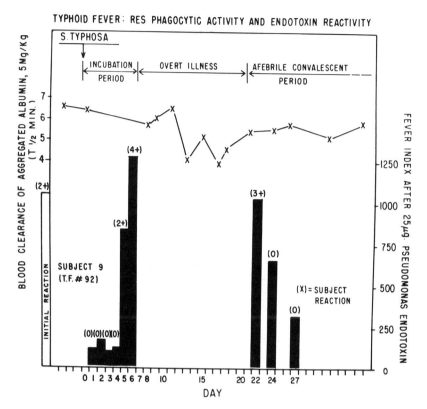

Figure 4. Typical pattern of response of informed volunteers to endotoxin during typhoid fever. Tolerance was induced before typhoid fever by daily intravenous injections of *Pseudomonas* endotoxin in increasing doses from 25 to 250 μg over a 30-day period. The initial control response is shown by the open bar on the left. Responses to 25 μg of the *Pseudomonas* endotoxin following oral infection with viable *S. typhosa* are shown by the solid bars. Note the hyperreactivity during the latter portion of the incubation period and its subsidence during early convalescence. The nonspecific phagocytic activity of the RES as measured by the blood clearance of radiolabeled aggregated human serum albumin showed no correlation with changes in endotoxin reactivity. Numbers in parenthesis above bars represent the intensity of subjective toxic responses (headache, chills, myalgia, nausea) graded from 0 to 4+.

II. Tolerance to Endotoxin Lethality

A. Evidence for RES Macrophages as a Major Endotoxin Target

As with pyrogenic tolerance, evidence is increasing that tolerance to the lethal effect of endotoxin is not mediated simply by accelerated RES uptake of circulating endotoxin which thereby protects other more susceptible tissues. Rather, as with fever, the RES macrophages per se appear to be a major target for endotoxin-induced lethality.

Heilman (1965) observed that cultured macrophages obtained from a variety of normal rabbit tissues were highly susceptible to the cytotoxic activity of endotoxins. This was confirmed and then extended by Glode *et al.* (1977) by the demonstration that macrophages from endotoxin-resistant C3H/HeJ mice exhibit marked resistance to the cytotoxic effects of endotoxin. Maier and Ulevitch (1981) recently demonstrated high levels of endotoxin-induced cytotoxicity for rabbit hepatic macrophages *in vitro* and *in vivo*.

Other studies by a number of laboratories have shown that when the phagocytic activity of the RES is stimulated with zymosan, triolein, glucan, graft-versus-host reactions, BCG, or *Brucella* infection, enhanced rather than diminished susceptibility develops to endotoxin lethality. Conversely, depression of RES phagocytic activity with ethyl stearate or methyl palmitate reduces, rather than augments, endotoxin lethality (Benacerraf *et al.*, 1959; Suter *et al.*, 1958; Cook *et al.*, 1980; Cooper and Stuart, 1961; Howard, 1961; Lemperle, 1966; Stuart and Cooper, 1962; Ueda *et al.*, 1966). From some of these observations, two decades ago, Cooper and Stuart (1961) concluded "Certainly it seems that the more phagocytes present in the liver and spleen the more sensitive are animals to endotoxin; on the other hand, decrease in cell numbers appears to have, if anything, a slightly beneficial effect on the resistance of the animals." More recent studies by Berendt *et al.* (1980) showed similar results. Mice bearing a syngeneic SA-1 sarcoma, or those treated with live mycobacteria or formalin-killed corynebacteria acquired a greatly increased susceptibility to the lethal effects of endotoxin that correlated with systemic activation of their macrophage system as measured by growth inhibition of *Listeria monocytogenes*.

Other lines of study also suggest that resistance of the RES to endotoxin, rather than enhanced RES clearance of circulating toxin, is the primary mechanism of tolerance to lethality. Normal rabbits were injected intravenously with an LD_{80} dose of endotoxin and 20 min later, after the RES had taken up approximately 50% of the toxin, the remainder in the circulation was rapidly reduced by exchange transfusion to levels seen in tolerant animals. Although such artificial removal of endotoxin from the circulation of the normal animals closely simulated the removal performed by the hyperphagocytic RES in tolerant animals (Fig. 5), mortality rates were not significantly reduced (Greisman and DuBuy, 1975). These findings have been confirmed by Gollan and McDermott (1979), who were unable to prevent mortality in dogs given endotoxin intravenously and exchange transfused 5 min later. These investigators concluded "Since the endotoxin particles are rapidly

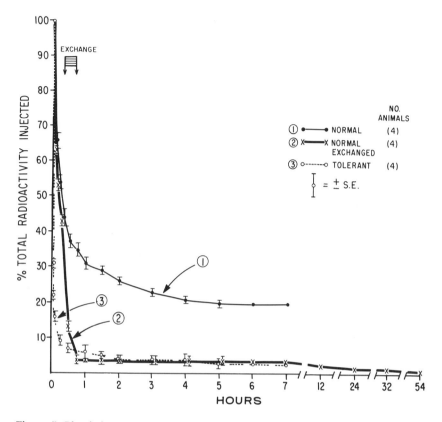

Figure 5. Blood clearance of 2500 μg of Cr[51]-tagged *E. coli* endotoxin in normal (curve 1). normal exchange-transfused (curve 2). and tolerant (curve 3). rabbits. Note the prolonged circulation of high concentrations of endotoxin in normal animals and the ability of exchange transfusions to simulate the endotoxin clearance curve of the tolerant animal. (From Greisman and DuBuy, 1975).

phagocytosed by the reticulo-endothelial system, even very early total blood exchange cannot dislodge them from their intracellular site." The above observations are also entirely consistent with those of Chedid *et al.* (1966), who reported that in contrast to endotoxins from smooth gram-negative bacteria, endotoxins from rough mutants were cleared equally rapidly from the blood of nontolerant and tolerant mice. It was concluded "It is, therefore, difficult to explain the increased resistance of the tolerant mouse against R (rough) endotoxin solely on the basis of its greater disappearance from the blood."

Agarwal *et al.* (1972) reported that following splenectomy, mice became more, rather than less, resistant to the lethal action of endotoxin.

Subsequently, Glode *et al.* (1976) demonstrated that C3H/HeN mice, after irradiation, became resistant to endotoxin lethality if reconstituted with histocompatible spleen cells from endotoxin-resistant C3H/HeJ mice. Conversely, histocompatible spleen cells from normal responder C3H/HeN mice, when used to reconstitute irradiated C3H/HeJ mice, enhanced their susceptibility to the lethal activity of endotoxin. These changes did not occur after transfer of spleen cells between identical strains. Subsequent studies with these chimera models using bone marrow cells instead of spleen cells yielded similar results (Michalek *et al.*, 1980). Macrophages appeared to best account for the cell type responsible for these effects (Rosenstreich *et al.*, 1977; Rosenstreich and Vogel, 1980).

B. Macrophage Mediator Production in Response to Endotoxin

That macrophages can be stimulated by endotoxin to produce and release a variety of mediators in addition to EP is now well established. Some that could exert detrimental effects include glucocorticoid-antagonizing factor (GAF) characterized by Moore *et al.* (1978), macrophage insulin-like activity factor (MILA), superoxides, lysosomal enzymes, procoagulant factor, and prostaglandins. Filkins (1980) suggested that while macrophage production of GAF in endotoxicosis may impair gluconeogenesis, MILA may augment tissue glucose uptake, the net effect resulting in the characteristic endotoxic hypoglycemia. Release of lysosomal enzymes from macrophages exposed to endotoxin has recently been suggested by Tanner *et al.* (1981) to contribute to hepatic parenchymal cell injury. Procoagulant factor release by macrophages exposed to endotoxin has been suggested to contribute to the disseminated intravascular coagulopathy (DIC) in typhoid fever (Hornick and Greisman, 1978). This concept has been extended by Maier and Ulevitch (1981) to include hepatic macrophage involvement in the initiation of DIC during endotoxemic shock. The latter investigators demonstrated that endotoxin is not only highly cytotoxic to isolated rabbit hepatic macrophages, but also stimulates selective increases in several cellular enzymes, including procoagulant factor. They concluded that "The combined effects of cytotoxicity and selective stimulation and release of mediators in LPS-stimulated hepatic macrophages may play a central role in the endotoxemic shock syndrome." Pabst and Johnston (1980) demonstrated that mouse macrophages exposed to endotoxins became primed to generate enhanced superoxide anion, and suggested that ". . . this same mechanism could also permit macrophages to mediate oxidative tissue injury." Rietschel *et al.* (1982) reviewed the endotoxin-induced macrophage mediators, addressing in particular the issue of prostaglandins. Peritoneal macrophages from endotoxin-tolerant mice were found refractory to prostaglandin production in response to endotoxin but not other stimuli (zymosan), suggesting that production of such mediators may contribute to endotoxin toxicity, and that tolerance may be related to macrophage resistance to such release.

C. Cellular Mechanisms Underlying Tolerance to Endotoxin Lethality

If, as with fever, the synthesis and release of macrophage mediators is an important contributing factor in endotoxin lethality, while tolerance is related, in part at least, to resistance to such mediator production, then the mechanisms underlying such resistance are critical. Agarwal and Berry (1968) demonstrated that both cellular and humoral mechanisms contribute to tolerance to endotoxin lethality in mice, and that activation of both these mechanisms could be prevented with the protein synthesis inhibitor actinomycin D. One of the cellular mechanisms of tolerance that develops after repetitive endotoxin injections, i.e., enhanced RES detoxification of endotoxin, has been considered earlier. DiLuzio and Crafton (1970) have presented considerable additional evidence that the altered susceptibility to endotoxin lethality produced by glucan and methyl palmitate is relatable to altered RES capacity to detoxify endotoxin and not to differences in RES uptake of the toxin. Whether the early phase of tolerance to endotoxin lethality that appears within 24 hr after an initial single intravenous injection of endotoxin in mice and guinea pigs is associated with enhanced RES detoxification remains to be determined. Of interest, such early lethal tolerance resembles early phase pyrogenic tolerance in that it exhibits no O specificity and has not been transferred with serum. Moreover, Urbaschek and Nowotny (1968) using a detoxified endotoxin (endotoxoid) in guinea pigs observed that such early lethal tolerance was not associated with detectable increments in antiendotoxin antibodies.

D. Humoral Mechanisms of Tolerance to Endotoxin Lethality

Boivin and Mesrobeanu (1938) initially reported that rabbit and horse antisera to Boivin endotoxins could passively transfer protection against lethality to mice and that such protection was strictly O specific. Zahl and Hutner (1944) observed that mice *actively* immunized with endotoxins from smooth gram-negative bacteria became resistant to heterologous endotoxin, but that protection against lethality was greatest against the homologous toxin. Wharton and Creech (1949) summarized their extensive studies on passive protection of mice against endotoxin lethality as follows: "It may be pointed out again that the gamma globulin fraction of rabbit antisera elicited by a series of injections of the polysaccharide (*Salmonella marcescens* endotoxin) showed strain specificity serologically and also protectively under the proper experimental conditions, although in the passive transfer, the use of an excessive amount of antibody toward one strain polysaccharide also protected mice to some extent against the lethal activity of the other strain polysaccharide." Abernathy (1957) observed two phases of tolerance to lethality after a single intraperitoneal injection of a Boivin-prepared endotoxin from *Brucella melitensis* into mice. The initial phase appeared within 24 hr, disappeared by day 3, then reappeared, reaching its maximum at 2 weeks; the tolerance to lethality persisted for at least 4 months, then declined substantially by 7 months. Such biphasic tolerance following a single intravenous injection of endotoxin, i.e., early and late phases, conforms to that described for pyrogenic tolerance as discussed earlier and is consistent with the concept of transient early cellular and delayed but more persistent antibody-mediated mechanisms of tolerance. Moreover, when late tolerance was tested at 1, 2, and 4 months after 3 weekly injections of a variety of Boivin endotoxin preparations, Abernathy observed that "protection against the homologous, or immunizing endotoxin persisted almost uniformly for two months, the only exception being mice immunized with shigella endotoxin" (wherein protection persisted 1 month). "In contrast, the range and degree of (late) heterologous protection was not impressive. . . ." Here, the importance of O-specific antibodies in late phase tolerance to endotoxin lethality appears clearly demonstrated. Freedman (1959) reported that rabbit antisera to Boivin-prepared endotoxins protected mice against homologous and heterologous endotoxin-induced lethality, but performed no quantitative studies on the relative levels of such protection and concluded that protection was not attributable to antibodies. Subsequently, Ziegler *et al.* (1973) reported that human antisera raised against heat-killed smooth gram-negative bacteria protected mice only against homologous endotoxin challenge; this contrasted with the earlier conclusion that O antisera prepared in rabbits provided protection to heterologous endotoxins that ". . . was always as good as that against the homologous endotoxin" (Davis *et al.*, 1969). McCabe *et al.* (1977), using lead-sensitized rats, observed that O-specific rabbit antisera provided the highest levels of passive protection against endotoxin lethality. Studies employing normal mice also indicate that passive protection against endotoxin lethality with rabbit antisera is consistently and significantly most marked with O-specific antiserum (Greisman, unpublished observations). Zaldivar and Scher (1979) were unable to protect mice against endotoxic lethality with murine-prepared O antiserum; however, their endotoxin challenge dose ($\geq 2\ LD_{100}$) may have been too overwhelming.

Considered collectively, the findings suggest that O antibodies are a major humoral protective factor in tolerance to lethality evoked by native endotoxins from smooth gram-negative bacteria but that antibodies to common antigens may also contribute.

Rabbit antisera to certain rough gram-negative bacterial mutants, e.g., Re 595 *S. minnesota* or J5 mutant of *E. coli* O111, have been reported capable of protecting mice against lethality from endotoxins prepared from many different smooth strains of gram-negative bacteria. Braude *et al.* (1977) have claimed that the latter antisera transfer protection equivalent to that afforded by O-specific antisera. This, however, has not been the experience of Greisman (unpublished observations) nor of McCabe *et al.* (1977). The importance of antibodies to lipid A in tolerance to endotoxin lethality when tolerance is induced with native endotoxins from smooth gram-negative bacteria remains to be defined, although McCabe *et al.* (1977) observed no passive protection using rabbit antisera to lipid A in lead-sensitized rats.

Although the Shwartzman reaction is not considered in the present review, the influence of antiendotoxin antibodies on this phenomenon will be mentioned, since their effect so closely parallels that described for the mitigation of the lethal and pyrogenic responses to endotoxin. Thus, the ability of endotoxins extracted from smooth gram-negative bacteria to prepare rabbit skin for the local Shwartzman reaction was preventable by pretreating these endotoxins with their homologous, but not with heterologous O-specific antisera (Radvany *et al.*, 1966). Similarly, when a "moderate" degree of endotoxin tolerance was actively induced by several daily intravenous injections of endotoxins from smooth gram-negative bacteria, tolerance to the local Shwartzman occurred that exhibited striking O specificity for the preparatory endotoxin (Kovats and Vegh, 1967). Employing O-specific antisera, protection against the local and the general Shwartzman reaction could be passively transferred; such protection exhibited both O and cross-specificity, the O-specific protection being more marked and of longer duration. Moreover, antisera against a rough mutant (J5) *E. coli* lacking O-specific side chains, also transferred protection (Braude and Douglas, 1972; Braude *et al.*, 1973). Ng *et al.* (1974) observed that anti-rough mutant sera to Re 595 *S. minnesota*, when preincubated with endotoxin, eliminated the Shwartzman skin preparatory activity of endotoxins extracted from the same or a closely related rough mutant strain of *Salmonella*; definite, but less marked neutralization occurred with endotoxin from the parent smooth *S. minnesota*.

That antibodies of any type are not essential for high levels of tolerance to endotoxin is perhaps best illustrated by the studies of Zaldivar and Scher (1979) wherein B-lymphocyte-defective (CBA/N) mice exhibited high levels of tolerance to endotoxin lethality when tested 8 days after a single intravenous endotoxin injection despite their inability to produce any detectable antiendotoxin antibodies. These investigators concluded that the findings ". . . strongly suggest that factors other than antibody play a critical role in the development of tolerance to the lethal effects of endotoxin." It would appear that following a single large dose of endotoxin (doubling of the initial tolerance-inducing endotoxin dose resulted in approximately 80% lethality), the nonimmunological tolerance mechanisms may remain activated for longer periods than occurs with the considerably smaller doses used for assay of pyrogenic tolerance.

E. Role of Accelerated RES Clearance of Circulating Endotoxin

Based upon the evidence available from many laboratories, emphasis in this review has been placed on the primary importance of resistance of the RES macrophages to the cytotoxic and mediator-releasing actions of endotoxin for the induction of pyrogenic and lethal tolerance. Accelerated blood clearance, with augmented uptake of endotoxin by the RES, appears to be a secondary protective mechanism that becomes effective during tolerance *only after resistance of the RES has been activated.* Moreover, its importance probably varies, depending upon the particular response to endotoxin under consideration. With fever, the responsible mediator, EP, appears to derive primarily from the RES macrophages per se, particularly the hepatic Kupffer cells. In contrast, the lethal activity of endotoxin involves extensive reactions outside the RES since in addition to its cytotoxic and mediator-releasing actions on RES macrophages, endotoxin initiates tissue injury by interacting with plasma proteins (e.g., complement, coagulation proteins) and numerous tissues (e.g., platelets, endothelium, neutrophils, and basophil/mast cells) (Morrison and Ulevitch, 1978). Therefore, while augmented RES clearance of circulating endotoxin appears to serve mainly an ancillary protective mechanism in pyrogenic tolerance (Greisman and Hornick, 1973), its contribution to lethal tolerance would be expected to be considerably greater.

III. Summary

The mechanisms underlying induction of tolerance to two effects of gram-negative bacterial endotoxins—fever and lethality—have been considered. Tolerance to the pyrogenic activity of circulating endotoxin appears to be based primarily upon resistance of RES macrophages, particularly hepatic Kupffer cells, to the EP-generating activity of endotoxin. Such resistance can be induced either by direct interaction of RES macrophages with endotoxin or by antiendotoxin antibodies, which protect the macrophages against the EP-stimulating activity of the toxin. The direct cellular effects responsible for tolerance are transient and require closely spaced or continuous exposure to endotoxin for maximum maintenance. At least two such mechanisms appear to be involved. One develops within hours and does not involve detectable enhancement of RES detoxification of endotoxin; this mechanism has been termed the *early pyrogenic refractory state*. The other develops later and is associated with enhanced RES inactivation of endotoxin. Antibody-mediated protection, though also delayed, is more enduring. Protective antibodies comprise those that react with O-specific antigens, and others that react with common core antigens. When endotoxins from smooth gram-negative bacteria are used to induce pyrogenic tolerance, O-specific antibodies appear to be the most important humoral protective component; when endotoxins from rough gram-negative mutants are administered, anticore antibodies comprise the major protective component. The latter antibodies convey broad-spectrum protection to heterologous endotoxins, but such protection is not as effective as O-specific protection. Antibodies to lipid A do not confer pyrogenic tolerance to normal recipients, but do so when combined with a recent injection of endotoxin. The importance of these antibodies when native endotoxins from smooth gram-negative bacteria are employed for tolerance induction remains to be determined. Accelerated clearance of circulating endotoxin, with increased uptake by the RES, appears to be an ancillary mechanism of pyrogenic tolerance that acts by diverting the toxin away from responsive macrophages outside the RES. This mechanism, however, becomes effective only after the RES macrophages per se have acquired resistance to the EP-generating activity of the toxin.

Tolerance to endotoxin lethality differs fundamentally from pyrogenic tolerance in that many more target tissues are involved, with generalized cytotoxicity and release of numerous mediators. There is, however, considerable evidence that as with EP release, RES macrophages constitute one of the major targets for endotoxin lethality, and that tolerance to lethality is primarily dependent upon induction of resistance of these RES macrophages to endotoxin cytotoxicity and mediator-releasing activity. The mechanisms underlying such RES resistance appear similar to those activated during pyrogenic tolerance, although the nature of the early phase of tolerance to lethality that develops within 24 hr has not been as well characterized. Accelerated clearance of circulating endotoxin, with increased uptake by the RES, probably constitutes a much more important secondary mechanism in lethal than in pyrogenic tolerance to endotoxin, since in addition to its actions on macrophages, endotoxin initiates injurious effects by interacting with plasma proteins (e.g., complement and coagulation factors) and numerous other cells (e.g., platelets, endothelium, neutrophils, and basophil/mast cells). However, as with pyrogenic tolerance, tolerance to lethality based solely upon diversion of endotoxin away from these plasma proteins and other target tissues probably would not be effective if the RES macrophages per se failed to acquire resistance to the toxin.

Tolerance is obviously a complex phenomenon and many areas remain unclear. Nevertheless, from the above considerations it is apparent that both immunologic and nonimmunologic mechanisms are activated during tolerance induction. The importance of each mechanism varies according to the nature of the response being assessed and also with the injection schedule of endotoxin, its antigenicity, dosage, and immunologic competency of the host. The diverse and often apparently conflicting observations on endotoxin tolerance may be explicable from analysis of the interplay of these various mechanisms.

IV. References

Abernathy, R. S., 1957, Homologous and heterologous resistance in mice given bacterial endotoxins, *J. Immunol.* **78**:387–394.

Agarwal, M. K., and Berry, L. J., 1968, Effect of actinomycin-D on RES and development of tolerance to endotoxin in mice, *J. Reticuloendothelial Soc.* **5**:353–367.

Agarwal, M. K., Parant, M., and Parant, F., 1972, Role of spleen in endotoxin poisoning and reticuloendothelial function, *Br. J. Exp. Pathol.* **53**:485–491.

Atkins, E., 1960, Pathogenesis of fever, *Physiol. Rev.* **40**:580–646.

Atkins, E., and Wood, W. B., Jr., 1955, Studies on the pathogenesis of fever. II. Identification of an endogenous pyrogen in the blood stream following the injection of typhoid vaccine, *J. Exp. Med.* **102**:499–516.

Atkins, E., Bodel, P., and Francis, L., 1967, Release of an endogenous pyrogen *in vitro* from rabbit mononuclear cells, *J. Exp. Med.* **126**:357–384.

Beeson, P. B., 1946, Development of tolerance to typhoid bacterial pyrogen and its abolition by reticulo-endothelial blockade, *Proc. Soc. Exp. Biol. Med.* **61**:248–250.

Beeson, P. B., 1947a, Tolerance to bacterial pyrogens. I. Factors influencing its development, *J. Exp. Med.* **86**:29–38.

Beeson, P. B., 1947b, Tolerance to bacterial pyrogens. II. Role of the reticuloendothelial system, *J. Exp. Med.* **86**:39–44.

Beeson, P. B., 1948, Temperature-elevating effect of a substance obtained from polymorphonuclear leucocytes, *J. Clin. Invest.* **27**:524 (Abstract).

Benacerraf, B., Thorbecke, G. J., and Jacoby, D., 1959, Effect of zymosan on endotoxin toxicity in mice, *Proc. Soc. Exp. Biol. Med.* **100**:796–799.

Berendt, M. J., Newborg, M. F., and North, R. J., 1980, Increased toxicity of endotoxin for tumor-bearing mice and mice responding to bacterial pathogens: Macrophage activation as a common denominator, *Infect. Immun.* **28**:645–647.

Bodel, P., and Atkins, E., 1967, Release of endogenous pyrogen by human monocytes, *N. Engl. J. Med.* **276**:1002–1008.

Boivin, A., and Mesrobeanu, L., 1938, Recherches sur les antigenes somatique et sur les endotoxines des bacteries. IV. Sur l'action anti-endotoxique de l'anticorps O, *Rev. Immunol.* **4**:40–52.

Braude, A. I., and Douglas, H., 1972, Passive immunization against the local Shwartzman reaction, *J. Immunol.* **108**:505–512.

Braude, A. I., Carey, F. J., and Zalesky, M., 1955, Studies with radioactive endotoxin. II. Correlation of physiologic effects with distribution of radioactivity in rabbits injected with lethal doses of *E. coli* endotoxin labelled with radioactive sodium chromate, *J. Clin. Invest.* **34**:858–866.

Braude, A. I., Zalesky, M., and Douglas, H., 1958, The mechanism of tolerance to fever, *J. Clin. Invest.* **37**:880–881 (Abstract).

Braude, A. I., Douglas, H., and Davis, C. E., 1973, Treatment and prevention of intravascular coagulation with antiserum to endotoxin, *J. Infect. Dis.* **128**(Suppl.):157–164.

Braude, A. I., Ziegler, E. J., Douglas, H., and McCutchan, J. A., 1977, Antibody to cell wall glycolipid of gram-negative bacteria: Induction of immunity to bacteremia and endotoxemia, *J. Infect. Dis.* **136**(Suppl.):167–173.

Chedid, L., Parant, F., Parant, M., and Boyer, F., 1966, Localization and fate of ^{51}Cr-labeled somatic antigens of smooth and rough salmonellae, *Ann. N.Y. Acad. Sci.* **133**: 712–726.

Cook, J. A., Dougherty, W. J., and Holt, T. M., 1980, Enhanced sensitivity to endotoxin induced by the RE stimulant, glucan, *Circ. Shock* **7**:225–238.

Cooper, G. N., and Stuart, A. E., 1961, Sensitivity of mice to bacterial lipopolysaccharide following alteration of activity of the reticulo-endothelial system, *Nature (London)* **191**:294–295.

Cremer, N., and Watson, D. W., 1957, Influence of stress on distribution of endotoxin in RES determined by fluorescein antibody technic, *Proc. Soc. Exp. Biol. Med.* **95**:510–513.

Davis, C. E., Brown, K. R., Douglas, H., Tate, W. J., III, and Braude, A. I., 1969, Prevention of death from endotoxin with antisera. I. The risk of fatal anaphylaxis to endotoxin, *J. Immunol.* **102**:563–572.

DiLuzio, N.R., and Crafton, C.G., 1970, A consideration of the role of the reticuloendothelial system (RES) in endotoxin shock, in: *Shock: Biochemical, Pharmacological, and Clinical Aspects* (A. Bertelli and N. Beck, eds.), pp. 27–57, Plenum Press, New York.

Dinarello, C. A., 1969, The role of the liver in the production of fever, Doctoral thesis, Yale University School of Medicine, New Haven, Conn.

Dinarello, C. A., Bodel, P. T., and Atkins, E., 1968, The role of the liver in the production of fever and in pyrogenic tolerance, *Trans. Assoc. Am. Physicians* **81**:334–344.

Farrar, W. E., Jr., Eidson, M., and Kent, T. H., 1968, Susceptibility of rabbits to pyrogenic and lethal effects of endotoxin after acute liver injury, *Proc. Soc. Exp. Biol. Med.* **128**:711–715.

Favorite, G. O., and Morgan, H. R., 1942, Effects produced by the intravenous injection in man of a toxic antigenic material derived from *Eberthella typhosa*: Clinical, hematological, chemical and serological studies, *J. Clin. Invest.* **21**:589–599.

Filkins, J. P., 1980, Endotoxin-enhanced secretion of macrophage insulin-like activity, *J. Reticuloendothelial Soc.* **27**:507–511.

Freedman, H. H., 1959, Passive transfer of protection against lethality of homologous and heterologous endotoxins, *Proc. Soc. Exp. Biol. Med.* **102**:504–506.

Freedman, H. H., 1960a, Passive transfer of tolerance to pyrogenicity of bacterial endotoxin, *J. Exp. Med.* **111**:453–463.

Freedman, H. H., 1960b, Reticuloendothelial system and passive transfer of endotoxin tolerance, *Ann. N.Y. Acad. Sci.* **88**:99–106.

Fukuda, T., and Murata, K., 1964, Hepatic pyrogen and cortisone-antipyresis in canine endotoxin fever, *Nature (London)* **204**:690–691.

Fukuda, T., and Murata, K., 1965, Species difference in the febrile and the leukocytic response to endotoxin: Endogenous hepatic pyrogen in dogs, *Jpn. J. Physiol.* **15**:169–179.

Galanos, C., Lüderitz, O., Rietschel, E. T., and Westphal, O., 1977, Newer aspects of the chemistry and biology of bacterial lipopolysaccharides, with special reference to their lipid A component, in: *Biochemistry of Lipids II* (T. W. Goodwin, ed.), *International Review of Biochemistry*, Vol. 14, pp. 239–335, University Park Press, Baltimore.

Glode, L. M., Mergenhagen, S. E., and Rosenstreich, D. L., 1976, Significant contribution of spleen cells in mediating the lethal effects of endotoxin *in vivo*, *Infect. Immun.* **14**:626–630.

Glode, L. M., Jacques, A., Mergenhagen, S. E., and Rosenstreich, D. L., 1977, Resistance of macrophages from C3H/HeJ mice to the *in vitro* cytotoxic effects of endotoxin, *J. Immunol.* **119**:162–166.

Gollan, F., and McDermott, J., 1979, Total hypothermic blood exchange in acute endotoxin shock, *Resuscitation* **7**:229–236.

Good, R. A., and Varco, R. L., 1955, A clinical and experimental study of agammaglobulinemia, *J. Lancet* **75**:245–271.

Good, R. A., and Zak, S. J., 1956, Disturbances in gamma globulin synthesis as "experiments of nature," *Pediatrics* **18**:109–149.

Greer, G. G., and Rietschel, E. T., 1978, Inverse relationship between the susceptibility of lipopolysaccharide (lipid A)-pretreated mice to the hypothermic and lethal effect of lipopolysaccharide, *Infect. Immun.* **20**:366–374.

Greisman, S. E., and DuBuy, B., 1975, Mechanisms of endotoxin tolerance. IX. Effect of exchange transfusion, *Proc. Soc. Exp. Biol. Med.* **148**:675–678.

Greisman, S. E., and Hornick, R. B., 1973, Mechanisms of endotoxin tolerance with special reference to man, *J. Infect. Dis.* **128**(Suppl.):265–276.

Greisman, S. E., and Woodward, W. E., 1965, Mechanisms of endotoxin tolerance. III. The refractory state during continuous intravenous infusions of endotoxin, *J. Exp. Med.* **121**:911–933.

Greisman, S. E., and Woodward, C. L., 1970, Mechanisms of endotoxin tolerance. VII. The role of the liver, *J. Immunol.* **105**:1468–1476.

Greisman, S. E., and Young, E. J., 1969, Mechanisms of endotoxin tolerance. VI. Transfer of the "anamnestic" tolerant response with primed spleen cells, *J. Immunol.* **103**:1237–1241.

Greisman, S. E., Carozza, F. A., Jr., and Hills, J. D., 1963, Mechanisms of endotoxin tolerance. I. Relationship between tolerance and reticuloendothelial system phagocytic activity in the rabbit, *J. Exp. Med.* **117**:663–674.

Greisman, S. E., Hornick, R. B., and Woodward, T. E., 1964a, The role of endotoxin during typhoid fever and tularemia in man. III. Hyperreactivity to endotoxin during infection, *J. Clin. Invest.* **43**:1747–1757.

Greisman, S. E., Wagner, H. N., Jr., Iio, M., and Hornick, R. B., 1964b, Mechanisms of endotoxin tolerance. II. Relationship between endotoxin tolerance and reticuloendothelial system phagocytic activity in man, *J. Exp. Med.* **119**:241–264.

Greisman, S. E., Young, E. J., and Woodward, W. E., 1966, Mechanisms of endotoxin tolerance. IV. Specificity of the pyrogenic refractory state during continuous intravenous infusions of endotoxin, *J. Exp. Med.* **124**:983–1000.

Greisman, S. E., Hornick, R. B., Wagner, H. N., Jr., Woodward, W. E., and Woodward, T. E., 1969a, The role of endotoxin during typhoid fever and tularemia in man. IV.

The integrity of the endotoxin tolerance mechanisms during infection, *J. Clin. Invest.* **48**:613–629.

Greisman, S. E., Young, E. J., and Carozza, F. A., Jr., 1969b, Mechanisms of endotoxin tolerance. V. Specificity of the early and late phases of pyrogenic tolerance, *J. Immunol.* **103**:1223–1236.

Greisman, S. E., Young, E. J., and DuBuy, B., 1973, Mechanisms of endotoxin tolerance. VIII. Specificity of serum transfer, *J. Immunol.* **111**:1349–1360.

Greisman, S. E., Young, E. J., Workman, J. B., Ollodart, R. M., and Hornick, R. B., 1975, Mechanisms of endotoxin tolerance: The role of the spleen, *J. Clin. Invest.* **56**:1597–1607.

Haeseler, F., Bodel, P., and Atkins, E., 1977, Characteristics of pyrogen production by isolated rabbit Kupffer cells *in vitro*, *J. Reticuloendothelial Soc.* **22**:569–581.

Hahn, H. H., Char, D. C., Postel, W. B., and Wood, W. B., Jr., 1967, Studies on the pathogenesis of fever. XV. The production of endogenous pyrogen by peritoneal macrophages, *J. Exp. Med.* **126**:385–394.

Hanson, D. F., Murphy, P. A., and Windle, B. E., 1980, Failure of rabbit neutrophils to secrete endogenous pyrogen when stimulated with staphylococci, *J. Exp. Med.* **151**:1360–1371.

Heilman, D. H., 1965, The selective toxicity of endotoxin for phagocytic cells of the reticuloendothelial system, *Int. Arch. Allergy Appl. Immunol.* **26**:63–79.

Herion, J. C., Walker, R. I., and Palmer, J. G., 1961, Endotoxin fever in granulocytopenic animals, *J. Exp. Med.* **113**:1115–1125.

Heyman, A., 1945, The treatment of neurosyphilis by continuous infusion of typhoid vaccine, *Vener. Dis. Inform.* **26**:51–57.

Hornick, R. B., and Greisman, S. E., 1978, On the pathogenesis of typhoid fever, *Arch. Intern. Med.* **138**:357–359.

Howard, J. G., 1961, Increased sensitivity to bacterial endotoxin of F1 hybrid mice undergoing graft-versus-host reaction, *Nature (London)* **190**:1122.

Kim, Y. B., and Watson, D. W., 1965, Modification of host responses to bacterial endotoxins. II. Passive transfer of immunity to bacterial endotoxin with fractions containing 19S antibodies, *J. Exp. Med.* **121**:751–759.

Kovats, T. G., and Vegh, P., 1967, Shwartzman reaction in endotoxin-resistant rabbits induced by heterologous endotoxin, *Immunology* **12**:445–453.

Lemperle, G., 1966, Effect of RES stimulation on endotoxin shock in mice, *Proc. Soc. Exp. Biol. Med.* **122**:1012–1015.

Lindberg, A. A., Svenson, S. B., and Greisman, S. E., 1982, in preparation.

McCabe, W. R., Bruins, S. C., Craven, D. E., and Johns, M., 1977, Cross-reactive antigens: Their potential for immunization-induced immunity to gram-negative bacteria, *J. Infect. Dis.* **136**(Suppl.):161–166.

Maier, R. V., and Ulevitch, R. J., 1981, The response of isolated rabbit hepatic macrophages (H-Mφ) to lipopolysaccharide LPS), *Circ. Shock* **8**:165–181.

Michalek, S. M., Moore, R. N., McGhee, J. R., Rosenstreich, D. L., and Mergenhagen, S. E., 1980, The primary role of lymphoreticular cells in the mediation of host responses to bacterial endotoxin, *J. Infect. Dis.* **141**:55–63.

Milner, K. C., 1973, Patterns of tolerance to endotoxin, *J. Infect. Dis.* **128**(Suppl.):237–245.

Moore, R. N., Goodrum, K. J., Couch, R., Jr., and Berry, L. J., 1978, Factors affecting macrophage function: Glucocorticoid antagonizing factor, *J. Reticuloendothelial Soc.* **23**:321–332.

Morgan, H. R., 1948, Tolerance to the toxic action of somatic antigens of enteric bacteria, *J. Immunol.* **59**:129–134.

Morrison, D. C., and Ulevitch, R. J., 1978, The effects of bacterial endotoxins on host mediation systems, *Am. J. Pathol.* **93**:527–617.

Mulholland, J. H., Wolff, S. M., Jackson, A. L., and Landy, M., 1965, Quantitative studies of febrile tolerance and levels of specific antibody evoked by bacterial endotoxin, *J. Clin. Invest.* **44:**920–928.

Muller-Ruchholtz, W., and Sprockhoff, H., 1969, Spezifitat der toleranz bakterieller pyrogene bei kaninchen, *Z. Immunitaetsforsch. Allerg. Klin. Immunol.* **136:**50–61.

Nelson, M. O., 1931, An improved method of protein fever treatment in neurosyphilis, *Am. J. Syph.* **15:**185–189.

Ng, A. K., Chang, C. M., Chen, C. H., and Nowotny, A., 1974, Comparison of the chemical structure and biological activities of the glycolipids of *Salmonella minnesota* R595 and *Salmonella typhimurium* SL1102, *Infect. Immun.* **10:**938–947.

Noll, H. J., and Braude, A. I., 1961, Preparation and biological properties of a chemically modified *Escherichia coli* endotoxin of high immunogenic potency and low toxicity, *J. Clin. Invest.* **40:**1935–1951.

Nowotny, A., Radvany, R., and Neale, N., 1965, Neutralization of toxic bacterial O-antigens with O-antibodies while maintaining their stimulus on non-specific resistance, *Life Sci.* **4:**1107–1114.

Ogata, S., and Kanamori, M., 1978, Effects of homologous O-antibody on host responses to lipopolysaccharide from *Yersinia enterocolitica.*Neutralization of its pyrogenicity, *Microbiol. Immunol.* **22:**485–494.

Ogata, S., Yamamoto, A., and Kanamori, M., 1975, Effect of antiserum against bacterial endotoxin on febrile response to lipopolysaccharide in host, *Jpn. J. Med. Sci. Biol.* **28:** 312–315.

Pabst, M. J., and Johnson, R. B., 1980, Increased production of superoxide anion by macrophages exposed *in vitro* to muramyl dipeptide or lipopolysaccharide, *J. Exp. Med.* **151:**101–114.

Page, A. R., and Good, R. A., 1957, Studies on cyclic neutropenia, *Am. J. Dis. Child.* **94:** 623–661.

Radvany, R., Neale, N., and Nowotny, A., 1966, Relation of structure to function in bacterial O-antigens. VI. Neutralization of endotoxic O-antigens by homologous O-antibody, *Ann. N.Y. Acad. Sci.* **133:**763–786.

Ralovich, B., Emody, L., and Lang, C., 1974, Problems of anti-endotoxemic immunity, *J. Hyg. Epidemiol. Microbiol. Immunol.* **18:**439–446.

Rietschel, E. T., and Galanos, C., 1977, Lipid A antiserum-mediated protection against lipopolysaccharide and lipid A-induced fever and skin necrosis, *Infect. Immun.* **15:**34–49.

Rietschel, E. T., Schade, U., Jensen, M., Wollenweber, H. W., Lüderitz, O., and Greisman, S. E., 1982, Bacterial endotoxins: Chemical structure, biological activity and role in septicaemia, *Scand. J. Infect. Dis.* **31**(Suppl.):8–21.

Rosenstreich, D. L., and Vogel, S. N., 1980, Central role of macrophages in the host response to endotoxin, in: *Microbiology 1980* (D. Schlessinger, ed.), pp. 11–15, American Society for Microbiology, Washington, D.C.

Rosenstreich, D. L., Glode. L. M., Wahl, L. M., Sandberg, A. L., and Mergenhagen, S. E. 1977, Analysis of the cellular defects of endotoxin-unresponsive C3H/HeJ mice, in: *Microbiology 1977* (D. Schlessinger, ed.), pp. 314–320, American Society for Microbiology, Washington, D.C.

Rutenburg, S. H., Rutenburg, A. M., Smith, E. E., and Fine, J., 1965, On the nature of tolerance to endotoxin, *Proc. Soc. Exp. Biol. Med.* **118:**620–623.

Stuart, A. E., and Cooper, G. N., 1962, Susceptibility of mice to bacterial endotoxin after modification of reticuloendothelial function by simple lipids, *J. Pathol. Bacteriol.* **83:** 245–254.

Sultzer, B. M., 1968, Genetic control of leucocyte responses to endotoxin, *Nature (London)* **219:**1253–1254.

Suter, E., Ullman, G. E., and Hoffman, R. G., 1958, Sensitivity of mice to endotoxin after vaccination with BCG (Bacillus Calmette-Guérin), *Proc. Soc. Exp. Biol. Med.* **99**:167–169.

Tanner, A., Keyhani, A., Reiner, R., Holdstock, G., and Wright, R., 1981, Proteolytic enzymes released by liver macrophages may promote hepatic injury in rat model of hepatic damage, *Gastroenterology* **80**:647–654.

Trejo, R. A., and DiLuzio, N. R., 1972, Influence of endotoxin tolerance on detoxification of *Salmonella enteritidis* endotoxin by mouse liver and spleen, *Proc. Soc. Exp. Biol. Med.* **141**:501–505.

Ueda, K., Ueda, Y., and Imaizumi, K., 1966, Studies on relation between endotoxin hyperreactivity and phagocytic activity of the reticuloendothelial system, *Jpn. J. Med. Sci. Biol.* **19**:127–135.

Urbaschek, B., and Nowotny, A., 1968, Endotoxin tolerance induced by detoxified endotoxin (endotoxoid), *Proc. Soc. Exp. Biol. Med.* **127**:650–652.

Westphal, O., 1963, Discussion Section, Part VII, in: *Bacterial Endotoxins* (M. Landy and W. Braun, eds.), p. 620, Rutgers University Press, New Brunswick, N.J.

Westphal, O., 1975, Bacterial endotoxins, *Int. Arch. Allergy Appl. Immunol.* **49**:1–43.

Wharton, D. R. A., and Creech, H. J., 1949, Further studies of the immunological properties of polysaccharides from *Serratia marcescens (Bacillus prodigiosus)*. II. Nature of the antigenic action and the antibody response in mice, *J. Immunol.* **62**:135–153.

Wiznitzer, T., Better, N., Rachlin, W., Atkins, N., Frank, E. D., and Fine, J., 1960, *In vivo* detoxification of endotoxin by the reticulo-endothelial system, *J. Exp. Med.* **112**:1157–1166.

Wolff, S. M., Mulholland, J. H., and Ward, S. B., 1965a, Quantitative aspects of the pyrogenic response of rabbits to endotoxin, *J. Lab. Clin. Med.* **65**:268–275.

Wolff, S. M., Mulholland, J. H., Ward, S. B., Rubenstein, M., and Mott, P. D., 1965b, Effect of 6-mercaptopurine on endotoxin tolerance, *J. Clin. Invest.* **44**:1402–1409.

Zahl, P. A., and Hutner, S. H., 1944, Protection against the endotoxins of some gram-negative bacteria conferred by immunization with heterologous organisms, *Am. J. Hyg.* **39**:189–196.

Zaldivar, N. M., and Scher, I., 1979, Endotoxin lethality and tolerance in mice: Analysis with the B-lymphocyte-defective CBA/N strain, *Infect. Immun.* **24**:127–131.

Ziegler, E. J., Douglas, H., and Braude, A. I., 1973, Human antiserum for prevention of the local Shwartzman reaction and death from bacterial lipopolysaccharides, *J. Clin. Invest.* **52**:3236–3238.

8

Effect of LPS on Nonspecific Resistance to Bacterial Infections

Monique Parant

I. Introduction

Bacterial endotoxins may affect the pathogenicity of a wide variety of infectious agents, not only gram-negative and gram-positive bacteria, but also fungi, parasites, and viruses. Originally demonstrated by Rowley (1956) using bacterial cell walls, and shortly later by several workers with partially purified endotoxins (Landy and Pillemer, 1956; Dubos and Schaedler, 1956), this capacity to nonspecifically enhance the host's resistance to a bacterial infection created more excitement about these substances (reviewed in Cluff, 1971). Subsequently, during the 1960s, chemical and physical methods of modifying these complex toxins without destroying their protective properties were developed (reviewed in Sultzer, 1971). A significant but transient increase in resistance to infection has been observed when endotoxin is given several hours before initiation of infection, and very low doses of endotoxin were found to be effective (see Cluff, 1971). However, depending upon the dose and route of administration, and upon the time interval prior to the infectious challenge, a transient increase in susceptibility to infection was also reported in the first papers (Rowley, 1956; Landy and Pillemer, 1956; Dubos and Schaedler, 1956). This negative phase has been produced when endotoxin is injected at the time of challenge or shortly thereafter.

Identification of the biologically active parts of the lipopolysaccharide (LPS) and dissociation between the multiple biological properties

Monique Parant • Immunothérapie Expérimentale, Institut Pasteur, 75724 Paris Cedex 15, France.

have not yet been achieved in spite of the excellent insight into the chemistry of endotoxin components (see Westphal, 1975; Lüderitz, 1977), and the chemical alteration studies (see Nowotny, 1969; Sultzer, 1971). Much of the work leading to the knowledge that toxicity of endotoxin is a property of the lipid A region has been pioneered and carried out by Westphal, Lüderitz, and their associates (Lüderitz, 1977). However, the various responses elicited by endotoxins are not uniformly and strictly correlated with the toxicity (Milner et al., 1971). The great decrease in pyrogenicity and toxicity of the phthalylated derivative of LPS (SPLPS) without any demonstrable loss of fatty acid residues from the lipid moiety is still unexplained (McIntire et al., 1976). Since SPLPS retained the potency of LPS for the adjuvant effect, the data are consistent with a clear separation of toxicity from adjuvanticity (McIntire et al., 1976), but SPLPS had simultaneously lost the capacity to protect mice against a bacterial infection (Chedid et al., 1975). Moreover, evidence is accumulating that the polysaccharide chain (PS) of LPS may not only function as a solubilizing carrier of lipid A, but may modulate its actions and also have some activity of its own (Nowotny et al., 1975; Lüderitz, 1977). Both the nontoxic derivatives, SPLPS and PS preparations, maintained their ability to protect mice against lethal irradiation (Chedid et al., 1975; Nowotny et al., 1975). By using acetic anhydride and anhydrous sodium acetate, Freedman and Sultzer obtained a detoxified endotoxin derivative still capable of stimulating host resistance to a variety of bacterial infections (see Sultzer, 1971). A reduction in toxicity of at least 10-fold has also been demonstrated by various procedures that accomplished cleavage of ester bonds without decreasing the protective effect against infectious challenges. However, comparative dose assays were not always performed, making it uneasy to determine whether quantitative differences would exist (see Sultzer, 1971; Cluff, 1971). In addition, when using solubilized lipid A in complex form with BSA, it was found almost as toxic as the original LPS and equally able to protect mice against an unrelated bacterial infection (Galanos, 1975; Parant et al., 1976).

Endotoxin-induced stimulation of resistance to infections occurs after the same time interval as the hyporesponsiveness to the toxic effects of endotoxins themselves, classically termed endotoxin tolerance since the study of the febrile response reported by Beeson (1947a,b). Both responses are antibody independent at least in an early phase, and nonspecific activation of phagocytic cells was considered as playing a major role (reviewed in Chedid and Parant, 1971; Cluff, 1971; Parant et al., 1967). Subsequently, a number of additional characteristics of endotoxin–macrophage interactions have been extensively documented (reviewed in Morrison and Ryan, 1979). Most of the collected data suggest that humoral mediators are involved in LPS-induced macrophage activation (Männel *et al.*, 1980; Nowakowski *et al.*, 1980). The protective activity of tumor necrosis serum (TNS) against several bacterial infections also suggests that endotoxin may exert its influence through the production of humoral mediators (Parant *et al.*, 1980).

The discovery and characterization of the C3H/HeJ mouse strain, which is refractory to essentially all biological effects of LPS, has proved to be a powerful model for studying the mechanisms underlying responsiveness to endotoxins (see Berry, 1977; Morrison and Ryan, 1979). Moreover, LPS-induced activation of macrophages is impaired in the genetically resistant C3H/HeJ strain (Chedid *et al.*, 1976; Ruco and Meltzer, 1978; Russo and Lutton, 1977; Rosenstreich and Vogel, 1980). Clearly, some of the LPS unresponsiveness of this mouse strain is related to the activation defect that their macrophages possess, but their B lymphocytes are also defective in their response to LPS, and more particularly to the lipid A moiety free of contaminating protein (Watson and Riblet, 1974; Skidmore *et al.*, 1977). Therefore, defective activation could be a secondary expression of lymphocyte unresponsiveness (Glode *et al.*, 1977; Nowakowski *et al.*, 1980). In several experimental models, the refractoriness to LPS stimulation can be completely compensated for by the transfer of closely related histocompatible high-responder bone marrow cells (Glode *et al.*, 1976a; Galelli *et al.*, 1979). Concerning LPS-increased resistance to infection, it can be assumed that both radioresistant and radiosensitive cells are defective in the low-responder C3H/HeJ mouse strain (Galelli *et al.*, 1979).

LPS responsiveness in high- and low-responder progeny shows some obvious discrepancies according to the assay. Depending mainly upon the method used to assess endotoxin effect, investigators concluded that the gene expression was either dominant or codominant (see Morrison and Ryan, 1979). While the mitogenic effect of LPS on B lymphocytes has been clearly demonstrated as being under codominant gene control (Coutinho, 1976; Glode *et al.*, 1976b; Watson *et al.*, 1978), adjuvant and immunogenic responses in F_1 hybrids have been described as closely resembling the high-responder parental strain (Watson and Riblet, 1974; Skidmore *et al.*, 1975), or as being intermediate (Watson *et al.*, 1978); McGhee *et al.*, 1979). The inflammatory response and endotoxic shock in intact mice appear to follow a pattern of multifactorial quantitative inheritance with an intermediate responsiveness in F_1 and F_2 hybrids (Sultzer, 1972; Sultzer and Goodman, 1977; McGee *et al.*, 1979). However, after removal of adrenals, which renders LPS high-responder mice highly susceptible to the lethal effect of endotoxins or of lipid A, LD_{50} of a nonmitogenic LPS administered intravenously was clearly found to have a similar pattern as in the high-responder parent (Parant *et al.*, 1977). Discrepancies between the results in F_1 hybrids may perhaps be related to differences in the high-responder parent, to variations in immunizing antigens, to the lack of dose–effect responses in most reports, to the route of administration, and to the possibility that some endotoxin preparations contained any contaminating mitogenic factors. In contrast to the numerous data briefly mentioned, inheritance of LPS-enhanced nonspecific resistance to infection has not been fully investigated. In a previous study, experiments involving survival and *in vivo* counts of viable bacteria suggested a dominance of the high-responder phenotype in C3H progeny (Parant *et al.*, 1977). Additional data have forced us to modify our position and will be described later in this chapter. Nevertheless, in the defective C3H/HeJ mice like in newborn mice in which the lack of competent cells may explain their unresponsiveness to LPS stimulation, endotoxin-induced serum factor from BCG-infected mice (TNS) was fully protective against bacterial challenge with *Klebsiella* or *Listeria* organisms (Parant *et al.*, 1980).

II. Activity of Detoxified Endotoxins

Most attempts to eliminate or reduce toxicity of endotoxins have often resulted in diminution or complete loss of their immunostimulating activities (see Nowotny, 1969; Sultzer, 1971). After treatment with phthalic anhydride, the derivative SPLPS differs greatly from the untreated LPS in its acute toxicity and pyrogenicity (Chedid *et al.*, 1975; Pistole, 1975; McIntire *et al.*, 1976). The degree of detoxification is related to the number of phthalyl or succinyl groups introduced, but with equal numbers of groups per LPS molecule, SULPS (sodium succinyl LPS) is 10 times as pyrogenic and toxic as SPLPS (McIntire *et al.*, 1976). In a previous study, we confirmed that sodium phthalate derivatives of several endotoxin preparations were at least 10,000 times less toxic than the original ones when injected in adrenalectomized mice, whereas SULPS was only 1,000 times less toxic (Chedid *et al.*, 1975). Both detoxified preparations retained their ability to act as an immunologic adjuvant and to induce blastic transformation of mouse splenic B-dependent lymphocytes (Chedid *et al.*, 1975; McIntire *et al.*, 1976). The nontoxic alkylated LPS preparations maintained also the capacity for protecting mice against irradiation (Chedid *et al.*, 1975), although they were less active than the polysaccharide-rich nontoxic preparation obtained by Nowotny *et al.* (1975). Notwithstanding its immunostimulant properties, SPLPS had lost the ability to protect mice against a *Klebsiella* infection (Chedid *et al.*, 1975), nor did the nontoxic polysaccharide increase nonspecific resistance to this infectious challenge (Parant *et al.*, 1976).

Studies were further pursued with a purified LPS, the reference bacterial endotoxin of the Bureau of Biologics of the FDA prepared from *E. coli* (Rudbach *et al.*, 1976). The phthalylated derivative was effectively detoxified, and comparative study with the parent compound demonstrated a reduced toxicity in adrenalectomized mice (LD_{50} instead of 0.02 μg per mouse). Pyrogenicity in the rabbit by the intravenous route was decreased by a factor of 10^4 (minimal pyrogenic dose 25 μg/kg instead of 1.65 ng/kg) and by a factor of about 3×10^3 when administered by the intracerebroventricular route (respectively 30 ng instead of 10 pg). Similarly, using normal human volunteers, the intravenous injection of 0.5 μg/kg did not induce any significant change in temperature (the minimal pyrogenic dose of the parent endotoxin is 0.5 ng/kg) nor modification in leukocyte counts, cortisol or growth hormone concentrations (Elin *et al.*, 1981). All the doses of SPLPS are expressed as amounts of LPS.

Against a severe infectious challenge with *Klebsiella pneumoniae* organisms, *E. coli* SPLPS was found inactive as in the previous study (Chedid *et al.*, 1975). To evaluate the end points and perform a comparative assay with LPS and SPLPS, mice were infected with a smaller bacterial inoculum. The effectiveness of *E. coli* LPS, administered 24 hr before the challenge, was clearly demonstrated since 0.001 μg was still able to protect significantly treated animals whereas almost all the untreated controls died. To obtain such a protection with SPLPS, a much higher dose (about 3×10^3-fold) was necessary (Table 1). Thus, a similar relationship was found between the toxicity and the protective activity against infection when the detoxified SPLPS derivative was used.

Another type of infectious challenge was used to evaluate the capacity of the detoxified derivative to produce a negative phase when injected simultaneously with the bacteria. The minimal effective dose of LPS is 10 μg per mouse by the intravenous route. As shown in Table 1, this treatment strongly decreased the resistance of mice to an intramuscular challenge with *Pseudomonas* (the minimal lethal dose by this route is 8×10^6 organisms). Within the limits of this assay (Table 1), SPLPS did not render mice highly susceptible to the bacterial challenge at doses that have been shown to be adjuvant-active (Chedid *et al.*, 1975). *In vitro* assays confirmed that the production of lymphocyte-activating factor (LAF, IL 1), which is an important mediator of LPS adjuvancy, was obtained with monocytes or macrophages incubated either with LPS or with SPLPS, and within the same range of doses (Damais and Parant, unpublished results). It, therefore, appears that

Table 1. Influence of LPS and SPLPS on Nonspecific Resistance to Bacterial
Infections in Mice When Given Either as a Previous or as a Simultaneous Injection

Infectious challenge[a]	Treatment (i.v.)			Mortality on day +15	
	Agent[b]	Dose (μg)	Time	Dead/total	p[c]
2×10^4	Saline	—	Day −1	23/24	
K. pneumoniae	LPS	0.001	Day −1	11/24	< 0.01
i.v.		0.01	Day −1	2/24	< 0.01
		0.1	Day −1	2/24	< 0.01
	SPLPS	0.1	Day −1	16/16	
		1	Day −1	13/16	
		10	Day −1	6/24	< 0.01
		100	Day −1	1/24	< 0.01
2×10^5	Saline	—	Day 0	0/16	
P. aeruginosa	LPS	10	Day 0	10/16	< 0.01
i.m.	SPLPS	10	Day 0	0/16	
		100	Day 0	0/16	
8×10^5	Saline	—	Day 0	0/16	
P. aeruginosa	LPS	10	Day 0	14/16	< 0.01
i.m.	SPLPS	10	Day 0	0/16	
		100	Day 0	3/16	

[a] Klebsiella pneumoniae of capsular type 2 and Pseudomonas aeruginosa (strain A237) were obtained from the Institut Pasteur Collection.
[b] E. coli LPS and its sodium phthalyl derivative; in SPLPS, the phthalyl moiety constitutes 35% by weight, and thus doses are expressed as amount of LPS.
[c] As calculated by the adjusted χ^2 method.

enhancement of nonspecific resistance to bacterial infection is not strictly related to adjuvant activity. However, detoxification of LPS, which decreases the induction of a negative phase of hypersusceptibility to infection, abolishes also its stimulating effect on resistance.

III. Activity of LPS in Immunocompromised Mice

The susceptibility to bacterial infections of animals undergoing general immunosuppression is well known and these agents often cause severe infection due to organisms rarely encountered in other circumstances. Natural resistance in mice to Pseudomonas aeruginosa or K. pneumoniae was decreased at least 10-fold with a single dose of cyclophosphamide or hydrocortisone. By contrast, a nonimmunosuppressive and nonsteroid anti-inflammatory drug such as indomethacin produced a slight increase in resistance when given alone and did not lessen the protective activity of LPS (unpublished results). In animals treated with the immunosuppressive drugs, a previous injection of endotoxin was always capable of stimulating their resistance, delaying the mortality, and often inducing a definitive protection (Parant *et al.*, 1976). When given shortly after a lethal irradiation, endotoxin was still able to increase the survival time of animals challenged with *Klebsiella* but failed to ensure a definitive survival. However, when thymectomized mice were lethally irradiated and reconstituted with anti-Θ-treated bone marrow cells, their responsiveness to LPS stimulation was fully recovered. Like in nude mice, LPS-enhanced resistance of these T-cell-deprived animals was quite similar to that observed in normal mice (Parant *et al.*, 1976); Galelli *et al.*, 1979). These studies indicate that enhancement of natural immunity by LPS requires a first step involving radioresistant cells followed by a second stage of complete destruction of infecting organisms that does not occur in the absence of radiosensitive cells.

Patients, mainly children, who have undergone splenectomy are known to often suffer from severe infections with microorganisms possessing polysaccharide capsules. Splenectomized rats and mice show an increased susceptibility to small doses of pneumococci (see Van Wyck *et al.*, 1976). Neonatal splenectomy was reported to affect the thymus cell cooperative capacity but did not affect the bone marrow cell population (see Auerbach, 1978). In our experimental model, animals were splenectomized at 7 days of age and were challenged after 5 or 6 weeks in comparison with sham-operated controls. No mortality and normal development were observed in these mice. However, these adult neonatally splenectomized mice were highly susceptible to *Klebsiella* infection whereas the mimimal lethal dose of a *Pseudomonas* strain was unchanged (Table 2). When LPS was injected under optimal conditions, it was

Table 2. Protective Activity of LPS against Bacterial Infections in Neonatally Splenectomized Mice

Infectious challenge[a]	Treatment[b] (time)	Mortality (dead/total)			p[c]
		Day +3	Day +6	Day +9	
10 *K. pneumoniae* i.v.	Saline (day −1)	25/32	30/32	30/32	
	LPS 1 μg (day −1)	1/32	17/32	17/32	< 0.01
8 × 10⁶ *P. aeruginosa* i.m.	Saline (day −7)	14/16	14/16	14/16	
	LPS 10 μg (day −7)	0/16	0/16	0/16	< 0.01

[a] In normal adult DBA/2 mice, the minimal lethal doses are respectively 8×10^3 *K. pneumoniae* by the i.v. route and 10^7 *P. aeruginosa* by the i.m. route. *P. aeruginosa* was a clinical isolate resistant to antibiotics and kindly provided by Dr. Pouillart, Paris.

[b] By the i.v. route, LPS was *E. coli* reference endotoxin.

[c] As calculated by the adjusted χ^2 method.

demonstrated to be very efficient in stimulating resistance against both challenges (Table 2).

During early postnatal life, mice have been found to be immunologically deficient and to be extremely susceptible to bacterial infections. Their poor immune status may be related to suppressor cells and also to the lack of maturity of macrophages and lymphocytes (see Parant *et al.*, 1978). Regarding the usual *Klebsiella* infection already described, the onset of resistance is around 21 days of age in the several mouse strains tested. Little is known concerning the neonate's capacity to respond to nonspecific immunostimulants except that LPS is devoid of adjuvant activity. Previous studies on nonspecific resistance to *Klebsiella* infection have shown that endotoxin stimulates them weakly although it can increase their phagocytic activity (Parant and Sacquet, 1966). By contrast, it was later shown that a synthetic immunoadjuvant, muramyl dipeptide (MDP), is able to confer a good protection to 7-day-old mice subsequently infected with *Klebsiella* (Parant *et al.*, 1978). In these experiments, both the treatment and the infectious challenge were administered by the subcutaneous route, and under these conditions endotoxin was completely devoid of protective effect. However, when *Klebsiella* organisms were injected by the intravenous route, death was always delayed for a short time by endotoxin administration, but no definitive protection has ever been observed even with a small bacterial inoculum. As depicted in Fig. 1, mice challenged with *Klebsiella* could be protected by endotoxin from 14 days of age. Therefore, younger animals provided a good experimental model to demonstrate the effectiveness either of synthetic adjuvant (reviewed in Parant, 1979) or of TNS (Parant *et al.*, 1980).

IV. Influence of LPS on Resistance to Infections in Low-Responder Mice and Progeny

The inability of C3H/HeJ mice to be stimulated by LPS is not a characteristic of all C3H strains since the control C3HeB/Fe strain used in our studies was fully protected against a *Klebsiella* infection. Even a TCA-extracted endotoxin was found ineffective in the low-responder strain showing that B-cell activation would not be the only mechanism (Chedid *et al.*, 1976). Transfer of cells from the histocompatible C3HeB/Fe Orl mice into thymectomized and irradiated C3H/He Orl recipients showed that bone marrow cells completely restored the capacity of low-responder mice to be stimulated by LPS and fully protected against the *Klebsiella* challenge. Conversely, high-responder mice similarly reconstituted 1 month before with low-respond-

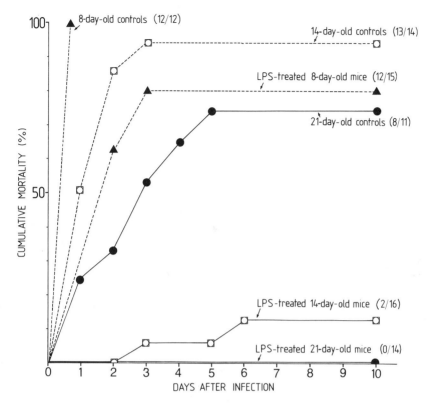

Figure 1. Influence of age on nonspecific resistance to a *Klebsiella* infection induced by LPS. Swiss mice were treated with 1 μg of *S. enteritidis* LPS by the subcutaneous route 1 day before being infected intravenously with 10^3 *K. pneumoniae* organisms. Numbers of dead animals/total are indicated in parentheses.

er bone marrow cells could not be definitively protected but retained their capacity to clear and destroy *Klebsiella* organisms during a few hours after injection. Therefore, several cell types could be incriminated and LPS unresponsiveness appears to be related to a defect of one radio resistant and long-lived cell type, and another one radiosensitive (Galelli *et al.*, 1979).

In vivo infection with *Mycobacterium bovis*, BCG strain, has been reported to render C3H/HeJ mice sensitive to some biological effects of LPS, particularly lethality, by compensating their profound defect through artificial activation of their macrophages (Vogel *et al.*, 1980). This type of experiment has to be reinvestigated in view of the results reported by Peavy *et al.* (1979). These latter authors did not find any change in the 50% lethal dose by using the intravenous route for both

BCG organisms and LPS administration. In the former work, describing a significant enhancement of sensitivity, the intraperitoneal route was chosen for previous infection and LPS challenge. Thus, in BCG-infected mice, the capacity of an additional stimulation with LPS was evaluated in low-responder mice, i.e., under conditions that induce the production of TNS in high-responder animals (see Hoffmann *et al.*, 1978). Briefly, 2 weeks after intravenous infection with BCG, two groups of C3H/He Orl mice received either 1 or 25 μg of LPS. In these mice, BCG alone induced a slight delay in mortality when subsequently infected with *Klebsiella* organisms but none of the animals survived. By contrast, the C3HeB/Fe Orl control mice were completely protected by BCG against the infectious challenge but simultaneously became extremely susceptible to the lethal effect of LPS like other high-responder animals ($LD_{100} < 1$ μg). In C3H/He mice, BCG did not induce an increased sensitivity to LPS and no more protection against *Klebsiella* was obtained after an additional injection of LPS. Surprisingly, serum obtained from a group of BCG-treated low-responder mice 2 hr after an injection of LPS was shown to induce a significant protection against the *Klebsiella* challenge after passive transfer in 8-day-old mice like TNS from other mouse strains (Parant *et al.*, 1980). However, protection was limited by comparison with data previously reported since serum (from C3H/He mice) did not protect young animals against a *Listeria* challenge. Therefore, the role of humoral factors in LPS-enhanced nonspecific resistance to bacterial infections would certainly be clarified by further purification of endogenous factors contained in TNS and comparative study of their activity in various test systems.

In an attempt to extend investigations on nonspecific resistance to bacterial infections in C3H mouse sublines and their progeny, several infectious challenges were used after considering the innate susceptibility in various groups. All the C3H mice were extremely susceptible to *K. pneumoniae* infection, more than the other mouse strains tested with the same infection, but *in vivo* bacterial enumerations indicated a still great sensitivity of C3H/He Orl mice. When challenged with *P. aeruginosa*, the LPS low-responder parental strain was found slightly less resistant than the others (minimal intravenous lethal dose = 2×10^7 cfu instead of 6×10^7 in C3HeB/Fe mice, and 4×10^7 in F_1 hybrids). Similar results were shown with the intramuscular route for infection. By contrast, C3H/He mice challenged with *Listeria monocytogenes* organisms exhibited a greater resistance than the other groups of C3H, and much more than Swiss mice. However, results reported in Table 3 clearly demonstrate that no improvement in survival was observed in C3H/He groups treated with LPS whatever the bacterial challenge was. Moreover, F_1 hybrids were completely protected like the high-responder pa-

Table 3. Influence of LPS on Resistance to *Listeria* or *Pseudomonas* Infection in C3H Mouse Strains

Infectious challenge, dose (CFU)[a]	Treatment (i.v.)		Mortality on day +15[b]		
	Agent, dose	Time	eB	He	(eB × He)F$_1$
Listeria i.v.					
6 × 10^4	Saline	Day −1	8/8	8/8	8/8
	LPS 1 μg	Day −1	1/8	8/8	1/8
10^4	Saline	Day −1	7/8	4/8	16/17
	LPS 0.1 μg	Day −1	0/8	5/8	0/8
	1 μg	Day −1	0/8	4/8	2/17
Pseudomonas i.v.					
6 × 10^7	Saline	Day −1	14/16	—	16/16
	LPS 10 μg	Day −1	4/16	—	6/16
3 × 10^7	Saline	Day −1	—	16/16	17/24
	LPS 10 μg	Day −1	—	15/16	5/24
Pseudomonas i.m.					
2 × 10^6	Saline	Day 0	0/16	9/16	2/16
	LPS 10 μg	Day 0	16/16	8/16	11/16
2 × 10^5	Saline	Day 0	0/16	0/16	0/16
	LPS 10 μg	Day 0	16/16	0/16	2/16

[a] *Listeria monocytogenes* was obtained from Dr. McKaness and injected i.v. *Pseudomonas aeruginosa* A237 was obtained from the Institut Pasteur Collection and injected either i.v. or i.m.
[b] In the following mouse strains: eB for C3HeB/Fe Orl, He for C3H/He Orl, and (eB × He)F$_1$ for their F$_1$ hybrids.

rental strain. In another type of experiment, the same three groups of animals were infected intramuscularly with *Pseudomonas* and some of them simultaneously received an intravenous injection of LPS. No negative phase was obtained in the low-responder animals whereas the other C3H subline was rendered more susceptible to the infectious challenge. In F$_1$ hybrids, results suggest the possibility of an intermediate responsiveness (Table 3).

A previous study had shown that *Corynebacterium granulosum* failed to significantly protect (C3HeB/Fe × C3H/He)F$_1$ hybrids against a *Klebsiella* challenge (like the low-responder strain) although the same treatment prevented mortality induced by *Listeria* infection (like the high-responder strain) (Parant *et al.*, 1977). In the same paper it was shown that the F$_1$ hybrids responded like the C3HeB/Fe parental strain when treated with LPS before the *Klebsiella* challenge. Even against a very small number of bacteria—the minimal lethal dose is about 10 *Klebsiella* organisms in the three groups of mice—LPS failed to protect the C3H/He mice. However, the high-responder strain was fully protected by the same treatment when the bacterial inoculum was increased up to 10^5 cfu. The protection conferred in F_1 hybrids by LPS was less marked as the challenge became more severe (see Table 4). Mortality in LPS-treated hybrids infected with *Klebsiella* could represent 80% of the mice with a much higher dose whereas the high-responder animals were still fully protected (Fig. 2). As shown in Fig. 2, a delay in mortality of several days was often observed in hybrids. Similarly, pretreatment with *Corynebacterium* could protect F_1 hybrids when they were infected with a small number of *Klebsiella* organisms, and this stimulation was still less effective than LPS (Table 4). The high-responder mouse strain was still protected when the number of viable organisms was increased (up to 3×10^4) whereas no F_1 hybrids survived (Table 4). Therefore, results reported in Table 4 show that the inability of C3H/He mice to respond to endotoxin does not depend on the virulence of the bacterial strain. The data obtained with *Corynebacterium* suggest, however, that unresponsiveness to LPS-enhanced resistance to infection may be related to a genetic defect in the capacity of C3H/He cells to respond to activation stimuli rather than to their unresponsiveness only to LPS as has also been proposed by Ruco and Meltzer

Table 4. Comparative Responsiveness to LPS or to *Corynebacterium* Stimulation in F_1 Hybrids and in the Parental C3H Mouse Strains Challenged with a *Klebsiella* Infection

K. pneumoniae challenge (i.v.)	Pretreatment[a] (i.v.)	Mouse strain[b]	Mortality on day $+15$ Treated	(Controls)	p
10^5 CFU	LPS	eB	0/8	(8/8)	< 0.01
	LPS	(eB × He)F_1	11/24	(24/24)	< 0.01
3×10^4 CFU	LPS	eB	0/8	(8/8)	< 0.01
	LPS	(eB × He)F_1	4/16	(16/16)	< 0.01
	C. granulosum	eB	8/16	(16/16)	< 0.01
	C. granulosum	(eB × He)F_1	24/24	(24/24)	—
10^4 CFU	LPS	eB	0/8	(8/8)	< 0.01
	LPS	(eB × He)F_1	0/8	(8/8)	< 0.01
	C. granulosum	eB	5/16	(16/16)	< 0.01
	C. granulosum	(eB × He)F_1	10/24	(23/24)	< 0.01
10^3 CFU	LPS	eB	0/8	(8/8)	< 0.01
	LPS	(eB × He)F_1	0/8	(8/8)	< 0.01
	C. granulosum	eB	1/8	(8/8)	< 0.01
	C. granulosum	(eB × He)F_1	1/8	(8/8)	< 0.01
10 CFU	LPS	eB	0/16	(16/16)	< 0.01
	LPS	He	16/16	(16/16)	—
	C. granulosum	eB	1/8	(8/8)	< 0.01
	C. granulosum	He	8/8	(8/8)	—

[a] Either 1 μg of LPS on day -1 or 300 μg of heat-killed *C. granulosum* on day -7.
[b] eB for C3HeB/Fe Orl, He for C3H/He Orl, and (eB × He)F_1 for their F_1 hybrids.

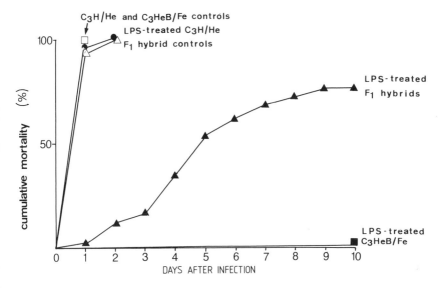

Figure 2. LPS-induced resistance in C3H/He Orl, C3HeB/Fe mice and their F_1 hybrids infected intravenously with 3×10^5 *K. pneumoniae* organisms. Data represent cumulative results of three identical experiments with groups of eight mice each.

(1978) and by Rosenstreich and Vogel (1980). Thus, neither the previous results (Parant *et al.*, 1977) nor the present ones permit a definite answer on the inheritance of LPS-induced resistance to infection. In such experiments studying resistance to a systemic challenge, both the responsiveness to LPS and the innate ability to clear and destroy invading bacteria are involved. Further analysis of this complex phenomenon requires additional *in vivo* and *in vitro* investigations with various bacterial challenges.

V. Conclusions

The relationships between the biochemical nature of endotoxins and of other constituents of the bacterial cell wall, and their biological properties have been the source of numerous investigations resulting in the synthesis of compounds having structural analogy with natural molecules. Moreover, as was underlined by Morrison and Ryan in their excellent review (1979), "defining the interaction of bacterial endotoxins with immunologically active cells has been one of the major factors contributing to current knowledge of the immune system and of its ability to function as our more potent defense against disease." Although it is not yet possible to attribute all the biological effects of this unique substance to a single entity, some interesting dissociations have already been described. It is of significant interest that nontoxic and nonpyrogenic LPS derivatives could retain potent adjuvant activity (McIntire et al., 1976; Chedid et al., 1975; Behling and Nowotny, 1977) and still induce CSF activity in mice (Nowotny et al., 1975). Modified endotoxins were also shown to be capable of initiating a lymphocyte proliferative response (see Morrison and Ryan, 1979) although this response could be attributed to lipid A, and was often linked to the adjuvanticity of endotoxin (Skidmore et al., 1975). The nonspecific increase in resistance to infections was also related to the presence of lipid A (Galanos, 1975) and was diminished or completely abolished in detoxified preparations of endotoxin as was demonstrated with SPLPS. However, SPLPS and experiments performed in C3H/HeJ mice indicate that when LPS is devoid of toxic effects, it is no longer able to induce a hypersusceptibility to infections by simultaneous administration. Further investigations of the relationships between structure and activity probably will lead to a better dissociation of biological properties of endotoxin as has been obtained with synthetic analogs of MDP (see Parant, 1979).

Because of the diversity of the biological activities of endotoxin on the host, it has been difficult to reveal precisely the mechanisms by which it produces any given effect. However, the expression of a common locus would appear to be a necessary component of many diverse endotoxic responses mediated by different cell types (Watson et al., 1978; Rosenstreich and Vogel, 1980). The involvement of cells is best understood for those effects that are clearly immunologic in nature, e.g., immunogenicity and adjuvanticity (Watson and Riblet, 1974; Coutinho, 1976; Glode et al., 1976b; Skidmore et al., 1975; Rosenstreich et al., 1977). The nature of the cellular involvement of other LPS-mediated effects such as lethality, resistance to infections, and tumor necrosis is still unclear (Glode et al., 1976a; Möller et al, 1978; Galelli et al., 1979; Männel et al., 1979). The amount of progress in the field of endotoxin research has been due in large part to the identification of the C3H/HeJ mouse strain (Heppner and Weiss, 1965; Sultzer, 1972; Vas et al., 1973). The importance of macrophages in endotoxin responsiveness (Bianco and Edelson, 1977; Ryan et al., 1979; O'Brien et al., 1980; Rosenstreich and Vogel, 1980) cannot rule out the participation of any radiosensitive bone marrow-derived cell (Glode et al., 1977; Galelli et al., 1979; Nowakowski et al., 1980). It is also pertinent to note that a variety of biologically active soluble factors or mediators are produced in vivo and in vitro by endotoxin. Some of them commonly associated with endotoxemia have already been involved in stimulation of nonspecific immunity (Freedman, 1960; Kampschmidt and Pulliam,

1975; Hoffmann *et al.*, 1978; Parant *et al.*, 1980). A number of these activities were found in TNS, and various factors could probably be purified. TNS was firstly described as causing a hemorrhagic necrosis of various tumors in mice (Carswell *et al.*, 1975). It has also been reported to possess properties of colony-stimulating factor and glucocorticoid-antagonizing factor (see Berry, 1977), lymphocyte-activating factor (IL 1) (Männel *et al.*, 1980) and various helper factors (Hoffmann *et al.*, 1978), interferon, cytotoxic and necrotizing factors (Kull and Cuatrecasas, 1981). The capacity of TNS to repair and to enhance the natural immunity of a genetically deficient or immature host (Parant *et al.*, 1980) suggests that soluble endogenous mediators, independent of humoral antibodies, are essential for the expression of nonspecific LPS-induced resistance to infections.

ACKNOWLEDGMENT

The author's work is supported by the Centre National de la Recherche Scientifique.

VI. References

Auerbach, R., 1978, Ontogeny of immune responsiveness in the absence of the spleen: a review, in: *Developmental and Comparative Immunology* Vol. 2, pp. 219–228, Pergamon Press, Elmsford, N.Y.

Beeson, P. B., 1947a, Tolerance to bacterial pyrogens. I. Factors influencing its development, *J. Exp. Med.* **86**:29.

Beeson, P. B., 1947b, Tolerance to bacterial pyrogens. II. Role of the reticuloendothelial system, *J. Exp. Med.* **86**:39.

Behling, U. H., and Nowotny, A., 1977, Immune adjuvancy of lipopolysaccharide and a non-toxic hydrolytic product demonstrating oscillating effects with time, *J. Immunol.* **118**:1905.

Berry, L. J., 1977, Bacterial toxins, *CRC Crit. Rev. Toxicol.* **1977**:239.

Bianco, C., and Edelson, P. J., 1977, Macrophage activation in C3H/HeJ mice, *Fed. Proc.* **36**:1263.

Carswell, E. A., Old, L. J., Kassel, R. L., Green, S., Fiore, N., and Williamson, B., 1975, An endotoxin-induced serum factor that causes necrosis of tumors, *Proc. Natl. Acad. Sci. USA* **72**:3666.

Chedid, L., and Parant, M., 1971, Hypersensitivity and tolerance to endotoxins, in: *Microbial Toxins*, Vol. V (S. Kadis, G. Weinbaum, and S. J. Ajl, eds.), pp. 415–459, Academic Press, New York.

Chedid, L., Audibert, F., Bona, C., Damais, C., Parant, F., and Parant, M., 1975, Biological activities of endotoxins detoxified by alkylation, *Infect. Immun.* **12**:714.

Chedid, L., Parant, M., Damais, C., Parant, F., Juy, D., and Galelli, A., 1976, Failure of endotoxin to increase nonspecific resistance to infection of lipopolysaccharide low-responder mice, *Infect. Immun.* **13**:722.

Cluff, L. E., 1971, Effects of lipopolysaccharides (endotoxins) on susceptibility to infections, in: *Microbial Toxins*, Vol. V (S. Kadis, G. Weinbaum, and S. J. Ajl, eds.), pp. 399–413, Academic Press, New York.

Coutinho, A., 1976, Genetic control of B-cell responses. I. Identification of the spleen B-cell defect in C3H/HeJ mice, *Scand. J. Immunol.* **5**:129.

Dubos, R. J., and Schaedler, R. W., 1956, Reversible changes in the susceptibility of mice to bacterial infections, *J. Exp. Med.* **104**:53.

Elin, R. J., Wolff, S. M., McAdam, K. P. W. J., Chedid, L., Audibert, F., Bernard, C., and Oberling, F., Properties of reference *Escherichia coli* endotoxin and its phthalylated derivative in humans, *J. Infect. Dis.* **144**:329.

Freedman, H. H., 1960, Passive transfer of tolerance to pyrogenicity of bacterial endotoxin, *J. Exp. Med.* **111**:453.

Galanos, C., 1975, Physical state and biological activity of lipopolysaccharides, toxicity and immunogenicity of the lipid A component, *Z. Immunitaetsforsch.* **149**:214.

Galelli, A., LeGarrec, Y., and Chedid, L., 1979, Transfer by bone marrow cells of increased natural resistance to *Klebsiella pneumoniae* induced by lipopolysaccharide in genetically deficient C3H/HeJ mice, *Infect. Immun.* **23**:232.

Glode, L. M., Mergenhagen, S. E., and Rosenstreich, D. L., 1976a, Significant contribution of spleen cells in mediating the lethal effects of endotoxin *in vivo*, *Infect. Immun.* **14**:626.

Glode, L. M., Scher, I., Osborne, B., and Rosenstreich, D. L., 1976b, Cellular mechanism of endotoxin unresponsiveness in C3H/HeJ mice, *J. Immunol.* **116**:454.

Glode, L. M., Jacques, A., Mergenhagen, S. E., and Rosenstreich, D. L., 1977, Resistance of macrophages from C3H/HeJ mice to the *in vitro* cytotoxic effects of endotoxin, *J. Immunol.* **119**:162.

Heppner, G., and Weiss, D. W., 1965, High susceptibility of strain A mice to endotoxin and endotoxin–red blood cell mixtures, *J. Bacteriol.* **90**:696.

Hoffman, M., Oettgen, H., Old, L., Mittler, R., and Hämmerling, U., 1978, Induction and immunological properties of tumor necrosis factor, *J. Reticuloendothelial Soc.* **23**:307.

Kampschmidt, R. F., and Pulliam, L. A., 1975, Stimulation of antimicrobial activity in the rat with leukocytic endogenous mediator, *J. Reticuloendothelial Soc.* **17**:162.

Kull, F. C., and Cuatrecasas, P., 1981, Preliminary characterization of the tumor cell cytotoxin in tumor necrosis serum, *J. Immunol.* **126**:1279.

Landy, M., and Pillemer, L., 1956, Increased resistance to infection and accompanying alteration in properdin levels following administration of bacterial lipopolysaccharides, *J. Exp. Med.* **104**:383.

Lüderitz, O., 1977, Endotoxins and other cell wall components of gram-negative bacteria and their biological activities, in: *Microbiology 1977* (D. Schlessinger, ed.), pp. 239–246, American Society for Microbiology, Washington, D.C.

McGhee, J. R., Michalek, S. M., Moore, R. N., Mergenhagen, S. E., and Rosenstreich, D. L., 1979, Genetic control of *in vivo* sensitivity to lipopolysaccharide: Evidence for codominant inheritance, *J. Immunol.* **122**:2052.

McIntire, F. C., Hargie, M. P., Schenck, J. R., Finley, R. A., Sievert, H. W., Rietschel, E. T., and Rosenstreich, D. L., 1976, Biologic properties of nontoxic derivatives of a lipopolysaccharide from *Escherichia coli* K235, *J. Immunol.* **117**:674.

Männel, D. N., Rosenstreich, D. L., and Mergenhagen, S. E., 1979, Mechanism of lipopolysaccharide-induced tumor necrosis: Requirement for lipopolysaccharide-sensitive lymphoreticular cells, *Infect. Immun.* **24**:573.

Männel, D. N., Farrar, J. J., and Mergenhagen, S. E., 1980, Separation of a serum-derived tumoricidal factor from a helper factor for plaque-forming cells, *J. Immunol.* **124**:1106.

Milner, K. C., Rudbach, J. A., and Ribi, E., 1971, General characteristics of endotoxin, in: *Microbial Toxins*, Vol. V (S. Kadis, G. Weinbaum, and S. J. Ajl, eds.), pp. 1–66, Academic Press, New York.

Möller, G. R., Terry, L., and Snyderman, R., 1978, The inflammatory response and resistance to endotoxin in mice, *J. Immunol.* **120**:116.

Morrison, D. C., and Ryan, J. L., 1979, Bacterial endotoxins and host immune responses, *Adv. Immunol.* **28**:293.

Nowakowski, M., Edelson, P. J., and Bianco, C., 1980, Activation of C3H/HeJ macrophages by endotoxin, *J. Immunol.* **125**:2189.

Nowotny, A., 1969, Molecular aspects of endotoxic reactions, *Bacteriol. Rev.* **33**:72.

Nowotny, A., Behling, U. H., and Chang, H. L., 1975, Relation of structure to function in bacterial endotoxins. VIII. Biological activities in a polysaccharide-rich fraction, *J. Immunol.* **115**:199.

O'Brien, A., Rosenstreich, D. L., Scher, I., Campbell, G. H., MacDermott, R. P., and Formal, S. B., 1980, Genetic control of susceptibility to *Salmonella typhimurium* in mice: Role of the LPS gene, *J. Immunol.* **124**:20.

Parant, M., 1979, Biological properties of a new synthetic adjuvant, muramyl dipeptide, *Semin. Immunopathol.* **2**:101.

Parant, M., and Sacquet, E., 1966, Augmentation de la résistance à l'infection du souriceau conventionnel ou axénique après une injection d'endotoxine, *C. R. Acad. Sci. Ser. D* **262**:1914.

Parant, M., Parant, F., Chedid, L., and Boyer, F., 1967, On the nature of some nonspecific host responses in endotoxin-induced resistance to infection, in: *The Reticuloendothelial System and Atherosclerosis* (N. R. DiLuzio and R. Paoletti, eds.), pp. 275–284, Plenum Press, New York.

Parant, M., Galelli, A., Parant, F., and Chedid, L., 1976, Role of B-lymphocytes in nonspecific resistance to *Klebsiella pneumoniae* infection of endotoxin-treated mice, *J. Infect. Dis.* **134**:531.

Parant, M., Parant, F., and Chedid, L., 1977, Inheritance of lipopolysaccharide-enhanced nonspecific resistance to infection and of susceptibility to endotoxic shock in lipopolysaccharide low-responder mice, *Infect. Immun.* **16**:432.

Parant, M., Parant, F., and Chedid, L., 1978, Enhancement of the neonate's nonspecific immunity to *Klebsiella* infection by muramyl dipeptide, a synthetic immunoadjuvant, *Proc. Natl. Acad. Sci. USA* **75**:3395.

Parant, M. A., Parant, F. J., and Chedid, L. A., 1980, Enhancement of resistance to infections by endotoxin-induced serum factor from *Mycobacterium bovis* BCG-infected mice, *Infect. Immun.* **28**:654.

Peavy, D. L., Baughn, R. E., and Muscher, D. M., 1979, Effects of BCG infection on the susceptibility of mouse macrophages to endotoxin, *Infect. Immun.* **24**:59.

Pistole, T. G., 1975, Biological activity of phthalated endotoxin, *Can. J. Microbiol.* **21**:1291.

Rosenstreich, D. L., and Vogel, S. N., 1980, Central role of macrophages in the host response to endotoxin, in: *Microbiology 1980* (D. Schlessinger, ed.), pp. 11–15, American Society for Microbiology, Washington, D.C.

Rosenstreich, D. L., Glode, L. M., and Mergenhagen, S. E., 1977, Action of endotoxin on lymphoid cells, *J. Infect. Dis.* **136**(Suppl.):239.

Rowley, D., 1956, Rapidly induced changes in the level of nonspecific immunity in laboratory animals, *Br. J. Exp. Pathol.* **37**:223.

Ruco, L. P., and Meltzer, M. S., 1978, Defective tumoricidal capacity of macrophages from C3H/HeJ mice, *J. Immunol.* **120**:329.

Rudbach, J. A., Akiya, F. I., Elin, R. J., Hochstein, H. D., Luoma, M. K., Milner, E. C. B.,

Milner, K. C., and Thomas, K. R., 1976, Preparation and properties of a national reference endotoxin, *J. Clin. Microbiol.* **3**:21.

Russo, M., and Lutton, J. D., 1977, Decreased *in vivo* and *in vitro* colony stimulating activity responses to bacterial lipopolysaccharide in C3H/HeJ mice, *J. Cell. Physiol.* **92**:303.

Ryan, J. L., Glode, L. M., and Rosenstreich, D. L., 1979, Lack of responsiveness of C3H/HeJ macrophages to lipopolysaccharide: The cellular basis of LPS-stimulated metabolism, *J. Immunol.* **122**:932.

Skidmore, B. J., Chiller, J. M., Morrison, D. C., and Weigle, W. O., 1975, Immunologic properties of bacterial lipopolysaccharide (LPS): Correlation between the mitogenic, adjuvant and immunogenic activities, *J. Immunol.* **114**:770.

Skidmore, B. J., Chiller, J. M., and Weigle, W. O., 1977, Immunologic properties of bacterial lipopolysaccharide (LPS). IV. Cellular basis of the unresponsiveness of C3H/HeJ mouse spleen cells to LPS-induced mitogenesis, *J. Immunol.* **118**:274.

Sultzer, B. M., 1971, Chemical modification of endotoxin, in: *Microbial Toxins*, Vol. V (S. Kadis, G. Weinbaum, and S. J. Ajl, eds.), pp. 91–126, Academic Press, New York.

Sultzer, B. M., 1972, Genetic control of host responses to endotoxin, *Infect. Immun.* **5**:107.

Sultzer, B. M., and Goodman, G. W., 1977, Characteristics of endotoxin-resistant low-responder mice, in: *Microbiology 1977* (D. Schlessinger, ed.), p. 304, American Society for Microbiology, Washington, D.C.

Van Wyck, D. B., Witte, M. H., Witte, C. L., and Strunck, R. C., 1976, Immunologic effects of partial and total splenectomy, in: *Immuno-aspects of the Spleen* (J. R. Battisto and J. W. Streilein, eds.), pp. 239–248, Elsevier/North-Holland, Amsterdam.

Vas, S. I., Roy, R. S., and Robson, H. G., 1973, Endotoxin sensitivity of inbred mouse strains, *Can. J. Microbiol.* **19**:767.

Vogel, S., Moore, R. N., Sipe, J. D., and Rosenstreich, D. L., 1980, BCG-induced enhancement of endotoxin sensitivity in C3H/HeJ mice. I. *In vivo* studies, *J. Immunol.* **124**:2004.

Watson, J., and Riblet, R., 1974, Genetic control of responses to bacterial lipopolysaccharides in mice. I. Evidence for a single gene that influences mitogenic and immunogenic responses to lipopolysaccharide, *J. Exp. Med.* **140**:1147.

Watson, J., Largen, M., and McAdam, K. P. W. J., 1978, Genetic control of endotoxic responses in mice, *J. Exp. Med.* **147**:39.

Westphal, O., 1975, Bacterial endotoxins, *Int. Arch. Allergy Appl. Immunol.* **49**:1.

9

Possibilities for Use of Endotoxin in Compromised Subjects

Richard I. Walker

I. Introduction—Approaches to Using Endotoxin in Medicine

The basic idea suggesting that pyrogenic agents may be beneficial to medicine is not new. Around 500 B.C., Parmenides, a Greek physician, stated, "If only I had the means of creating fever artificially, I should be able to cure all illnesses" (Westphal *et al.*, 1977). Over 2000 years later, the pyrogenic principle was found to be endotoxin. Fever therapy using whole gram-negative microorganisms had been in use for about 50 years when Boivin *et al.* (1933) described a method to purify the endotoxic material.

In the early 1940s, Morgan and his co-workers adapted the Boivin procedure to isolation and purification of an antigenic material from *Salmonella typhi* that, in sublethal amounts, induced fever and leukopenia in rabbits (Morgan, 1940, 1941; Cundiff and Morgan, 1941; Morgan *et al.*, 1943). He tested the idea that endotoxin might induce responses in man similar to those seen in animals and, therefore, be of use for inducing therapeutic febrile reactions for treatment of asymptomatic neurosyphilis (Favorite and Morgan, 1942). Injections containing 1–2 μg of endotoxin caused fevers from 103 to 105°F in patients

Richard I. Walker ● Naval Medical Research Institute, Bethesda, Maryland 20014. The opinions and assertions contained herein are the private ones of the author and are not to be construed as official or as reflecting the views of the Navy Department or the naval service at large. The experiments reported herein were conducted according to the principles set forth in the "Guide for the Care and Use of Laboratory Animals," Institute of Laboratory Animal Resources, National Research Council, DHEW, Publ. No. 74–23.

(Favorite and Morgan, 1942, 1946). This dose was relatively safe for humans, but had to be increased three- to fivefold in subsequent treatments to achieve similar results due to the development of endotoxin tolerance (Morgan, 1948).

Several approaches to further realization of the medical potential of purified endotoxin are available. One is to reduce the toxic aspects of the molecule but leave those necessary to achieve beneficial effects intact. Various biochemical and physical treatments have been used in attempts to obtain this goal. Many of these approaches will reduce toxicity with variable retention of beneficial properties (Johnson and Nowotny, 1964; Martin and Marcus, 1966; Sultzer, 1971; Chedid et al., 1975; Ribi et al., 1979; Elin, 1979; Raziuddin, 1980).

Emulsification of endotoxin in oil and water has been found to have a depot effect on the toxin, thereby reducing toxicity but retaining the capability to induce resistance to certain viral, bacterial, and parasitic infections as well as cancer implants (Wagner et al., 1959; Goble and Singer, 1960; Prigal, 1961, 1969). Cuevas et al. (1974) found that one intramuscular injection of emulsified endotoxin increased the resistance of immunosuppressed mice to subsequent E. coli peritonitis. $FeCl_3$ used to monitor emulsions was found to be capable of detoxifying endotoxin (Prigal, 1967; Prigal et al., 1973).

A promising physical method of endotoxin detoxification was developed by Previte et al. (1967). They found that ionizing radiation (20 Mrad ^{60}Co) increased the LD_{50} of S. typhimurium endotoxin from 1.1 mg/20 g mouse to more than 9.5 mg. This treatment did not inactivate pyrogenic components of endotoxin nor its ability to protect the animals against gram-negative infection. Bertók and his colleagues extended these studies and found that irradiated endotoxin was not only less toxic but also caused less activation of complement and depression of blood pressure (Füst et al., 1977). However, irradiated endotoxin administered 5 days previously protected rats against septic shock induced by the injection of colonic contents as well as against hemorrhagic shock (Balogh and Bertók, 1974). These and other beneficial effects of irradiated endotoxin were recently reviewed by Bertók (1980). The only beneficial property of endotoxin found to be reduced by radiation was the capacity to mobilize stem cells.

Another means to more safely use endotoxin treatments is to find preparations of toxin that are naturally attenuated. Endotoxins of some microorganisms are significantly less toxic than those from many other flora. These naturally attenuated toxins could be relatively safer to administer, but retain significant immunopotentiating effects. For example, the endotoxin of P. aeruginosa is less toxic for mice than that of E.

coli (Pavlovskis *et al.*, 1975). A commercial preparation of *Pseudomonas* polysaccharide is available that enhances granulopoiesis (Windle and Wilcox, 1950) and leukocytosis (Soylemozoglu and Wells, 1951) and activates the reticuloendothelial system (Windle *et al.*, 1950). Although not tested for its beneficial effects, the endotoxin of *Yersinia pestis* is also less toxic than many others (Davies, 1956). Recently, Paquet *et al.* (1978) found that LPS from *Coxiella burnetii* induces nonspecific resistance to *Candida albicans*, but may be less toxic than *E. coli* endotoxin. Another gram-negative microorganism whose endotoxin deserves further study as an immunomodulator is *Campylobacter*. Fumarola *et al.* (1981) found that endotoxic reactions (e.g., *Limulus* test reactivity) elicited by heat-killed *Campylobacter* are quantitatively less than those obtained with a similar preparation of *E. coli*.

Whether endotoxin is administered in its natural or attenuated form, pharmacologic means will have to be available to control potentially destructive responses that could develop in a compromised host. Although a thorough understanding of mechanisms of endotoxin action is not yet available, sufficient data have been accumulated to propose means necessary to attenuate detrimental host responses to endotoxin.

Endotoxin interacts with numerous humoral factors and cell types. However, the consequence of many of these interactions only amplify endotoxin effects and are not determinative for survival. Some investigators traced the distribution in animals of labeled endotoxin in order to learn more of determinative events. After interacting with platelets, most injected endotoxin became localized in the liver (Braude *et al.*, 1955), and considerable work has been done to demonstrate that endotoxin has a direct effect on this and other organs in which it is found.

The concept of a direct determinative action of endotoxin on liver or other organs is cast in doubt, however, by the observation that endotoxin administered intramuscularly is also lethal, although amounts associated with toxicity are never found in the liver (Noyes *et al.*, 1959). This finding prompted more recent studies, which showed that lethality and other expected metabolic events could be obtained after subcutaneous administration of endotoxin although 85% of the material never left the site of injection (Petschow and Moon, 1981). These data are consistent with those of Berry (1971) and more recently by Feuerstein and Ramwell (1981), which indicate a mediated response to endotoxin. These two studies obtained different responses to endotoxin by liver and lung tissue respectively when the toxin was first given to the intact animal rather than added directly to the tissue *in vitro*.

The major cell type responsible for mediating determinative effects of endotoxin is apparently of marrow origin. Reconstitution of X-irradiated C3HeB/FeJ mice with marrow from the C3H/HeJ mouse, a strain genetically resistant to endotoxin, confers increased resistance on the recipient animals (Michalek *et al.*, 1980). The macrophage of the C3H/HeJ mouse is refractory to endotoxin (Sultzer and Goodman, 1977). The studies with genetically resistant mice confirm the report by Peavy and Brandon (1980) that increases and decreases in the susceptibility of mice to the lethal effects of endotoxin are paralleled by similar changes in susceptibility of their macrophages to endotoxin *in vitro*.

The determinative lethal effect mediated by macrophages may occur in the microcirculation. A basic event during endotoxin shock is inadequate capillary perfusion (Hardaway, 1980), and Fine and Minton (1966) suggested that endotoxin kills via destruction of the integrity of blood vessels in the splanchnic area. Such findings are consistent with those of Urbaschek (1971), who used cinematography to study microcirculatory events in living animals challenged with lethal amounts of endotoxin. He saw that the lumen of the capillaries becomes narrowed due to swelling of the endothelial cells and the presence of wall-adherring blood cells. Blood flow continues to decrease with time and eventually stasis occurs, capillary wall permeability increases, and microbleeding occurs.

Abnormal production of vasoactive substances, including prostaglandins, may account for the mediated failure to maintain capillary perfusion and thereby cause subsequent metabolic events associated with endotoxemia. When cell membranes are perturbed by physical or biochemical means, arachidonic acid at the membrane surface is converted into various prostaglandin endoperoxides. Some of these form thromboxane A_2 (TXA$_2$), which contributes to platelet aggregation and vasoconstriction. Specific inhibition of thromboxane synthesis alleviates some of the detrimental effects of endotoxin (Wise *et al.*, 1980a,b; Smith *et al.*, 1980). Normal vascular endothelial cells can convert these same endoperoxides into prostacyclin (PGI$_2$), which inhibits platelet aggregation caused by factors other than direct interaction with endotoxin (Fletcher and Ramwell, 1978) and is a vasodilator. Maintenance of a favorable ratio between PGI$_2$ and TXA$_2$ may be determinative to survival.

II. Experimental Studies—Endotoxin in Compromised Hosts

Numerous studies have described beneficial effects obtained when endotoxin is administered prior to compromising situations and these phenomena may be important clinically if applied to certain individuals facing high risk of infection or trauma. A major use of endotoxin as a beneficial agent, however, will probably be in individuals who have already been compromised (e.g., burned, infected). The possibility that these subjects may be even more sensitive than normal to the lethal effects of endotoxin must be taken into consideration.

Injury induced by ionizing radiation can be used to illustrate the association of endotoxin with events in a compromised host. Mice (B6CBF$_1$) given a lethal dose of X-rays (850 rad) demonstrated a biphasic pattern of endotoxin involvement (Walker et al., 1975; Walker, 1979). Livers of these animals were sampled at intervals and the presence or absence of bacteria or Limulus amebocyte lysate-reactive material determined. During the first phase, which lasted from days 1 to 4 after radiation, up to 80% of the mice were positive for endotoxin, but no animals died or showed visible signs of distress. No bacteria were detected at this time. This initial period was followed by a phase lasting until days 7–8 in which no endotoxin or bacteria were detected. After this time, endotoxin was again detectable in liver homogenates. Gram-negative bacteria were also detected and infection preceded death, which occurred between 11 and 14 days after radiation.

Detection of endotoxin in liver homogenates after radiation may be due to disruption of tight junctional barriers (zonula occludens) between adjacent epithelial cells of the intestine (Walker and Porvaznik, 1978). The biological significance of this phenomenon is open to question. Fine and his colleagues have argued that increased intestinal permeability to endotoxin, which occurs following many types of injury, contributes to mortality if unrelieved (Fine and Minton, 1966; Caridis et al., 1973). However, the early presence of endotoxin in livers of irradiated animals elicited no observable ill effects, and could actually be beneficial through stimulation of host defenses.

A biphasic increase in sensitivity to injected endotoxin occurs in irradiated mice and coincides with the periods during which disruption of ileal tight junctional barriers was observed (Walker, 1979). The early increase in sensitivity of B6CBF$_1$ mice to 300 μg of endotoxin occurred on day 3 postradiation, as observed previously by Smith et al. (1963). This dose of endotoxin killed no mice prior to radiation, but 80% mortality was achieved on day 3. Six days after radiation, no mice died from a similar challenge, but 100% mortality resulted from 300 μg of S. typhi endotoxin administered on days 8–10.

The factors responsible for sensitivity of irradiated mice to endotoxin are unknown. Endotoxin of intestinal origin already present in host tissues (Meter, 1974; Walker et al., 1975) could contribute to toxic effects of additional exogenous endotoxin. Alternatively, the presence of mediators such as histamine that disrupt tight junctions (M.

Provaznik, personal communication) could sensitize mice to endotoxin (Urbaschek, 1966) or macrophages may be hyperresponsive to endotoxin (Peavy and Brandon, 1980) at the times of peak sensitivity. The finding that increased sensitivity to endotoxin occurring late after radiation could be reduced by treatment of the mice with orally administered antibiotics suggests that endotoxin can synergistically enhance infection (Galley et al., 1975).

The negative phase wherein small amounts of endotoxin enhance infection by otherwise sublethal numbers of microorganisms has been described by others (Sanarelli, 1924; Dubos and Schaedler, 1956; Suter, 1964; Flynn and McEntegart, 1973). Little is known regarding the mechanism of this phenomenon (Cluff, 1971), but it obviously increases the risk associated with endotoxin therapy of infected patients. Enhanced microbial virulence following endotoxin administration is dose dependent (Dubos and Schaedler, 1956). However, means to eliminate the transient immunosuppression induced by any dose of endotoxin must be available before use of the substance can be considered for infected subjects.

The foregoing does not mean that endotoxin cannot be used for compromised subjects in whom infection has not yet occurred. The problem of enhanced microbial virulence could be avoided if the endotoxin is administered shortly after injury. Opportunistic infections such as described in irradiated mice (Walker et al., 1975) do not occur until sometime after the host is compromised. For example, in American troops injured in Vietnam, P. aeruginosa was a major cause of infection, but not until after the fifth hospital day (Tong, 1972). The lag period can be due to disturbances among the normal colonization resistant flora and subsequent overgrowth in the intestine by opportunistic pathogens (van der Waaij, 1979). For example, we found that numbers of both facultative organisms and segmented filamentous microflora (SFM) decrease dramatically in ilea of rats given sublethal or lethal radiation (Porvaznik et al., 1979). However, in sublethally irradiated animals the SFM return to normal levels by days 3–4 postexposure followed by a slower increase in facultative flora back towards normal numbers. In lethally irradiated animals, the SFM population never recovers, but a massive overgrowth of the ileum by gram-negative facultative flora such as Proteus and E. coli occurs and correlates with the appearance of microorganisms in the liver, an event that precedes death.

Some of the reasons given above for increased sensitivity to endotoxin may also apply to tumor-bearing animals. C57BL/6 mice engrafted with the Lewis lung (3LL) carcinoma are more sensitive than normal to 300 μg of S. typhi endotoxin when it is administered on days

3 or 14 but not day 7 after engraftment (Walker *et al.*, 1980). In contrast to irradiated mice (Walker, 1979), no contribution of intestinal microorganisms to this sensitivity was detected. Specifically, we found no evidence for disruption of tight junctional barriers between epithelial cells of the intestine nor did SFM disappear from the ileum as seen in irradiated animals. This latter finding was consistent with an absence of cultivatable microorganisms in liver samples removed from tumor-bearing mice. Early sensitivity to endotoxin, as in irradiated animals, could be due to circulating mediators or activated macrophages (Berendt *et al.*, 1980). Also, at day 14, breakdown products from the large tumor could be toxic (Havas *et al.*, 1960).

Enhanced sensitivity to endotoxin in compromised hosts may be reduced by using detoxified preparations. We used $FeCl_3$ to treat endotoxin from *S. typhi* and found that, although less toxic, it was equal to normal endotoxin in conferring radioprotection and endotoxin tolerance (Galley *et al.*, 1975). Also, this preparation was a more effective B-cell mitogen than untreated toxin and it could be used to treat CBA donor mice to preclude lethal graft-vs.-host disease when their spleen cells were engrafted into irradiated $B6CBF_1$ mice. However, 25 μg of $FeCl_3$-treated endotoxin killed all mice given 850 rad X-radiation 8 days previously. This toxic effect could be avoided by intestinal decontamination, thus indicating that the toxin was enhancing low-level bacterial infections.

Chromium chloride is also effective in reducing the toxicity of endotoxin (Prigal *et al.*, 1973), but it has not been tested in compromised subjects. Studies in our laboratory demonstrated that the ability to induce B-cell mitogenicity, colony-stimulating factor, dermal Shwartzman reactions, or radioprotective effects is reduced by chromium treatments (Snyder *et al.*, 1978), but the capacity to disrupt tight junctions is normal and Hageman factor activation is enhanced (Walker *et al.*, 1979). These same studies, however, showed that $CrCl_3$-treated endotoxin is as effective as untreated endotoxin in enhancing nonspecific resistance to *K. pneumoniae* infections.

Recent studies in this laboratory (Walker *et al.*, 1982) have tested the safety of irradiated endotoxin in compromised animals with encouraging results. When $B6CBF_1$ mice receiving 1000 rad ^{60}Co 3 or 7 days previously were challenged with 300 μg of *E. coli* endotoxin, 60 and 44% mortality was obtained respectively. No animals died, however, when given irradiated (20 Mrad) endotoxin at these times. Similarly, C57BL/6 mice engrafted with 5×10^6 Lewis lung carcinoma cells 3 days previously showed 60% mortality when given 300 μg of unirradiated endotoxin, but no animals receiving irradiated endotoxin died. Irradiated and unirradiated endotoxin preparations were also given simultaneously with *K. pneumoniae*. One hundred micrograms of either endotoxin preparation enhanced mortality from infection in NMRI mice when given with 10^5 organisms. No such effect was seen with 50 μg of irradiated toxin, but unirradiated endotoxin at this concentration continued to enhance virulence.

We were able to demonstrate that the less toxic irradiated preparation still retained beneficial properties. When 100 μg of irradiated endotoxin was used to treat mice 24 hr prior to challenge with 10^6 or 10^7 *K. pneumoniae*, it not only conferred some protection against infection, but also delayed that mortality that was obtained.

Recent studies indicate that the host response to endotoxin may soon be manipulated to ensure safety from detrimental effects of the toxin. An association has been found between PGI_2/TXA_2 ratios and resistance of C3H/HeJ mice to endotoxin (Walker *et al.*, 1982). C3H/HeJ mice are unusual in that they are generally unresponsive to endotoxin, but, in contrast to their endotoxin-sensitive paired C3H/HeN or C3HeB/FeJ strain mice, are extremely sensitive to some bacterial infections. For example, as little as 10 *K. pneumoniae* will kill all C3H/HeJ mice, but 10^6 organisms are necessary to kill C3HeB/FeJ mice (Walker and Fletcher, 1980). When hematopoietic effects of endotoxin on C3H/HeJ mice were studied, we found that above a threshold dose, they respond to the endotoxin in a way qualitatively similar to C3H/HeN animals (MacVittie *et al.*, 1983). If the complex was pure relative to associated protein, then the defect in the C3H/HeN mice was absolute with reference to hematopoietic effects.

We studied microcirculatory events in endotoxin-resistant animals in order to detect critical events that might be manipulated for the benefit of the host, thereby increasing the safety of endotoxin usage. In a previous study, endotoxin-sensitive C3HeB/FeJ and endotoxin-resistant C3H/HeJ mice were administered 1 mg of *S. typhi* endotoxin, a dose lethal for all sensitive animals, but no resistant mice (Walker and Parker, 1980). At intervals after challenge, damage to the pulmonary microcirculation [expressed as number of polymorphonuclear leukocytes (PMNL) observed per $400\times$ microscope field] was assessed. At 1 hr after challenge, similar degrees of leukocyte sequestration were observed in lungs of both strains of mice. However, by 6 hr after challenge, leukocyte numbers in lung capillaries of sensitive mice remained at a high level (60 PMNL/$400\times$ field) whereas they had almost returned to normal in C3H/HeJ mice (10 PMNL/$400\times$ field) examined 6 hr after challenge.

Leukocyte sequestration data for resistant and responder mice were compared to results obtained when PGI_2/TXA_2 ratios were determined in cultured lung slices removed from C3H/HeJ and C3H/HeN mice at 1 or 6 hr after challenge with 0.8 mg of endotoxin. This dose was toxic for all C3H/HeN mice. C3H/HeJ mice maintained a high PGI_2/TXA_2 ratio at 1 and 6 hr (21.1 and 24.3, respectively) in contrast to sensitive animals, which had only normal ratios at these times (10.3–12.2). The high ratio obtained with lung slices from C3H/HeJ mice seemed to be due to a low level of thromboxane at 1 hr, but at 6 hr excessive production of prostacyclin was noted. The exact nature of a possible macrophage contribution to the abnormal prostaglandin responses seen in C3H/HeJ mice remains to be determined.

The evidence presented in the foregoing suggests that mediated damage to microcirculatory function can be detrimental to survival and could result from a failure to maintain an optimal PGI_2/TXA_2 ratio. Pharmacologic measures are now available to manipulate this ratio in favor of the host. Animals have been protected against the lethal effects of endotoxin with general inhibitors of prostaglandin synthesis such as indomethacin and ibuprofen (Hilton and Wells, 1976; Fletcher and Ramwell, 1978; Wise et al., 1980a). Some protection has also been obtained with drugs that block synthesis of thromboxane (Wise et al., 1980b; Smith et al., 1980) or promote synthesis of prostacyclin (Fletcher and Ramwell, 1979). Prostacyclin infusion has even been used to prevent lethal consequences after shock has begun to occur (Krausz et al., 1981).

Taken together, many studies suggest that reduction of macrophage responsiveness to endotoxin can contribute to reduction of microvascular damage leading to irreversible shock. A major event to be controlled is prostaglandin synthesis. Use of selective inhibitors of prostaglandin synthesis or infusion of prostacyclin can counter the deleterious events associated with endotoxin, but could leave activation of beneficial processes intact. This latter possibility remains to be determined.

III. Beneficial Effects of Endotoxin

The use of endotoxin therapy has tremendous potential for compromised hosts. Enhancement of nonspecific resistance is important to such hosts because they are subject to infectious complications initiated by a variety of opportunistic pathogens. Although endotoxin initially increases susceptibility to infection, as more time passes between treatment and challenge with microorganisms, resistance to infection with a variety of bacteria (Rowley, 1955; Prigal, 1961; Margherita and Friedman, 1965; Berger and Fukui, 1963; Sultzer, 1968; Galelli et al., 1981), protozoan parasites (Goble and Singer, 1960; McGregor et al., 1969; Singer et al., 1964), viruses (Wagner et al., 1959; Gledhill, 1959; Finkelstein, 1961), and fungi (Kimball *et al.*, 1968; Wright *et al.*, 1969; Kobayashi *et al.*, 1969) is enhanced.

Microbial agents that nonspecifically modulate immune responses should not only offer protection against a variety of pathogens, but also may contribute to repair of trauma and thereby reduce tissue damage and the potential for subsequent infection. This principle was recently demonstrated by Stinnett *et al.* (1981), who improved survival in beagles given a 33% total body surface third-degree flame burn by administration of *Corynebacterium parvum* vaccine at intervals after injury. Similarly, Urbaschek and Nowotny (1969) used detoxified *Serratia marcescens* endotoxin to prevent burn-induced damage to the microcirculation. More recently, normal entotoxin was used to reduce burn severity in mice when given immediately after injury (Spillert *et al.*, 1981). A polysaccharide material from *Pseudomonas* (Windle *et al.*, 1950) can improve vascularity in areas traumatized by frostbite and thereby reduce tissue loss (Wexler and Tryczynski, 1955). Endotoxin also stimulates hematopoiesis (Windle and Wilcox, 1950), which would be important to marrow recovery after radiation. Some controversial evidence indicates that endotoxin can contribute to repair of damaged nerves (Windle and Chambers, 1950; Clemente *et al.*, 1951; Matinian and Andreasian, 1976).

Another effect of endotoxin that could be significant to compromised hosts, particularly those subjected to radiation, is regulation of granulopoiesis. Endotoxin produced by intestinal microflora contributes to this function even in normal hosts. For example, when dogs are treated orally with the poorly adsorbed antibiotics bacitracin and neomycin, enteric gram-negative microorganisms become undetectable in the feces and levels of a serum granulopoietic factor (CSF) are reduced to 50% of normal values (Walker *et al.*, 1978; MacVittie and Walker, 1978). If these same animals are allowed to become recolonized, then CSF levels return to normal.

Endotoxin has long been known to be an effective adjuvant (Johnson *et al.*, 1956). More recently, it has been used to enhance the adjuvant effect of liposomes to protein antigens (van Rooijen and van Nieuwmegen, 1980). Antibody titers to human serum albumin were greater and antibody could be detected earlier when endotoxin was incorporated in the protein–liposome complexes.

IV. Conclusion

Immunomodulator agents such as endotoxin hold tremendous promise for medicine. Endotoxin could emerge as an exceptionally useful agent of this type due to its long evolutionary association with multicellular animals. Unfortunately, endotoxin can cause shock and death when administered in amounts that trick the body into executing self-destructive activities that, in moderation, would be beneficial. However, means are becoming available to attenuate either the endotoxin molecule or the hosts' response to it. To finally harness the beneficial effects of endotoxin to combat infectious agents and promote healing processes will bring the realization of a dream already 2500 years old.

ACKNOWLEDGMENT

Portions of research cited herein were supported by NMR&DC work unit No. MR041.05.01.0057.

V. References

Balogh, A., and Bertók, L., 1974, Prevention of septic and haemorrhagic shock by detoxified endotoxin pretreatment, *Acta Chir. Acad. Sci. Hung.* **15**:21.
Berendt, M. J., Newborg, M. F., and North, R. J., 1980, Increased toxicity of endotoxin for tumor-bearing mice and mice responding to bacterial pathogens: Macrophage activation as a common denominator, *Infect. Immun.* **28**:645.
Berger, F. M., and Fukui, G. M., 1963, Endotoxin induced resistance to infections and tolerance, *Proc. Soc. Exp. Biol. Med.* **114**:780.
Berry, L. J., 1971, Metabolic effects of bacterial endotoxins, in: *Microbial Toxins*, Vol. V, *Bacterial Endotoxins* (S. Kadis, G. Weinbaum, and S. J. Ajl, eds.), pp. 165–208, Academic Press, New York.
Bertók, L., 1980, Radio-detoxified endotoxin as a potent stimulator of nonspecific resistance, *Perspect. Biol. Med.* **1980**:61.
Boivin, A., Mesrobeanu, J., and Mesrobeanu, L., 1933, Technique pour la préparation des polysaccharides microbiens spécifiques, *C. R. Soc. Biol.* **113**:490.
Braude, A. I., Carey, F. J., and Zalesky, M., 1955, Studies with radioactive endotoxin. II. Correlation of physiological effects with distribution of radioactivity in rabbits injected with lethal doses of *E. coli* endotoxin labelled with radioactive sodium chromate, *J. Clin. Invest.* **34**:858.
Caridis, D. J., Ishiyama, M., Woodruff, P. W. H., and Fine, J., 1973, Role of the intestinal flora in clearance and detoxification of circulating endotoxin, *J. Reticuloendothelial Soc.* **14**:513.
Chedid, L., Audibert, F., Bona, C., Damais, C., Parant, F., and Parant, M., 1975, Biological activities of endotoxins detoxified by alkylation, *Infect. Immun.* **12**:714.
Clemente, C. D., Chambers, W. W., Greene, L., Mitchell, S. Q., and Windle, W. F., 1951, Regeneration of the transected spinal cord of adult cats, *Anat. Rec.* **109**:280.
Cluff, L. E., 1971, Effects of lipopolysaccharides (endotoxins) on susceptibility to infections, in: *Microbial Toxins*, Vol. V, *Bacterial Endotoxins* (S. Kadis, G. Weinbaum, and S. J. Ajl, eds.), pp. 399–413, Academic Press, New York.
Cuevas, P., Fine, J., and Monaco, A. P., 1974, Successful induction of increased resistance to gram-negative bacteria and to endotoxin in immunosuppressed mice, *Am. J. Surg.* **127**:460.

Cundiff, R. J., and Morgan, H. R., 1941, The inhibition of the bactericidal power of human and animal sera by antigenic substances obtained from organisms of the typhoid–salmonella group, *J. Immunol.* **42**:361.

Davies, D. A. I., 1956, A specific polysaccharide of Pasteurella pestis, *Biochem.* **63**:103.

Dubos, R. J., and Schaedler, R. W., 1956, Reversible changes in the susceptibility of mice to bacterial infections. I. Changes brought about by injection of pertussis vaccine or of bacterial endotoxins, *J. Exp. Med.* **104**:53.

Elin, R. J., 1979, Non-specific resistance to infection, in: *Immunological Aspects of Infectious Diseases* (G. Dick, ed.), pp. 1–19, University Park Press, Baltimore.

Favorite, G. O., and Morgan, H. R., 1942, Effects produced by the intravenous injection in man of a toxic antigenic material derived from *Eberthella typhosa*: Clinical, hematological, chemical and serological studies, *J. Clin. Invest.* **21**:589.

Favorite, G. O., and Morgan, H. R., 1946, Therapeutic induction of fever and leucocytosis using a purified typhoid pyrogen, *J. Lab. Clin. Med.* **6**:672.

Feuerstein, N., and Ramwell, P., 1981, *In vivo* and *in vitro* effects of endotoxin on prostaglandin release from rat lung, *Br. J. Pharmacol.* **73**:511.

Fine, J., and Minton, R., 1966, Mechanism of action of bacterial endotoxin, *Nature (London)* **210**:97.

Finkelstein, R. A., 1961, Alteration of susceptibility of embryonated eggs to Newcastle disease virus by *Escherichia coli* and endotoxin, *Proc. Soc. Exp. Biol. Med.* **106**:481.

Fletcher, J. R., and Ramwell, P. W., 1978, *E. coli* endotoxin shock in the dog: Treatment with lidocaine or indomethacin, *Br. J. Pharmacol.* **64**:185.

Fletcher, J. R., and Ramwell, P. W., 1979, Lidocaine treatment following baboon endotoxin shock improves survival, *Adv. Shock Res.* **2**:291.

Flynn, J., and McEntegart, M. G., 1973, Gonococcal enhancement of staphylococcal virulence for the mouse, *J. Med. Microbiol.* **6**:371.

Fumarola, D., Miragliotta, G., Jirillo, D., and Monno, R., 1981, Endotoxin-like activity associated with heat-killed organisms of the genus *Campylobacter*, International Workshop on *Campylobacter* Infections, Reading, England (March 1981), P/No. 28.

Füst, G., Bertók, L., and Juhász-Nagy, S., 1977, Interactions of radio-detoxified *Escherichia coli* endotoxin preparations with the complement system, *Infect. Immun.* **16**:26.

Galelli, A., LeGarrec, Y., and Chedid, L., 1981, Increased resistance and depressed delayed-type hypersensitivity to *Listeria monocytogenes* induced by pretreatment with lipopolysaccharide, *Infect. Immun.* **32**:88.

Galley, C. B., Walker, R. I., Ledney, G. D., and Gambrill, M. R., 1975, Evaluation of biologic activity of ferric chloride-treated endotoxin in mice, *Exp. Hematol.* **3**:197.

Gledhill, A. W., 1959, Sparing effect of serum from mice treated with endotoxin upon certain murine virus diseases, *Nature (London)* **183**:185.

Goble, F. C., and Singer, L., 1960, The reticuloendothelial system in experimental malaria and trypanosomiasis, *Ann. N.Y. Acad. Sci.* **88**:199.

Hardaway, R. M., 1980, Endotoxemic shock, *Dis. Colon Rectum* **23**:597.

Havas, H. F., Donnelly, A. J., and Levine, S. I., 1960, Mixed bacterial toxins in the treatment of tumors. III. Effect of tumor removal on the toxicity and mortality rates in mice, *Cancer Res.* **20**:393.

Hilton, J. G., and Wells, C. H., 1976, Effects of indomethacin and nicotinic acid on *E. coli* endotoxin shock in anesthetized dogs, *J. Trauma* **16**:968.

Johnson, A. G., and Nowotny, A., 1964, Relationship of structure to function in bacterial O-antigens. III. Biological properties of endotoxoids, *J. Bacteriol.* **87**:809.

Johnson, A. G., Gaines, S., and Landy, M., 1956, Studies on the O antigen of *Salmonella typhosa*. V. Enhancement of antibody response to protein antigens by the purified lipopolysaccharide, *S. Exptl. Med.* **103**:225.

Kimball, H. R., Williams, T. W., and Wolff, S. M., 1968, Effect of bacterial endotoxin on experimental fungal infections, *J. Immunol.* **100**:24.

Kobayashi, H., Yasuhira, K., and Uesaka, I., 1969, Effect of *Escherichia coli* and its endotoxin on the resistance of mice to experimental cryptococcal infection, *Jpn. J. Microbiol.* **13**:223.

Krausz, M. M., Utsunomiya, T., Feuerstein, G., Wolfe, J. H. N., Shepro, D., and Hechtman, H. B., 1981, Prostacyclin reversal of lethal endotoxemia in dogs, *J. Clin. Invest.* **67**:1118.

McGregor, R. R., Sheagren, J. N., and Wolff, S. M., 1969, Endotoxin-induced modification of *Plasmodium berghei* infection in mice, *J. Immunol.* **102**:131.

MacVittie, J. J., and Walker, R. I., 1978, Canine granulopoiesis: Alterations induced by suppression of gram-negative flora, *Exp. Hematol.* **6**:639.

MacVittie, J. J., Walker, R. I., and Patchen, M., 1983, LPS-induced, dose-dependent, hematopoietic responses in LPS resistant C3H/HEJ and C57BL/105cN mutant mice, *J. Cell. Physiol.*, in press.

Margherita, S. S., and Friedman, H., 1965, Induction of non-specific resistance by endotoxin in unresponsive mice, *J. Bacteriol.* **89**:277.

Martin, W. J., and Marcus, S., 1966, Detoxified bacterial endotoxins. II. Preparation and biological properties of chemically modified crude endotoxins from *Salmonella typhimurium*, *J. Bacteriol.* **91**:1750.

Matinian, L. A., and Andreasian, A. S., 1976, Enzyme therapy in organic lesions of the spinal cord, Brain Information Service, UCLA.

Meter, I. D., 1974, Endotoxemia in lethally irradiated animals, *Radiobiologya* **14**:98.

Michalek, S. M., Moore, R. N., McGhee, J. R., Rosenstreich, D. L., and Mergenhagen, S. E., 1980, The primary role of lymphoreticular cells in the mediation of host responses to bacterial endotoxin, *J. Infect. Dis.* **141**:55.

Morgan, H., 1940, Preparation of antigenic material inducing leucopenia from *Eberthella typhosa* cultured in synthetic medium, *Proc. Soc. Exp. Biol. Med.* **43**:529.

Morgan, H. R., 1941, Immunologic properties of an antigenic material isolated from *Eberthella typhosa*, *J. Immunol.* **41**:161.

Morgan, H. R., 1948, Resistance to the action of the endotoxins of enteric bacilli in man, *J. Clin. Invest.* **22**:706.

Morgan, H. R., Favorite, G. O., and Horneff, J. A., 1943, Immunizing potency in man of a purified antigenic material isolated from *Eberthella typhosa*, *J. Immunol.* **46**:301.

Noyes, H. E., McInturf, C. R., and Blahuta, G. J., 1959, Studies on distribution of *Eschericchia coli* endotoxin in mice, *Proc. Soc. Exp. Biol. Med.* **100**:65.

Paquet, A., Jr., Rael, E. D., Klassen, D., Martinez, I., and Baca, O. G., 1978, Mitogenic and protective activity associated with lipopolysaccharide from *Coxiella burnetii*, *Can. J. Microbiol.* **24**:1616.

Pavlovskis, O. R., Callahan, L. T., III, and Pollack, M., 1975, *Pseudomonas aeruginosa* exotoxin, in: *Microbiology 1975* (D. Schlessinger, ed.), pp. 252–256, American Society for Microbiology, Washington, D.C.

Peavy, D. L., and Brandon, C. L., 1980, Macrophages: Primary targets for LPS activity, in: *Bacterial Endotoxins and Host Response* (M. K. Agarwal, ed.), pp. 299–309, Elsevier/North-Holland, Amsterdam.

Petschow, B. W., and Moon, R. J., 1981, Murine response and tissue localization of subcutaneously administered *Salmonella typhimurium* endotoxin, *Abstr. Ann. Meet. ASM* (1981) p. 18, B23.

Porvaznik, M., Walker, R. I., and Gillmore, J. D., 1979, Reduction of the indigenous filamentous microorganisms in rat ilea following γ-radiation, *Scanning Electron Microsc.* **3**:15.

Previte, J. J., Chang, Y., and El-Bisi, H. M., 1967, Detoxification of *Salmonella typhimurium* lipopolysaccharide by ionizing radiation, *J. Bacteriol.* **93:**1607.

Prigal, S. J., 1961, Development in mice of prolonged non-specific resistance to sarcoma implant and staphylococcus infection following a repository injection of lipopolysaccharide, *Nature (London)* **191:**1111.

Prigal, S. J., 1967, The ferric chloride spot test for the evaluation and standardization of emulsion of allergens, *J. Allergy* **39:**1.

Prigal, S. J., 1969, The induction of prolonged resistance against a lethal staphylococcic infection in mice with a single injection of an emulsified lipopolysaccharide, *J. Allergy* **44:**176.

Prigal, S. J., Herp, A., and Gerstein, S., 1973, The detoxification of lipopolysaccharides derived from bacterial endotoxins by ferric chloride, *J. Reticuloendothelial Soc.* **14:**250.

Raziuddin, S., 1980, Biological activities of chemically modified endotoxins from *Vibrio cholerae, Biochim. Biophys. Acta* **620:**193.

Ribi, E., Parker, R., Strain, S. M., Mizuno, Y., Nowotny, A., Von Eschen, K. B., Cantrell, J. L., McLaughlin, C. A., Hwang, K. M., and Goren, M. B., 1979. Peptides as a requirement for immuno-therapy of the guinea pig line-10 tumor with endotoxins, *Cancer Immunol. Immunother.* **7:**43.

Rowley, D., 1955, Stimulation of natural immunity to *Escherichia coli* infections: Observations on mice, *Lancet* **1:**232.

Sanarelli, G., 1924, De la pathogenie du cholera: Le cholera experimental, *Ann. Inst. Pasteur* **38:**11.

Singer, I., Kimble, E. T., III, and Ritts, R. E., Jr., 1964, Alterations of the host–parasite relationship by administration of endotoxin to mice with infections of trypanosomes, *J. Infect. Dis.* **114:**243.

Smith, E. F., III, Tabas, J. H., and Lefer, A. M., 1980, Beneficial actions of imidazole in endotoxin shock, *Prostaglandins Med.* **4:**215.

Smith, W. W., Alderman, I. M., Schneider, C., and Cornfield, J., 1963, Sensitivity of irradiated mice to bacterial endotoxin, *Proc. Soc. Exp. Biol. Med.* **113:**778.

Snyder, S. L., Walker, R. I., MacVittie, T. J., and Sheil, J. M., 1978, Biological properties of bacterial lipopolysaccharides threated with chromium chloride, *Can. J. Microbiol.* **24:**495.

Soylemozoglu, B., and Wells, J. A., 1951, Comparison of leukocyte response to ACTH and bacterial pyrogen, *Proc. Soc. Exp. Biol. Med.* **77:**43.

Spillert, C. R., Ghuman, S. S., McGovern, P. J., Jr., and Lazaro, E. J., 1981, Effects of endotoxin in murine burns, *Adv. Shock Res.* **5:**163.

Stinnett, J. D., Alexander, J. W., Morris, M. J., Dreffer, R. L., Craycroft, T. R., Anderson, P. E., Ogle, C. R., and MacMillan, B. G., 1981, Improved survival in severely burned animals using intravenous *Corynebacterium parvum* vaccine post injury, *Surgery* **89:**237.

Sultzer, B. M., 1968, Endotoxin-induced resistance to staphylococcal infection: Cellular and humoral responses compared in two mouse strains, *J. Infect. Dis.* **118:**340.

Sultzer, B. M., 1971, Chemical modification of endotoxin and inactivation of its biological properties, in: *Microbial Toxins*, Vol. V, *Bacterial Endotoxins* (S. Kadis, G. Weinbaum, and S. J. Ajl, eds.), pp. 91–126, Academic Press, New York.

Sultzer, B. M., and Goodman, G. W., 1977, Characteristics of endotoxin-resistant low-responder mice, in: *Microbiology 1977* (D. Schlessinger, ed.), pp. 304–309, American Society for Microbiology, Washington, D.C.

Suter, E., 1964, Hyperreactivity to endotoxin in infection, in: *Bacterial Endotoxins* (M. Landy and W. Braun, eds.), pp. 435–447, Rutgers University Press, New Brunswick, N.J.

Tong, M. J., 1972, Septic complications of war wounds, *J. Am. Med. Assoc.* **219:**1044.

Urbaschek, B., 1966, The histamine-sensitizing effect of endotoxins, *Ann. N.Y. Acad. Sci.* **133**:709.

Urbaschek, B., 1971, Addendum—The effects of endotoxin in the microcirculation, in: *Microbial Toxins*, Vol. V, *Bacterial Endotoxins* (S. Kadis, G. Weinbaum, and S. J. Ajl, eds.), pp. 261–275, Academic Press, New York.

Urbaschek, B., and Nowotny, A., 1969, Endotoxin tolerance induced by endotoxoid, in: *La structure et les effets biologiques des produits bactériéns provenants de germes Gram-négatifs* (L. Chedid, ed.), pp. 357–365, Colloq. Int. sur les Endotoxines, CNRS No. 174, Paris.

van der Waaij, D. (ed.), 1979, The colonization resistance of the digestive tract in experimental animals and its consequences for infection prevention, acquisition of new bacteria and the prevention of spread of bacteria between cage mates, in: *New Criteria for Antimicrobial Therapy: Maintenance of Digestive Tract Colonization Resistance*, pp. 43–53, Excerpta Medica, Amsterdam.

van Rooijen, N., and van Nieuwmegen, R., 1980, Endotoxin enhanced adjuvant effect of liposomes, particularly when antigen and endotoxin are incorporated within the same liposome, *Immunol. Commun.* **9**:742.

Wagner, R. R., Snyder, R. M., Hook, E. W., and Luttrell, C. N., 1959, Effect of bacterial endotoxin on resistance of mice to viral encephalitides, *J. Immunol.* **83**:87.

Walker, R. I., 1979, Hematologic contributions to increases in resistance or sensitivity to endotoxin, in: *Experimental Hematology Today—1979* (S. J. Baum and G. D. Ledney, eds.), pp. 55–62, Springer-Verlag, Berlin.

Walker, R. I., and Fletcher, J. W., 1980, Evidence that injury to platelets can contribute to the diversity of host responses to endotoxin, in: *Bacterial Endotoxins and Host Response* (M. K. Agarwal, ed.), pp. 327–338, Elsevier/North-Holland, Amsterdam.

Walker, R. I., and Parker, G. A., 1980, Clearance of inflammatory cells from the microcirculation of endotoxin-resistant C3H/HeJ mice, *Can. J. Microbiol.* **26**:725.

Walker, R. I., and Porvaznik, M. J., 1978, Association of leukopenia and intestinal permeability with radiation-induced sensitivity to endotoxin, *Life Sci.* **23**:2315.

Walker, R. I., Ledney, G. D., and Galley, C. B., 1975, Aseptic endotoxemia in radiation injury and graft-vs-host disease, *Radiat. Res.* **62**:242.

Walker, R. I., MacVittie, T. J., Sinha, B. L., Ewald, P. E., Egan, J. E., and McClung, G. L., 1978, Antibiotic decontamination of the dog and its consequences, *Lab. Anim. Sci.* **28**:55.

Walker, R. I., Porvaznik, M., Egan, J. E., and Miller, A. M., 1979, Hageman factor activation and tight junction disruption in mice challenged with attenuated endotoxin, *Experienta* **35**:759.

Walker, R. I., Ledney, G. D., Exum, E. D., Porvaznik, M., and Merrell, B. R., 1980, Responses to endotoxin of mice bearing the Lewis lung (3LL) carcinoma, *Toxicon* **18**:573.

Walker, R. I., Casey, L., Ramwell, P., and Fletcher, J. R., 1982, Association of prostacyclin production with resistance of C3H/HeJ mice to endotoxin shock. *Adv. Shock Res.* **7**:125.

Walker, R. I., Ledney, G. D., and Bertok, L., 1983, Reduced toxicity of irradiated endotoxin in mice compromised by irradiation, tumors, or infections, *J. Trauma*, in press.

Westphal, O., Westphal, U., and Sommer, T., 1977, The history of pyrogen research, in: *Microbiology 1977* (D. Schlessinger, ed.), pp. 221–235, American Society for Microbiology, Washington, D.C.

Wexler, B. C., and Tryczynski, E., 1955, Effect of a bacterial polysaccharide (piromen) in preventing tissue loss in rabbits due to gangrene induced by frostbite, *J. Lab. Clin. Med.* **45**:296.

Windle, W. F., and Chambers, W. W., 1950, Regeneration in the spinal cord of the cat and dog, *J. Comp. Neurol.* **93**:241.

Windle, W. F., and Wilcox, H. H., 1950, Extramedullary hemopoiesis in rabbit and cat induced by bacterial pyrogens, *Am. J. Physiol. Soc.* **163**:762.

Windle, W. F., Chambers, W. W., Ricker, W. A., Ginger, L. G., and Koenig, H., 1950, The reaction of tissues to administration of pyrogenic preparation from a *Pseudomonas* species, *Am. J. Med. Sci.* **219**:422.

Wise, W. C., Cook, J. A., Halushka, P. V., and Knapp, D. R., 1980a, Protective effects of thromboxane synthetase inhibitors in rats in endotoxic shock, *Circ. Res.* **46**:854.

Wise, W. C., Cook, J. A., Eller, T., and Halushka, P. V., 1980b, Ibuprofen improves survival from endotoxic shock in the rat, *J. Pharmacol. Exp. Ther.* **215**:160.

Wright, L. J., Kimball, H. R., and Wolff, S. M., 1969, Alterations in host responses to experimental *Candida albicans* infections by bacteria endotoxin, *J. Immunol.* **103**:1276.

10

Stimulation of Nonspecific Resistance by Radiation-Detoxified Endotoxin

Lóránd Bertók

I. Introduction

Many years ago, some investigators observed that within 24 hr of intravenous inoculation with killed typhoid bacilli, animals developed increased resistance to the live culture. Later, other workers identified the endotoxic lipopolysaccharides as the active material in such vaccines. The effects of this LPS-induced enhanced resistance have been demonstrated in infections induced by parasites, fungi, viruses, gram-positive and gram-negative bacteria, and mycobacteria. It is also well known that repeated parenteral injections of small doses of LPS will produce endotoxin tolerance. Among the several factors that may be involved in inducing this tolerance, bacterial LPS seems to play the major role in stimulating the host's natural defenses (Milner *et al.*, 1971; Pusztai-Markos, 1975). This tolerance has been manifested in an immunologically nonspecific manner, where endotoxic shock could be prevented by previous applications of small doses of endotoxin or endotoxin derivatives.

The undesirable side effects of endotoxin (such as pyrogenicity, hypotensive, abortifacient, etc.) made the application of endotoxin in the induction of the above beneficial effects unacceptable (Berczi *et al.*, 1966; Bertók, 1978, 1980).

Several investigators have sought to decrease the lethality of LPS while retaining the beneficial properties. The idea that such detoxification may be possible arose from successful detoxification of bacterial

Lóránd Bertók ● "Frèdèric Joliot-Curie" National Research Institute for Radiobiology and Radiohygiene, Budapest, Hungary.

protein toxins (e.g., tetanus or diphtheria anatoxins or toxoids). Various physical, chemical, and immunochemical procedures were applied in the attempt to produce detoxification (for review see Sultzer, 1971). One of the first procedures was the application of deacylation of endotoxin by potassium methylate (Nowotny, 1963, 1964, 1969). Phthalylation with O-phthalic anhydride gave excellent results (McIntire *et al.*, 1976). We employed the use of ionizing radiation (Previte *et al.*, 1967; Bertók *et al.*, 1973). In this chapter are summarized the results obtained with the radiation-detoxified LPS (RD-LPS) in experiments aiming to induce elevated nonspecific resistance (NSR).

II. Preparation, Biological and Chemical Properties of the RD-LPS

The parent LPS was isolated by the phenol–water method of Westphal *et al.* (1952) from culture of *E. coli* and purified by ultracentrifugation. For detoxification, the material was suspended in distilled water at a concentration of 10 mg/ml and irradiated with different doses (50, 100, 150, and 200 kGy) in a ^{60}Co source (Noratom-Gamma).

The most important biological properties of the RD-LPS irradiated with 50 kGy are summarized in Table 1. The listed results demonstrate that the noxious properties of the parent LPS were reduced. However, the endotoxin tolerance-inducing, the shock-preventing, the radioprotective, the NSR-enhancing, and the immunoadjuvant capacity of the starting material were preserved (Bertók, 1978, 1980). It should be pointed out here that the toxic properties of the irradiated preparations could be further reduced by increasing the dose of radiation (Füst *et al.*, 1977). Since a detailed review has been published recently on these results (Bertók, 1980), Table 1 only summarizes the highlights of this research.

Regarding the chemical changes in the structure of LPS induced by radiation, only limited information is available. Radiation decreases the KDO and glycosamine content and has some effect on the fatty acid composition of the lipid fraction (El-Sabbagh *et al.*, 1982). The amino acids of the preparation are also altered by irradiation and this effect seems also to be dose dependent.

III. RD-LPS as a Stimulator of NSR

A. Endotoxic Shock

Endotoxins enter the circulatory system in various forms of shock, and the importance of their hemodynamic effect in the development and irreversibility of shock has been described (Fine *et al.*, 1959; Fine,

Table 1. Comparison of Various Biological Effects of Parent and RD-LPS (50 kGy) Preparation

Effect	LPS	RD-LPS
LD_{50} (mg/kg) (rat)[a]	20	50
Complement decrease *in vivo* (%) (rabbit)[b]	35	10
Anticomplementary activity *in vitro* (μg)[a]	30	174
Hypotension (mm Hg) (dog)[a]	< 60	> 10
Leukocyte decrease (%) (rabbit)[b]	66	45
Platelet decrease (%) (rabbit)[b]	80	50
Stem cell mobilization (CFU-S) (mouse)[c]	521	247
Fibrinogen decrease (%) (rabbit)[b]	40	17
Pyrogenicity (°C) (rabbit)[d]	2.4	1.2
Local Shwartzman activity (mm^2) (rabbit)[b]	1200	600
Fetopathy (%) (rat)[e]	100	2
Spermatozoa agglutination *in vitro* (swine)[f]	+	−
Interferon induction (mouse)[g]	+	+
Hyperlipidemia[h]	+	−

[a] Füst *et al.* (1977).
[b] Szilágyi *et al.* (1978).
[c] Gidáli *et al.* (1978) cited in Bertók (1980). For determination of stem cell mobilization, five male (BALB/c × CBA)F$_1$ hybrid mice were injected i.v. (500 μg of LPS or RD-LPS). Five days later, the colony-forming (spleen) unit (CFU-S) of blood samples was determined in the same five hybrid mice.
[d] Antal and Bertók (1978) cited in Bertók (1980). For determination of pyrogenicity, 10 male New Zealand rabbits weighing 2500 g were injected i.v. (1 μg of LPS or RD-LPS). Difference between baseline and maximum temperature is expressed in °C.
[e] Csordás *et al.* (1978).
[f] Sutka and Bertók (1978) cited in Bertók (1980). For spermatozoa agglutination, boar spermatozoa were used.
[g] Tálas and Bertók (1978) cited in Bertók (1980). For detection of interferon-inducing capacity, 10 male and female BALB/c mice weighing 20–25 g were injected i.v. (25 μg of LPS or RD-LPS).
[h] Gaál *et al.* (1981). For measurement of the hyperlipidemic effect, 10 male and female C57BL mice weighing 20 g were injected i.p. (1–2 mg/kg). Blood samples were taken from 1 to 144 hr. The VLDL, LDL, and HLDL fractions were determined.

1965). Examples of endotoxic shocks (enteroendotoxemia) are the so-called "coli-dyspepsia" of newborns (human baby, calf, baby pig), edematous disease of pigs, septic shock and peritonitis, etc. (Bertók, 1977, 1980).

Experimental endotoxic shock can be provoked by intravenous or intraperitoneal injection of toxic doses of LPS in endotoxin-sensitive animal species (Berczi *et al.*, 1966). The RD-LPS pretreatment (i.v., i.p., or s.c.) in rats, mice, hamsters, guinea pigs, dogs, pigs, and horses can induce complete endotoxin tolerance. The effective doses are 25–50 μg in mice, 1–2 mg in pigs, and 2–4 mg in dogs. This tolerance persists in animals for a few weeks (Bertók, 1980).

The protective effect of the RD-LPS can be demonstrated by measuring a few basic parameters of the cardiovascular system of dogs in acute endotoxin shock. A single dose of 4 mg of RD-LPS was given i.p., 48 hr before the induction of shock by 2 mg/kg LPS given i.v. The 2 mg/kg dose of RD-LPS 50 kGy did not elicit any significant change in blood pressure or in cardiac output. Table 2 shows the difference between the values of blood pressure and cardiac output of the control and pretreated animals, respectively. In control animals, LPS diminished the blood pressure and cardiac output by about 50% in 2 min. However, RD-LPS pretreatment considerably reduced these manifestations of the LPS administration. The comparison of blood pressure and cardiac output permitted the quantitative assessment of this protective effect (Balogh *et al.*, 1973, Balogh *et al.*, 1981).

B. Fecal Peritonitis and Septic Shock

Prevention of fecal peritonitis and sepsis is an important problem in surgery. A suitable model of this (i.p. injection of 30 mg colonic contents) has been worked out in rats.

If 1 mg RD-LPS 50 kGy was given i.p. 5 days before the challenge, fatal peritonitis and septic shock was prevented in 90% of the cases (see Table 3). All control animals perished within 12 hr. From the blood of these animals, *Proteus vulgaris* was isolated. It is worth mentioning that in a few percent of RD-LPS-pretreated survivors, local (sequestered) peritonitis or periorchitis was demonstrated. It is likely that the enhancement of NSR by RD-LPS can localize the diffuse peritonitis in these cases. Based on these experiments, we believe that RD-LPS pretreatment may be useful in the preparation of abdominal (gastrointestinal) operations (Balogh *et al.*, 1973; Balogh and Bertók, 1974).

Table 2. Effect of LPS (2 mg/kg i.v.) on the Systemic Circulation of Untreated (A, $N = 5$) and RD-LPS-Pretreated (B, $N = 5$) Dogs[a]

| | | | Change (min after endotoxin administration) | | | | | |
		Control	2	5	15	30	45	60
Mean arterial	A	89.2	− 42.2*	− 46.8*	− 49.0*	− 28.2	− 30.6*	− 35.0*
blood pressure		± 6.1	± 18.8	± 14.1	± 11.4	± 11.8	± 10.2	± 8.2
(mm Hg)	B	97.6	− 24.2	− 16.6	− 3.8	− 4.4	− 10.0	− 10.1
		± 12.6	± 12.6	± 8.6	± 5.1	± 2.9	± 7.2	± 10.9
Cardiac output	A	104.6	− 51.8*	− 54.9*	− 54.8**	− 38.5**	− 41.7*	− 53.8**
(ml/min/kg)		± 7.1	± 15.6	± 12.8	± 10.2	± 6.0	± 10.1	± 7.3
	B	103.2	− 23.7	− 16.2	− 4.2	− 3.6	− 10.1	− 3.1
		± 11.4	± 12.5	± 9.8	± 9.1	± 5.1	± 4.3	± 12.3

[a] Asterisks refer to significant changes (*$p < 0.05$, **$p < 0.01$) from the control value.

Table 3. Protective Effect of RD-LPS (50 kGy) Pretreatment in Experimental Fecal Peritonitis and Septic Shock of Rats[a,b]

Pretreatment	Lethality (dead/total)
—	200/200
LPS 1 mg i.p.	21/200
RD-LPS 1 mg i.p.	20/200

[a] Data from Balogh *et al.* (1973).
[b] Challenge (+ 120 hr), colonic contents (30 mg) i.p.

C. Hemorrhagic Shock

Fine *et al.* (1959; Fine, 1965) proposed that hemorrhagic shock among others is also an endotoxemic state. We induced hemorrhagic shock (hypotension) in dogs by bleeding and demonstrated protective value of RD-LPS in this type of shock, too. One i.p. injection of the RD-LPS (50 kGy) can protect the majority of dogs against lethal hemorrhagic shock. Table 4 shows the results of such an experiment.

D. Intestinal Ischemic Shock Induced by Occlusion of the Superior Mesenteric Artery

It is well known that occlusion of the superior mesenteric artery (SMAO) induces intestinal ischemia, so-called "vascular-ileus." In this form of ileus, the fatal shock cannot be prevented by revascularization. Occlusion of the cranialis mesenteric artery is a suitable animal model for the SMAO. LPS absorption from the intestinal canal is the most important factor in the pathogenesis of irreversible SMAO shock (Fine, 1965; Sófalvi *et al.*, 1977). We attempted to prevent the intestinal isch-

Table 4. Protective Effect of RD-LPS (50 kGy) Pretreatment in Experimental Hemorrhagic Shock of Dogs[a,b]

Pretreatment	Lethality dead/total
—	8/12
RD-LPS 4 mg i.p.	2/12

[a] Data from Balogh and Bertók (1974).
[b] Challenge (+48 hr), hypotonia (35 mm Hg 150 min).

emic shock in rats by RD-LPS pretreatment. The RD-LPS pretreatment protected 74% of the animals in experimental intestinal eschemia (Table 5).

E. Tourniquet Shock

The tourniquet is a good and accepted model for the study of the shock induced by limb ischemia. Orban et al. (1978) demonstrated that in the pathogenesis of tourniquet shock, endotoxin is absorbed from the intestinal tract. For this reason the tourniquet seemed to be suitable for demonstration of the protective effect of RD-LPS in this form of shock. Table 6 summarizes the results of an experiment. The RD-LPS pretreatment can protect the majority of animals from the consequences of endotoxic shock induced by limb ischemia.

F. Fetopathy and Abortion Induced by LPS

Urinary tract infections have often been blamed for premature deliveries. These are caused by gram-negative bacteria, and their abortifacient effect is assumed to be due to their endotoxin content (Good and Thomas, 1953). The abortifacient effect and the Shwartzman reactivity of bacterial endotoxins were also demonstrated in animal experiments (Apitz, 1935; Chedid et al., 1962; Dzvornyak et al., 1970; Parant and Chedid, 1964; Rieder and Thomas, 1960). Endotoxin increases $PGF_{2\alpha}$ synthesis in the endometrium, and thus may play a role in the abortifacient effect. Various pharmaceutical compounds cannot prevent this effect of LPS (Skarnes and Harper, 1972). We tested the efficiency of RD-LPS pretreatment in the following experimental system. Physiological saline solution was given to one group of pregnant rats. A second group was treated with RD-LPS. Neither maternal nor fetal damages were observed in the first group and only four fetuses out of 202 were found dead in the second group. A third group of animals was injected

Table 5. Protective Effect of RD-LPS (50 kGy) Pretreatment in SMAO Shock of Rats[a,b]

Pretreatment	Lethality (dead/total)
—	15/15
RD-LPS 100 μg i.v.	4/15

[a] Data from Kisida et al. (1974).
[b] Challenge (+24 hr), SMA ligature (150 min).

Table 6. Protective Effect of RD-LPS (50 kGy) Pretreatment in Experimental Tourniquet Shock (Limb Ischemia) of Rats[a,b]

Pretreatment	Lethality (dead/total)
—	23/35
RD-LPS 200 μg i.v.	6/35

[a] Data from Orbán *et al.* (1978).
[b] Challenge (+48 hr), tourniquet (270 min).

with 1 mg LPS. Twenty-four hours later, all mothers were found alive; on the other hand, all fetuses were dead in the uterus, and abortion started soon thereafter. A fourth group of animals was pretreated with RD-LPS 24 hr before the LPS injection. In this group, none of the mothers died, and only 21 fetuses out of 207 were found dead. Accordingly, RD-LPS gave 90% protection against the abortifacient effect of LPS (Table 7).

Histological examinations supported the above-mentioned results. Control animals (first group) did not show any abnormality. In the livers of the second group of pregnant animals, a swelling of the Kupffer cells was visible. In the giant cell layer of the placentas, focal necroses and moderate degenerations could be observed. Light focal necroses were found also in the labyrinth layer and the decidua. In the third group, the livers of all pregnant animals showed extensive midzonal necroses. In the livers of 50% of the fetuses, singular cellular necroses could be detected. However, extensive necroses were seen in the laby-

Table 7. Effect of RD-LPS (50 kGy) Pretreatment on Endotoxin-Induced Fetopathy and Abortion in Pregnant Rats[a]

Pretreatment	Challenge (+24 hr)	Lethality (dead/total) Parent	Lethality (dead/total) Fetus
Physiological NaCl solution	—	0/20	0/210
RD-LPS 200 μg i.v.	—	0/20	4/202
—	1 mg LPS	0/20	198/198*
RD-LPS 200 μg i.v.	1 mg LPS	0/20	21/207*

[a] Data from Csordás *et al.* (1978).

rinth layer and in the giant cell layer where the necroses involved both clear and giant cells. Decidual necroses were extensive, and focal necroses were also present in the amnionic epithelium. So-called hyaline thrombi were formed in the maternal blood spaces and in damaged areas of the labyrinths. In the fourth group of animals, we found that all necrotic phenomena had become less severe. Singular cell necroses were detectable in the livers of the pregnant animals, and Kupffer cells were swollen. In the placentas, a mild degeneration of the labyrinth-trophoblast was evident; however, the clear cells and the amnionic epithelium revealed scattered focal necroses (Csordás et al., 1978).

G. Radiation Disease

Endotoxins play a very important role in the pathogenesis of the so-called intestinal syndrome of radiation (Bertók and Kocsár, 1974). On the other hand, it is well known that a subtoxic dose of parenterally given LPS is radioprotective in both intestinal and bone marrow syndromes. For this reason, we used RD-LPS as a protector against radiation-induced disease. As indicated in Table 8, both the RD-LPS and the parent LPS saved on average 70% of the irradiated rats (Bertók, 1980).

H. Protective Effect in Various Experimental Infections

LPS pretreatment can prevent lethal bacterial infections through elevated NSR. NSR plays a significant role both in the pathogenesis of some infectious diseases and in the ability of the body to recover from them. Facultative pathogenic bacteria may flourish and cause disease when specific and nonspecific resistance are impaired. For example, *Pasteurella multiseptica* cannot cause a fatal illness in normal rats, but if introduced following irradiation of the animals with X-rays, a classical

Table 8. Radioprotective Effect of
RD-LPS (50 kGy) in X-Irradiated
(800 R) Rats[a,b]

Pretreatment	Lethality dead/total
—	25/25
LPS 100 μg i.v.	8/25
RD-LPS 100 μg i.v.	7/25

[a] Data from Bertók (1980).
[b] Irradiation (+24 hr), X-rays (800 R) (total body).

pasteurellosis with 100% mortality ensues. This model was used by us to assay the capacity of preparations to induce NSR. It has been shown that the pretreatment of rats with RD-LPS can protect all animals from the *Pasteurella* infection. This pretreatment will also protect 90–100% of rats inoculated with *Klebsiella pneumoniae* or *Proteus vulgaris* (Bertók, 1980).

LPS is an interferon inducer and the RD-LPS has the same effect. The lethality of experimental Aujeszky disease (pseudorabies virus, herpes group) was reduced from 100% to 30% by this means (Bertók, 1980).

I. Protection of Antilymphocyte Serum-Treated (Immunosuppressed) Animals by RD-LPS

The absence of NSR in immunosuppressed patients is a major problem of transplantation procedure. It is well known that the majority of transplant patients die of common septicemia because of their inability to mount an efficient immune response. Antilymphocyte serum (ALS) is a potent immunosuppressant commonly used in organ transplantation. For this reason, the possibility of augmenting a patient's NSR and induction of endotoxin tolerance after ALS treatment has a great importance. In the experiments to be reported, induction of endotoxin tolerance in animals treated with ALS is described.

RD-LPS was given to ALS-treated rats to see whether it could increase their resistance to lethal dose of LPS. Results are shown in Table 9. It is evident that in spite of the suppression of thymus function by ALS, induction of tolerance to LPS was normal.

Table 9. Induction of Endotoxin Tolerance by RD-LPS (50 kGy) in ALS-Treated (Immunosuppressed) Rats[a,b]

Pretreatment		Lethality
ALS[c]	RD-LPS	dead/total
—	—	18/20
—	100 μg i.v.	0/20
2 × 2 ml/100 g body wt i.p. (−72 and −24 hr)	100 μg i.v.	0/20

[a] Data from Bertók *et al.* (1979).
[b] Challenge (+24 hr), LPS (5 mg) i.v.
[c] ALS was prepared in rabbits by injecting rat thymic lymphocytes. The cytotoxic activity of ALS was tested by dye exclusion and ^{51}Cr release. The immunosuppressive effect of ALS was demonstrated on the immune response to SRBC in rats.

IV. RD-LPS as Immunoadjuvant

A. Adjuvant Effect

Endotoxins are well known to exert a marked adjuvant activity on the humoral immune response of mammals. Data are scarce and controversial concerning the adjuvant effect of detoxified LPS preparations and it seems to depend closely on the detoxification method (Sultzer, 1971). Practically only the potassium methylate (Nowotny, 1964) and the RD-LPS preparations retained their adjuvant activity. The RD-LPS preparations (50, 100, 150, and 200 kGy) in a dose-dependent manner have a good immunoadjuvant effect in rats using SRBC as antigen. In the case of 50 and 100 kGy LPS preparations, the relative spleen weight, the splenic PFC (direct and indirect) count, and the hemolysin (MER and MES) titer produced by them are mostly identical with the parent LPS. In the case of 150 and 200 kGy LPS preparations, the majority of these parameters are decreased (Elekes and Bertók, 1979).

The RD-LPS (50 kGy) also shows a good adjuvant effect in mice using inactivated food-and-mouth-disease virus as antigen (Sólyóm and Bertók, 1983).

B. Immunoregeneration in Irradiated Animals

Ionizing radiation decreases all immune functions. For this reason, it is important to look for possibilities of immunoregeneration. It has been demonstrated that RD-LPS preparations (50, 100, 150, and 200 kGy) can regenerate the immune functions in sublethally irradiated rats. The relative spleen weight, the PFC (direct and indirect) count, and the serum hemolysin level are elevated 21 days after irradiation (Elekes et al., 1978). In the regeneration of the immune system, both T and B lymphocytes are necessary. It is likely that the relative radioresistant T-helper cell population has an important role in the above regenerating effect of RD-LPS (Temesi et al., in press).

V. Conclusion

As illustrated above, the RD-LPS preparations have a wide range of beneficial effects (summarized in Table 10). The most important question arising from these results deals with the possible mechanisms resulting in enhanced NSR. In the pathogenesis of the majority of the experimental models we used, the damage of lysosomal membranes is very important. In the development of endotoxin tolerance or NSR en-

Table 10. Protective Value of RD-LPS (50 kGy) Pretreatment in Various
Shocks, Radiation Disease, and Infections[a]

Experimental intervention (species)	Protective value (%)
Endotoxin shock	100
(rat, pig)	
Radiation disease (X-rays or ^{60}Co-γ)	70
(rat)	
Septic shock (induced by fecal peritonitis)	90
(rat)	
Tourniquet shock (limb ischemia)	60
(rat)	
Intestinal ischemic shock (induced by SMAO)	74
(rat)	
Hemorrhagic shock (hypotension)	70
(dog)	
Abortus (LPS induced)	90
(rat)	
Pasteurella infection (in X-irradiated animals)	100
(rat)	
Klebsiella infection	90
(rat)	
Aujeszky-virus infection	70
(rat)	
Immunosuppression/LPS-tolerance (in ALS-treated animals)	100
(rat)	

[a] From Bertók (1980).

hancement, the so-called lysosomal membrane "condensation" may play an important role (Bertók, 1978; Kutas *et al.*, 1969). RD-LPS pretreatment (endotoxin tolerance) prevented lysosomal enzyme (β-glucosaminidase, cathepsin D) liberation from lysosomes challenged by parent (toxic) LPS in rabbits (Szilágyi *et al.*, 1982).

At the same time, the RD-LPS preparations preserve many beneficial effects such as RES activation, phagocytosis promotion, macrophage activation, natural antibody and properdin level elevation, and some others. According to Chedid and Parant, "tolerance to the lethal effect of endotoxin could be mediated by several nonspecific mechanisms, such as increased stability of lysosome membranes, corticoid secretion, stimulation of the reticuloendothelial system, and, finally, biological degradation and detoxification of the antigen" (Chedid *et al.*, 1978). As the above results indicate, ionizing radiation detoxifies the LPS while maintaining its capacity to induce tolerance to toxic effects and to mobilize the host defenses in an immunologically nonspecific fashion.

VI. References

Apitz, K., 1935, A study of generalized Shwartzman phenomenon, *J. Immunol.* **29**:255.

Balogh, A., and Bertók, L., 1974, Experimentelle Abwehr des septischen und hämmorrhagischen Shocks durch Vorbehandlung mit detoxifiziertem endotoxin, *Acta Chir. Acad. Sci. Hung.* **15**:21.

Balogh, A., Bertók, L., and Kocsár, L., 1973, Prevention of experimental peritonitis (septic shock) by detoxified endotoxin pretreatment [Hungarian], *Kiserl. Orvostud.* **25**:157.

Berczi, I., Bertók, L., and Bereznai, T., 1966, Comparative studies on the toxicity of *Escherichia coli* lipopolysaccharide endotoxin in various animal species, *Can. J. Microbiol.* **12**: 1070.

Bertók, L., 1977, Physico-chemical defense of vertebrate organisms: The role of bile acids in defense against bacterial endotoxins, *Perspect. Biol. Med.* **21**:70.

Bertók, L., 1978, Immunological properties of detoxified lipopolysaccharides, in: *Immunology 1978* (J. Gergely, G. Medgyesi, and Z. Hollan, eds.), pp. 463–470, Akadémiai Kiadó, Budapest.

Bertók, L., 1980, Radio-detoxified endotoxin as a potent stimulator of nonspecific resistance, *Perspect. Biol. Med.* **24**:61.

Bertók, L., and Kocsár, L., 1974, Die Rolle des Endotoxins beim intestinalen Syndrom der Strahlenkrankheit, *Wiss. Z. Karl Marx Univ. Leipzig Math. Naturwiss. Reihe* **23**:65.

Bertók, L., Kocsár, L., Várterész, V., Bereznai, T., and Antoni, F., 1973, Procedure for preparation and use of radio-detoxified bacterial endotoxin [Hungarian], Patent 162,973, Budapest.

Bertók, L., Elekes, E., and Merétey, K., 1979, Endotoxin tolerance in rats treated with antilymphocyte serum, *Acta Microbiol. Acad. Sci. Hung.* **26**:135.

Chedid, L., Boyer, F., and Parant, M., 1962, Etude de l'action abortive des endotoxines injectees a la souris gravide normale, castree ou hypohysectomisee, *Ann. Inst. Pasteur Paris* **102**:77.

Chedid, L., Parant, M., and Parant, F., 1978, Recent aspects of nonspecific resistance and adjuvants, in: *Immunology 1978* (J. Gergely, G. Medgyesi, and Z. Hollán, eds.), pp. 453–462, Akadémiai Kiadó, Budapest.

Csordás, T., Bertók, L., and Csapó, Z., 1978, Experiments on prevention of the endotoxin-abortifacient effect by radio-detoxified endotoxin pretreatment in rats, *Gynecol. Obstet. Invest.* **9**:57.

Dzvornyák, L., Ruzicska G., Deseö, G., and Csaba, B., 1970, Distribution of P³²-labeled *E. coli* endotoxin in the tissues of pregnant rabbits and their fetuses, *Am. J. Obstet. Gynecol.* **106**:721.

Elekes, E., and Bertók, L., 1979, Effect of decreased toxicity endotoxin preparations induced by radiation on the immune response of normal and irradiated rats [Hungarian], *Izotoptechnika* **23**:73.

Elekes, E., Bertók, L., and Merétey, K., 1978, Adjuvant activity of endotoxin preparations in normal and irradiated rats, *Acta Microbiol. Acad. Sci. Hung.* **25**:17.

El Sabbagh, M. C., Galanos, C., Bertók, L., Füst, G., and Lüderitz, O., 1982, Effect of ionizing radiation on chemical and biological properties of *Salmonella minnesota* R595 lipopolysaccharide, *Acta Microbiol. Acad. Sci. Hung.*, **29**:321.

Fine, J., 1965, Shock, in: *Handbook of Physiology*, Section 2, *Circulation* (W. F. Hamilton and P. Dow, eds.), Vol. III, p. 2031, American Physiological Society, Washington, D.C.

Fine, J., Frank, E. D., Ravin, H. A., and Rutenburg, S. H., 1959, The bacterial factor in traumatic shock, *N. Engl. J. Med.* **260**:214.

Füst, G., Bertók, L., and Juhász-Nagy, S., 1977, Interactions of radio-detoxified *Escherichia coli* endotoxin preparations with the complement system, *Infect. Immun.* **16**:26.

Gaál, H. D., Kremmer, T., Balint, Zs., Holczinger, L., Bertok, L., and Nowotny, A., 1983, Effects of bacterial endotoxins and their detoxified derivatives on lipid metabolism in mice, *Appl. Toxicol. Appl, Pharmacol.*, in press.

Good, R. A., and Thomas, L., 1953, Studies on the generalized Shwartzman reaction, *J. Exp. Med.* **97**:871.

Kisida, E., Bertók, L., and Karika, G., 1974, Toxicity of peritoneal fluid from dogs with an occluded superior mesenteric artery, *J. Reticuloendothelial Soc.* **15**:13.

Kisida, E., Bertók, L., and Karika, G., 1975, Role of endotoxin in the pathogenesis of superior mesenteric artery occlusion-induced shock [Hungarian], *Magy. Sebész.* **28**:313.

Kutas, V., Bertók, L., and Szabó, L. D., 1969, Effect of endotoxin on the serum ribonuclease activity in rats, *J. Bacteriol.* **100**:550.

McIntire, F. C., Hargie, M. P., Schenck, J. R., Finley, R. A., Sievert, H. W., Rietschel, E. T., and Rosenstreich, D. L., 1976, Biologic properties of nontoxic derivatives of a lipopolysaccharide from *Escherichia coli* K235, *J. Immunol.* **117**:674.

Milner, K., Rudbach, J. A., and Ribi, E., 1971, General characteristics, in: *Microbial Toxins*, Vol. IV (G. Weinbaum, S. Kadis, and S. J. Ajl, eds.), pp. 1–65, Academic Press, New York.

Nowotny, A., 1963, Endotoxoid preparations, *Nature (London)* **197**:721.

Nowotny, A., 1964, Chemical detoxification of bacterial endotoxins, in: *Bacterial Endotoxins* (M. Landy and W. Braun, eds.), p. 29, Rutgers University Press, New Brunswick, N.J.

Nowotny, A., 1969, Molecular aspects of endotoxic reactions, *Bacteriol. Rev.* **33**:72.

Orbán, I., Bertók, L., Regös, J., and Gazsó, L., 1978, Die Rolle der bakteriellen Endotoxine beim Tourniquet Shock der Ratten, *Acta Chir. Acad. Sci. Hung.* **19**:339.

Parant, M., and Chedid, L., 1964, Protective effect of chlorpromazine against endotoxin induced abortion, *Proc. Soc. Exp. Biol. Med.* **106**:906.

Previte, J. J., Chang, Y., and El-Bisi, H., 1967, Detoxification of *Salmonella typhimurium* lipopolysaccharide by ionizing radiation, *J. Bacteriol.* **93**:1607.

Pusztai-Markos, Z., 1975, The role of humoral and cellular factors in the stimulation of nonspecific resistance induced by endotoxin, in: *Gram-Negative Bacterial Infections and Mode of Endotoxin Action* (B. Urbaschek, R. Urbaschek, and E. Neter, eds.), pp. 147–154, Springer-Verlag, Berlin.

Rieder, R., and Thomas, L., 1960, Studies of the mechanisms involved in the production of abortion by endotoxin, *J. Immunol.* **84**:189.

Skarnes, R. C., and Harper, K. J., 1972, Relationship between endotoxin induced abortion and synthesis of prostaglandin F, *Prostaglandins* **1**:191.

Sófalvi, A., Bertók, L., Kisida, E., and Karika, G., 1977, Local Sanarelli–Shwartzman phenomenon provoked by superior mesenteric artery occlusion in rabbits [Hungarian], *Kiserl. Orvostud.* **29**:613.

Sólyom, A., and Bertók, L., 1983, Application of radio-detoxified endotoxin (TOLERIN²) as adjuvant for experimental foot-and-mouth disease vaccine, *Bull. Internat. Epizool. (Paris)*, in press.

Sultzer, B. M., 1971, Chemical modification of endotoxin and inactivation of its biological properties, in: *Microbial Toxins*, Vol. V (S. Kadis, G. Weinbaum, and S. J. Ajl, eds.), pp. 91–126, Academic Press, New York.

Szilágyi, T., Csernyánszky, H., Gazdy, E., and Bertók, L., 1978, Haematological effect and Shwartzman reactivity of radio-detoxified endotoxin, *Acta Microbiol. Acad. Sci. Hung.* **25**:159.

Szilágyi, T., Csernyánszky, H., Gazdy, E., and Bertók, L., 1980, The procoagulant activity of leukocytes pretreated with radio-detoxified endotoxin, *Acta Microbiol. Acad. Sci. Hung.* **27**:191.

Szilágyi, T., Cjernyánstky, H., Gazdy, E., and Bertók, L., 1982, The lysosomal enzyme liberation and tolerance inducing effects of readio-detoxified endotoxins, *Acta Microbiol. Acad. Sci. Hung.* **29**:155.

Temesi, A., Bertók, L., and Pellet, S., 1983, Stimulation of human peripheral lymphocytes by endotoxin and radio-detoxified endotoxin. Role of various sera, *Acta Microbiol. Acad. Sci. Hung.*, in press.

Westphal, O., Lüderitz, O., and Bister, F., 1952, Über die Extraktion der Bakterien mit Phenol-Wasser, *Z. Naturforsch.* **7b**:148.

11

Lymphocyte Activation by Endotoxin and Endotoxin Protein

The Role of the C3H/HeJ Mouse

Barnet M. Sultzer

I. Introduction

A perusal of the literature on endotoxins in the past decade makes it clear that a renewed interest has developed to explain the multiple pathophysiological and immunological host responses to lipopolysaccharide endotoxin (LPS). In great measure this was due to the recognition that many of these responses were under genetic control. The genetic approach taken by numerous workers arose from the discovery of the mutant C3H/HeJ mouse strain, which is highly resistant to the toxic effects of LPS and which displays cellular responses both *in vivo* and *in vitro* that are diametrically opposite to those from normally susceptible mice (Sultzer, 1968, 1969). The availability of this inbred strain gave rise to a variety of comparative studies so that a rather extensive literature on this subject has accumulated and continues to do so. However, within the allotted space for this article, I will only attempt to highlight those aspects of the work with this strain that have dealt with the activation of lymphocytes by LPS, and to discuss in some detail the newer studies originating from the use of the C3H/HeJ mouse that have established that various proteins associated with the LPS in the outer membrane of gram-negative bacteria are potent activators of immunocompetent cells. A more comprehensive review of the

Barnet M. Sultzer • Department of Microbiology and Immunology, State University of New York Downstate Medical Center, Brooklyn, New York 11203.

literature on the various responses of the C3H/HeJ mouse to LPS may be found in the recent article by Morrison and Ryan (1979).

II. Lymphocyte Activation by LPS

A. B-Cell Responses to LPS

Bacterial LPS was first described to be a potent mitogen for mouse splenic lymphocytes in culture by Peavy *et al.* (1970). Applying their system with some modifications to the culture of C3H/HeJ spleen cells, we found that DNA synthesis was some five times less in C3H/HeJ cells as compared to endotoxin-susceptible A/HeJ spleen cells in the presence of a Boivin-type LPS (Sultzer, 1972). When C3H/HeJ spleen cells were cultured with Westphal-type LPS, which contained less than 1% protein, the response was essentially nil (Sultzer and Nilsson, 1972). At about the same time, Gery *et al.* (1972) and Andersson *et al.* (1972a) reported that LPS was a mitogen for murine B cells. Furthermore, immunoglobulin and antibody synthesis could be induced by LPS in cultured mouse spleen cells to a variety of antigens, so that in effect LPS was characterized as the first polyclonal B-cell activator (Andersson *et al.*, 1972b). In further studies on the mitogenic and immune responses to LPS by the C3H/HeJ mouse (Sultzer, 1973), DNA synthesis was found to be normal in spleen cells cultured with the T-cell mitogen concanavalin A (Con A). In addition, the PPD-tuberculin stimulation of DNA synthesis in lymphocytes from normal and *Mycobacterium bovis* (BCG)-sensitized C3H/HeJ mice was similar to that obtained with LPS-responder CBA/J and A/HeJ mice. The fact that the antibody response to LPS was diminished in C3H/HeJ mice, whereas their immune response to sheep erythrocytes was normal, led to the conclusion that the B cells of C3H/HeJ mice were deficient in their capacity to be activated by LPS (Sultzer, 1973). Indeed, this deficiency led to the suggestion that the C3H/HeJ resistance to the toxicity of endotoxin possibly was related to the hyporeactivity of their cells to LPS (Sultzer, 1972), a notion that, although somewhat simplistic, has gained some experimental support (Rosenstreich *et al.*, 1977). The apparent inability of C3H/HeJ B cells to respond to LPS led to a series of studies from many groups that definitively characterized this rather selective defect. Thus, the failure of C3H/HeJ B cells to respond to LPS was found from cell transfer experiments not to be due to the absence of helper cells or the presence of suppressor cells (Watson and Riblet, 1975; Slowe and Waldmann, 1975; Sultzer, 1976; Coutinho, 1976; Glode *et al.*, 1976; Skidmore *et al.*, 1977); rather, it was the result of the interaction between LPS and the lymphocytes. Binding of the LPS to

C3H/HeJ cells did not differ from that observed with cells from normal mice (see below); however, the responses subsequent to the initial binding event, which normally consisted of proliferation and differentiation, did not occur to any appreciable extent. Furthermore, the C3H/HeJ mouse failed to maintain a sustained high-titer primary IgM antibody response to LPS, failed to be primed for an enhanced secondary immune response to LPS, failed to respond to the adjuvant effect of LPS, and was unable to reverse the induction of tolerance to a protein antigen under the influence of LPS (Watson and Riblet, 1974; Rudbach and Reed, 1977; Goodman et al., 1978).

More extensive characterization of the B lymphocytes of C3H/HeJ mice has led to other findings that have contributed to our understanding of B-cell maturation and function in the mouse. For example, Melchers (1977), studying B-lymphocyte development in fetal liver, concluded that the differentiation of fetal liver cells to immunoglobulin-secreting antibody-producing cells occurs in two stages. The first conversion of a pre-B to a B cell is independent of externally added mitogen (such as LPS), and the second differentiation of a B cell to an antibody-producing cell requires the addition of the mitogen. Because the C3H/HeJ cells are nonresponsive to LPS, it was possible to test and verify by appropriate matings with responder strains that the antibody-producing responses developed in vitro with LPS were really due to the fetal liver cells of the embryos rather than contaminating cells of the mother. In other studies aimed at defining the frequencies of mitogen-reactive B cells in the mouse at different ages, Andersson et al. (1977) were unable to find LPS-reactive B cells in the spleens of new-born or 6- to 8-week-old C3H/HeJ mice; in contrast, various other strains of mice showed frequencies of 1 in 3 to 1 in 10 in young adults and 10% of their number in newborns. However, C3H/HeJ spleens at birth and later showed the normal frequencies of lipoprotein-reactive B cells, confirming the selective genetic deficiency of a subpopulation of LPS-reactive B cells in this strain. Hoffmann et al. (1977) were able to distinguish different pathways for the induction of mitosis and antibody production resulting from the B-cell activation by LPS by the use of the C3H/HeJ mouse. Thus, C3H/HeJ spleen cells could produce antibody to hapten-conjugated autologous mouse erythrocytes only if LPS and macrophages from LPS-responder C3HeB/FeJ mice or LPS-induced tumor necrosis serum were added to the culture system. The mitotic response, however, was not restored, suggesting that adjuvanticity and mitogenicity represent distinct pathways of B-cell activation by LPS, probably subject to different mechanisms of regulation. In this regard, studies directed at the adjuvant activity of LPS for thymus-independent antigens such as type III pneumococcal polysaccharide (S3) turned up the unexpected finding that LPS has a strong suppressive effect on antibody responses to this antigen in BALB/c mice (Braley-Mullen *et al.*, 1977). This was not the case in the C3H/HeJ strain, which led these authors to suggest that suppression was dependent on the B-cell mitogenic activity of the LPS. This suppression, by the way, was presumably directed against virgin B cells, since the development of IgG memory responses or the response of primed mice to S3 were not affected.

In other studies on the differentiation of B lymphocytes, Watson (1977) examined the induction of surface Ia antigen on bone marrow cells by LPS (Hämmerling *et al.*, 1975) and showed that lipid A fails to induce the expression of these antigens in C3H/HeJ, but not C3H/DiSn cells. However, protein-containing LPS preparations or the lipid A-associated protein (LAP) of Morrison *et al.* (1976) could induce Ia antigen expression in C3H/HeJ immature B cells, confirming that a subpopulation of C3H/HeJ immature B cells lack the LPS-response pathway of differentiation.

The possible involvement of lymphoid cells activated by LPS in what may be termed secondary immunologic reactions has also called upon the use of the C3H/HeJ mouse for analysis of the basis of these reactions. More specifically, Männel *et al.* (1979) were able to show by adoptive transfer of bone marrow cells from LPS-responder histocompatible C3H/HeN mice to their C3H/HeJ counterparts that the necrosis of a grafted fibrosarcoma on the C3H/HeJ mouse occurred when it was injected with LPS. This did not occur if the C3H/HeN mouse received C3H/HeJ bone marrow cells, suggesting that LPS produces necrosis of tumors by activating host lymphoreticular cells. In similar adoptive transfer experiments using bone marrow cells and the same mouse strains, Rosenstreich and McAdam (1979) found that LPS-induced serum amyloid A (SAA) protein occurred in C3H/HeJ mice that had received C3H/HeN cells, whereas this did not occur in those C3H/HeJ mice receiving their autologous cells. These authors suggested that lymphoreticular cells may be involved in the LPS-induced SAA protein synthesis and speculate that B cells and/or macrophages, therefore, may underlie the relationship between chronic infection or inflammation and the development of secondary amyloidosis. That B lymphocytes may also act in a regulatory role is supported in a recent report by Vallera *et al.* (1980), who found that LPS suppressed the development of cell-mediated cytotoxicity *in vitro* in a murine mixed lymphocyte culture (MLC). From various experiments, including the use of the unresponsiveness of C3H/HeJ cells in the MLC, these authors concluded that the proliferation of B lymphocytes was necessary for expression of this LPS-induced suppression, and that most likely a factor was released from these B cells that had a modulatory effect on the T cells involved in the generation of this cell-mediated cytotoxicity. Finally, it should be mentioned that there are instances where C3H/HeJ mouse lymphocytes respond in a normal or elevated fashion to LPS. An interesting example reported by Levy and Edgington (1980) is the LPS-induced procoagulant activity (PCA) in C3H/HeJ mice. It is known that lymphocyte collaboration is required for the induction of macrophage PCA by LPS in mice. Unexpectedly, purified LPS can induce PCA in C3H/HeJ lymphoid cells, indicating that, although the lymphocytes of these mice cannot be stimulated to proliferate by LPS, they can be triggered to the degree that they can induce macrophages to produce PCA.

B. T-Cell Responses to LPS

From all available evidence, LPS has not been shown to directly activate T cells of mice or humans to proliferate, in contrast to the well-documented direct effect of LPS on B cells. However, LPS has been found to stimulate T-cell differentiation, and the adjuvant effect of LPS *in vivo* has been claimed to involve T cells. Furthermore, T-cell activity also has been implicated in LPS effects on *in vitro* antibody production and B-cell mitogenesis.

In this regard, the C3H/HeJ mouse has found use in examining the interaction of LPS and T cells. For example, Norcross and Smith (1977) reported that when LPS-unresponsive immunoincompetent thymus cells are combined with a minimal number of peripheral lymphoid cells in the presence of LPS, synergistic DNA synthesis occurs mostly as a result of peripheral cell proliferation. The Con A-unresponsive thymus cell may be from the C3H/HeJ strain, but the peripheral B lymphocyte must be from an LPS-responsive strain, suggesting that, under certain circumstances, proliferative responses to LPS may be regulated by a subclass of thymic T cells.

In those splenic T cells that do respond to Con A, Bick *et al.* (1977) found, however, that LPS-nonresponder (C3H/HeJ) cells respond, but do so poorly to low concentrations of Con A in comparison with the LPS-responder C3H/Tif cells. Injection of LPS 2 days before culture inhibited the response to low doses of Con A in cultures of C3H/Tif spleen cells, but had no inhibitory effect on the dose–response profile of C3H/HeJ spleen cells. Furthermore, the low-dose Con A response of splenic T cells was found to be dependent upon the presence of an LPS-sensitive, Ia-positive cell, which is absent or not activated in the C3H/HeJ mouse.

When submitogenic doses of Con A were used in the experiments of Tanabe and Nakano (1979a,b), DNA synthesis in thymocytes from LPS-responder BALB/c mice was significantly enhanced by adding

LPS. However, the DNA synthesis of thymocytes from C3H/HeJ mice was not enhanced by adding LPS. In addition, the thymocytes from C3H/HeJ mice responded to a lesser extent than other strains of mice to the mitogenic stimulation of rabbit anti-mouse thymocyte serum. Both of these deficiencies could be restored by the addition of a supernatant factor derived from LPS-responder murine spleen cells cultured with LPS. C3H/HeJ spleen cells did not produce this factor, which was estimated to have a molecular weight of about 20,000, was heat labile, and sensitive to trypsin but not to DNase or RNase.

Another expression of activated T cells induced by Con A relates to the helper function of these cells. For the induction of thymus-dependent immune responses in B cells, helper T cells are fundamental. At least one effector mechanism of activated T cells in the induction of B cells is the release of soluble factors. In this regard, Primi *et al.* (1979) have recently reported that Con A-activated T cells secrete factors with polyclonal B-cell activating properties. Again, this was achieved with concentration of Con A lower than those optimal for induction of DNA synthesis. While spleen cells from nude mice were not activated, C3H/HeJ spleen cells were activated to polyclonal antibody synthesis by Con A. The authors concluded, therefore, that the LPS-responsive subset of B cells in normal mice is not the target, to a large extent, for the Con A-induced polyclonal B-cell activation. However, the results also suggest that C3H/HeJ helper T cells do respond to Con A, distinguishing this subset of T cells from those that proliferate in the presence of low concentration of Con A and depend on the presence of an Ia-positive cell or endotoxin-induced cell factors.

Further investigations into the helper effect of T cells on polyclonal B-cell responses also have been reported by Goodman and Weigle (1979). By admixture of separated B- and T-cell populations, these workers demonstrated that normal fresh splenic T cells were able to augment polyclonal B-cell responsiveness to LPS up to severalfold, provided the two cell types were cocultured in equal numbers, a circumstance not apparent in normal spleen cell cultures. Interestingly, further emphasis on the phenomenon of LPS activation of T cells was provided in these studies when it was found that T cells from C3H/HeJ mice were deficient in the capacity to enhance polyclonal B-cell responsiveness of B cells derived from LPS-responder (C3H/HeN) mice. This result carries the implication, as the authors point out, that LPS may act directly on T cells and in the C3H/HeJ mouse this T-cell response is deficient.

That T lymphocytes can be involved in a deleterious immune response, such as experimental autoimmune thyroiditis, has also been documented. Esquivel *et al.* (1978) showed that the injection of soluble mouse thyroglobulin (MTg) in conjunction with LPS abrogates self-tolerance to MTg in mice and the severity of the resulting thyroiditis is regulated by T cells according to their *H-2* haplotype. These results also imply that LPS activates (directly or indirectly) specific MTg-reactive T cells in the presence of MTg. However, in addition to the T-cell regulation, Kong *et al.* (1978) reported that, in the response to MTg, the responsiveness to LPS is a prerequisite. This was determined by using C3HeB/FeJ and C3H/HeJ mice, which have the good responder ($H-2^k$) haplotype to MTg but differ in B-cell LPS responsiveness. After appropriate treatment of the mice with MTg, followed by injection of phenol-extracted LPS, high levels of antibody and thyroid infiltration were observed only in the LPS-responder C3HeB/FeJ mice and not in the C3H/HeJ mice. Nevertheless, the transfer of C3H/HeJ thymocytes did enable C3H/HeN nude mice to respond to MTg and LPS adjuvant with high MTg antibody titers and thyroid lesions. The authors concluded, therefore, that the induction of thyroiditis by MTg plus LPS depends upon both an *Ir-Tg* gene expressed by T cells and genetic LPS responsiveness reflected in B cells. However, based on the previous report by these same workers, the possibility exists that C3H/HeJ T cells (MTg-reactive) are activated by LPS, in contrast to the reported deficiencies of C3H/HeJ T-helper cells to be activated by LPS.

C. The Receptor Hypothesis

Soon after the recognition of the deficient C3H/HeJ B-cell response to LPS (lipid A), two questions of paramount interest were extensively investigated. The first dealt with what kind of genetic control was exercised over lymphocyte activation by LPS, and the second concerned the nature of the binding of LPS to low-responder and high-responder cells.

Genetic experiments from a number of laboratories, which involved the breeding of the C3H/HeJ mouse to various inbred strains, have led to a consensus that the mitogenic responsiveness to LPS is governed by a single gene locus composed of codominant alleles (Coutinho, 1976; Glode and Rosenstreich, 1976; Sultzer, 1976; Watson *et al.*, 1977). Watson *et al.* (1978a) determined that this locus, termed *Lps*, was on chromosome 4 and linked to the major urinary protein locus (*Mup-1*). In other studies, Watson *et al.* (1980) reported that the *Lps* locus regulated a number of immunological reactions to lipid A including polyclonal and adjuvant responsiveness, and further claimed that some nonimmunological responses such as hypothermia, elevation of serum amyloid protein and colony-stimulating factor levels were also controlled by this locus (Watson *et al.*, 1978b). Nevertheless, it is important to recognize that many reactions to LPS are complex phenomena undoubtedly involving different cell types and cell products. These interactions may be triggered as secondary or tertiary events subject to gene control independent of the *Lps* locus (e.g., lethality and the inflammatory response to LPS), so that the final phenotypic expression may not obey simple Mendelian genetics (Sultzer, 1972).

The matter of the binding of LPS to lymphocytes was approached with the obvious possibility that the aberrant behavior of C3H/HeJ B cells to LPS was due to a deficiency in their binding of LPS or lipid A. Again, the results of experiments from many laboratories using a variety of techniques led to the generally accepted conclusion that the binding of LPS to the lymphocytes of C3H/HeJ mice does not differ to any appreciable extent from that obtained with lymphocytes from high-responder strains (Gregory *et al.*, 1980; Kabir and Rosenstreich, 1977; Symons and Clarkson, 1979; Sultzer, 1976; Watson and Riblet, 1975). Thus, binding per se is not considered responsible for the triggering of murine B lymphocytes into DNA synthesis, but rather that a secondary mitogen receptor may exist and is lacking or somehow incompetent in the plasma membrane of C3H/HeJ cells. There is one exception to the reports listed above. Nygren *et al.* (1979) claimed that the binding of LPS horseradish peroxidase conjugate to lymphocytes from the LPS high-responder C3H/Tif strain was greater at low concentrations of LPS (0.7, 7.0 μg/ml) than that seen with C3H/HeJ cells, although there was no difference at high concentrations of LPS (70 μg/ml). In view of the fact that the differences obtained represented 2–4 cells at 0.7 μg/ml and 10–20 cells at 7 μg/ml, one might be cautious in accepting these preliminary findings as conclusive.

As previously mentioned, the concept of a secondary mitogen receptor on B cells was a natural outgrowth of considerations drawn from the results of both the genetic and binding experiments. The first reported experiments to identify this receptor were by Forni and Coutinho (1978) and Coutinho *et al.* (1978), who developed an anti-LPS receptor antiserum by injecting rabbits with responder C3H/TiF spleen cells and then extensively absorbing the serum with tissue from the C3H/HeJ strain. From numerous and varied control tests, this antiserum was concluded to be specific for the mitogen receptor; however, Watson *et al.* (1980) have failed to reproduce these results. Antisera were produced using spleen cells from C3H/DiSn, C3H/Bi, and C3H.SW responder mice that were extensively absorbed both *in vitro* and *in vivo* using C3H/HeJ tissue. None of the antisera so made were found to preferentially bind LPS-responder-strain spleen cells as determined by fluorescence or mitogen assays. Although the C3H/TiF strain was not used, Watson points out that an antiserum raised against any

LPS-responder spleen cell population and absorbed with C3H/HeJ tissue should bind to a cell surface antigen that is related to LPS responsiveness. Interestingly, the antiserum prepared by Forni and Coutinho (1978) resulted in the identification of the C57BL/10.ScCr strain as an LPS nonresponder (Coutinho and Meo, 1978). Nevertheless, one is still faced with the conflicting results described above, so that although it is still attractive to consider that the *Lps* locus codes for a cell surface receptor for lipid A, it is possible that the defect in C3H/HeJ cells may be failure of the transfer of the LPS triggering signal from the cell surface receptor to the nucleus due to a component controlled by the *Lps* locus.

In this regard, Watanabe and Ohara (1981) reported the application of a method for separating mouse splenic lymphocytes into karyoplasts (minicells) and cytoplasts (enucleated cells) with high purity to study the basis for LPS unresponsiveness. These authors reasoned that a third possibility existed in that abnormal events may occur in the nucleus of C3H/HeJ cells in response to LPS. By combining polyethylene glycol with poly-L-arginine, the fusion of irradiated or mitomycin C-treated C3H/HeN responder splenic lymphocytes with intact karyoplasts (nuclei) from C3H/HeJ spleen cells was accomplished. The resulting hybrid cells responded well to the mitogenic effect of LPS, clearly indicating that no abnormality existed in C3H/HeJ nuclei. Apparently, LPS nonresponsiveness still may be attributable to either cell-surface or cytoplasmic events.

III. Lymphocyte Activation by Endotoxin Protein

As mentioned earlier, our initial observations on the ability of C3H/HeJ splenic lymphocytes to respond to the activating effects of endotoxin were obtained using TCA-extracted endotoxins that were known to contain protein complexed to the LPS. Essentially protein-free LPS preparations made by the phenol method were inactive on C3H/HeJ spleen cells but TCA extracts were active, although to a lesser extent than for normal mice (Sultzer, 1972). That certain microbial proteins could nonspecifically activate C3H/HeJ spleen cells and those from other strains as well was revealed when we set about to demonstrate the specificity of the C3H/HeJ deficiency for LPS by sensitizing the mice to PPD-tuberculin. Spleen cells from such mice were stimulated to DNA synthesis by PPD; however, spleen cells from normal control mice also were activated by PPD. Subsequently, it was shown that PPD stimulated the proliferation of normal B cells from C3H/HeJ mice and other strains as well (Sultzer and Nilsson, 1972). Numerous studies confirmed and extended this finding. Indeed, tuberculin was shown to activate a different subpopulation of B cells than LPS (Sultzer *et al.*, 1977), to also act as a polyclonal activator of antibody production without the need for division (Nilsson *et al.*, 1973), to stimulate migration inhibition factor in normal guinea pigs (Yoshida *et al.*, 1973), and to act as an adjuvant of specific antibody production (Kreisler and Möller, 1974).

From consideration of the activity of PPD, our attention returned to the activating properties of endotoxin containing protein for C3H/HeJ lymphocytes. While it was generally accepted that the protein component was superfluous for the toxic and basic pathophysiological effects of LPS and at best acted as an inert carrier of the other biologically active components, the possibility occurred to us either that the presumably inert protein might restore the response of C3H/HeJ B cells to LPS appropriately presenting the lipid A to the lymphocyte surface, or that the protein itself was active. By hot-phenol extraction of the LPS–protein complex, we found that the protein, which was solubilized in the phenol phase and recovered by ethanol precipitation, was highly active on C3H/HeJ spleen cells (Sultzer and Goodman, 1976). Furthermore, this protein, which we designated *endotoxin protein* (EP) in keeping with the earlier literature, did not stimulate the T cells of mice to DNA synthesis, but did activate the spleen cells of congenitally athymic mice. In addition, EP stimulated C3H/HeJ splenic lymphocytes to produce antibody plaque cells against hapten-conjugated sheep erythrocytes (Sultzer and Goodman, 1976).

Extension of these initial experiments confirmed the original suggestion that EP was a B-cell activator in mice. EP was active without accessory T cells or macrophages and was a more potent stimulator than LPS or PPD-tuberculin. Of particular interest was the finding that EP stimulated DNA synthesis in lymphocytes from rats, rabbits, guinea pigs, and human peripheral blood. This stands in contrast to purified LPS, which is at best a weak activator of lymphocytes from these species (Goodman and Sultzer, 1979a).

The finding that EP was a mitogen for human peripheral blood lymphocytes (HPBL) called for further study, which revealed that human B lymphocytes in the absence of T cells were stimulated to DNA synthesis. EP was also able to induce HPBL to produce antibodies in culture against both sheep erythrocytes and hapten-conjugated sheep erythrocytes (Goodman and Sultzer, 1979b). Interestingly, EP mitogenesis but not polyclonal activation was inhibited by human serum, adding further evidence to the proposition that these events were dissociable.

Work from other laboratories on the activation of C3H/HeJ lymphocytes also revealed that the ability of C3H/HeJ spleen cells to be stimulated to mitogenesis by LPS was dependent on the method employed in extraction of the LPS (Skidmore *et al.*, 1975). LPS extracted with the phenol–water technique was ineffective, whereas LPS extracted by the butanol–water technique of Morrison and Leive (1975) proved to be active, although quantitatively less in C3H/HeJ cells than in the counterpart high-responder strain. While it was speculated that the phenol treatment might have degraded the lipid A in some fashion to lessen its activity, as detected by the C3H/HeJ spleen cell response, it remained for Morrison *et al.* (1976) to demonstrate by aqueous phenol extraction that the protein present in the butanol-extracted LPS was the active component (Betz and Morrison, 1977). The protein separated by Morrison, designated lipid A-associated protein (LAP), has been found to be active on nonlymphoid cells as well. In a number of *in vitro* situations, LAP has been found to be active in inducing cytotoxic macrophages, increasing glucose utilization in fibroblasts, and unlike LPS or the LPS–LAP complex, can cause the release of histamine and serotonin from rat peritoneal mast cells without cytolysis taking place (Morrison and Betz, 1977). On the other hand, LAP had no direct effect on platelets or polymorphonuclear neutrophils, although the serum–complement-mediated lysis of platelets activated by LPS was significantly enhanced by the presence of LAP complexed with LPS (Morrison *et al.*, 1980).

At this juncture, it is important to compare EP and LAP from several points of view, since the obvious question is whether these materials are the same. At first glance, it appeared that the proteins isolated were probably very similar or the same since the products obtained were both active on C3H/HeJ spleen cells (Sultzer and Goodman, 1976; Morrison *et al.*, 1976). However, on closer inspection, several differences appear. First and foremost, EP preparations activate DNA synthesis and polyclonal antibody in HPBL, although varying in effectiveness depending on the source, i.e., EP from *S. typhi* is the most active, EP from *S. typhimurium* is active but somewhat less, and EP from *E. coli* O127:B8 shows a minimum of activity. The LAP preparations derived from various serotypes of *E. coli* and *S. typhimurium*, on the other hand, do not stimulate HPBL to DNA synthesis (Morrison, personal communication). However, EP has not been tested for activity on nonlymphoid cells such as fibroblasts, platelets, and neutrophils as yet so that no comparisons can be drawn concerning these biological properties. Nevertheless, recent work, which will be discussed, has shown EP can activate macrophages in culture to be cytotoxic for tumor cells, as is the case with LAP (Sultzer, LeGarrec, and Chedid, unpublished observation).

A preliminary characterization of the chemical properties of EP has recently been reported, which revealed that the protein was a complex of numerous polypeptide species that varied depending upon the organism from which it was isolated and also the bacterial culture conditions used (Goodman and Sultzer, 1979c). By SDS-polyacrylamide gel electrophoresis, EP from *E. coli* consisted of five major protein bands of approximate molecular weights of 35,000, 17,000, 14,000, 12,000, and 10–11,000. EP from *S. typhi* showed four major protein bands of approximate molecular weights of 35,000, 32,000, 13–13,500, and 10–11,000. It is important to note that although there were some similar protein bands in both EP preparations (those at 35,000 and 10–11,000), there were major differences. The protein bands present at molecular weights of 17,000 and 12,000 in the *E. coli* EP were present as only minor components in the *S. typhi* EP, whereas the 32,000-molecular-weight polypeptide seen in the *S. typhi* EP was not found in the *E. coli* EP. It should be emphasized at this point that *S. typhi* EP preparations of the type described here are quite active on HPBL, whereas the *E. coli* EP preparations are weakly active. They cannot be distinguished based on their activity on mouse lymphocytes. This is also true of an *S. typhi* EP preparation made from bacteria grown in fermentation tank culture, which shows a loss of most of the higher-molecular-weight polypeptides, as opposed to *S. typhi* EP preparations made from shaker flask cultures. The former is inactive on HPBL whereas the latter is stimulatory. Our tentative hypothesis at present is that the specific polypeptides present in *S. typhi* EP including the 32,000- and 13–13,500-molecular-weight species may be associated with activity for HPBL.

This brings into question the LAP, which also was characterized by SDS-PAGE (Morrison *et al.*, 1980). In their report, Morrison *et al.* showed that the LAP from *E. coli* O111:B4 was heterogeneous. It consisted of a major protein component at an approximate molecular weight of 42,000 with other prominent bands at 28,000, 22,000, 16,000, and 8000. The 8000-molecular-weight component appeared to comigrate with the peptidoglycan-associated lipoprotein of Braun (1977). Clearly, the EP from *E. coli* and the LAP from *E. coli* may be similar but do show differences in their polypeptide profiles. Whether these differences arise from differences in the *E. coli* strains, in the bacterial culture conditions, in the butanol versus the trichloroacetic acid extraction, or the less likely differences in the analytical PAGE system used in both laboratories cannot be determined at present; however, both *E. coli* EP and LAP are weakly active or inactive on HPBL, which emphasizes to us the significance of the different *S. typhi* polypeptide components of EP, which are active on HPBL. Further studies on separation of the polypeptide components of EP now in progress should help to define the components that are active on HPBL as well as the lymphocytes and nonlymphoid cells from other species.

IV. Additional Stimulating Properties of Endotoxin Protein

A. Activation of Cytotoxic Macrophages

In any event, studies have continued on characterizing the biological potential of EP in its present state. This approach appears both feasible and necessary so that the biological properties of the individual polypeptides can be compared with the aggregated complex. From recent work at the Institut Pasteur in collaboration with Y. LeGarrec and L. Chedid, we have found that EP can activate murine macrophages to be cytotoxic for tumor cells *in vitro*. As shown in Table 1, peritoneal macrophages from CBA mice when stimulated with EP in culture will cause the cytolysis of P815 mastocytoma cells, as measured by the release of ^{51}Cr from the labeled targets. With a ratio of macrophages to target cells of 5 : 1, maximum specific release was obtained. These results, in general, agree with those obtained with LAP (for review, see Morrison and Ryan, 1979).

B. Induction of Interferon

Additional studies with LeGarrec and Chedid were directed at defining whether EP could stimulate interferon *in vivo* and *in vitro* as is the case with LPS. As shown in Table 2, interferon activity was found

Table 1. Cytolytic Effect of Endotoxin Protein-Stimulated
Macrophages on Tumor Cells

Target cell concentration[a]	Mean % specific release ^{51}Cr[b]
2×10^4	50.8
4×10^4	41.7
2×10^5	44.2
4×10^5	79.6

[a] Thioglycolate-induced peritoneal macrophages (2×10^6) from CBA mice cultured as monolayers. Macrophages were stimulated with 50 μg/ml of EP 24 hr prior to the addition of ^{51}Cr-labeled P815 mastocytoma cells.

[b] Specific release = [(experimental release − macrophage control release)/(total releasable counts − macrophage control release)] $\times$ 100.

Table 2. Induction of Circulating Interferon in Mice by Endotoxin Protein

Inducer[b]	Titer (U/ml serum)[a]			
	CBA	C57BL/6	C3HeB/FeJ	C3H/He (Orl)
LPS (*E. coli*)	64	512	ND[c]	64
LPS (*S. typhi*)	ND	2048	2048	ND
EP (*E. coli*)	128	1024	ND	1024
EP (*S. typhi*)	ND	1024	512	32

[a] Interferon titrations were performed by the cytopathogenic inhibition test on L-cell cultures in microplates using vesicular-stomatitis virus challenge. Titers are expressed in international reference units.
[b] The inducers were injected intravenously at a dose of 200 μg per mouse. Blood was taken for serum 2 hr after the injection.
[c] ND, not done.

in the serum of mice 2 hr after the intravenous injection of EP. By 7 hr, the activity was diminished or no longer apparent, closely resembling the results obtained with LPS. The interferon activity obtained with EP was found to be acid resistant and heat labile, resembling that type of interferon induced by LPS *in vivo*. It should be noted that interferon activity was found in C3H/He (Orl) mice, which have been shown by Chedid's group to be low responders to the mitogenicity of LPS (Chedid and LeGarrec, 1980). On the other hand, when spleen cells were cultured from high-responder CBA mice and low-responder C3H/HeJ mice in the presence of 100 μg of LPS or EP, interferon activity could be detected only from CBA cell cultures (Table 3). Whether this lack of response of C3H/HeJ cells in culture is quantal or quantitative remains to be determined.

Table 3. Induction of Interferon in Mouse Spleen Cells by Endotoxin Protein

Inducer[b]	Titer (U/ml supernatant)[a]	
	CBA	C3H/HeJ
Control medium	0	0
LPS (*E. coli*)	16	0
LPS (*S. typhi*)	16	0
EP (*E. coli*)	32	0
EP (*S. typhi*)	64	8

[a] Interferon assay by cytopathogenic inhibition test. Titers are expressed in international reference units.
[b] Spleen cells were cultured at 1×10^7 cells/ml for 24 hr at 37°C in the presence of 100 μg/ml of each stimulant.

C. Adjuvanticity

Since EP can stimulate both proliferation and polyclonal antibody production *in vitro*, as well as activate macrophages, it seemed reasonable to investigate whether EP could act as an adjuvant in a manner similar to LPS endotoxin. For this purpose, we have adopted a mouse protection test developed by Dr. John P. Craig for assaying the efficiency of adjuvants to enhance protection against the lethality of cholera enterotoxin. In this assay, mice immunized by intraperitoneal injection of a single dose of toxoid are challenged 42 days later by approximately 3 LD_{50} of cholera enterotoxin given intravenously. The dose–response curve for intravenous cholera toxin is precise and steep; 3 LD_{50} kills all mice within 2–3 days. Therefore, the 50% protective dose (PD_{50}) of glutaraldehyde cholera toxoid in control mice can be readily compared with the PD_{50} obtained in mice receiving both the toxoid and adjuvant. In collaboration with J. P. Craig, we have tested the ability of EP from *S. typhi* to enhance the protection afforded by cholera toxoid. An example of the results is shown in Table 4. Ten micrograms of EP can enhance the protection of the Wyeth cholera toxoid by approximately 16-fold. In other experiments, EP from *Bordetella pertussis* can enhance protection by about 27-fold when given at a dose of 10 μg. These preliminary results are encouraging and further experiments are in progress to characterize the degree of protection, the levels of serum neutralizing antitoxin, and whether EP can act as an adjuvant for cholera toxoid with lymphocytes and accessory cells in culture.

D. Protective Immunogenicity

From the close association of EP with LPS, it is reasonable to conclude that the polypeptides of the protein complex are intrinsic components of the outer membrane of gram-negative bacteria. Also, it is

Table 4. Immunoadjuvant Effect of Endotoxin Protein on Cholera Toxoid Protection in CF-1 Mice

Vaccine	Cholera-toxoid PD_{50}[a]
Toxoid[b]	0.435 μg
Toxoid + EP (10 μg)[c]	0.027 μg

[a] Cholera toxin (Wyeth 1100) challenge (2.8 LD_{50}) given i.p. 6 weeks postvaccination.
[b] Cholera toxoid (Wyeth 11201) given i.p. in eight dilutions, six mice per group.
[c] Endotoxin protein derived from *S. typhi*.

conceivable that at least part of these molecules may be exposed at the cell surface and exposed in greater measure upon disintegration of the organism *in vivo*. Therefore, it is not unlikely that EP or its component polypeptides may act as an immunogen and induce a protective immune response by the host of a humoral and/or cellular nature. Furthermore, there is the possibility that outer-membrane protein immunogens may circumvent the toxicity problem inherent to vaccines for gram-negative bacteria in which the O antigens present on the LPS are the principle immunodeterminants. For these reasons, an investigation of the possible application of EP as a protective antigen in salmonellosis of mice began some time ago in collaboration with Dr. T. K. Eisenstein.

For these experiments, EP was prepared in the usual manner (Sultzer and Goodman, 1976) from *S. typhimurium* W118-2 cells by isolation from TCA-extracted LPS. This strain of *S. typhimurium* has been well characterized as a challenge organism for infecting outbred and inbred strains of mice (Eisenstein and Angerman, 1978). At the outset, the intraperitoneal LD_{50} of this organism was determined in various strains of mice including the CD-1, C3H/HeNCr1Br, C3H/HeJ, and C3HeB/FeJ to serve as a basis for immunization experiments. As expected, the CD-1 mouse was normally resistant with an LD_{50} 1×10^4 cells and the C3H/HeNCr1Br strain was somewhat less resistant with an LD_{50} of 1.2×10^3 cells. However, the C3HeB/FeJ mouse, like the C3H/HeJ strain, was hypersusceptible to *Salmonella* infection with a theoretical lethal dose approaching a single cell (Eisenstein *et al.*, 1980). Of particular significance here is that the C3HeB/FeJ mouse is sensitive to the toxic effects of LPS and is a high responder to the mitogenic and polyclonal activating effects of LPS in contrast to the LPS-resistant, low-responder C3H/HeJ mouse. Clearly, *Salmonella* susceptibility is not associated with the defective LPS mitogenic responsiveness controlled by the Lps^d allele.

When these strains of mice were vaccinated with various preparations including purified LPS, TCA–LPS containing 10–12% protein, and EP, another interesting picture emerged. Briefly, CD-1 mice are readily protected (90% survival at 60 days) against a challenge of 500 LD_{50} of *S. typhimurium* when given two doses of 50 and 25 μg of *S. typhimurium* EP, which is equivalent to or better than protection afforded by LPS (Eisenstein and Sultzer, 1981). As seen in Table 5, C3H/HeNCr1Br mice likewise are protected by *S. typhimurium* EP. However, when C3H/HeJ and C3HeB/FeJ mice were vaccinated with LPS, EP, and TCA–LPS and challenged with 6 to 24 cells of *S. typhimurium*, protection was poor. LPS and EP could not stimulate resistance. The TCA–LPS prolonged the mean time to death but survival diminished below

Table 5. Protection against *S. typhimurium* Infection in C3H/HeNCr1Br Mice Immunized with Endotoxin Protein

Vaccine	Dose (μg)[a]		60-day survival[b]	
	1	2	%	Alive/total
LPS	50	25	90	9/10
	100	100	63	5/8[c]
TCA–LPS	50	25	100	10/10
	100	100	100	4/4[d]
EP	50	25	100	10/10
	100	100	100	10/10
PBS[e]	—	—	11	1/9

[a] Mice immunized i.p. with dose 1, and 14 days later with dose 2.
[b] Mice infected i.p. 21 days after dose 2 with 2.3 × 10^4 cells of *S. typhimurium* W118-2 (= 19 LD$_{50}$ doses).
[c] Two mice died from toxicity of immunization.
[d] Six mice died from toxicity of immunization.
[e] PBS, phosphate-buffered saline.

statistically significant levels at 60 days (Sultzer *et al.*, 1980; Eisenstein *et al.*, 1980). The basic conclusion we draw from these results simply is that highly susceptible mice are difficult to immunize against salmonellosis using immunogens of the type described, whereas more inherently resistant mice can be well protected. In this instance, the *S. typhimurium* EP was shown to be highly protective without eliciting any toxicity, as compared to the *S. typhimurium* LPS, which killed as many as 20–60% of the mice during the immunization procedure (see Table 5). Clearly, outer-membrane proteins from gram-negative bacteria offer the potential of being effective immunogens. In fact, Kuusi *et al.* (1979) also have provided evidence that "porin" protein may be protective or at least play an important role in anti-*Salmonella* immunity.

V. Conclusions

Needless to say, some of the more intriguing aspects of the many biological phenomena induced by endotoxin have been those nonspecific responses of the host that may be labeled beneficial. The more detailed knowledge we have now about the activation of lymphocytes by LPS (lipid A), briefly highlighted in this chapter, and the greater insight we have into the genetic control of these and other host responses to LPS have provided a better understanding of what we call nonspecific resistance; however, the problem of LPS toxicity has not been alleviated. Perhaps detoxified derivatives or nontoxic polysaccharide components discussed elsewhere may find application in the future. In this regard, the eventual application of what we call endotoxin protein may be feasible. Strictly speaking, the outer-membrane protein we find complexed to the LPS when various extraction procedures are used should not be considered a derivative of the endotoxin. Nevertheless, the intriguing aspect of EP is that, in its present form, this material can induce the nonspecific activation of immunocompetent cells to a greater extent and in species that do not react well with LPS without gross toxicity. At the same time, EP extracted from pathogenic organisms, such as *Salmonella*, may contain immunological determinants that can induce specific immunity. In this context, perhaps it is even conceivable that EP may act as an adjuvant for itself. Certainly, separation and purification of the constituent polypeptides should reveal which components act as specific immunogens, which serve to activate lymphocytes and macrophages, and which may serve to induce interferon.

ACKNOWLEDGMENT

The work on endotoxin protein was supported by Grant AI-16782 from the United States Public Health Service.

VI. References

Andersson, J., Möller, G., and Sjöberg, O., 1972a, Selective induction of DNA synthesis in T and B lymphocytes, *Cell. Immunol.* **4**:381.

Andersson, J., Sjöberg, O., and Möller, G., 1972b, Induction of immunoglobulin and antibody synthesis *in vitro* by lipopolysaccharides, *Eur. J. Immunol.* **2**:349.

Andersson, J., Coutinho, A., and Melchers, F., 1977, Frequencies of mitogen-reactive B cells in the mouse. I. Distribution in different lymphoid organs from different inbred strains of mice at different ages, *J. Exp. Med.* **145**:1511.

Betz, S. J., and Morrison, D. C., 1977, Chemical and biological properties of a protein rich fraction of bacterial lipopolysaccharides. I. The *in vitro* murine lymphocyte response, *J. Immunol.* **119**:1475.

Bick, P. H., Persson, U., Smith, E., Möller, G., and Hammarström, L., 1977, Genetic control of lymphocyte activation: Lack of response to low doses of concanavalin A in lipopolysaccharide-nonresponder mice, *J. Exp. Med.* **146**:1146.

Braley-Mullen, H., Hayes, N., and Sanders, M., 1977, Suppression of IgM responses to type III pneumococcal polysaccharide by lipopolysaccharide (LPS), *Cell. Immunol.* **30**:300.

Braun, V., 1977, Lipoprotein from the outer membrane of *Escherichia coli* as an antigen, immunogen, and mitogen, in: *Microbiology 1977* (D. Schlessinger, ed.), pp. 251–261, American Society for Microbiology, Washington, D.C.

Chedid, L., and LeGarrec, Y., 1980, Transfer of cells or of serum factors in nonspecific resistance to infection, in: *Microbiology 1980* (D. Schlessinger, ed.), pp. 19–24, American Society for Microbiology, Washington, D.C.

Coutinho, A., 1976, Genetic control of B-cell responses. II. Identification of the spleen B-cell defect in C3H/HeJ mice, *Scand. J. Immunol.* **5:**129.

Coutinho, A., and Meo, T., 1978, Genetic basis for unresponsiveness to lipopolysaccharide in C57B1/10/Cr mice, *Immunogenetics* **7:**17–24.

Coutinho, A., Forni, L., and Watanabe, T., 1978, Genetic and functional characterization of an antiserum to the lipid-A-specific triggering receptor on murine B-cells, *Eur. J. Immunol.* **8:**63.

Eisenstein, T. K., and Angerman, C. R., 1978, Studies on the protective capacity and immunogenicity of lipopolysaccharide, acetone-killed cells, and a ribosome-rich extract of *Salmonella typhimurium* in C3H/HeJ and CD-1 mice, *J. Immunol.* **121:**1010.

Eisenstein, T. K., and Sultzer, B. M., 1981, *Salmonella* vaccines: Protection by endotoxin protein and lipopolysaccharide in two different mouse strains, in: *Bacterial Vaccines,* Vol. 4.(J. Robbins, J. Hill, and J. Sadoff, eds.), pp. 423–428, Thieme-Stratton, New York.

Eisenstein, T. K., Deakins, L. W., and Sultzer, B. M., 1980, The C3HeB/FeJ mouse, a strain in the C3H lineage which separates *Salmonella* susceptibility and immunizability from mitogenic responsiveness to lipopolysaccharide, in: *Genetic Control of Natural Resistance to Infection and Malignancy* (E. Shamere, P. A. R. Kongsharer, and M. Landy, eds.), pp. 115–120, Academic Press, New York.

Esquivel, P. S., Kong, Y. M., and Rose, N. R., 1978, Evidence for thyroglobulin-reactive T cells in good responder mice, *Cell. Immunol.* **37:**14.

Forni, L., and Coutinho, A., 1978, An antiserum which recognizes lipopolysaccharide-reactive B cells in the mouse, *Eur. J. Immunol.* **8:**56.

Gery, I., Kruger, J., and Spiesel, S. Z., 1972, Stimulation of B-lymphocytes by endotoxin: Reactions of thymus-deprived mice and karyotypic analysis of dividing cells in mice bearing T_6T_6 thymus grafts, *J. Immunol.* **108:**1088.

Glode, L. M., and Rosenstreich, D. L., 1976, Genetic control of B cell activation by bacterial lipopolysaccharide is mediated by multiple distinct genes or alleles, *J. Immunol.* **117:**2061.

Glode, L. M., Mergenhagen, S. E., and Rosenstreich, D. L., 1976, Significant contribution of spleen cells in mediating the lethal effects of endotoxin *in vivo, Infect. Immun.* **14:**626.

Goodman, G. W., and Sultzer, B. M., 1979a, Further studies on the activation of lymphocytes by endotoxin protein, *J. Immunol.* **122:**1329.

Goodman, G. W., and Sultzer, B. M., 1979b, Endotoxin protein is a mitogen and polyclonal activator of human B-lymphocytes, *J. Exp. Med.* **149:**713.

Goodman, G. W., and Sultzer, B. M., 1979c, Characterization of the chemical and physical properties of a novel B-lymphocyte activator, endotoxin protein, *Infect. Immun.* **24:**685.

Goodman, M. G., and Weigle, W. O., 1979, T-Cell regulation of polyclonal B-cell responses. I. Helper effects of T-cells, *J. Immunol.* **122:**2548.

Goodman, M. G., Parks, D. E., and Weigle, W. O., 1978, Immunologic responsiveness of the C3H/HeJ mouse: Differential ability of the butanol-extracted lipopolysaccharide (LPS) to evoke LPS-mediated effects, *J. Exp. Med.* **147:**800.

Gregory, S. H., Zimmerman, D. H., and Kern, M., 1980, The lipid A moiety of lipopolysaccharide is specifically bound to B-cell subpopulations of responder and nonresponder animals, *J. Immunol.* **125:**102.

Hämmerling, I., Chin, A. F., Abbot, J., and Scheid, M. C., 1975, The ontogeny of murine

B-lymphocytes. I. Induction of phenotypic conversion of Ia$^-$ to Ia$^+$ lymphocytes, *J. Immunol.* **115**:1425.

Hoffman, M. K., Galanos, C., Koenig, S., and Oettgen, H. F., 1977, B-Cell activation by lipopolysaccharide: Distinct pathways for induction of mitosis and antibody production, *J. Exp. Med.* **146**:1640.

Kabir, S., and Rosenstreich, D. L., 1977, Binding of bacterial endotoxin to murine spleen lymphocytes, *Infect. Immun.* **15**:156.

Kong, Y. M., El-Rehewy, M., Giraldo, A. A., Esquivel, P. S., Jeffries, C. D., Rose, N. R., and David, C. S., 1978, The effect of LPS responsiveness on induction of autoimmune thyroiditis, *Fed. Proc.* **37**:1380.

Kreisler, J. M., and Möller, G., 1974, Effect of PPD on the specific immune response to heterologous red cells *in vitro*, *J. Immunol.* **112**:151.

Kuusi, N., Nurminen, M., Saxen, H., Valtonen, M., and Makela, P. H., 1979, Immunization with major outer membrane proteins in experimental salmonellosis of mice, *Infect. Immun.* **25**:857.

Levy, G. A., and Edgington, T. S., 1980, Lymphoid procoagulant activity and mitogenesis in the C3H/HeJ mouse: Discordant response to lipopolysaccharide stimulation, *J. Immunol.* **124**:2665.

Männel, D. N., Rosenstreich, D. L., and Mergenhagen, S. E., 1979, Mechanism of lipopolysaccharide-induced tumor necrosis: Requirement for lipopolysaccharide sensitive lymphoreticular cells, *Infect. Immun.* **24**:573.

Melchers, F., 1977, B lymphocyte development in fetal liver. I. Development of reactivities to B-cell mitogens *in vivo* and *in vitro*, *Eur. J. Immunol.* **7**:476.

Morrison, D. C., and Betz, S. J., 1977, Chemical and biological properties of a protein rich fraction of bacterial lipopolysaccharide. II. The *in vitro* peritoneal mast cell response, *J. Immunol.* **119**:1790.

Morrison, D. C., and Leive, L. 1975, Fractions of lipopolysaccharide from *Escherichia coli* O111:B4 prepared by two extraction procedures, *J. Biol. Chem.* **250**:2911.

Morrison, D. C., and Ryan, J. L., 1979, Bacterial endotoxins and host immune responses, *Adv. Immunol.* **28**:294.

Morrison, D. C., Betz, S. J., and Jacobs, D. M., 1976, Isolation of a lipid A bound polypeptide responsible for "LPS-initiated" mitogenesis of C3H/HeJ spleen cells, *J. Exp. Med.* **144**:840.

Morrison, D. C., Wilson, M. E., Razuiddin, S., Betz, S. J., Curry, B. J., Dades, Z., and Munkenbeck, P., 1980, Influence of lipid A-associated protein on endotoxin stimulation of nonlymphoid cells, in: *Microbiology 1980* (D. Schlessinger, ed.), pp. 30–35, American Society for Microbiology, Washington, D.C.

Nilsson, B. S., Sultzer, B. M., and Bullock, W. W., 1973, Purified protein derivative of tuberculin induces immunoglobulin production in normal mouse spleen cells, *J. Exp. Med.* **137**:127.

Norcross, M. A., and Smith, R. T., 1977, Regulation of B-cell proliferative responses to lipopolysaccharide by a subclass of thymus T-cells, *J. Exp. Med.* **145**:1299.

Nygren, H., Dahlen, G., and Möller, G., 1979, Bacterial lipopolysaccharides bind selectively to lymphocytes from lipopolysaccharide high-responder mouse strains, *Scand. J. Immunol.* **10**:555.

Peavy, D. L., Adler, W. H., and Smith, R. T., 1970, The mitogenic effects of endotoxin and staphylococcal enterotoxin B on mouse spleen cells and human peripheral lymphocytes, *J. Immunol.* **105**:1453.

Primi, D., Hammarström, L., Möller, G., Smith, C. I. E., and Uhr, J., 1979, Con-A-activated T-cells secrete factors with polyclonal B-cell-activating properties, *Scand. J. Immunol.* **9**:467.

Rosenstreich, D. L., and McAdam, K. P. W. J., 1979, Lymphoid cells in endotoxin-induced production of the amyloid-related serum amyloid A protein, *Infect. Immun.* **23**: 181.

Rosenstreich, D. L., Glode, L. M., Wahl, L. M., Sandberg, A. L., and Mergenhagen, S. E., 1977, Analysis of the cellular defects of endotoxin-unresponsive C3H/HeJ mice, in: *Microbiology 1977* (D. Schlessinger, ed.), pp. 314–320, American Society for Microbiology, Washington, D.C.

Rudbach, J. A., and Reed, N. D., 1977, Immunological responses of mice to lipopolysaccharide: Lack of secondary responsiveness by C3H/HeJ mice, *Infect. Immun.* **16**:513.

Skidmore, B. J., Morrison, D. C., Chiller, J. M., and Weigle, W. O., 1975, Immunologic properties of bacterial lipopolysaccharide (LPS). II. The unresponsiveness of C3H/HeJ mouse spleen cells to LPS-induced mitogenesis is dependent on the method used to extract LPS, *J. Exp. Med.* **142**:1488.

Skidmore, B. J., Chiller, J. M., and Weigle, W. O., 1977, Immunologic properties of bacterial lipopolysaccharide (LPS). IV. Cellular basis of the unresponsiveness of C3H/HeJ mouse spleen cells to LPS-induced mitogenesis, *J. Immunol.* **118**:274.

Slowe, A., and Waldmann, H., 1975, The "intrinsic adjuvanticity" and immunogenicity of trinitrophenylated lipopolysaccharide, *Immunology* **29**:825.

Sultzer, B. M., 1968, Genetic control of leucocyte responses to endotoxin, *Nature (London)* **219**:1253.

Sultzer, B. M., 1969, Genetic factors in leucocyte responses to endotoxin: Further studies in mice, *J. Immunol.* **103**:32.

Sultzer, B. M., 1972, Genetic control of host responses to endotoxin, *Infect. Immun.* **5**:107.

Sultzer, B. M., 1973, Mitogenic and immune responses to endotoxin: A deficiency in C3H/HeJ mice, *Abstr. Annu. Meet. Am. Soc. Microbiol.* p. 85, M69.

Sultzer, B. M., 1976, Genetic analysis of lymphocyte activation by lipopolysaccharide endotoxin, *Infect. Immun.* **13**:1579.

Sultzer, B. M., and Goodman, G. W., 1976, Endotoxin protein: A B-cell mitogen and polyclonal activator of C3H/HeJ lymphocytes, *J. Exp. Med.* **144**:821.

Sultzer, B. M., and Nilsson, B. S., 1972, PPD-tuberculin—A B-cell mitogen, *Nature New Biol.* **240**:198.

Sultzer, B. M., Nilsson, B. S., and Kirschenbaum, D., 1977, Nonspecific stimulation of lymphocytes by tuberculin, *Infect. Immun.* **15**:799.

Sultzer, B. M., Goodman, G. W., and Eisenstein, T. K., 1980, Endotoxin protein as an immunostimulant, in: *Microbiology 1980* (D. Schlessinger, ed.), pp. 61–65, American Society for Microbiology, Washington, D.C.

Symons, D. B. A., and Clarkson, C. A., 1979, The binding of LPS to the lymphocyte surface, *Immunology* **38**:503.

Tanabe, M. J., and Nakano, M., 1979a, Dysfunction of thymocytes of C3H/HeJ mice in blastogenic response to antithymocyte serum *in vitro* and its repair with lipopolysaccharide induced mediator, *Microbiol. Immunol.* **23**:287.

Tanabe, M. J., and Nakano, M., 1979b, Lipopolysaccharide-induced mediators assisting the proliferative response of C3H/HeJ thymocytes to concanavalin A, *Microbiol. Immunol.* **23**:1097.

Vallera, D. A., Gamble, C. E., and Schmidtke, J. R., 1980, Lipopolysaccharide-induced immunomodulation of the generation cell-mediated cytotoxicity. II. Evidence for the involvement of a regulatory B lymphocyte, *J. Immunol.* **124**:641.

Watanabe, T., and Ohara, J., 1981, Functional nuclei of LPS-nonresponder C3H/HeJ mice after transfer into LPS-responder C3H/HeN cells by cell fusion, *Nature (London)* **290**:58.

Watson, J., 1977, Differentiation of B lymphocytes in C3H/HeJ mice: The induction of Ia antigens by lipopolysaccharide, *J. Immunol.* **118**:1103.

Watson, J., and Riblet, R., 1974, Genetic control of responses to bacterial lipopolysaccharide in mice. I. Evidence for a single gene that influences mitogenic and immunogenic responses to lipopolysaccharides, *J. Exp. Med.* **140**:1147.

Watson, J., and Riblet, R., 1975, Genetic control of responses to bacterial lipopolysaccharides in mice. II. A gene that influences the activity of a membrane receptor for lipopolysaccharides, *J. Immunol.* **114**:1462.

Watson, J., Riblet, R., and Taylor, B., 1977, The response to lipopolysaccharides in recombinant inbred lines of mice, *J. Immunol.* **118**:1604.

Watson, J., Kelly, K., Largen, M., and Taylor, B. A., 1978a, The genetic mapping of defective LPS response gene in C3H/HeJ mice, *J. Immunol.* **120**:422.

Watson, J., Largen, M., McAdam, K. P. W. J., and Taylor, B. A., 1978b, Genetic control of endotoxic responses in mice, *J. Exp. Med.* **147**:39.

Watson, J., Kelly, K., and Whitlock, C., 1980, Genetic control of endotoxin sensitivity, in: *Microbiology 1980* (D. Schlessinger, ed.), pp. 4–10, American Society for Microbiology, Washington, D.C.

Yoshida, T., Hidekichi, S., and Cohen, S., 1973, The protection of migration inhibition factor by B and T cells of the guinea pig, *J. Exp. Med.* **38**:784.

12

Adjuvant Action of Bacterial Endotoxins on Antibody Formation

A Historical Perspective

Arthur G. Johnson

One of the most dramatic of the potentially beneficial actions of the endotoxins derived from gram-negative bacteria has been their capacity to elevate antibody levels to unrelated antigens in animals. This truly profound effect of the isolated lipopolysaccharide component (LPS) has been known for over 25 years (Johnson *et al.*, 1956), and has found application in many experimental immunology laboratories throughout the world.

A number of early observations and studies of the adjuvant action occurring following combined immunization with bacterial cells in conjunction with unrelated antigens preceded definition of LPS as the active component. The use of typhoid vaccine for nonspecific immunization was exemplary in this respect. Herrmann (cited by Petersen, 1922) observed that typhoid vaccine stimulated production of specific anti-streptococcic agglutinins in rabbits that had failed to respond with antibody when cultures of streptococci alone were used. It was reported also by Landy and Webster (1952) that rabbits immunized with a purified Vi antigen frequently showed higher than normal Vi antibody levels if alcohol-killed *Salmonella typhosa* O901 cells were given in conjunction with the Vi antigen. Cultures of other gram-negative bacteria, Friedlander's bacillus or *Serratia marcescens*, were found by Pawlowsky (1930) to favorably influence the resistance of rabbits infected with *Ba-*

Arthur G. Johnson ● Department of Medical Microbiology/Immunology, University of Minnesota, Duluth, Minnesota 55812.

cillus anthracis. Immunization of patients against typhoid fever or cholera was noted by Kutscher (cited by Petersen, 1922) to cause a rise in the titer of *Shigella dysenteriae* agglutinins. Similarly, the injection of *E. coli*, *S. dysenteriae*, or *Corynebacterium diphtheriae* into rabbits previously immunized against typhoid bacilli, also was shown to cause a rise in typhoid agglutinins (Conradi and Bieling, 1916). Killed cultures of such organisms as *Staphylococcus aureus, Serratia marcescens,* or *B. anthracis* were known to have been employed in the early days of vaccine therapy with the objective of increasing the natural resistance of animals. Acetic acid extracts of bacterial cultures were shown to have the same effect (Hahn, 1929). Stuhl (1918) observed that infected wounds healed more rapidly following inoculation of typhoid vaccine. The favorable effect of the injection of unrelated organisms, either living or dead, was postulated by these early workers to occur as a result of the local inflammatory reaction induced by the bacteria, the cellular response aiding in the disposal of the microorganisms subsequently injected.

Several studies have shown that incorporation of certain suspensions of bacteria, such as *Hemophilus pertussis,* with one or more immunizing agents, will increase the antigenic effectiveness of the latter (Bell, 1948; Ordman and Grasset, 1948). Greenberg and Fleming (1947) found that this enhancement was related to the dose of *H. pertussis,* and in a further classical study Fleming *et al.* (1948) recorded higher antitoxin titers in humans receiving *H. pertussis* vaccine along with diphtheria and tetanus toxoids than those in individuals who received each toxoid singly. The addition of pertussis vaccine enhanced antibody response to diphtheria regardless of whether alum was present or not (Ungar 1952a,b).

The component of the *Hemophilus* bacillus responsible for this enhancing activity was attributed in 1956 to the LPS portion of the cell wall (Johnson *et al.,* 1956) and later specifically to the lipid moiety (Abdelnoor, 1969; Chiller *et al.,* 1973). Recently, the polysaccharide per se also has been reported to exert an adjuvant action (Behling and Nowotny, 1977). The initial findings were a spin-off from research being done at Walter Reed Army Medical Center under the direction of Dr. Maurice Landy, in which the objective was to purify the immunogenic components of typhoid vaccine. Purification of the O antigen resulted in an LPS with only 0.6% nitrogen (Webster *et al.,* 1955), which possessed most if not all of the endotoxic properties of gram-negative organisms (Landy and Johnson, 1955). When this LPS was tested for its capacity to enhance antibody formation to diphtheria toxoid in rabbits, the dramatic results shown in Table 1 were seen.

A remarkable enhancing action again was observed when using purified proteins as antigens (Table 2). Both IgM and IgG were elevated

Table 1. Enhancement of Antibody Response in Rabbits to Diphtheria
Toxoid by LPS

Primary immunization: 3 injections i.v. at 3-day intervals	Mean antitoxin units by rabbit intradermal neutralization		
	10 days	29 days	7 days postsecondary[a]
C. diphtheriae toxoid (20 Lf)	0.065	0.013	1.0
C. diphtheriae toxoid (20 Lf) + LPS (5 μg)	1.9	0.145	32.0

[a] A second injection of 5 Lf toxoid or 5 Lf toxoid + 5 μg LPS given i.v. on day 29 to top and bottom groups, respectively.

and adjuvanticity was exerted on a variety of antigens in a number of animal species with endotoxins isolated from several bacterial genera (Merritt and Johnson, 1964; Johnson et al., 1956). Similar results were obtained by Condie et al. (1955) using dilutions of meningococcal endotoxin. Susceptibility to the toxic action of the LPS appeared in early experiments to be required for adjuvanticity to be exerted (Johnson et al., 1956; Condie and Good, 1956). However, later studies revealed the endotoxin could be detoxified by several different procedures with retention of some of its adjuvant action (Johnson and Nowotny, 1964; Schenck et al., 1969; Chedid et al., 1975).

Titration of the time of administration of endotoxin relative to antigen revealed a surprising immunomodulatory property of the LPS in that suppression of the response occurred when LPS was given a day or two prior to antigen (Kind and Johnson, 1959). This was explored in depth by Franzl and McMaster (1966) and later by others (reviewed

Table 2. Enhancement of Antibody Response in Rabbits to
Bovine Serum Albumin by LPS

Immunization: 3 injections i.v. at 3-day intervals	Mean antibody nitrogen/ml serum, day 7 (μg)
Bovine serum albumin (10 mg)	4
Bovine serum albumin (10 mg) + LPS (10 μg)	162

in Morrison and Ryan, 1979). Further illustration of the complexity of the mechanism of action of these molecules is found in the recent reports documenting their effects on many different cells with immune potential (e.g., polymorphonuclear leukocytes, macrophages, T and B cells). This is suggestive of a common event initiated by the LPS at the membrane of a variety of cells.

As may be deduced, utilization of the LPS as an adjuvant component to vaccines destined for human or veterinary usage requires much more knowledge. Answers to the following questions appear to be fundamental:

1. Are any toxic manifestations of this adjuvant required for its action?
2. What is the chemical definition of the ligand(s) responsible for enhancement and suppression of antibody synthesis to unrelated antigens?
3. What are the cellular subset targets and the nature of the receptor receiving the enhancing and suppressing stimuli of the LPS?
4. Is the adjuvant and suppressive action mediated through release of endogenous cytokines and if so, how can they be selectively controlled?
5. What new safety tests are needed to assess any dangers associated with a potent stimulus of the immune response?

References

Abdelnoor, A., 1969, Relationship of chemical composition of endotoxins to their biological properties, Ph.D. thesis, University of Michigan, Ann Arbor.

Behling, U. H., and Nowotny, A., 1977, Immune adjuvancy of lipopolysaccharide and a nontoxic hydrolytic product demonstrating oscillating effects with time, *J. Immunol.* **118**:1905.

Bell, J. A., 1948, Diphtheria immunization: Use of alum precipitated mixture of pertussis vaccine and diphtheria toxoid, *J. Am. Med. Assoc.* **137**:1009.

Chedid, L., Audibert, F., Bona, C., Damais, C., Parant, F., and Parant, M., 1975, Biological activities of endotoxins detoxified by alkylation, *Infect. Immun.* **12**:1712.

Chiller, J. M., Skidmore, B. J., Morrison, D. C., and Weigle, W. O., 1973, Relationship of the structure of bacterial lipopolysaccharides to its function in mitogenesis and adjuvanticity, *Proc. Natl. Acad. Sci. USA* **70**:2129.

Condie, R. M., and Good, R. A., 1956, Inhibition of immunological enhancement by endotoxin in refractory rabbits: Immunochemical study, *Proc. Soc. Exp. Biol. Med.* **91**:414.

Condie, R. M., Zak, S. J., and Good, R. A., 1955, Effect of meningococcal endotoxin on the immune response, *Proc. Soc. Exp. Biol. Med.* **90**:355.

Conradi, H., and Bieling, R., 1916, Ueber Fehler-quellen der Gruberwidalschen Reaktion, *Dtsch. Med. Wochenschr.* **42**:1280.

Fleming, D. S., Greenberg, L., and Beith, E. M., 1948, The use of combined antigens in the immunization of infants, *Can. Med. Assoc. J.* **59:**101.

Franzl, R. E., and McMaster, P. D., 1966, The enhancement and suppression of hemolysin production by a bacterial endotoxin, *J. Exp. Med.* **127:**1087.

Greenberg, L., and Fleming, D. S., 1947, Increased efficiency of diphtheria toxoid when combined with pertussis vaccine, *Can. J. Public Health* **38:**279.

Hahn, M., 1929, Naturliche immunitat (resistenz), in: *Handbuch Im Pathologick Mikroorganismen*, 3rd ed. Kolle, Kraus Uhlenhuth, Vol. 1, p. 742, G. Fischer, Jena.

Johnson, A. G., and Nowotny, A., 1964, Relationship of structure to function in bacterial O-antigens. III. Biological properties of endotoxoids, *J. Bacteriol.* **87:**809.

Johnson, A. G., Gaines, S., and Landy, M., 1956, Studies on the O antigen of *Salmonella typhosa*. V. Enhancement of antibody response to protein antigens by the purified lipopolysaccharide, *J. Exp. Med.* **103:**225.

Kind, P. D., and Johnson, A. G., 1959, Studies on the adjuvant action of bacterial endotoxins. I. Time limitation of enhancing effect and restoration of antibody formation in x-irradiated rabbits, *J. Immunol.* **82:**415.

Landy, M., and Johnson, A. G., 1955, Studies on the O antigen of *Salmonella typhosa*. IV. Endotoxic properties of the purified antigen, *Proc. Soc. Exp. Biol. Med.* **90:**57.

Landy, M., and Webster, M. E., 1952, Studies on Vi antigen. III. Immunological properties of purified Vi antigen derived from *Escherichia coli* 5396/38, *J. Immunol.* **69:**143.

Merritt, K., and Johnson, A. G., 1964, Type of antibody elevated by 5-fluoro-2-deoxyuridine and endotoxin during the primary response of the mouse, *Proc. Soc. Exp. Biol. Med.* **115:**1132.

Morrison, D. C., and Ryan, J. L., 1979, Bacterial endotoxins and host immune responses, *Adv. Immunol.* **28:**293.

Ordman, D., and Grasset, E., 1948, Combined pertussis–diphtheria prophylactic antigens: Experimental study to determine specific immunizing value of these antigens used in combination, *J. Hyg.* **46:**117.

Pawlowski, A. D., 1930, Zur Frage der Infektion und der Immunitat, *Z. Hyg. Infektionskr.* **33:**261.

Petersen, W. F., 1977, *Protein Therapy and Non-specific Resistance*, Macmillan Co., New York.

Schenck, J. R., Hargie, M. P., Brown, M. S., Ebert, D. S., Yoo, A. L., and McIntire, F. C., 1969, The enhancement of antibody formation by *Escherichia coli* lipopolysaccharide and detoxified derivative, *J. Immunol.* **102:**1411.

Stuhl, C., 1918, Typhus Schutzimpfung und Phagozytose, *Muench. Med. Wochenschr.* **65:** 614.

Ungar, J., 1952a, Recent advances in pertussis immunization, *Irish J. Med. Sci.* **316:**145.

Ungar, J., 1952b, Combined prophylactics, *Proc. R. Soc. Med.* **45:**674.

Webster, M. E., Sagin, J. F., Landy, M., and Johnson, A. G., 1955, Studies on the O antigen of *Salmonella typhosa*. I. Isolation and purification of the antigen, *J. Immunol.* **74:** 455.

13

In Vitro Adjuvant Effect of Endotoxin
Review and Experiments

Masayasu Nakano and Takehiko Uchiyama

I. Introduction

Mishell and Dutton (1967) devised a method that initiates an immune response in cultures of free spleen cells from unimmunized mice. This method allows the investigation of biochemical and cellular events associated with induction and development of the immune response and enables the investigator to use an intrinsically simpler system for analysis than is true when intact animals are employed.

The action of endotoxin on the immune system *in vivo* and *in vitro* has been demonstrated by many investigators (reviewed in Morrison and Ryan, 1979). *In vitro* demonstration of the adjuvant potency of endotoxin was first reported by Ortiz-Ortiz and Jaroslow (1970). These investigators demonstrated that if lipopolysaccharide (LPS) was added to the spleen cell suspensions from unimmunized mice at 12 to 24 hr after sheep red blood cell (SRBC) antigen, it gave a severalfold increase in the number of plaque-forming cells (PFC) after 6 days of culture in comparison with controls in the presence of SRBC, while there was little or no enhancement when LPS was given at other times during the 6 days of culture. They also found that when the suspension was incubated with LPS before the addition of SRBC, the response was depressed. Thus, the effect of endotoxin on antigen-dependent response *in vitro* was immunoregulatory. It works as an adjuvant under some conditions, but as an immunosuppressant under others.

Masayasu Nakano ● Department of Microbiology, Jichi Medical School, Tochigi 329-04, Japan. Takehiko Uchiyama ● Department of Microbiology, Kitasato University School of Medicine, Kanagawa 228, Japan.

Endotoxin itself has the capacity to initiate nonspecifically the synthesis and secretion of antibody by immunopotent B lymphocytes in the absence of specific antigens—polyclonal B-cell activation (Möller, 1975). The antigen-dependent immune adjuvant activity of endotoxin is quite prominent. It is obviously a synergistic effect of endotoxin and antigen, and not an additive effect of both of these immunostimulants, since the magnitude of the response cannot be explained by the simple sum of the polyclonal B-cell activation by endotoxin and the response to antigen alone.

Haptens conjugated with endotoxin are immunogenic *in vitro* at very low concentration (Coutinho *et al.*, 1974; Jacobs and Morrison, 1975). Coutinho *et al.* (1974, 1975) demonstrated that the 4-hydroxy-3,5-dinitrophenyl hapten coupled to LPS (NNP–LPS) activated polyclonal as well as specific anti-NNP antibody synthesis in cultures of murine spleen cells, but the optimal concentrations for induction of specific anti-NNP cells were several orders of magnitude lower than the concentrations required for polyclonal activation. These low concentrations failed to activate nonspecific cells, but they induced specific response of high-avidity NNP-specific cells with typical kinetics of antigenic response *in vitro*. These findings indicate that the adjuvant action of LPS conjugated to hapten is different from its property of polyclonal activation.

Adjuvant effect of endotoxin *in vitro* does not always reflect the status *in vivo*. Simultaneous injections of endotoxin and antigen elicit obvious adjuvant effect *in vivo* (Johnson *et al.*, 1956; Neter, 1969; Nakano *et al.*, 1971), while addition of antigen and LPS together to *in vitro* cell cultures at the same time can hardly induce the adjuvant effect (Ortiz-Ortiz and Jaroslow, 1970; Koenig and Hoffmann, 1979). Andersson *et al.* (1972) demonstrated that the adjuvant effect of LPS could be observed even in the cultured spleen cells of mice tolerant to LPS, which had been injected repeatedly with detoxified LPS and had lost the ability to respond to LPS *in vivo*. The usefulness of the *in vitro* culture system, however, should not be suspect because of the discrepancy between *in vivo* and *in vitro* experiments, for cells in culture may reveal their actual *in vivo* responses but without the modifications that are always inevitable *in vivo*.

II. Endotoxin as Adjuvant for T-Cell-Dependent Antigens

T cells are needed for antibody production carried out by B cells against certain antigens [thymus-dependent (TD) antigens]. The experiments of Armerding and Katz (1974) provided the indication of T-cell participation in the *in vitro* adjuvant effect of endotoxin on TD antigens. They demonstrated that spleen cells from mice primed to 2,4-dinitrophenyl (DNP)–keyhole limpet hemocyanin (KLH) conjugate develop considerably enhanced *in vitro* anti-DNP antibody response when exposed to LPS in the presence of antigen, provided the antigen employed is the DNP coupled to the original immunizing carrier, and spleen cells from mice that were previously primed to carrier protein develop augmented primary response to DNP coupled to the original immunizing carrier in the presence of LPS. Another decisive evidence for an obligate requirement for T lymphocytes in endotoxin-induced adjuvanticity on the primary response against particulate antigens [SRBC or SRBC coated with 2,4,6-trinitrophenyl (TNP)] was reported by McGhee *et al.* (1979) (see Chapter 17).

Hoffmann *et al.* (1975) reported nonspecific perturbation of T-cell function by endotoxin on T-cell-dependent response in spleen cell cultures; the SRBC response was significantly suppressed by the addition of endotoxin at the initiation of culture, while significant enhancement ensued if it was added at a late time. Uchiyama and Jacobs (1978a,b) have extended these studies and demonstrated that endotoxin acting on separate T-cell subpopulations can both enhance and suppress the *in vitro* immune response to hapten-coupled erythrocytes (TNP–SRBC). The endotoxin-induced enhancement was due to the effect of one cell population (helper T cells), whereas another population (suppressor T cells) depressed the PFC response to TNP–SRBC in response to endotoxin.

Goodman and Weigle (1981) demonstrated that polyclonal activation of murine splenic B lymphocytes by LPS is subject to regulation by helper and suppressor influences from T lymphocytes, and the soluble fraction of splenic T-cell sonicate exerts both helper and suppressor regulatory influences. The helper and suppressor activities are distinct molecular entities derived from distinct splenic T-lymphocyte subpopulations. However, participation of these regulatory T cells (which have been observed in LPS-induced polyclonal response) in the adjuvant effect of LPS has not as yet been studied.

Jacobs (1979) demonstrated that T-cell replacing factor (TRF) markedly enhances the ability of endotoxin to stimulate the response of unfractionated spleen cells, B cells from normal mice, and *nu/nu* spleen cells to SRBC antigen under conditions where endotoxin alone is barely stimulatory. Furthermore, Hoffmann and Watson (1979), Hoffman *et al.* (1979a,b), and Hoffman (1980) found that endotoxin-activated macrophages produced an active factor [interleukin 1 (IL 1)] that exhibited synergistic effects with TRF on the induction of immune response to erythrocyte antigens in macrophage-depleted spleen cell cultures.

These studies suggest that LPS-induced IL 1 acts on antigen-reactive B cells and then B cells convert into antibody-secreting cells (PFC) under the regulation by helper T cells or by a mediator (TRF) that they release.

Adjuvant effect of endotoxin for TD antigen declines with aging. Kishimoto *et al.* (1976) demonstrated that the anti-TNP response of young spleen cells of mice to TNP–SRBC was markedly enhanced by the addition of LPS to the cultures, whereas no or little enhancement of the response was induced in the aged spleen cells even in the presence of high concentration of LPS. Their results, however, did not clarify what kinds of cells—B cells, T cells, or other accessory cells—in the aged spleen cells are responsible for the impaired adjuvant effect of LPS.

Mice of the CBA/N subline that bear an X-chromosome-linked defect in B-lymphocyte lineage are deficient in responses to a number of immunologic stimuli. Spleen cells with CBA/N-genetic defect are affected in antibody responses to thymus-independent (TI) antigens, but can respond to TD antigens. The CBA/N spleen cells respond poorly to the mitogenic and polyclonal effect of LPS. Campbell and Kind (1979) examined the adjuvant effect of endotoxin on the CBA/N spleen cells to TD antigens (SRBC and goat erythrocytes), and showed that the CBA/N spleen cells developed higher numbers of PFC when cultured with LPS.

III. Adjuvant Effect of Endotoxin in the Absence of T Lymphocytes

Spleen cells of *nu/nu* or thymectomized, irradiated, and bone marrow cell-reconstituted (AT × BM) mice cannot respond to TD antigens. However, in the presence of LPS, these cells do respond to TD antigens. Sjöberg *et al.* (1972) and Watson *et al.* (1973) demonstrated that spleen cells treated with anti-θ antibody and complement or the cells of nude mice respond to SRBC antigen in the presence of LPS. Similar results were also demonstrated by Schrader (1973, 1974) using a soluble TD antigen where antibody responses to fowl γ-globulin could be elicited in the presence of endotoxin and the absence of T-helper cells. Another approach to examine the substitution of helper T cells by endotoxin was conducted by several investigators who prepared covalent conjugates of low-molecular-weight haptens, e.g., NNP (Coutinho *et al.*, 1974) or TNP (Jacobs and Morrison, 1975) with endotoxin. These complexes were found, when tested in the spleen cells of AT × BM or *nu/nu* mice, to yield immune responses that shared the T-independent property of endotoxin carrier.

IV. Endotoxin as Adjuvant for T-Cell-Independent Antigen

Some antigens, such as hapten-conjugated endotoxin, hapten-conjugated *Brucella* (*Brucella* organisms contain endotoxin in their cell wall), hapten-conjugated Ficoll, and pneumococcal polysaccharide, are capable of inducing an immune response in the absence of mature T lymphocytes. These TI antigens can be classified as TI-1 and TI-2 antigens based on the immune responses of CBA/N-defective mice, which are profoundly affected in the immune responses to TI-2 antigens such as TNP–Ficoll, and less affected, but still considered to be abnormal, are responses to TI-1 antigens such as hapten-conjugated LPS.

The responses to hapten-conjugated LPS (TI-1) have been described in the previous sections. The lymphocyte population in the CBA/N-defective mice is either entirely devoid or extremely depleted of Lyb5$^+$ B cells. Mosier *et al.* (1976) demonstrated that CBA/N-defective F$_1$ male spleen cells respond to TNP–LPS *in vitro*, though the magnitude of the PFC response is obviously poor when compared with that of phenotypically normal F$_1$ female littermate cells. Moreover, their responses differ from those of female cells in that more TNP–LPS is required to elicit the peak anti-TNP response and that the anti-TNP antibody secreted by male cells is of lower avidity than that of female cells. More precise analysis on the interaction between CBA/N-defective cells and antigens was investigated by Boswell *et al.* (1980b). They examined the anti-TNP PFC response in CBA/N-defective spleen cell cultures to TI-1 (hapten-coupled endotoxin), TI-2, or TD antigens, and found that accessory cells (macrophages) and/or T cells are necessary for the induction of antibody synthesis of B cells by TD, TI-2, and low-dose TI-1 antigens. However, the response of the CBA/N B cells (Lyb5$^-$ subset), as well as normal cells (mixed population of Lyb5$^-$ and Lyb5$^+$ subsets), to high-dose TI-1 antigen does not require these supplemental cells. These results suggest that the hapten-coupled endotoxin is capable of triggering directly antigen-specific antibody-producing B cells without cooperation of T cells or macrophages under some conditions.

There is strong evidence supporting the existence of separate subpopulations of B cells that respond to TD and TI antigens. Diaz-Espada *et al.* (1978) studied the susceptibility of hapten-primed spleen cells to TNP–LPS *in vitro* and found this susceptibility to be dependent upon the carrier used in the immunization (whether TD or TI) and the time elapsed from priming up to the preparation of the cultures. As the immunization of mice with DNP–ovalbumin reduced the responsiveness to TNP–LPS *in vitro*, they concluded that part, but not all, of a TD–anti-DNP population evolved after antigen challenge to a stage of LPS

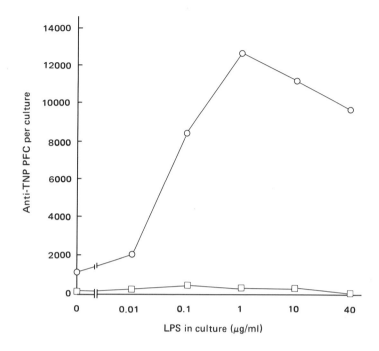

Figure 1. Adjuvant effect of endotoxin on anti-TNP PFC response of murine spleen cell cultures in the presence of thymus-independent TNP–Ficoll antigen. Spleen cells (1.5 × 10[7]) of C57BL/6 mice were cultured with (o) or without (□) TNP–Ficoll (1 ng/ml). LPS at graded concentrations was added to cultures at 48 hr after initiation of culture. Numbers of anti-TNP PFC were determined on day 4.

unresponsiveness, leaving intact a TI–anti-DNP population susceptible to LPS.

Endotoxin has an adjuvant effect on the response to TI-2 antigen. As shown in Fig. 1, spleen cell cultures in the presence of endotoxin and TNP–Ficoll yielded many anti-TNP PFC after 4 days' culture, and the numbers of PFC in the cultures were much greater than those in control cultures with TNP–Ficoll or endotoxin.

V. The Role of Macrophages on the Adjuvant Effect of Endotoxin

Evidence for a pivotal role for macrophages in the endotoxin-induced enhancement of the immune response *in vitro* was first provided by the experiments of Hoffmann *et al.* (1977). They examined the *in vi-*

tro response to TNP-conjugated mouse erythrocytes which were capable of bypassing T-cell help in the presence of endotoxin. When LPS and macrophages from LPS-responder C3HeB/FeJ mice were added to LPS-nonresponder C3H/HeJ spleen cell cultures, enhanced antibody response was observed. Similar experiments by McGhee *et al.* (1979) have examined the adjuvant effects of endotoxin on the *in vitro* immune response of subpopulations of endotoxin-responsive and -nonresponsive cells. Their experiments indicated that both macrophages and T lymphocytes were required to be endotoxin responsive in order to manifest enhanced immune response to SRBC.

Butler *et al.* (1979) found that the stimulation of immunocytes with LPS induced the release of a helper factor that appeared to be a monokine, and the immunological mechanism of the primary *in vitro* antibody response to SRBC antigens involved this soluble immunomodulatory factor. Hoffmann and his collaborators (1979a,b, 1980) extended their studies and demonstrated that immune enhancing effect of macrophages can be replaced by a soluble factor (IL 1) that is produced by endotoxin-activated macrophages.

VI. Adjuvant Effect of LPS on Bone Marrow Cell Culture

Bone marrow is the nursery of primitive precursors of hemopoietic and lymphoid cells. B and T lymphocytes and macrophages arise from these stem cells, and migrate into the peripheral lymphoid organs and other parts in the body. As the lymphoid cells in bone marrow are still premature and lack immunocompetency, they can hardly respond to antigen *in situ*. However, endotoxin, instead of antigens, has some ability to stimulate them into proliferative response and polyclonal antibody synthesis.

Adjuvant effect of endotoxin on bone marrow cell cultures can be seen with TNP–Ficoll (TI antigen). We prepared bone marrow cells from femurs of BALB/c mice and these cells were cultured in the presence of TNP–Ficoll or TNP–KLH for 5 days. LPS was added to these cultures at 30 min after initiation of culture. As shown in Fig. 2, tremendous numbers of anti-TNP PFC were generated in the cultures with TNP–Ficoll and LPS. The number of PFC in the cultures with TNP–KLH (TD antigen) and LPS was the same as that with LPS alone. These antigens themselves did not induce any significant PFC response. However, endotoxin-induced adjuvant effect on bone marrow cells to TD particulate antigen (TNP–SRBC) was also absent. Even though the bone marrow cell cultures were supplemented with both

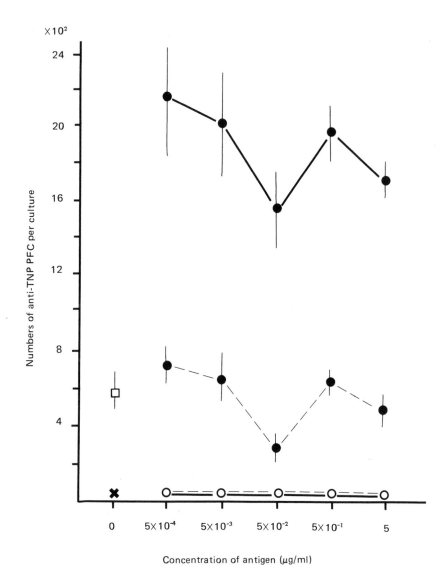

Figure 2. Adjuvant effect of endotoxin on anti-TNP PFC response of murine bone marrow cell cultures. Bone marrow cells (2×10^6) were cultured with or without antigen (TNP–Ficoll or TNP–KLH). LPS (10 μg/ml) was added into cultures 30 min later. Anti-TNP PFC responses of cultures were determined on day 5. (○ — ○) TNP–Ficoll; (○ ---- ○) TNP–KLH; (●—●) TNP–Ficoll and LPS; (●----●) TNP-KLH and LPS; (☐) LPS; (✗) background PFC. Mean of triplicate cultures and standard deviation.

macrophage and T-lymphocyte populations prepared from spleen, adjuvant effect of endotoxin on the response in cultures to these TD antigens could not be observed (data not shown). These results suggest that only the premature B cells that are destined to respond to TI antigens are ready to receive the signals of endotoxin and antigen, and that when both signals are received, they can turn into the specific antibody-secreting cells, though the antigenic signal itself cannot trigger them to become immunocompetent antibody-producing cells.

These premature B cells responsive to TNP–Ficoll-dependent adjuvant effect of LPS seem to be generated from the cells lacking Ia marker on their cell surfaces, which is the product of Ir gene in *H-2* gene complexes in mice, whereas the cells with Ia surface markers are also necessary for the elicitation of adjuvant effect. In order to eliminate Ia-positive cells, we incubated the bone marrow cells in the presence of anti-Ia serum and complement at 37°C for 30 min and then the cells were washed. These Ia-negative cells were cultured and the adjuvant effect of LPS was examined. As shown in Table 1, adjuvant effect was evident in these Ia-negative cell cultures as well as in the cultures of bone marrow cells not pretreated with anti-Ia serum and complement. However, when anti-Ia serum was added to the cultures of bone marrow cells from ATL and ATH mice, the adjuvant effect and the polyclonal effect of LPS were completely abolished in the Ia-matched ATL cell cultures but not in the Ia-unmatched ATH cell cultures (Table 2). These findings suggest that the Ia-positive cell population in bone marrow cells is responding to LPS, but the Ia-positive cell population is generated from Ia-negative bone marrow cells during the culture.

Table 1. Adjuvant Effect of LPS on Anti-TNP PFC Response in Ia-Negative Bone Marrow Cell Cultures against TNP–Ficoll Antigen[a]

Addition to culture	No. of anti-TNP PFC per culture (mean ± S.E.)	
	Bone marrow cells	Ia-negative bone marrow cells
None	28 ± 5	52 ± 19
TNP–Ficoll (5 μg/ml)	111 ± 44	23 ± 9
TNP–Ficoll + LPS (10 μg/ml)	1906 ± 447	1474 ± 450
LPS	1124 ± 85	981 ± 210

[a] Bone marrow cells were obtained from ATL mice. Ia-negative cells were prepared by treatment of the bone marrow cells with anti-Ia (ATH anti-ATL) serum (final dilution, 1:400) and rabbit complement at 37°C for 30 min. The cells were cultured (2 × 10⁶ cells/2 ml/well) in the presence or absence of TNP–Ficoll. LPS was added 3 hr after initiation of culture. PFC numbers were determined on day 5.

Table 2. Effect of Anti-Ia^k (ATH Anti-ATL) Serum on Adjuvant Effect of LPS in Bone Marrow Cell Cultures^a

	No. of anti-TNP PFC per culture (mean ± S.E.)	
Addition to culture	ATL cells	ATH cells
None	43 ± 16	21 ± 3
TNP–Ficoll (5 μg/ml)	58 ± 22	165 ± 45
TNP–Ficoll + anti-Ia^k serum	30 ± 2	21 ± 3
TNP–Ficoll + LPS (10 μg/ml)	2240 ± 327	3758 ± 548
TNP–Ficoll, LPS, + anti-Ia^k serum	26 ± 10	4525 ± 800
LPS	1110 ± 351	2015 ± 641
LPS + anti-Ia^k serum	41 ± 10	4090 ± 1128

^{a} Bone marrow cells obtained from ATL or ATH mice were cultured (2 × 10^6 cells/2 ml/well) in the presence or absence of TNP–Ficoll. Some of cultures received anti-Ia^k serum (final dilution, 1:400). LPS was added 3 hr after initiation of culture. PFC numbers were determined on day 5.

VII. Immune Suppression by Endotoxin

Endotoxin, primarily recognized as an immune adjuvant, also induces immune suppression, depending on the experimental conditions, especially on the time interval between the adjuvant application and immunization (Kind and Johnson, 1959; Franzl and McMaster, 1968; for review see Neter, 1969; Morrison and Ryan, 1979). Behling and Nowotny (1977) demonstrated oscillating effects on immune adjuvancy of endotoxin *in vivo*: an alternating enhancement and suppression of PFC response as a function of time in mice given LPS at various days before challenge with SRBC antigen.

The suppressive effect of endotoxin in cultures is dependent on animal strains used, cell concentrations, and the time of LPS administration with respect to antigen (Koenig and Hoffmann, 1979). Decrease of the cell concentration of spleen cells in cultures resulted in enhancement of antibody production by LPS rather than suppression and it seemed that the LPS was inhibitory only in cultures undergoing a vigorous immune response. The immune response was suppressed only when LPS was added before or together with antigen.

Baltz and Rittenberg (1977) showed that pretreatment of spleen cells with nanogram quantities of endotoxin leads to inhibition of primary response to hapten-conjugated carriers. In this case, the inhibition of response appears to be rapidly induced. Exposure of spleen cells to LPS (or very low dose of antigen) for 1 hr at 4°C is enough to induce the unresponsiveness against immunogenic DNP–dextran. The authors suggest that this may represent a 'down regulation' by which B cells discriminate antigenic signals in determining the response to antigens in the environment; in other words, down regulation by low-dose specific or nonspecific signals provides a means of distinguishing background noise from the true antigenic stimuli.

Persson (1977) reported that spleen cells from LPS-injected mice could suppress normal mouse spleen cells in their response to the antigen (SRBC), and the suppressor cells contained in the LPS-activated spleen cells were found to be inhibitory after removal of θ -positive T cells and macrophages. Koenig and Hoffmann (1979) demonstrated that the suppressor cell (as mediator of LPS-induced immune suppression) was a B cell with the cell surface antigen phenotype of Ig^+, Ia^+, CR^+, $Lyb2^+$, and $PC1^-$, and macrophages or T cells were not required for the generation of suppressor B cells by LPS. Furthermore, Kempf and Rubin (1979) found that polyclonal stimulation of normal mouse spleen cells by LPS resulted in the generation of a factor (molecules of $\sim$ 24,000 daltons) that was capable of suppressing the humoral immune response *in vitro*, and B cells were essential for production of the suppressive mediator. They suggested that the suppressive activities seen by Koenig and Hoffmann (1979) and Persson (1977), generated by the addition of LPS at initiation of culture, may be mediated by this factor.

On the other hand, another kind of suppressor cell in the LPS-mediated enhancement of the T-cell-dependent immune response was reported by Walker and Weigle (1978) and Uchiyama and Jacobs (1978a,b). Walker and Weigle demonstrated that LPS could be highly suppressive to the secondary antibody response to soluble protein antigens (human or turkey γ-globulin) *in vitro*. LPS-induced suppression of the primary antibody response was usually seen when LPS was added before or together with antigen, while the most suppressive effect of LPS on their secondary response was observed when LPS was added 12 hr after culture with antigen. As removal of B cells not bearing specificity for the primary antigen did not reduce suppression by LPS, they thought that the suppression was not due to the B cell responsive to LPS. Uchiyama and Jacobs characterized their suppressor cells as T cells. They demonstrated that the TNP–PFC response of normal spleen cells to TNP–SRBC antigen was inhibited by adding spleen cells from the mice injected with LPS previously. The suppressor cells were irradiation sensitive, adherent to nylon wool, and sensitive to anti-T-cell serum. McGhee *et al.* (1980) presented evidence for the existence of T-suppressor cells that interfere with the immune response to TNP–LPS *in vitro*. They found that splenic T cells from conventional mice, but not from germfree mice, suppressed TNP–LPS response irrespective of whether B cells were obtained from germfree or conventional spleens. Therefore, the authors presumed that the normal gram-negative gut flora induces the production of a T-cell population that regulates B-cell response to bacterial endotoxin.

Immune-suppression by LPS may be related with dysfunction of helper T cells. Portnoï *et al.* (1981) found that helper activity of T lymphocytes obtained from LPS-injected mice decreased and that the production of TRF in concanavalin A-stimulated spleen cells from these mice was defective. Therefore, they concluded that the inhibition of the immune response resulted from a defect in helper T-cell differentiation signal(s) necessary for B precursors to initiate antibody secretion.

There is considerable evidence to indicate a suppressive effect of macrophages on endotoxin-induced adjuvant effect of the immune response under some conditions. We obtained spleen cells from

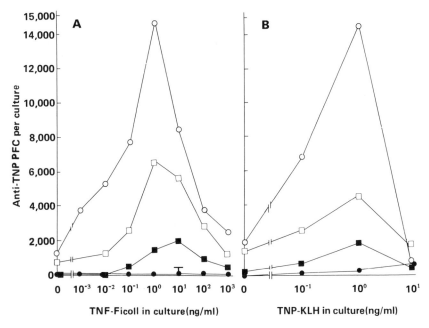

Figure 3. Augmentation of adjuvant effect of LPS by macrophage removal. Whole spleen cells and macrophage-depleted spleen cells of C57BL/6 mice were cultured with graded concentrations of TNP–Ficoll (A) or TNP–KLH (B). LPS (1 μg/ml) was added into cultures 48 hr after initiation of culture. Numbers of anti-TNP PFC in cultures were assessed on day 4. (■) Whole spleen cell cultures in the absence of LPS; (□) whole spleen cell cultures in the presence of LPS; (●) macrophage-depleted spleen cell cultures in the absence of LPS; (○) macrophage-depleted spleen cell cultures in the presence of LPS.

Table 3. Adjuvant Effect of LPS on *in Vitro* **Antibody Response to TNP–Ficoll or TNP–KLH in the Presence or Absence of T Cells and Macrophages**[a]

Spleen cells	Antigen: TNP–Ficoll (3 ng) or TNP–KLH (1 μg)	LPS (1 μg)	No. of anti-TNP PFC per culture				
			Response to TNP–Ficoll			Response to TNP-KLH	
			Expt. 1	Expt. 2	Expt. 3	Expt. 1	Expt. 2
Whole spleen	−	−	140	220	125	125	140
cells	+	−	1,300	2,135	1,610	1,900	1,580
	−	+	1,360	800	1,080	1,040	1,360
	+	+	5,500	10,480	11,440	4,480	6,800
T-cell-depleted	−	−	190	80	150	115	190
spleen cells	+	−	540	1,700	420	90	250
	−	+	1,400	880	880	880	1,400
	+	+	4,800	9,680	8,480	1,640	2,640
Macrophage-de-	−	−	75	200	115	115	175
pleted spleen	+	−	300	320	280	425	530
cells	−	+	1,360	1,860	960	960	1,360
	+	+	13,200	17,640	12,000	6,600	10,640
Macrophage- and	−	−	130	150	15	115	130
T-cell-depleted	+	−	240	300	120	150	290
spleen cells	−	+	1,560	1,600	800	800	2,000
	+	+	10,280	17,360	14,040	3,360	7,920

[a] LPS was added at 48 hr after initiation of culture. PFC numbers were determined on day 4.

C57BL/6 mice. Macrophage-depleted cells were prepared by passage of the spleen cells over a column of Sephadex G-10. The cells were cultured in the presence of varying doses of TNP–Ficoll (TI) or TNP–KLH (TD) antigen. LPS was added to the cultures 48 hr after initiation. The numbers of anti-TNP PFC in the cultures were determined on day 4. As shown in Fig. 3, adjuvant effect of LPS was prominent in the macrophage-depleted cell cultures in comparison with whole spleen cell cultures. In order to confirm this finding, whole spleen cells, macrophage-depleted cells, T-cell-depleted cells (anti-BAT serum and complement-treated spleen cells), and both macrophage- and T-cell-depleted cells (Sephadex G-10 column-passed and anti-BAT serum plus complement-treated spleen cells) were prepared. These cells were cultured with or without antigens and/or LPS. As shown in Table 3, adjuvant effect of endotoxin was greater in the cultures of macrophage-depleted cells than that of whole spleen cells. The effect of T cell depletion could be seen on the response to TNP–KLH, but not to TNP–Ficoll. The response to TNP–Ficoll is macrophage dependent (Boswell *et al.*, 1980a). The response to TNP–Ficoll in macrophage-depleted cell cultures as well as both macrophage- and T-cell-depleted cell cultures diminished obviously in comparison with that in whole spleen cells. However, when LPS was added to macrophage-depleted cell cultures, anti-TNP responses of the cultures were hightened, regardless of the presence of T cells. Furthermore, whole spleen cells, macrophages (plastic dish-adherent spleen cells), and macrophage-depleted spleen cells were prepared from LPS-responder C3H/HeN or nonresponder C3H/HeJ mice. The macrophage-depleted cells were reconstituted by supplementation of macrophages from either strain. These cells were cultured in the presence or absence of TNP–KLH and/or LPS. As shown in Table 4, positive effect of macrophage-depletion on endotoxin-induced immune enhancement could be seen in the cultures of C3H/HeN spleen cells, but not in those of C3H/HeJ spleen cells. Addition of C3H/HeN macrophages to macrophage-depleted C3H/HeN spleen cells diminished the adjuvant effect of LPS, while C3H/HeJ macrophages instead of C3H/HeN macrophages did not cause such a diminution in the C3H/HeN spleen cell cultures. These results indicate

Table 4. Regulation of LPS-Induced Adjuvant Effect on *in Vitro* Antibody Response to TNP–KLH by Macrophages[a]

				No. of anti-TNP PFC per culture	
Spleen cells for culture	Macrophage supplementation	Antigen: TNP–KLH (1 μg)	LPS (1 μg)	C3H/HeN cell culture	C3H/HeJ cell culture
Whole spleen cells	None	–	–	20	30
		–	+	730	50
		+	–	770	580
		+	+	2240	710
Macrophage-depleted spleen cells	None	–	–	60	20
		–	+	1600	30
		+	–	260	90
		+	+	6480	100
	Macrophages from C3H/HeN spleen cells	–	–	30	20
		–	+	1560	120
		+	–	440	410
		+	+	2800	730
	Macrophages from C3H/HeJ spleen cells	–	–	80	10
		–	+	1320	180
		+	–	540	410
		+	+	7040	710

[a] LPS was added at 48 hr after initiation of culture. PFC numbers were determined on day 4.

that LPS responder macrophages may regulate negatively the adjuvant effect of LPS on B lymphocytes.

VIII. Chemical Component of Endotoxin Responsible for Adjuvant Effect

Endotoxin is an integral component of the outer membrane of gram-negative bacteria. Endotoxic activity of endotoxin is attributed to the lipid moiety in the molecular complexes. Lipid A itself prepared by hydrolysis from LPS has a potent ability to induce adjuvant effect on the *in vitro* immune response of murine spleen cells to TD and TI antigens (unpublished observation). However, there are some other components in endotoxin that are capable of inducing adjuvant activity. Frank *et al.* (1977) demonstrated that a nontoxic, relatively low-molecular-weight, polysaccharide-rich preparation (PS), free of detectable lipid, prepared from endotoxic LPS of *Serratia marcescens* by hydrolysis in 1 N HCl at 100°C for 30 min was stimulatory on the *in vitro* anti-SRBC immune response of murine splenocytes, though a chemically detoxified preparation (MexB) prepared from the endotoxic LPS by treatment with potassium methylate had no stimulatory activity. Such nontoxic adjuvant preparations obtained from endotoxin may bring great improvements for the practical use of microbial adjuvants.

IX. Conclusion

Endotoxin itself is a potent immunomodulator both *in vivo* and *in vitro*. Direct interaction of endotoxin with B lymphocytes leads to synthesis and secretion of antibodies directed not only against antigenic determinants on the endotoxin itself, but also with specificities characteristic of the complete repertoire of variable-region gene products. Macrophages stimulated with endotoxin increase their phagocytic or pinocytic capability of engulfing antigens and produce immunopotentiating factors such as IL 1. Evidence on the direct action of endotoxin for T lymphocytes is obscure. However, LPS can support the mitogenic response of T lymphocytes by concanavalin A (Fradet *et al.*, 1976) probably via a mediator produced by LPS-stimulated macrophages (Tanabe and Nakano, 1979).

In this chapter, we have reviewed the studies by others and ourselves concerning the adjuvant action of LPS on *in vitro* antibody response under the stimulation of specific antigens. Accumulated knowledge about the cellular mechanisms of the adjuvant effect underlines the multiple-pathway nature of the response. The action of endotoxin on immunocompetent cells in the presence of antigens is apparently different from that in the absence of antigens. Thymic dependency of antigens is also concerned with the adjuvant effect. Especially, premature B lymphocytes in bone marrow respond only to the TI antigen in endotoxin supplementation, suggesting that the subpopulation responsive to TI antigens differs from that responsive to TD antigens.

Adjuvant effect of endotoxin is under the control of B and T lymphocytes and macrophages. Moreover, recent studies indicate that some of the cell populations stimulated by endotoxin produce active mediators that regulate the adjuvant action of endotoxin (Butler et al., 1979). The regulatory mechanisms are complex, and the precise roles of these cells and mediators are as yet unclear. Endotoxin is one of the most powerful immunomodulators and it is a ubiquitous substance in nature. The analysis of the regulatory mechanisms in the presence of antigens may provide useful procedures for the clinical and therapeutic application of such immunomodulators.

X. References

Andersson, J., Sjöberg, O., and Möller, G., 1972, Induction of immunoglobulin and antibody synthesis in vitro by lipopolysaccharides, Eur. J. Immunol. 2:349.

Armerding, D., and Katz, D. H., 1974, Activation of T and B lymphocytes in vitro. I. Regulatory influence of bacterial lipopolysaccharide (LPS) on specific T-cell helper function, J. Exp. Med. 139:24.

Baltz, M., and Rittenberg, M. B., 1977, Down regulation in B lymphocytes: Low dose signals, Eur. J. Immunol. 7:218.

Behling, U. H., and Nowotny, A., 1977, Immune adjuvancy of lipopolysaccharide and a nontoxic hydrolytic product demonstrating oscillating effects with time, J. Immunol. 118:1905.

Boswell, H. S., Ahmed, A., Scher, I., and Singer, A., 1980a, Role of accessory cells in B cell activation. II. The interaction of B cells with accessory cells results in the exclusive activation of an Lyb5+B cell subpopulation, J. Immunol. 125:1340.

Boswell, H. S., Nerenberg, M. I., Scher, I., and Singer, A., 1980b, Role of accessory cells in B cell activation. III. Cellular analysis of primary immune response deficits in CBA/N mice: Presence of an accessory cell–B cell interaction defect, J. Exp. Med. 152:1194.

Butler, R. C., Nowotny, A., and Friedman, H., 1979, Macrophage factors that enhance the antibody response, Ann. N.Y. Acad. Sci. 332:564.

Campbell, P. A., and Kind, P. D., 1979, Enhancement by LPS of B-cell function in genetically deficient mice, Cell. Immunol. 42:240.

Coutinho, A., Gronowicz, E., Bullock, W. W., and Möller, G., 1974, Mechanism of thymus-independent immunocyte triggering: Mitogenic activation of B cells results in specific immune responses, J. Exp. Med. 139:74.

Coutinho, A., Gronowicz, E., and Möller, G., 1975, Mechanism of B-cell activation and paralysis by thymus-independent antigens: Additive effects between NNP–LPS and LPS in the specific response to the hapten, *Scand. J. Immunol.* **4**:89.

Diaz-Espada, F., Martinez-Alonso, C., and Bernabe, R. R., 1978, Development of B cell subsets: Effect of priming on *in vitro* TNP-LPS responsiveness, *J. Immunol.* **121**:13.

Fradet, Y., Roy, R., and Daguillard, F., 1976, Regulation of *in vitro* lymphocyte responses. I. Adjuvant effect of lipopolysaccharide (LPS) on low-zone unresponsiveness to concanavalin A, *Cell. Immunol.* **27**:94.

Frank, S., Specter, S., Nowotny, A., and Friedman, H., 1977, Immunocyte stimulation *in vitro* by nontoxic bacterial lipopolysaccharide derivatives, *J. Immunol.* **119**:855.

Franzl, R. E., and McMaster, P. D., 1968, The primary immune response in mice. I. The enhancement and suppression of hemolysin production by a bacterial endotoxin, *J. Exp. Med.* **127**:1087.

Goodman, M. G., and Weigle, W. O., 1981, T cell regulation of polyclonal B cell responsiveness. III. Overt T helper and latent T suppressor activities from distinct subpopulations of unstimulated splenic T cells, *J. Exp. Med.* **153**:844.

Hoffmann, M. K., 1979, Control of B-cell differentiation by macrophages, *Ann. N.Y. Acad. Sci.* **332**:557.

Hoffmann, M. K., 1980, Macrophages and T cells control distinct phases of B cell differentiation in the humoral immune response *in vitro*, *J. Immunol.* **125**:2076.

Hoffmann, M. K., and Watson, J., 1979a, Helper T cell-replacing factors secreted by thymus-derived cells and macrophages: Cellular requirements for B cell activation and synergistic properties, *J. Immunol.* **122**:1371.

Hoffmann, M. K., Weiss, O., Koenig, S., Hirst, J. A., and Oettgen, H. F., 1975, Suppression and enhancement of the T cell-dependent production of antibody to SRBC *in vitro* bacterial lipopolysaccharide, *J. Immunol.* **114**:738.

Hoffmann, M. K., Galanos, C., Koenig, S., and Oettgen, H. F., 1977, B-Cell activation by lipopolysaccharide: Distinct pathways for induction of mitosis and antibody production, *J. Exp. Med.* **146**:1640.

Hoffmann, M. K., Koenig, S., Mittler, R. S., Oettgen, H. F., Ralph, P., Galanos, C., and Hämmerling, U., 1979b, Macrophage factor controlling differentiation of B cells, *J. Immunol.* **122**:497.

Jacobs, D. M., 1979, Synergy between T cell-replacing factor and bacterial lipopolysaccharide (LPS) in the primary antibody response *in vitro*: A model for lipopolysaccharide adjuvant action, *J. Immunol.* **122**:1421.

Jacobs, D. M., and Morrison, D. C., 1975, Stimulation of a T-independent primary anti-hapten response *in vitro* by TNP-lipopolysaccharide (TNP-LPS), *J. Immunol.* **114**:360.

Johnson, A. J., Gaines, S., and Landy, M., 1956, Studies on the 0 antigen of *Salmonella typhosa*. V. Enhancement of antibody response to protein antigens by purified lipopolysaccharide, *J. Exp. Med.* **103**:225.

Kempf, K. E., and Rubin, A. S., 1979, Generation by lipopolysaccharide of a late-acting soluble suppressor of antibody synthesis, *Cell. Immunol.* **43**:30.

Kind, P., and Johnson, A. J., 1959, Studies on the adjuvant action of bacterial endotoxins on antibody formation. I. Time limitation of enhancing effect and restoration of antibody formation in X-irradiated rabbits, *J. Immunol.* **82**:415.

Kishimoto, S., Takahama, T., and Mizumachi, H., 1976, *In vitro* immune response to the 2,4,6-trinitrophenyl determinant in aged C57BL/6J mice: Changes in the humoral immune response to, avidity for the TNP determinant and responsiveness to LPS effect with aging, *J. Immunol.* **116**:294.

Koenig, S., and Hoffmann, M. K., 1979, Bacterial lipopolysaccharide activates suppressor B lymphocytes, *Proc. Natl. Acad. Sci. USA* **76**:4608.

McGhee, J. R., Farrar, J. J., Michalek, S. M., Mergenhagen, S. E., and Rosenstreich, D. L., 1979, Cellular requirements for lipopolysaccharide adjuvanticity: A role for both T lymphocytes and macrophages for *in vitro* responses to particulate antigens, *J. Exp. Med.* **149**:793.

McGhee, J. R., Kiyono, H., Michalek, S. M., Babb, J. L., Rosenstreich, D. L., and Mergenhagen, S. E., 1980, Lipopolysaccharide (LPS) regulation of the immune response: T lymphocytes from normal mice suppress mitogenic and immunogenic responses to LPS, *J. Immunol.* **124**:1603.

Mishell, R. I., and Dutton, R. W., 1967, Immunization of dissociated spleen cell cultures from normal mice, *J. Exp. Med.* **126**:423.

Möller, G., 1975, One non-specific signal triggers B lymphocytes, *Transplant. Rev.* **23**:126.

Morrison, D. C., and Ryan, J. L., 1979, Bacterial endotoxins and host immune responses, *Adv. Immunol.* **28**:293.

Mosier, D. E., Scher, I., and Paul, W. E., 1976, *In vitro* responses of CBA/N mice: Spleen cells of mice with an X-linked defect that precludes immune responses to several thymus-independent antigens can respond to TNP-lipopolysaccharide, *J. Immunol.* **117**: 1363.

Nakano, M., Shimamura, T., and Saito, K., 1971, Cellular mechanisms of adjuvant action of bacterial lipopolysaccharide in anti-sheep red blood cell antibody response, *Jpn. J. Microbiol.* **15**:149.

Neter, E., 1969, Endotoxins and the immune response, *Curr. Top. Microbiol. Immunol.* **47:** 82.

Ortiz-Ortiz, L., and Jaroslow, B. N., 1970, Enhancement by the adjuvant, endotoxin, of an immune response induced *in vitro, Immunology* **19**:387.

Persson, U., 1977, Lipopolysaccharide-induced suppression of the primary immune response to a thymus-dependent antigen, *J. Immunol.* **118**:789.

Portnoï, D., Motta, I., and Truffa-Bachi, P., 1981, Immune unresponsiveness of spleen cells from lipopolysaccharide-treated mice to particulate thymus-dependent antigen. I. Evidence for differentiation signal defect, *Eur. J. Immunol.* **11**:156.

Schrader, J. W., 1973, Mechanism of activation of the bone marrow-derived lymphocyte. III. A distinction between a macrophage-produced triggering signal and the amplifying effect on triggered B lymphocytes of allogenic interactions, *J. Exp. Med.* **138**: 1466.

Schrader, J. W., 1974, The mechanism of bone marrow-derived (B) lymphocyte activation. II. A "second signal" for antigen-specific activation provided by flagellin and lipopolysaccharide, *Eur. J. Immunol.* **4**:20.

Sjöberg, O., Andersson, J., and Möller, G., 1972, Lipopolysaccharide can substitute for helper cells in the antibody response *in vitro, Eur. J. Immunol.* **2**:326.

Tanabe, M. J., and Nakano, M., 1979, Lipopolysaccharide-induced mediators assisting the proliferative response of C3H/HeJ thymocytes to concanavalin A, *Microbiol. Immunol.* **23**:1097.

Uchiyama, T., and Jacobs, D. M., 1978a, Modulation of immune response by bacterial lipopolysaccharide (LPS): Multifocal effects of LPS-induced suppression of the primary antibody response to a T-dependent antigen, *J. Immunol.* **121**:2340.

Uchiyama, T., and Jacobs, D. M., 1978b, Modulation of immune response by bacterial lipopolysaccharide (LPS): Cellular bases of stimulatory and inhibitory effects of LPS on the *in vitro* IgM antibody response to a T-dependent antigen, *J. Immunol.* **121**: 2347.

Walker, S. M., and Weigle, W. O., 1978, Effect of bacterial lipopolysaccharide on the *in vitro* secondary antibody response in mice. I. Description of the suppressive capacity of lipopolysaccharide, *Cell. Immunol.* **36**:170.

Watson, J., Epstein, R., Nakoinz, I., and Ralph, P., 1973, The role of humoral factors in the initiation of *in vitro* primary immune responses. II. Effects of lymphocyte mitogens, *J. Immunol.* **110**:43.

14

Stimulation of Immunomodulatory Factors by Bacterial Endotoxins and Nontoxic Polysaccharides

Herman Friedman, Steven Specter, and R. Christopher Butler

I. Introduction

Immune responses are the result of a complex series of cellular and molecular interactions following exposure of lymphoid cells to specific antigens. Much is now known concerning the nature and mechanisms of antibody responses and cell-mediated immunity at both the humoral and the cellular level. The magnitude and range of the immune response to a particular antigen is regulated by a wide variety of factors, including soluble factors derived from various cell types that appear to serve as additional "signals" to the immune response mechanism (Schimpl and Wecker, 1975; Meltzer and Oppenheim, 1977). In this regard, components of microorganisms, including those derived from either gram-negative or gram-positive bacteria, are known to exert marked adjuvant effects on the immune response of animals and man, both *in vivo* and *in vitro* (Friedman, 1979; Garzelli *et al.*, 1982). Microbial products have been shown to either enhance or suppress the immune response, as well as modify the nature of the response. In this regard, many studies have been performed concerning the adjuvant effects of cell wall-derived lipopolysaccharides (LPS) from gram-negative bacteria. Various laboratories have shown that LPS can both stimulate

Herman Friedman and Steven Specter ● Department of Microbiology and Immunology, University of South Florida College of Medicine, Tampa, Florida 33612. **R. Christopher Butler** ● The Arlington Hospital, Arlington, Virginia 22205.

and inhibit the immune response, possibly by modifying macrophage activity as well as activity of responding immunocytes. In this regard, studies in this laboratory have shown that not only the intact LPS moiety from gram-negative bacteria influences the immune response, but also a lower-molecular-weight lipid A-free polysaccharide (PS) derivative. Nontoxic PS derived from LPS, when injected into experimental animals or added to spleen cell cultures *in vitro*, influence the antibody response to a wide variety of antigens, including sheep red blood cells (SRBC). As indicated below, both LPS and nontoxic PS show marked immunomodulatory effects on murine lymphoid cells. This effect appears due to stimulation of antibody helper factors when lymphoid cells are exposed to these materials.

II. General Experimental Methods

In order to examine immunomodulation by bacterial LPS and nontoxic PS, a murine model system was utilized with SRBC as test antigen. For this purpose, BALB/c mice were obtained from The Jackson Laboratory, Bar Harbor, Maine. LPS was obtained by acid digestion procedures from *Serratia marcescens* cultures, as described previously (Nowotny *et al.*, 1966). The phenol–water extract contained less than 0.5% nitrogen but was rich in both the PS and the lipid A moiety. The endotoxin had the usual characteristics associated with LPS, including the ability to stimulate antibody forming cells *in vivo* and *in vitro*, Shwartzman reactivity in rabbits, pyrogenicity, etc. A nontoxic PS derivative was obtained by acid hydrolysis of the endotoxin exactly as described (Chang *et al.*, 1974). Control *E. coli* endotoxin preparations were obtained commercially from Difco Laboratories.

Adjuvant activity of the preparations was determined by injecting graded quantities of LPS or PS intraperitoneally into groups of 8- to 10-week-old mice followed by challenge immunization at various times thereafter by i.p. injection of graded amounts of SRBC. The antibody response was determined at various times thereafter by obtaining individual blood specimens by retroorbital plexus puncture from individual mice and harvesting the serum. Micro hemagglutination and hemolysin tests were performed using 0.025-ml volumes of twofold dilutions of serum. In addition, groups of three to six mice were sacrificed at specified time intervals after immunization, their spleens obtained at autopsy, and dispersed single-cell suspensions prepared by teasing with needles and forceps in sterile medium 199 containing fetal calf serum. The numbers of individual hemolytic antibody plaque-forming cells (PFC) per cell suspension were determined by the standard hemolytic plaque assay using a micro procedure (Cunningham and Szenberg, 1968). The number of individual PFCs was calculated from the number of antibody plaques from three or more per cell sample, and calculated per million spleen cells tested or per cell suspension. In addition, a completely *in vitro* culture system was used in which 5×10^6 normal viable spleen cells were cultured in microtiter plates in 0.25-ml volumes of medium to which were added graded concentrations of LPS or PS and sheep RBC as the antigen (Kamo *et al.*, 1976). The numbers of hemolytic PFCs per well were determined at various times thereafter by the Cunningham plaque assay (Cunningham and Szenberg, 1968).

In order to examine the possibility that soluble lymphoid cell-derived factors were involved, culture supernatants obtained from mouse spleen cells incubated with graded concentrations of either LPS or PS were added to suspensions of normal spleen cells immunized *in vitro* with SRBC. The culture supernatants were characterized by various treatments, such as incubation with enzymes or treatment at different temperatures.

III. Results

Injection of intact *Serratia* LPS into mice enhanced markedly the immune response to challenge immunization with SRBC (Table 1). A chemically detoxified LPS (i.e., Mex B) had little if any significant effect on the response. However, the acid hydrolysis-prepared PS was almost as effective as the intact LPS in enhancing the immune response (Table 1). Similarly, LPS was a marked adjuvant for the antibody response to SRBC *in vitro*, as was PS (Table 1). The nontoxic Mex B had no en-

Table 1. Comparative Adjuvanticity of *Serratia* LPS and Detoxified LPS and PS for Anti-SRBC Responses

Bacterial preparation[a]	Serum titer[b]	PFC response[c]	
		In vivo[d]	*In vitro*[d]
None (control)	1:256	962 ± 130	1560 ± 320
LPS	1:1024	3580 ± 295	4530 ± 320
Mex B	1:256	1015 ± 150	1310 ± 370
PS	1:1024	2961 ± 175	4120 ± 295

[a] Indicated *Serratia* product, in 10- to 20-μg doses, injected i.p. into mice or added to cultures of 5×10^6 normal mouse spleen cells *in vitro*.
[b] Average hemagglutination titer 10 days after immunization.
[c] Average PFC response for 3–5 cultures per group 4–5 days after immunization.
[d] Mice injected i.p. with 4×10^8 SRBC, cultures immunized *in vitro* with 2×10^6 SRBC.

hancing effect. The stimulation of the *in vivo* immune response to SRBC was evident during both the primary and the secondary response (Table 2). Both LPS and PS had a similar enhancing effect on the secondary IgG response, indicating that the T-dependent switch to the 7 S class of immunocompetent cells was stimulated. The enhanced IgG response was even greater than that observed for the IgM response, both after primary and secondary immunizations with SRBC (Table 2).

In vitro models of antibody formation are useful in dissecting the cell types involved in antibody formation, as well as separation of those parameters involving nonimmune systems from those by lymphoid cells per se. As is evident in Table 3, graded doses of PS enhanced the immune response over a 2- to 7-day period *in vitro*, similar to the enhancement observed with the intact LPS. Optimum enhancement occurred with both preparations on the fourth or peak day of the response, but activity was evident even at earlier and later times during the course of the *in vitro* immune response. Approximately 10–20 μg of either LPS or PS was optimal for enhancing this response. However, as shown in Table 4, addition of either LPS or PS to the cultures during the first day of incubation was optimal for maximum immune enhancement. Addition of LPS or PS 48 hr after culture initiation resulted in lower enhancement, while addition on day 4 had essentially no or little effect. Previous experiments had shown that a concentration of 10^8 SRBC stimulated an optimum immune response. However, many model systems concerning adjuvants utilize suboptimal doses of antigen. Thus, it was of interest to determine whether different concentrations of SRBC would reveal differences between LPS and PS effects. As is

Table 2. Comparative Effects of Bacterial LPS and Detoxified LPS and PS on Primary vs. Secondary Antibody Response of Mouse Spleen Cells to SRBC

	Antibody response[b]		
	Primary	Secondary	
Bacterial preparation[a]	IgM	IgM	IgG
None (control)	738 ± 110	530 ± 180	1865 ± 210
LPS	1975 ± 395	1040 ± 260	4870 ± 560
Mex B	820 ± 170	580 ± 150	1910 ± 320
PS	2100 ± 370	1130 ± 280	4150 ± 370

[a] Indicated *Serratia* preparation injected in 10- to 20-μg dose i.p. into mice immunized i.p. with 4×10^8 SRBC.
[b] Average PFC response ± S.E. for 3–6 mice per group 5 days after primary or secondary immunization with SRBC.
[c] Mice primed 4 weeks earlier with 4×10^8 SRBC i.p.

Table 3. Effect of LPS and PS on Antibody Response of Mouse
Spleen Cells *in Vitro* to SRBC

Bacterial product[a]	Antibody response[b]		
	Day +2	Day +4	Day +7
None (control)	310	1240	610
LPS 0.1 μg	361	1350	680
1.0 μg	425	1730	1130
10.0 μg	586	2870	1870
20.0 μg	520	1530	950
PS 0.1 μg	360	1310	730
1.0 μg	395	1950	970
10.0 μg	520	2930	2110
20.0 μg	650	2760	2000

[a] Indicated *Serratia* LPS or PS added to cultures of 5×10^6 normal mouse spleen cells immunized *in vitro* with 2×10^6 SRBC.
[b] Average PFC response for three or more cultures per group on day indicated after *in vitro* immunization.

evident in Table 5, the lowest dose of SRBC resulted in lower PFC responses. Both LPS and PS enhanced this response in a dose-related manner. Furthermore, the optimum doses of the bacterial products, i.e., 10–20 μg, produced a greater percent increase in the PFC response for the lowest SRBC dose as compared to the increases observed with the optimum or even superoptimum doses. Furthermore, it is evident that even in the absence of SRBC, both LPS and PS induced a clonallike activation of the background antibody response in an equal manner.

Table 4. Effect of Time of Addition of LPS and PS to Cultures
on Antibody Enhancement

Time of addition to cultures (hr)[a]	Antibody response[b]	
	LPS	PS
None	1170 ± 210	1030 ± 170
0	3760 ± 410	3100 ± 250
18	3100 ± 370	3250 ± 350
48	1970 ± 295	2010 ± 310
96	1230 ± 175	1340 ± 250

[a] 10–20 μg LPS or PS added to cultures of 5×10^6 normal spleen cells immunized *in vitro* with 2×10^6 SRBC.
[b] Average PFC response ± S.E. for 3–5 cultures per group 5 days after *in vitro* immunization.

Table 5. Effect of Antigen Concentration on Immunostimulation by LPS and PS of Antibody Responsiveness of Mouse Spleen Cells to SRBC

Bacterial product added to cultures[a]	SRBC concentration for immunization[b]			
	None	2×10^4	2×10^6	2×10^8
None	58 ± 16	197 ± 30	785 ± 130	930 ± 250
LPS 1.0 μg	117 ± 25	710 ± 65	1360 ± 190	1550 ± 320
10.0 μg	238 ± 39	930 ± 130	2250 ± 360	1600 ± 320
20.0 μg	310 ± 74	1150 ± 280	2030 ± 410	2100 ± 250
PS 1.0 μg	110 ± 36	360 ± 85	1196 ± 210	1240 ± 510
10.0 μg	295 ± 45	765 ± 130	270 ± 370	1760 ± 340
20.0 μg	260 ± 58	970 ± 190	2300 ± 280	1530 ± 240

[a] Indicated dose of LPS or PS added to cultures of 5×10^6 normal mouse spleen cells *in vitro*.
[b] Indicated dose of SRBC added to cultures and average PFC response ($\pm$ S.E.) determined for 3–4 cultures per group 4 days later.

The results presented in Table 6 show that a soluble factor may be involved in the immune enhancement induced by the endotoxin, as well as the nontoxic PS. For these experiments, optimum amounts of LPS or PS were injected into normal nontreated mice that were bled 2 hr later to obtain serum as a source of post-LPS or post-PS serum. Addition of graded quantities of these sera to cultures of mouse spleen cells resulted in a dose-dependent enhancement of the immune response to SRBC. These results suggested development of soluble mediators as immunoenhancement factors.

Table 6. Effect of Post-LPS and Post-PS Serum on Antibody Response of Mouse Spleen Cells to SRBC

Serum added to cultures (ml)[a]	PFC response[b]
None (control)	765 ± 210
Post-LPS 0.005	810 ± 200
0.01	1170 ± 320
0.05	1460 ± 190
0.1	1530 ± 760
Post-PS 0.005	960 ± 150
0.01	1230 ± 210
0.05	1270 ± 320
0.1	1360 ± 240

[a] Indicated volume of posttreatment serum added to 5×10^6 mouse spleen cells immunized *in vitro* with 2×10^6 SRBC; serum donor injected 2 hr earlier with 10 μg LPS or PS.
[b] Average PFC response of 5–6 cultures per group 4 days after *in vitro* immunization.

Table 7. Effect of LPS, PS, and Posttreatment Serum on Antibody Response of FLV Suppressed Spleen Cells

Culture treatment[a]	Addition to cultures	PFC response[b]
None (control)		1160 ± 210
FLV	None	275 ± 37
	LPS	865 ± 410
	Post-LPS serum	910 ± 320
	PS	945 ± 196
	Post-PS serum	830 ± 240

[a] Cultures of 5×10^6 spleen cells either untreated, as control, or incubated with 100 LD_{50} FLV plus LPS or PS (10.0 μg or 0.1 μg) 2-day posttreatment mouse serum.
[b] PFC response $\pm$ S.E. for 4–5 cultures 4 days after *in vitro* immunization.

In many sytems, it has been observed that endotoxins may reverse or abrogate the tumorigenic potential of a wide variety of carcinogens and especially oncogenic viruses. Furthermore, postendotoxin serum has been considered a major source of tumor-necrotizing factor and exhibits potent antitumor activity both *in vivo* and *in vitro*. In earlier studies, it was observed that infection of susceptible mice with a leukemogenic virus, i.e., Friend leukemia virus (FLV), markedly depresses the antibody response. Lipopolysaccharides, as well as other similar substances, reverse or inhibit some of the immunosuppressive effects of this leukemia virus. As is evident in Table 7, both LPS as well as PS, when added to cultures of spleen cells from FLV-infected mice, markedly enhanced the otherwise depressed antibody response. Similarly, post-LPS and post-PS sera had similar immunoenhancing effects for spleen cells from FLV-infected mice. Table 8 shows that cell-free supernatants from normal mouse spleen cell cultures incubated for 3–5 days with either LPS or PS also had similar enhancing activity for both virus-infected splenocytes as well as normal spleen cells *in vitro*. The optimum enhancement occurred with supernatants from 5-day-treated cultures, indicating that the effect was not due to a possible carryover of LPS or PS with the small amount of supernatant added to the cultures being immunized with SRBC. It was evident that the immunoenhancing supernatants from LPS- and PS-treated cultures had a greater effect on normal spleen cells, but also markedly enhanced the response of spleen cells from the virus-infected mice.

IV. Discussion and Conclusions

Studies concerning the immunomodulatory effects of bacterial endotoxins have revealed that these microbial products exhibit a wide variety of activities on the immune response system (Skidmore *et al.*,

Table 8. Effect of Culture Supernatants from LPS- and PS-Treated Spleen Cells on Antibody Responsiveness of Normal and FLV-Infected Spleen Cells Immunized *in Vitro* with SRBC

Supernatant from cell cultures[a]	Days of culture treatment	Antibody response[b]	
		Normal	Virus infected
Untreated		962 ± 130	240 ± 38
LPS-treated	+1	1070 ± 240	510 ± 260
	+3	2760 ± 370	870 ± 240
	+5	2200 ± 310	950 ± 160
PS-treated	+1	1130 ± 210	310 ± 140
	+3	2640 ± 420	730 ± 190
	+5	2530 ± 370	810 ± 220

[a] Cell-free culture supernatants from normal or treated spleen cell cultures added in 0.1-ml volume to indicated spleen cell cultures (5×10^6) immunized *in vitro* with 2×10^6 SRBC; treated cultures incubated for indicated time with 10 μg LPS or PS.
[b] Average PFC response ± S.E. for 3–5 cultures per group 5 days after *in vitro* immunization.
[c] Spleen cell donors injected with 100 LD_{50} FLV 10–15 days earlier.

1975). Many models have been used, including antibody formation *in vivo* as well as immune responses *in vitro*, utilizing newer immunobiologic and cultural methods. Previous studies in this laboratory have indicated that not only intact bacterial LPS but also lipid A-free PS derivatives could enhance the antibody response to SRBC, as well as other antigens (Frank *et al.*, 1977). It is evident from results presented in this report that graded doses of *Serratia* LPS and PS enhance the antibody response of mouse spleen cells *in vitro* after *in vivo* immunization with SRBC, as well as after *in vitro* challenge with the same antigen. The high-molecular-weight detoxified Mex B preparation, shown to have some biologic activity but devoid of immunoenhancing activity, had no such effect either *in vivo* or *in vitro*. Thus, it is not the molecular weight or form of the endotoxin preparation per se that is important in immunoenhancement, since the PS derivative free of detectable lipid A is a relatively small molecule, at least in comparison to the intact LPS preparation.

Since earlier data had suggested that LPS affects the immune response by stimulating interleukin production by macrophages, it was of interest to determine whether PS may function in a similar manner (Meltzer and Oppenheim, 1977). Indeed, as shown in this study, the PS had the equivalent ability to induce and/or stimulate an enhanced immune response *in vitro*, regardless of the dose of antigen used and/or the time of addition to the cultures. Furthermore, serum obtained from mice injected 2 hr earlier with LPS or PS had immunoenhancing activity for other spleen cell cultures immunized *in vitro* with SRBC. Finally, spleen cell cultures exposed to either LPS or PS resulted in supernatants that had immunoenhancing activity for both normal spleen cells, as well as leukemia virus-suppressed splenocytes. These results indicate that the PS is active in stimulating immunoenhancing soluble factors presumably interleukins. There have been no unequivocal results to date showing that lipid A per se has similar activities. Thus, the lipid A-free PS immunoenhancement appears due to the PS, as shown by both direct and indirect means. For example, stimulation of intermediatory substances such as interleukins would account for most if not all of the immunoenhancement induced by these substances. The PS moiety could be the active agent in this respect.

It is of interest that even immunosuppressed spleen cell preparations, when exposed to either the LPS, the PS, or small quantities of sera from mice treated 2 hr earlier with these preparations, respond in an enhanced manner to a challenge immunization with SRBC. This suggests that even though the immunosuppressed spleen cell cultures have fewer immunocompetent cells capable of producing detectable antibody to SRBC following incubation with these bacterial preparations and/or the induced soluble factors in serum or culture supernatants, increased numbers of antibody-forming cells develop. This suggests that the effects noted may be due to the enhancement of cellular differentiation or clonal expansion. Interleukins are known to have such effects in a wide variety of model systems. It seems apparent that the PS has the ability to stimulate interleukin activities similar to that induced by whole LPS and other interleukin-stimulating materials. Since the PS is nontoxic and can be administered in relatively large doses, it seems plausible that this material may be useful not only for studying the mechanisms involved in immune enhancement by bacterial cell wall components, but also may have some practical value in a wide variety of systems where immunoenhancement is desirable.

V. Summary

Endotoxin derived from *S. marcescens*, as well as a nontoxic PS derivative, were studied in terms of the nature and mechanism of immunoenhancement *in vivo* and *in vitro*. These bacterial products, when injected into normal mice or added to cultures of spleen cells, enhance the antibody response to SRBC. The effects were dose and time dependent. The PS had similar enhancing activities to the intact LPS. Injection of these materials into mice resulted in development of post-LPS or post-PS serum that, in turn, had immunoenhancing properties when added in relatively small quantities to cultures of normal mouse spleen cells immunized with SRBC. Furthermore, spleen cell cultures exposed to these materials *in vitro* for several days developed immunoenhancing supernatant active on normal mouse spleen cell cultures. These bacterial preparations, including the posttreatment sera and supernatants, also had immunoenhancing activity for immunosuppressed spleen cell cultures derived from leukemia virus-infected mice. These results indicate that a nontoxic, relatively low-molecular-weight PS derived from LPS may be useful not only for studying the nature and mechanism of immunoenhancement induced by bacterial cell wall components, but also may have some practical value in reversing virus-induced immunosuppression at the cellular level.

ACKNOWLEDGMENTS

The capable technical assistance of Mr. Chandu Patel and Mrs. Leony Mills in these studies is acknowledged.

VI. References

Chang, H., Thompson, J. J., and Nowotny, A., 1974, Release of colony stimulating factor (CSF) by nonendotoxin breakdown products of bacterial lipopolysaccharides, *Immunol. Commun.* **3**:401.

Cunningham, A. J., and Szenberg, A., 1968, Further improvements in the plaque technique for detecting single antibody-forming cells, *Immunology* **14**:599.

Frank, S. J., Specter, S., Nowotny, A., and Friedman, H., 1977, Immunocyte stimulation *in vitro* by nontoxic bacterial lipopolysaccharide derivatives, *J. Immunol.* **119**:855.

Friedman, H., 1979, Subcellular factors in immunity, *Ann. N.Y. Acad. Sci.* **332**:16–25.

Garzelli, C., Campa, M., Colizzi, V., Benedettini, G., and Falcone, G., 1982, Evidence for autoantibody production associated with polyclonal B-cell activation by *Pseudomonas aeruginosa*, *Infect. Immun.* **35**:13.

Kamo, I., Pan, S.-H., and Friedman, H., 1976, A simplified procedure for *in vitro* immunization of dispersed spleen cell cultures. *J. Immunol. Methods* **11**:55.

Meltzer, M. S., and Oppenheim, J. J., 1977, Bidirectional amplification of macrophage–lymphocyte interactions: Enhanced lymphocyte activating factor production by activated adherent mouse peritoneal cells, *J. Immunol.* **118**:77.

Nowotny, A., Cundy, K., Neale, N. L., Nowotny, A. M., Radvany, R., Thomas, S. P., and Tripodi, D. J., 1966, Relation of structure to function in bacterial O-antigens. IV. Fractionation of the components, *Ann. N.Y. Acad. Sci.* **133**:586.

Schimpl, A., and Wecker, E., 1975, A third signal in B-cell activation given by TRF, *Transplant. Rev.* **23**:176.

Skidmore, B. J., Chiller, J. M., Morrison, D. C., and Weigle, W. O., 1975, Immunologic properties of bacterial lipopolysaccharides (LPS): Correlation between the mitogenic, adjuvant and immunogenic activity, *J. Immunol.* **114**:770.

15

Interleukin 1

Release from LPS-Stimulated Mononuclear Phagocytes

Lawrence B. Lachman

I. Introduction

Interleukin 1 (IL 1), previously known as lymphocyte-activating factor (LAF), is key mediator of macrophage function. Macrophages release IL 1 following stimulation with bacterial endotoxin (Gery and Waksman, 1972), substances that possess adjuvant properties (Oppenheim et al., 1980), or phagocytic stimuli (Rosenwasser et al., 1979). IL 1 can completely replace macrophages in some in vitro immunological assay systems (Maizel et al., 1981) and greatly reduce the number of macrophages required in other systems (Rosenberg and Lipsky, 1981). IL 1 probably replaces the soluble activating factor released by macrophages but cannot replace the macrophage function of antigen or lectin presentation (Rosenberg and Lipsky, 1981). IL 1 may also regulate nonimmunological responses that involve macrophages. Specifically, IL 1 may be identical to the fever-producing substance endogenous pyrogen (Murphy et al., 1980; Rosenwasser et al., 1979), and appears to stimulate prostaglandin and collagenase release from synovial cells (Mizel et al., 1981) and to release acute-phase reactants (Selinger et al., 1980).

Clearly, a mediator with the broad spectrum of biological activities of IL 1 has been the subject of intense research. This chapter will attempt to emphasize the cellular sources of IL 1, the stimulants required

Lawrence B. Lachman • Immunex Corporation, Seattle, Washington 98101.

for macrophages to release IL 1, and finally the purification and bio-logical properties of IL 1.

Prior to beginning the more detailed aspects of this chapter, a brief statement concerning the inherent difficulties of lymphokine and monokine research is required. Although lymphokines and monokines are easily prepared by culturing mononuclear cells with lectins or endo-toxin, there are few other easy aspects of this research. Almost all lymphokines and monokines are assayed in time-consuming, labor-intensive bioassay systems. The assay systems are difficult to standardize due to daily variation in the response of indicator cells. In addition, the assay systems are often adversely affected by chemicals and buffers used in the purification procedures. Lymphokines and monokines are notoriously difficult to purify. The major difficulty is in obtaining large enough quantities of starting material to compensate for the many low-yield purification steps required to purify a protein to homogeneity. The sensitivity of the assay systems can be extemely deceiving. Culture medium exhibiting strong biological activity may contain less than 1 ng/ml of the effector lymphokine. Thus, if the lymphokine could be obtained with 100% yield from 1 liter of culture medium, only 1 μg of protein would be obtained; hardly enough protein to begin detailed biochemical analysis even with the most sensitive amino acid analyzers and protein sequencers. Obtaining homogeneous preparations of lymphokines and monokines is of utmost importance to the unambiguous understanding of their biological properties. Also, homogeneous preparations will permit detailed chemical analysis, which could lead to the preparation of these valuable molecules through gene cloning. The following chapter will review many aspects of IL 1 preparation and pu-rification as well as its multiple biological activities.

II. IL 1 Release from Mononuclear Cells

A. IL 1 Assay

The most frequently used assay for IL 1 is stimulation of [^{3}H]thy-midine incorporation by mouse thymocytes. Briefly, thymocytes are cul-tured for 72 hr in the presence of 10% (v/v) IL 1 with a pulse of [^{3}H] thymidine for the final 18–24 hr of culture. There are two major varia-tions of this assay. The first variation measures IL 1 as a direct mitogen for unstimulated thymocytes (Lachman et al., 1980), and the second variation measures the ability of IL 1 to increase [^{3}H]thymidine incor-poration by PHA-stimulated thymocytes (Mizel et al., 1978). Although PHA is a poor mitogen for mouse thymocytes, PHA plus IL 1 demon-strates a strong synergistic response. A major drawback to the PHA-stimulated comitogen assay is that IL 1 is not the only substance that can stimulate lectin-pulsed thymocytes synergistically. Specifically, phorbol myristic acetate (PMA) also stimulates pronounced [³H]thymidine incorporation by PHA-stimulated T cells (Rosenstreich and Mizel, 1979). This would be of minor significance if it were not for the fact that PMA has been frequently and *erroneously* thought to stimulate IL 1 release from macrophages (Diamanstein *et al.*, 1981). Since lipopolysaccharide (LPS) is the most common stimulant of IL 1 release, thymocytes of the C3H/HeJ strain are most frequently used to assay for IL 1 (Mizel *et al.*, 1978). Because mouse thymocytes respond very poorly to endotoxin, other strains of mice, particularly the NIH Swiss, can be used satisfactorily to assay endotoxin-containing culture medium.

B. The Effect of Endotoxin-Contaminated Tissue Culture Medium upon IL 1 Release

Although IL 1 release was believed to result from LPS stimulation of monocytes, IL 1 was frequently detected in the medium of unstimulated human mononuclear cells (Wood and Cameron, 1978; Lachman and Metzgar, 1980a). The explanation for this finding was the ubiquitous presence of endotoxin in fetal calf serum (FCS), tissue culture medium, and Ficoll–Hypaque used to prepare mononuclear cells. Table 1 demonstrates a correlation between endotoxin-contaminated tissue culture medium and IL 1 release by buffy coat leukocytes. Similar results were obtained for FCS by Wood and Cameron (1978). Fortunately, many commercial preparations of Ficoll–Hypaque (Sigma Chemical Co.) and tissue culture media are negative in the *Limulus* lysate assay (LLA). The significance of these observations is to emphasize that only substances that are negative in the LLA can be used to prepare IL 1 or be evaluated as stimulants of IL 1 release. Many types of chemical and biological immunostimulants used in *in vitro* experiments are contaminated with endotoxin (Fumarola and Jirillo, 1979).

C. IL 1 Release by Peripheral Blood Monocytes

Many types of experiments have directly demonstrated that IL 1 is released by human monocytes (Gery and Waksman, 1972). Human polymorphonuclear (PMN) leukocytes have not been previously evaluated for IL 1 release. The results in Table 2 demonstrate that PMN leukocytes do not release IL 1.

Human monocytes demonstrate a remarkable sensitivity to LPS as expressed by IL 1 release. Figure 1 shows a dose–response curve for LPS stimulation of IL 1 from human mononuclear cells. Maxium IL 1

Table 1. The Effect of Endotoxin Contamination in Commercially Prepared Tissue Culture Media upon IL 1 Release

Tissue culture medium	Limulus lysate clot	Serum	Limulus lysate clot	Leukocytes	IL 1 release [³H]Thymidine incorporation by thymocytes (cpm/culture)
Microbiological Associates[a]	−	Freshly prepared human serum	−[c]	+[d]	229
				−[e]	199
Gibco[b]	+	Freshly prepared human serum	−	+	1931
				−	173

[a] Specially prepared and tested endotoxin-free minimal essential medium (MEM) with glutamine.
[b] MEM with glutamine or RPMI 1640 with glutamine.
[c] Normal human serum was diluted 1:20 in endotoxin-free water. Undiluted serum is inhibitory in the *Limulus* lysate assay. The final concentration of serum in the leukocyte cultures is 5%.
[d] Leukocytes were prepared by 800g centrifugation of heparinized whole human blood. All precautions to ensure endotoxin-free conditions were followed. These included the use of new bottles of heparin, pyrogen-free disposable needles and syringes, and plastic tissue culture flasks and pipets.
[e] Control cultures not containing leukocytes.

Table 2. Comparison of IL 1 Release by Mononuclear Cells
and PMN Leukocytes[a]

Leukocyte culture conditions	[^{3}H]Thymidine incorporation by thymocytes (cpm/culture)
Layer I (monocytes and lymphocytes)	
0.5 × 10^6/ml	450
0.5 × 10^6/ml + LPS (20 μg/ml)	3449
Layer II (PMN leukocytes)	
0.5 × 10^6/ml	307
0.5 × 10^6/ml + LPS	473
Control[b]	233
Control + LPS	473
Medium	250

[a] Monocytes and PMN leukocytes were prepared by centrifugation over an endotoxin-free Ficoll–Hypaque step gradient as described by English and Anderson (1974).
[b] Control culture not containing leukocytes.

release could be obtained in culture medium containing 0.1 ng/ml of LPS. Similar results have been reported by Wood and Cameron (1978). Thus, the sensitivity of human monocytes to endotoxin, as expressed by IL 1 release, parallels the sensitivity of the LLA and explains why whenever mononuclear cells are cultured without specific precautions for endotoxin contaminations, the medium will contain significant levels of IL 1 (Lachman and Metzgar, 1980a).

The effect of endotoxin upon IL 1 release can be blocked by preincubation of LPS with polymyxin B (PB) (Wood and Cameron, 1978). In the process of confirming this observation, it was found that PB could block the effect of LPS only if added prior to the addition of serum to the mixture (Table 3). When LPS was incubated for 1 hr with serum, PB could no longer inhibit IL 1 release. This activation of serum by LPS, or activation of LPS by serum (whichever way you care to look at it) occurred slightly better with heat-inactivated serum and was not inhibited by 10 mM EDTA. The activation of serum would appear to be mediated by the lipid A moiety of LPS since it can be blocked by PB (Morrison and Jacobs, 1976). An explanation for these observations may be that the interaction of serum with LPS generates a stimulatory component, such as a complement component (nonenzymatic since the reaction occurs in the presence of EDTA) which binds to monocytes and activates IL 1 release. Another possibility is that a serum enzyme may hydrolyze endotoxin to its actual stimulatory form. It should be emphasized that IL 1 is not endotoxin (Wood and Cameron, 1978). Also, the requirement for the interaction of endotoxin with serum could explain why in the absence of serum, only about 10% as much IL 1 can be generated as in the presence of serum.

To further investigate the question of serum activation by endotoxin, another stimulant of IL 1 release was evaluated. Since phagocytosis

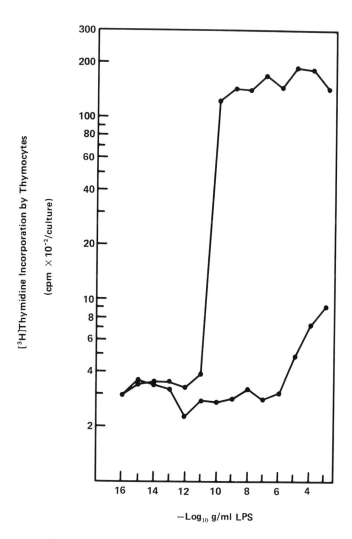

Figure 1. The effect of LPS upon IL 1 release by human leukocytes. Endotoxin-free buffy coat leukocytes were cultured in increasing concentrations of *E. coli* LPS for 24 hr and the levels of IL 1 were determined in the mouse thymocyte assay (upper curve). The effect of increasing concentrations of LPS upon the thymocyte assay (lower curve) was also measured by culturing control samples containing LPS but lacking leukocytes.

Table 3. Inhibition of LPS Activation of Human Serum by Polymyxin B

| | | | | IL 1 release |
| | Time of addition (hr)[a] | | | [^{3}H] Thymidine incorporation by mouse thymocytes (cpm/culture) |
	0	1	6	
Experimental	LPS + PB[b]	NHS	WBC	234
groups	LPS + PB[c]	NHS	WBC	121
	LPS + NHS	PB[b]	WBC	1256
	LPS + NHS	PB[c]	WBC	1201
	LPS + NHS		WBC	1216
Controls	PB[b]	NHS	WBC	130
	PB[c]	NHS	WBC	142
	NHS		WBC	229
	NHS			138
	LPS			174
	LPS	NHS		155

[a] Human buffy coat leukocytes and normal serum were prepared as described in text to maintain endotoxin-free conditions. The culture conditions were as follows: LPS, 1 ng/ml; normal human serum (NHS), 5%; white blood cells (WBC), 0.5 + 10^6/ml. The time-zero additions were added to 1.0 ml of endotoxin-free medium. IL 1 release was measured following 24 hr of culture.
[b] 2.5 μg/ml of polymyxin B (PB).
[c] 5.0 μg/ml of PB.

of staphylococci leads to IL 1 release by macrophages (Hanson *et al.*, 1980), the question was asked if phagocytosis was required or if the staphylococci could activate serum in a manner similar to LPS. It should be emphasized that endotoxin-free staphylococci were used for these experiments. As shown in Table 4, incubation of staphylococci with serum and removal of the bacteria by centrifugation and sterile filtering resulted in an activated serum that could stimulate IL 1 release from monocytes. Again the possibility cannot be ruled out that serum hydrolyzes a macrophage stimulatory substance from the bacterial cell wall. Isolation of the stimulatory component should allow identification of its origin and may provide very interesting insight into the signal requirement for IL 1 release from monocytes.

D. IL 1 Release from Acute Monocytic Leukemic Cells

Lachman *et al.* (1978) demonstrated that peripheral blood leukocytes from patients with acute monocytic leukemia (AMoL) or acute myelomonocytic leukemia (AMML) release IL 1 when cultured in the presence of endotoxin-containing medium. The leukemic cells do not

Table 4. IL 1 Release from Leukocytes Stimulated with Staphylococci-Activated Serum[a]

	[3H]Thymidine incorporation by mouse thymocytes (cpm/culture)	
Culture conditions[b]	− Leukocytes	+ Leukocytes
Staphylococci-activated unheated serum	1465	14,366
Staphylococci-activated heat-inactivated[c] serum	782	8,950
Heat-inactivated serum	512	975
Heat-inactivated serum + LPS[d]	691	16,332
Heat-inactivated serum + staphylococci[e]	617	10,680
Medium	547	550

[a] Normal human serum (NHS) was activated with staphylococci by mixing 1.0 ml of serum with 3.9 × 10^8 staphylococci in 1.0 ml of endotoxin-free MEM. The mixture was heated for 4 hr at 37°C. The activated serum was separated from the bacteria by centrifugation and sterile-filtration.
[b] The endotoxin-free culture conditions were 1.0 × 10^6 buffy coat leukocytes, 5% (v/v) NHS or 5% (v/v) activated serum, and MEM to a final volume of 1.0 ml. The cultures were incubated for 24 hr at 37°C and the supernatant solutions assayed for IL 1 activity. Control cultures contained all components except leukocytes.
[c] 56°C, 30 min.
[d] 1 μg/ml.
[e] 2.0 × 10^7 bacteria were added to a 1.0-ml leukocyte culture.

spontaneously release IL 1. Release of IL 1 from AMML cells correlates directly with the presence of nonspecific-esterase-positive cells (indicating monocytic origin) in the leukemic population. The AMoL cells are a nearly pure population of IL 1-producing cells and can be obtained in large quantity from patients receiving therapeutic leukophoresis. For example, leukophoresis of an AMoL patient with a peripheral white blood count of 300,000/mm^3 yielded 5 × 10^{11} leukemic cells. These cells were frozen in large test tubes in liquid nitrogen with serum and Me$_2$ SO as a cryopreservative. The AMoL cells were used at later times to produce extremely active preparations of IL 1 from a single donor. IL 1 derived from AMoL cells or human monocytes demonstrates exactly the same biochemical and biological characteristics (Lachman *et al.*, 1981). Interestingly, the serum and urine of these leukemic patients did not contain measurable quantities of IL 1.

E. IL 1 Release from Macrophage Cell Lines

Lachman *et al.* (1977) were the first to demonstrate IL 1 release from the then recently described continuous cultures of murine macrophages. The cell lines released IL 1 in response to LPS stimulation and could release IL 1 in the absence of serum (Lachman and Metzgar,

1980b). The continuous macrophage cell lines have become the universal source of murine IL 1.

Stimulation of IL 1 from the P388D$_1$ cell line has unfortunately created some unresolved questions. Mizel *et al.* (1978) reported that 0.5 µg/ml of PMA stimulated IL 1 release and later reported that exposure to 10 µg/ml of PMA for 5 hr did not stimulate IL 1 release (Mizel and Mizel, 1981). The explanation for these findings was briefly mentioned earlier. When PMA is continuously cultured with macrophages, the PMA is added along with the culture medium to the thymocyte assay. The PMA is stimulatory if the IL 1 assay contains both thymocytes and PHA, and gives the false impression that the culture medium contains IL 1 (Rosenstreich and Mizel, 1979). Also, concentrations of PMA in the range of 1 µg/ml can be cytotoxic to thymocytes, and thus give a spurious dose–response curve indicating that this concentration is nonmitogenic. When PMA in the range of 1 µg/ml is cultured with cells, the effective concentration in the medium is lowered due to absorption by the cells. The residual PMA in the culture medium may now be lowered to a range that is mitogenic for thymocytes, again giving the false impression that high doses of PMA are nonmitogenic but stimulate IL 1 release.

No human cell lines have been demonstrated to release IL 1. In particular, the U937 diffuse histiocytic lymphoma, which has many properties of human monocytes, does not release IL 1 spontaneously or in response to LPS (Lachman *et al.*, 1979). Recently, Larrick *et al.* (1980) demonstrated that the U937 cells could be activated by treatment with lymphokine preparations. Unfortunately, these activated cells also failed to release IL 1 (unpublished results).

F. Effect of Various Macrophage Stimulants upon IL 1 Release

The macrophage stimulants tuftsin, fMet-Leu-Phe, NaIO$_4$, and levamisole were all tested for their ability to stimulate IL 1 release from human monocytes. The agents were negative in the LLA clot assay. The results shown in Table 5 demonstrate that these compounds did not stimulate IL 1 release from monocytes and did not spuriously effect the IL 1 assay in the concentrations tested. The synthetic mycobacteria cell wall analog muramyl dipeptide (MDP) has been demonstrated to stimulate IL 1 release from human monocytes (Oppenheim *et al.*, 1980). MDP is a substance that can be obtained in endotoxin-free form since it is chemically synthesized.

A recent report by Moore *et al.* (1980) indicates that colony-stimulating factor (CSF) from L929 cells and from lectin-stimulated murine splenocytes stimulates mouse peritoneal macrophages to release IL 1.

Table 5. The Effect of Various Macrophages Stimulants upon IL 1 Release by Human Leukocytes

			IL 1 release	
			[³H]Thymidine incorporation by thymocytes (cpm/culture)	
Leukocyte culture conditions[a]			Control culture (− leukocytes)	Experimental culture (+ leukocytes)
	Stimulant	Dose (μg/ml)		
Expt. 1	LPS	20	1497	11463
	Tuftsin[b]	100	647	545
		1	741	606
		0.01	600	671
		—	597	634
Expt. 2	LPS	20	537	4214
	fMet-Leu-Phe[c]	1	333	144
		10^{-3}	213	117
		10^{-6}	211	145
		—	301	108
Expt. 3	LPS	20	150	6654
	NaIO₄, 5 min[d]	1	157	107
	20 min[d]	1	303	184
		—	134	145
Expt. 4	LPS	20	860	6453
	Levamisole[e]	200	517	842
		20	764	524
		2	660	690
		0.2	564	755
		0.02	680	800
		—	594	254

[a] Human buffy coat leukocytes and normal serum were prepared to maintain endotoxin-free conditions. The stimulant to be evaluated was added to (except in the case of NaIO₄) cultures of leukocytes (0.5×10^6/ml) containing 5% normal human serum. Control cultures not containing leukocytes, but containing the stimulant were also prepared. Cultures of leukocytes stimulated with LPS were prepared for each experiment.
[b] Tuftsin (Thr-Lys-Pro-Arg-OH) was purchased from Calbiochem–Behring Corp. and was found to be endotoxin-free at these concentrations. Culture supernatants were assayed for IL 1 activity before and after dialysis to determine if residual tuftsin could be inhibiting the response of thymocytes to IL 1. Both dialyzed and nondialzed samples were negative for IL 1 activity.
[c] N-Formyl-methionine-leucine-phenylalanine, prepared by Peninsula Laboratories, was found to be endotoxine-free at these concentrations.
[d] Peripheral blood leukocytes were treated with 5 mM NaIO₄ (1 μg/ml) for 5 or 20 min at 4°C in endotoxin-free PBS. The leukocytes were washed twice and cultured under the usual conditions. The viability of the leukocytes was 100% following the NaIO₄ treatment.
[e] Levamisole, prepared by Janssen R and D, New Brunswick, N.J., was kindly supplied by Dr. Ralph Snyderman, Department of Medicine, Duke University Medical School. Levamisole was found to be endotoxin-free at 10^{-3} M (200 μg/ml). Levamisole-containing cultures were dialyzed against Hepes-Cl for 24 hr prior to assay since solutions containing greater than 10^{-10} M levamisole were toxic to mouse thymocytes.

This very interesting finding may indicate a pathway by which CSF, a granulopoietic factor, may also amplify the immune response by stimulating IL 1 release from macrophages. Also, CSF, produced by fibroblasts or macrophages near the site of tumor growth or wound repair, could also amplify the local immune response.

IL 1 has been found in the serum of bacillus Calmette-Guérin (BCG)-infected mice following intravenous injection of LPS (Männel *et al.*, 1980). The serum was also demonstrated to contain a tumor cytostatic factor known as tumor-necrotizing factor. This report is the only demonstration of IL 1 activity produced *in vivo*. Control animals injected with LPS alone or BCG-treated animals did not contain IL 1 in their serum. These results may partially explain the role of LPS in tumor regression to be the generation of a cytotoxic factor for tumor cells and amplification of the immune response to the tumor mediated by a soluble macrophage-derived factor. It is of interest to note that serum and urine samples of patients with AMoL did not demonstrate measurable levels of IL 1 although the leukemic cells released large quantities of IL 1 when cultured in the presence of LPS or staphylococci.

III. Purification of Human IL 1

Human IL 1 for purification studies has been prepared from peripheral blood monocytes, AMoL cells, and AMML cells. No difference in the biochemical properties of the IL 1 has been noted from these three sources. The IL 1-containing medium is usually prepared by culturing the monocytes or leukemic cells in minimum essential medium (MEM) containing 1–5% normal serum and 10 μg/ml of LPS. The conditioned medium is harvested by centrifugation following 24–48 hr of culture at 37°C (Lachman *et al.*, 1980). IL 1 can be prepared in large quantities in spinner cultures and can also be prepared by pooling monocytes from many normal donors. A mixture of monocytic cells from normal donors is not sufficient to stimulate IL 1 release; LPS must also be present. The requirement for the presence of serum has been discussed earlier in this review. The presence of 1–5% serum is the major obstacle in the purification of IL 1. The purification is actually the separation of a low-molecular-weight protein (see below), present in ng per ml quantities, from a sea of serum proteins.

A. Molecular Weights of Human IL 1

Sephadex G-100 chromatography of concentrated IL 1-containing medium demonstrated a low-molecular-weight activity of approximately 15,000 and a high-molecular-weight activity of approximately 60,000

(Togawa *et al.*, 1979). The 60,000-molecular-weight activity chromato-graphed in the same region as albumin and immunoglobulin, the major serum proteins, while the 15,000-molecular-weight activity was separated from most of the serum proteins. It is for this reason that all attempts to purify human IL 1 have been directed toward the 15,000-molecular-weight activity. Prior to Sephadex chromatography of the IL 1-containing medium, time-consuming membrane ultrafiltration of the protein-rich medium was required to reduce the volume. An alternative method of separating the low-molecular-weight activity in a more rapid and automated fashion was sought. A hollow fiber device of 50,000-molecular-weight cutoff was found to rapidly and efficiently separate the 15,000-molecular-weight IL 1 activity from approximately 99% of the serum protein present in the original medium. The details of this technique have been described (Lachman *et al.*, 1980). The < 50,000-molecular-weight fraction could be concentrated by membrane ultrafiltration using a 10,000-molecular-weight cutoff membrane. This 10–50,000-molecular-weight fraction was further purified by a second passage through the 50,000-molecular-weight hollow fiber device. This procedure again removes 99% of the remaining protein and revealed a peculiar property of the IL 1, namely, that much of the IL 1 activity that was able to pass through the 50,000-molecular-weig ht hollow fiber cartridge in the first step of purification was no longer able to pass through the same hollow fiber device after concentration (Lachman *et al.*, 1980). This property of IL 1 has been investigated by Togawa *et al.* (1979) and is believed to be due to the formation of a high-molecular-weight complex between IL 1 and a serum protein (not identified). The reason serum protein is present in the < 50,000-molecular-weight fraction following the first hollow fiber step was that the cutoff of the hollow fiber device was not absolute, but allowed about 1% of serum protein of > 50,000 molecular weight to pass. A second passage of the IL 1 through the 50,000-molecular-weight hollow fiber cartridge reduced the yield of the desired 15,000 molecular weight activity but greatly increased the purity (Lachman *et al.*, 1980). The second < 50,000-molecular-weight sample contained about 30% of the starting low-molecular-weight IL 1 activity but only 0.01% of the original protein. The second < 50,000-molecular-weight fraction was concentrated to approximately 40 ml by membrane ultrafiltration with a 10,000-molecular-weight cutoff membrane and the sample was further purified by isoelectric focusing.

B. Isoelectric Focusing of the 10–50,000-Molecular-Weight Fraction

The 10–50,000-molecular-weight fraction containing the IL 1 activity was further purified by sucrose gradient isoelectric focusing (IEF). IL 1 activity was recovered as a sharp peak of activity (Fig. 2) with an average isoelectric point of pH 7.0. IEF of the 10–50,000-molecular-weight activity revealed very little charge heterogeneity for IL 1. Small peaks of IL 1 activity were found at pH 5.6 and 5.3 but they diluted to background following one 1:10 dilution, while the pH 7.0 peak demonstrated strong biological activity at a 1:100 dilution. The recovery of biological activity following IEF was disappointing. Only between 10 and 30% of the applied IL 1 activity was recovered. This poor recovery must be viewed in light of the ability of IEF to (1) separate IL 1 from albumin, the major protein in the 10–50,000-molecular-weight fraction and (2) the ability to separate IL 1 from contaminating endotoxin. The pH 7 peak was found to be negative in an LLA assay following dialysis and sterile filtration. The ability to remove endotoxin from the highly purified IL 1 activity allowed for several types of *in vivo* experiments to be performed (see below).

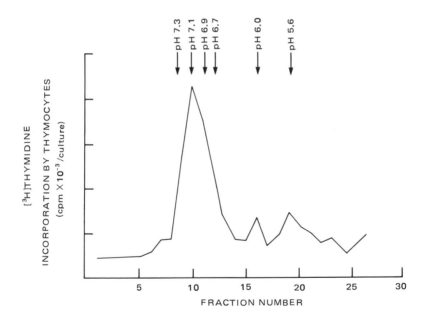

Figure 2. Isoelectric focusing of the 10–50,000-molecular-weight IL 1 activity. The 10–50,000-molecular-weight fraction of IL 1 activity was further purified by sucrose gradient IEF in a pH 8–4 Ampholine gradient. When a stable pH gradient had been reached (18 hr at 4°C, 1600-V constant power), 0.5-ml fractions were collected, the pH determined, and each fraction dialyzed to remove the sucrose and Ampholine. The amount of IL 1 in each fraction was determined in the [^{3}H]thymidine-incorporation assay using mouse thymocytes. The profile of activity shown above was obtained by adding the IEF fractions at a final concentration of 1% (v/v) to the thymocyte cultures.

C. Polyacrylamide Gel Electrophoresis of the IEF-Purified IL 1 and Evaluation of Purity

The IEF-purified IL 1 activity was further purified by two types of polyacrylamide gel electrophoresis (PAGE). Nondenaturing Tris–glycine PAGE in 7% gels resulted in excellent recovery of IL 1 activity (Fig. 3). Unfortunately, this procedure did not result in homogeneously purified IL 1. Transferrin was found to migrate in exactly the same region as the IL 1. In an attempt to separate the IL 1 from other proteins of higher molecular weight, SDS-PAGE was investigated. As shown in Fig. 4, SDS-PAGE was able to separate IL 1 from the other proteins in the IEF-purified sample. The SDS gel clearly demonstrates that the IEF-purified sample contains numerous proteins besides IL 1, but that IL 1 is the protein present in the largest quantity. It should be noted that this gel has been stained by the silver staining technique of Oakley *et al.* (1980), which is approximately 1000 times more sensitive than Coomassie blue staining. This gel experiment was the first time that a protein band that corresponded with the biological activity of IL 1 was visible on a gel (Lachman *et al.*, 1980). The molecular weight of the IL 1 was estimated from the SDS gel to be about 11,000, which was in good agreement with previous studies. The IL 1 activity was eluted from the gel and subjected to SDS-PAGE in another gel of greater per-

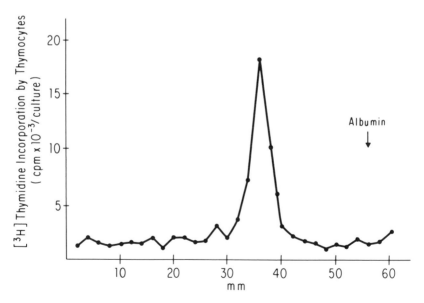

Figure 3. Preparative PAGE of IEF-purified IL 1. A 100-μl aliquot of IEF-purified IL 1 was applied to a 100-mm 7% Tris-glycine (pH 8.3) tube gel containing a 5-mm spacer gel. When the tracking dye had migrated 100 mm, the gel was cut into 2-mm slices and these were eluted and assayed for IL 1 activity.

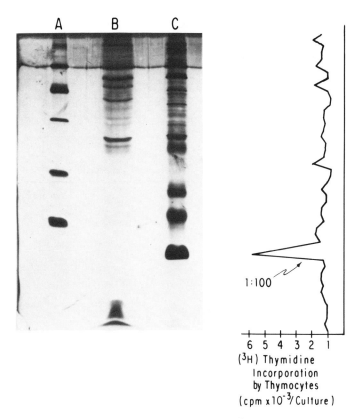

Figure 4. SDS-PAGE of IEF-purified IL 1. Two 1.0-ml samples of the IEF-purified IL 1 were dialyzed against water and lyophilized. One sample was applied to a 15% analytical slab gel and the other was applied to a 15% tube gel. The IEF-purified IL 1 (C) contained many protein bands, but in particular three bands of <20,000 molecular weight that were not present in an identically purified sample of control medium (B), which contained medium and serum but no cells. The molecular-weight standards (A) are 94K, 43K, 30K, 20.1K, and 14.4K. An [^{3}H]thymidine-incorporation assay of the tube gel slices demonstrated a single peak of IL 1 activity, which corresponded with the darkly staining band of approximately 11,000 molecular weight. The recovered IL 1 was active at a final concentration of 1% (v/v) in the thymocyte cultures and when tested again by SDS-PAGE in an 18% gel, demonstrated a single protein band. The AgNO$_3$ staining procedure of Oakley *et al.* (1980) was used to stain the slab gels.

cent acrylamide concentration. Silver staining of this gel demonstrated that the eluted IL 1 activity had been purified to homogeneity. The SDS gel-purified activity is suitable for the preparation of hybridoma antibody and amino acid analysis–protein sequencing, but contains too little biological activity to perform most of the experiments that need to be confirmed with homogeneous IL 1. For this reason, current experiments involve the use of nondenaturing gradient polyacrylamide gels to prepare homogeneous IL 1 that retains greater biological activity.

IV. Biochemical Properties of Human IL 1

Biochemical characterization of IL 1 has progressed through the indirect route of treatment with various enzymes, chemicals, or heat. As shown in Table 6, chymotrypsin and trypsin were able to inactivate IEF-purified IL 1, whereas neuraminidase and ribonuclease did not affect the biological activity. Treatment of IEF-purified IL 1 with cyano-

Table 6. Enzymatic or Chemical Hydrolysis of IEF-Purified IL 1 Activity

Hydrolysis conditions[a]		IL 1 release [³H]Thymidine incorporation by thymocytes (cpm/culture)	
Enzyme or chemical	Buffer	Undiluted	1:2
—	Hepes-NaCl	5519	3397
—	A[b]	6381	4284
Trypsin[b,c]	A	1297	635
—	B	6662	3727
Chymotrypsin	B	1331	767
—	C	4124	3703
Neuraminidase	C	4748	8041
Ribonuclease	C	7428	4775
Pronase	A	Toxic	
Deoxyribonuclease	D	Toxic	
—	E	3846	1915
BrCN[d]	E	410	414
Background		423	423

[a] To measure a reduction in IL 1 activity following enzymatic hydrolysis of IL 1 requires three control experiments for each enzyme: (1) The IL 1 activity must be stable when dialyzed against the buffer used for enzymatic hydrolysis and when finally dialyzed against the buffer (Hepes-NaCl) used during the assay; (2) due to the continuous presence of the enzyme in the thymocyte assay, the enzyme must not be toxic to the thymocytes; (3) the enzyme must not affect the response of the thymocytes to IL 1. Dialysis of IEF-purified IL 1 against the hydrolysis buffers and redialysis against the buffer used for the assay (Hepes-NaCl) did not significantly affect IL 1 activity. Control samples of the enzymes used were dialyzed against the hydrolysis buffer and the assay buffer and evaluated for their effect upon the background level of the thymocytes and the response of the thymocytes to IL 1 samples. Pronase and DNase were toxic to mouse thymocytes.

[b] Buffers and enzyme concentrations were as follows. Trypsin: 115 units/ml of IL 1–Buffer A (0.46 M Tris–HCl pH 8.1 containing 0.01 M CaCl$_2$). Chymotrypsin: 30 units/ml of IL 1–Buffer B (0.08 M Tris–HCl pH 7.8 containing 0.1 M CaCl$_2$). Neuraminidase: 0.75 unit/ml of IL 1–Buffer C (0.1 M NaOAc pH 5). Ribonuclease: 37 units/ml of IL 1–Buffer C. Pronase: 45 units/ml of IL 1–Buffer A. Deoxyribonuclease: 900 units/ml of IL 1–Buffer D (0.1 M NaOAc pH 5 containing 5 mM MgSO$_4$).

[c] The enzymatic hydrolysis was performed for 2 hr at 37°C, and the reaction was terminated by the addition of 10% human serum.

[d] IEF-purified IL 1, dialyzed overnight against E (17 nM K$_2$HPO$_4$ pH 6.3 + 0.3 N HCl), was treated with BrCN (7.5 mg/ml) for 2 hr at 25°C. The hydrolyzed sample and control were dialyzed overnight against Hepes-NaCl prior to assay.

gen bromide (BrCN) under acidic conditions also destroyed the biological activity of IL 1 (Table 6). It should be noted that IL 1 activity was able to withstand fairly well the strong acidic conditions of 0.3 N HCl in which the BrCN reaction was performed. Reduction and alkylation of IL 1 with iodoacetamide did not affect the biological activity of IL 1 (Table 7). Heat treatment at 100°C for 10 min inactivated IL 1 biological activity, whereas 56°C for 30 min reduced the activity by 50% (Table 8). The conclusion of all these experiments is that human IL 1 is a nonglycosylated protein that is quite stable to heat and extreme conditions of acid. IEF-purified IL 1 has been screened for protease activity in the very sensitive casein hydrolysate assay. The IEF-purified material did not hydrolyze [³H]casein to a measurable extent over a 1-hr period. The assay has been demonstrated to be sensitive to 1–2 ng trypsin/20-μl sample (Levin et al., 1976).

It should be noted that IL 1 is an extremely hydrophobic protein. IL 1 will bind nonspecifically to most types of sterile filters and ultrafiltration membranes and most types of chromatography medium. For this reason, sterile filtration must be done in the presence of serum or with Acrodisc membranes (Gelman Scientific Corp.) and ultrafiltration should only be performed with membranes of the YM series (Amicon Corp.). During experiments in which insolubilized enzymes were mixed with IEF-purified IL 1, it was found that all enzymes tested, including

Table 7. Reduction and Alkylation of IEF-Purified IL 1 Activity in Nondenaturing Solvent

Reduction and alkylation condition[a]			IL 1 release [³H]Thymidine incorporation by thymocytes (cpm/culture)	
2-Mercaptoethanol	Iodoacetamide	IL 1	Undiluted	1:2
−	−	+	6063	2833
+	−	+	5513	2621
+	+	+	5163	3055
−	+	+	5950	3369
+	+	−	493	

[a] IL 1 activity purified by IEF was dialyzed overnight against phosphate-buffered saline (PBS). The samples to be reduced (0.5-ml volumes) were incubated for 4 hr at 37°C following the addition of 5 μl of 1 M 2-mercaptoethanol. Control samples, lacking 2-mercaptoethanol, were incubated the same period of time. Following incubation, a reduced and a nonreduced sample of IL 1 activity were treated in the dark with 0.5 ml of 0.5 M iodoacetamide dissolved in PBS for 1 hr at 25°C. Control samples received 0.5 ml of PBS. The alkylation reaction was terminated by the addition of 250 μl of 1 M 2-mercaptoethanol. The control samples also received the same amount of 2-mercaptoethanol and the samples were dialyzed against Hepes-Cl for 24 hr with two changes of buffer prior to the thymocyte assay.

Table 8. Heat Treatment of IEF-Purified IL 1 Activity

		IL 1 release
Conditions		[³H]Thymidine incorporation by thymocytes (cpm/culture)
Temperature	Minutes	
Unheated	—	1359
56°C	10	986
	20	820
	30	729
100°C	2	544
	5	631
	10	263
Medium		198

DNase and RNase, were able to inactivate IL 1. Upon further examination, it was found that gentle shaking of Sepharose, dextran, or polyacrylamide, the insolubilizing agents for the enzymes, was able to remove IL 1 activity from the IEF-purified sample.

V. Biological Activities of Human IL 1

The introductory section of this review contained a brief summary of some of the general biological properties of IL 1. This section will discuss in greater detail some of the specific biological properties of IEF- and PAGE-purified human IL 1.

A. Human IL 1 Stimulates Both Mature and Immature Mouse Thymocyte Subpopulations

The ability to stimulate [³H]thymidine incorporation by mouse thymocytes was the assay used to monitor for IL 1 activity throughout the purification. The exact effect of IL 1 upon mouse thymocytes has been studied only briefly. Peanut agglutinin (PNA) has been demonstrated to distinguish mature from immature mouse thymocytes. Briefly, PNA⁺ thymocytes (~ 95% of the population) are the immature, cortisone-sensitive, cortical cells of the thymus, while PNA⁻ thymocytes represent the small population (~ 5%) of mature, cortisone-insensitive medullary thymus cells. Otterness et al. (1981) have demonstrated that PNA⁻ mature thymocytes respond better to IL 1 than the immature PNA⁺ cells.

B. Regulation of Interleukin 2 Release by IL 1

The demonstration by Morgan et al. (1976) that T-cell growth factor (TCGF), now known as interleukin 2 (IL 2), could stimulate the continuous *in vitro* growth of human and mouse T cells made this factor the focus of intense research interest. The requirements for IL 2 release were T lymphocytes, macrophages, and lectin. Lymphocytes depleted of macrophages were able to bind lectin but could not be stimulated to incorporate significant amounts of [³H]thymidine. Several groups simultaneously demonstrated that macrophages regulated the release of IL 2 from T lymphocytes and that the macrophages could be replaced by the soluble mediator IL 1 (Smith *et al.*, 1980); Larsson and Coutinho, 1980). Smith *et al.* (1980) further demonstrated that the immunosuppressive role of dexamethasone was in fact not due to an effect upon lymphocytes but rather upon macrophages. Dexamethasone inhibited IL 1 release from macrophages and thus prevented lectin-stimulated [³H]thymidine incorporation. When exogenous IL 1 was added to dexamethasone-containing cultures, no inhibition of [³H]thymidine incorporation by lymphocytes was detected. These findings indicate that certain immunodeficiency conditions may represent a defect at the macrophage, as opposed to the lymphocyte, level of cellular interaction and recognition.

C. The Effect of IL 1 on Human Thymocytes and T Lymphocytes

Human IL 1 is able to stimulate [³H]thymidine incorporation by human cells. Maizel *et al.* (1981) have demonstrated that IEF- and PAGE-purified human IL 1 can stimulate [³H]thymidine incorporation by human thymocytes and T lymphocytes. Unlike mouse thymocytes, human IL 1 cannot stimulate significant [³H]thymidine incorporation by human cells in the absence of either Con A or PHA. But, in the presence of lectin, IL 1 can stimulate significant [³H]thymidine incorporation by macrophage-depleted thymocytes or T lymphocytes. In fact, the level of [³H]thymidine incorporation in the presence of IL 1 is equal to the level of [³H]thymidine incorporation found when 5% human macrophages are added to the cultures of T lymphocytes, indicating the ability of IL 1 to completely replace macrophages in this system.

D. *In Vivo* Effects of Human IL 1

IEF-purified human IL 1 has been administered intravenously to rabbits and rats. As noted previously, IEF-purified human IL 1 is endotoxin-free as determined by the LLA assay. This is a fortuitous result, since no special precautions were taken during the purification to ensure endotoxin-free conditions. The availability of endotoxin-free IL 1 makes possible *in vivo* experiments that would be impossible to interpret if the IL 1 contained endotoxin.

Human IEF-purified IL 1 was administered intravenously to rabbits by Dr. Elisha Atkins and the rabbits were monitored for an increase in body temperature (Bernheim *et al.*, 1980). One-half milliliter of the sample was found to raise the body temperature of rabbits by 1°C. The observed fever was monophasic and occurred 30 min following injection and subsided within 1 hr. An identical fever response was observed in animals that had been recently tolerized to endotoxin, which confirms that the IL 1-induced fever was not due to endotoxin.

These results are in agreement with the findings of Rosenwasser *et al.* (1979) and Murphy *et al.* (1980) that IL 1 may be the same factor as the previously described monocyte endogenous pyrogen. Unfortunately, all these experiments suffer from the same basic problem; they have not been performed with homogeneously purified preparations of IL 1. These experiments will no doubt be rapidly repeated as homogeneous IL 1 becomes available.

Dr. Ralph Kampschmidt administered IEF-purified IL 1 to rats and monitored the animals for (1) rapid increase in the circulating levels of peripheral blood neutrophils and (2) 24-hr levels of plasma fibrinogen. The rationale for these experiments was that IL 1 exhibited many of the same properties as the previously described monokine leukocytic endogenous mediator (LEM) (Kampschmidt and Upchurch, 1980). Intravenous injection of 0.10 ml of IEF-purified IL 1 was found by Kampschmidt to increase circulating neutrophil levels by 8000 cells per mm^3 1 hr after injection. Also, a dose of 0.25 ml of IL 1 was found to increase 24-hr fibrinogen levels by 45 mg/100 ml plasma (unpublished results). These properties are identical to those described for LEM and again confirm that the IEF-purified IL 1 is free of contaminating endotoxin. As mentioned previously, the shortcoming of these experiments is that the IEF-purified IL 1 is not homogeneous. In any event, these experiments do indicate that IL 1 may be a central mediator of inflammation that can exert biological effects on the nervous system (fever), the bone marrow (margination of neutrophils), and the liver (secretion of acute-phase reactants such as fibrinogen).

VI. Summary and Conclusion

Macrophages stimulated with endotoxin or phagocytic stimulants release IL 1 into the culture medium. Very highly purified, but not yet homogeneous, preparations of human IL 1 have been demonstrated to stimulate [^{3}H]thymidine incorporation by mouse thymocytes as well as human thymocytes and peripheral T cells. IL 1 enhances the antibody response of B cells and stimulates the generation of cytotoxic T cells (Farrar *et al.*, 1979). Also, the essential role of macrophages in the lectin response of lymphocytes can be replaced by IL 1. IL 1 may affect the total response to bacterial antigens or tumor by stimulating fibroblasts to secrete prostaglandin and collagenase and liver cells to secrete acute-phase proteins.

The availability of homogeneous preparations of human IL 1 is of paramount importance. Homogeneous IL 1 can be used to prepare hybridoma antibody to IL 1, which will lead to rapid assays. Simplified assays can be used to determine IL 1 levels in serum or cell supernatants and thus allow exploration of the possibility that IL 1 release may be reduced in tumor-bearing patients and immunodeficient patients. This avenue of research may indicate that IL 1 could be used as an immunotherapeuthic agent, similar to interferon.

In conclusion, IL 1 research has centered upon the difficult task of purification. Homogeneous preparations of human IL 1 can be expected in the very near future and the interesting and exciting biological studies of IL 1 can begin.

VII. References

Bernheim, H. A., Block, L. H., Francis, L., and Atkins, E., 1980, Release of endogenous pyrogen-activating factor from concanavalin A-stimulated human lymphocytes, *J. Exp. Med.* **152**:1811.

Diamanstein, T., Klos, M., and Reiman, J., 1981, Studies on T-lymphocyte activation. I. Is competence induction in thymocytes by phorbol myristate acetate an accessory cell-independent event?, *Immunology* **43**:183.

English, D., and Anderson, B. R., 1974, Single-step separation of red blood cells, granulocytes and mononuclear leukocytes on discontinuous density gradients of Ficoll–Hypaque, *J. Immunol. Methods* **5**:249.

Farrar, J. J., Simon, P. L., Farrar, W. L., Koopman, W. J., and Fullar-Bonar, J., 1979, Role of mitogenic factor, lymphocyte-activating factor and immune interferon in the induction of humoral and cell-mediated immunity, *Ann. N.Y. Acad. Sci.* **332**:303.

Fumarola, D., and Jirillo, E., 1979, Endotoxin contamination of some commercial preparations used in experimental research, in: *Biomedical Applications of the Horseshoe Crab*, pp. 379–385, Liss, New York.

Gery, I., and Waksman, B. H., 1972, Potentiation of T-lymphocyte response to mitogens. II. The cellular source of potentiating mediators, *J. Exp. Med.* **136**:143.

Hanson, D. F., Murphy, P. A., and Windle, B. E., 1980, Failure of rabbit neutrophils to secrete endogenous pyrogen when stimulated with staphylococci, *J. Exp. Med.* **151**:1360.

Kampschmidt, R. F., and Upchurch, H. F., 1980, Neutrophil release after injection of endotoxin or leukocytic endogenous mediator into rats, *J. Reticuloendothelial Soc.* **28**:191.

Lachman, L. B., and Metzgar, R. S., 1980a, Purification and characterization of human lymphocyte activating factor, in: *Biochemical Characterization of Lymphokines* (A. L. deWeck, F. Kristensen, and M. Landy, eds.), pp. 397–398, 405–409, Academic Press, New York.

Lachman, L. B., and Metzgar, R. S., 1980b, Characterization of high and low molecular weight lymphocyte activating factor (interleukin 1) from P388D$_1$ and J774.1 mouse macrophage cell lines, *J. Reticuloendothelial Soc.* **27**:621.

Lachman, L. B., Hacker, M. P., Blyden, G. T., and Handschumacher, R. E., 1977, Preparation of lymphocyte activating factor from continuous murine macrophage cell lines, *Cell. Immunol.* **34**:416.

Lachman, L. B., Moore, J. O., and Metzgar, R. S., 1979, Preparation and characterization of lymphocyte activating factor (LAF) from acute monocytic and myelomonocytic leukemia cells, *Cell. Immunol.* **41**:199.

Lachman, L. B., Page, S. O., and Metzgar, R. S., 1980, Purification of human interleukin, 1, *J. Supramol. Struct.* **13**:457.

Lachman, L. B., George, F. W., IV, and Metzgar, R. S., 1981, Human interleukin 1 and 2: Purification and characterization, in: *Lymphokines and Thymic Factors* (A. L. Goldstein and M. A. Chirigos, eds.), pp. 21–32, Raven Press, New York.

Larrick, J. W., Fisher, D. G., Anderson, S. J., and Koren, H. S., 1980, Characterization of a human macrophage-like cell line stimulated *in vitro*: A model of macrophage functions, *J. Immunol.* **125**:6.

Larsson, E.-L., and Coutinho, A., 1980, Mechanism of T cell activation. I. A screening of "step one" ligands, *Eur. J. Immunol.* **10**:93.

Levin, N., Hatcher, V. B., and Lazarus, G. S., 1976, Proteinases of human epidermis: A possible mechanism for polymorphonuclear leukocyte chemotaxis, *Biochim. Biophys. Acta* **452**:458.

Maizel, A. L., Mehta, S. R., Ford, R. J., and Lachman, L. B., 1981, Effect of interleukin 1 on human thymocytes and purified human T cells, *J. Exp. Med.* **153**:470.

Männel, D. N., Farrar, J. J., and Mergenhagen, S. E., 1980, Separation of a serum-derived tumoricidal factor from a helper factor for plaque-forming cells, *J. Immunol.* **124**:1106.

Mizel, S. B., and Mizel, D., 1981, Purification to apparent homogeneity of murine interleukin 1, *J. Immunol.* **126**:834.

Mizel, S. B., Rosenstreich, D. L., and Oppenheim, J. J., Phorbol myristic acetate stimulates LAF production by the macrophage cell line P388D₁, *Cell. Immunol.* **40**:230.

Mizel, S. B., Dayer, J. M., Krane, S. M., and Mergenhagen, S. E., 1981, Stimulation or rhuematoid synovial cell collegenase and prostaglandin production by partially purified lymphocyte-activating factor (interleukin 1), *Proc. Natl. Acad. Sci. USA* **78**:2474.

Moore, R. N., Oppenheim, J. J., Farrar, J. J., Carter, C. S., Jr., Waheed, A., and Shadduck, R. K., 1980, Production of lymphocyte-activating factor (interleukin 1) by macrophages activated with colony-stimulating factors, *J. Immunol.* **125**:1302.

Morgan, D. A., Ruscetti, F. W., and Gallo, R. C., 1976, Selective *in vitro* growth of T lymphocytes from normal human bone marrows, *Science* **193**:1007.

Morrison, D. C., and Jacobs, D. M., 1976, Inhibition of lipopolysaccharide-initiated activation of serum complement by polymyxin B, *Infect. Immun.* **13**:298.

Murphy, P. A., Simon, P. L., and Willoughby, W. F., 1980, Endogenous pyrogens made by rabbit alveolar macrophages, *J. Immunol.* **124**:2498.

Oakley, B. R., Kirsch, D. R., and Morris, N. R., 1980, A simple ultra-sensitive silver stain for detecting proteins in polyacrylamide gels, *Anal. Biochem.* **105**:361.

Oppenheim, J. J., Togawa, A., Chedid, L., and Mizel, S., 1980, Components of mycobacteria and muramyl dipeptide with adjuvant activity induce lymphocyte activating factor, *Cell. Immunol.* **50**:71.

Otterness, I. G., Lachman, L. B., and Bliven, M. L., 1981, Effects of levamisole on the proliferation of thymic lymphocyte subpopulations, *Immunopharmacology* **3**:61.

Rosenberg, A. S., and Lipsky, P. E., 1981, The role of monocytes in pokeweed mitogen-stimulated human B cell activation: Separate requirements for intact monocytes and a soluble monocyte factor, *J. Immunol.* **126**:1341.

Rosenstreich, D. L., and Mizel, S. B., 1979, Signal requirements for T lymphocyte activation. I. Replacement of macrophage function with phorbol myristic acetate, *J. Immunol.* **123**:1749.

Rosenwasser, L. J., Dinarello, C. A., and Rosenthal, A. S., 1979, Adherent cell function in murine T-lymphocyte antigen recognition. IV. Enhancement of murine T-cell antigen recognition by human leukocytic pyrogen, *J. Exp. Med.* **150**:709.

Selinger, M. J., McAdams, K. P., Kaplan, M. M., Sipe, J. D., Vogel, S. N., and Rosenstreich, D. L., 1980, Monokine induced synthesis of serum amyloid A protein by hepatocytes, *Nature (London)* **285**:498.

Smith, K. A., Lachman, L. B., Oppenheim, J. J., and Favata, M. F., 1980, The functional relationship of the interleukins, *J. Exp. Med.* **151**:1551.

Togowa, A., Oppenheim, J. J., and Mizel, S. B., 1979, Characterization of lymphocyte-activating factor (LAF) produced by human mononuclear cells: Biochemical relationship of high and low molecular weight forms of LAF, *J. Immunol.* **122**:2112.

Wood, D. D., and Cameron, P. M., 1978, The relationship between bacterial endotoxin and human B cell-activating factor, *J. Immunol.* **121**:53.

16

Immunomodulatory Effects of Endotoxin

Diane M. Jacobs

I. Introduction and Background

Exposure of cells of the lymphoid system to endotoxin in conjunction with conventional antigens results in a dramatic modification of the specific immune response that would otherwise occur. Johnson *et al.* (1956) initially observed that endotoxin administered concomitantly with protein antigens resulted in markedly increased antibody titers. The effect could be obtained using endotoxin prepared from several species of gram-negative bacteria. The ability of endotoxin to thus act as a conventional adjuvant was found to depend on the time of administration relative to antigen, as demonstrated by Kind and Johnson (1959) and Franzl and McMaster (1968). They showed that enhanced antibody responses occurred only when endotoxin was administered at the same time or within a few days after antigen administration, but a depressed antibody response occurred if endotoxin preceded antigen. Thus, these early investigations demonstrated the dual nature of the effect of endotoxin on the immune response.

A new concept was introduced by Dresser (1961) in considering the nature of immunological adjuvants. He proposed that antigens that normally induce tolerance do so because they lack a characteristic termed *adjuvanticity*, but they act as immunogens in the presence of an external (conventional) adjuvant. He further suggested that a more stringent required characteristic for an agent to be considered an adjuvant would be its ability to prevent the induction of experimental toler-

Diane M. Jacobs ● Department of Microbiology, Schools of Medicine and Dentistry, State University of New York, Buffalo, New York 14214.

ance. Claman (1963) found that endotoxin administered to mice with normally tolerogenic deaggregated bovine gamma-globulin interfered with the induction of tolerance to this antigen. Similarly, Brooke (1965) demonstrated that endotoxin prevented the induction of tolerance to pneumococcal polysaccharide SIII. More extensive characterization of the ability of endotoxin to prevent tolerance to protein antigens was carried out by Golub and Weigle (1967).

These early studies described the basic immunomodulatory activities of endotoxin. Subsequent studies have relied heavily on our increased knowledge of the characteristics of lymphocyte classes and our understanding of the interactions that result in cell activity. Research on endotoxin has identified lipopolysaccharide (LPS) as the most important biologically active component in endotoxin preparations, and has elucidated its structure. The details of such studies on endotoxin are beyond the scope of this chapter and can be found elsewhere in this volume and in other publications (Galanos et al., 1977). It should be pointed out that preparations of endotoxin used in immunological studies often contain impurities with biological characteristics similar to those of purified LPS (Morrison et al., 1976; Sultzer and Goodman, 1976). To be strictly accurate, the term LPS should be reserved for those preparations known to be pure. However, both terms are often used interchangeably in the literature, albeit inaccurately at times. For the sake of simplicity and brevity, only LPS will be referred to in the rest of this chapter.

Based on new information of lymphocyte classes, a number of papers appeared that provided strong evidence that LPS directly influenced the activity of B cells. Gery et al. (1972) and Andersson et al. (1972a) found that LPS stimulated DNA synthesis in B cells in the apparent absence of either T cells or macrophages. Such B-cell mitogenicity was accompanied by differentiation of these cells to synthesize and secrete immunoglobulin (Andersson et al., 1972b). As secretion is independent of added antigen, and is of many different antibody specificities, this has been referred to as the polyclonal response. The data of Sjöberg et al. (1972) also showed that LPS appeared to substitute for T cells in vitro; when LPS was added to cultures of spleen cells depleted of T cells but containing sheep erythrocytes (SRBC), the specific plaque-forming cell (PFC) response to this antigen could be achieved.

The ability of LPS to stimulate B cells and its inability to have a measurable effect such as mitogenicity on T cells raised the possibility that its adjuvanticity in vivo might be attributable to its interaction with B cells. Several lines of evidence have suggested that LPS adjuvanticity in vivo can occur in the absence of functional T cells. Schmidtke and

Dixon (1972) were able to induce an antihapten response in mice injected with LPS and trinitrophenylated mouse erythrocytes (TNP–MRBC), an otherwise nonimmunogenic antigen. Since the carrier was a self-antigen, the inference was made that no self-reactive helper T cells were present, and that LPS acted independently of T cells. Chiller and Weigle (1973) showed that adjuvanticity *in vivo* could also be achieved in the absence of functional T cells. Adult-thymectomized bone marrow-reconstituted (ATxBM) mice did not respond to administration of the T-dependent antigen, aggregated human gamma-globulin (AHGG), but did respond when the antigen was administered with LPS. LPS–B-cell interaction could also be involved in the prevention as well as the termination of tolerance. Louis *et al.* (1973) induced tolerance in mice by administration of deaggregated human gamma-globulin (DHGG). LPS administered within a few hours of antigen prevented tolerance; mice produced antibody when challenged subsequently with the immunogenic form of the same antigen. AHGG. Analysis of the status of both T and B cells in the responding animals by passive transfer experiments revealed that animals primed with tolerogen and LPS possessed primed B cells but tolerant T cells. Furthermore, mice rendered tolerant to DHGG produced antibody when LPS was administered with aggregated gamma-globulin (AHGG) if the challenge was carried out at a time when B cells had recovered from tolerance although T cells had not recovered (Chiller and Weigle, 1973).

Other lines of evidence suggest that the antibody response achieved during LPS-induced adjuvanticity does depend on the presence of helper T cells. The earliest suggestion that T cells are required for expression of adjuvanticity came from Allison and Davies (1971), who found that mice depleted of T cells by neonatal thymectomy or treatment with antilymphocyte serum gave poor immune respones to T-dependent antigens even when these antigens were administered with LPS. Adjuvanticity could be restored if hosts were reconsituted with a source of T cells.

Parks *et al.* (1977) described a dichotomy between T-cell requirements for adjuvanticity and the prevention of tolerance. They found that congenitally athymic nude mice would not produce antibody to AHGG when the antigen was administered with LPS, indicating a T cell requirement for adjuvanticity in these mice. In contrast, the B cells from nude mice injected with the tolerogen DHGG and LPS did become primed by this procedure; antibody was produced when cells from these hosts were transferred to irradiated primed heterozygous littermates, which served as a source of helper T cells. Although the authors inferred from these results that nude mice may have a B cell deficiency, additional interpretations are possible. Previous results

(Chiller and Weigle, 1973) demonstrating adjuvanticity in ATxBM mice may have depended on a small number of residual T cells in these animals. It may be the LPS does not circumvent the T cell requirement for a T-dependent antigen *in vivo* but improves the response of specific B cells when they cooperate with required helper T cells.

Studies by Katz and colleagues provided evidence that LPS interacts with T cells. Hamaoka and Katz (1973) evaluated the anti-DNP responses of irradiated mice that received an adoptive transfer of spleen cells from DNP–KLH-primed donors and were subsequently immunized with DNP–BGG alone or with LPS. Animals receiving LPS produced higher antibody titers than recipients of antigen alone, but titers were markedly depressed in both groups if spleen cells were depleted of T cells with anti-θ serum before transfer. In a similar adoptive transfer system using both carrier-primed T cells and hapten-primed B cells, Newburger *et al.* (1974) found that development of carrier-specific helper cells in donor mice was improved when priming was carried out with both LPS and antigen. Armerding and Katz (1974) found an influence of LPS in T-helper function *in vitro*. Spleen cells from mice primed with a protein carrier or a haptenated protein and cultured with hapten–carrier conjugates gave higher antihapten responses in the presence of LPS, but only when the stimulating carrier was homologous with the immunizing carrier. As the effect of LPS was carrier specific, it would seem that the activity of carrier-specific T cells is modulated by LPS.

Additional evidence that T cells are required in LPS-mediated adjuvanticity and that they must interact with LPS has come from McGhee *et al.* (1979a). These investigators examined the response to SRBC of cultures of purified subpopulations of splenic lymphoid cells taken from C3H/HeN (LPS responder) mice and the syngeneic LPS-nonresponder C3H/HeJ strain. They found that the enhanced PFC response to antigen engendered in the presence of LPS required T cells from responder animals, and adjuvanticity could also be expressed by nonresponder B cells cultured in the presence of responder T cells. An additional requirement for B cells of either type was the presence of macrophages from LPS-responder mice. This macrophage requirement was also demonstrated by Hoffmann *et al.* (1977), who were able to induce an anti-TNP response to TNP–MRBC in macrophage-depleted spleen cells of either LPS responders or nonresponders when the cells were cultured with macrophages from the LPS-responder strain.

A role for macrophages in mediating adjuvanticity by presentation of antigen to responding cells had earlier been proposed by Unanue *et al.* (1969) and Spitznagel and Allison (1970). It now appears more likely that a role for macrophages in LPS-mediated adjuvanticity is dependent on their release of lymphocyte-activating factor (LAF) in response to LPS stimulation (Rosenstreich and Mizel, 1978). The active component in the serum of mice injected with LPS that potentiates the response of LPS-nonresponder B cells (Hoffmann *et al.*, 1977) is also likely to be LAF.

The suppressive activity of LPS on the induction of antibody synthesis has been reexamined in an effort to determine the cellular basis for this phenomenon and its relationship, if any, to the adjuvanticity of LPS. High doses of LPS added to cultures of normal spleen cells at the time of antigen stimulation suppress the primary response to SRBC (Hoffmann *et al.*, 1975). Spleen cells taken from mice injected with LPS cultured in the presence of antigen are also depressed in their ability to give a primary response (Persson, 1977). Spleen cells from such treated mice, as well as normal spleen cells incubated *in vitro* for 2 days, were capable of actively suppressing the *in vitro* primary response to SRBC. The cells responsible for this activity appeared to be B cells (Persson, 1977). Secondary *in vitro* antibody responses are also susceptible to inhibition by addition of LPS. Walker and Weigle (1978) demonstrated this effect in cultures of spleen cells from mice primed *in vivo* to HGG and turkey gamma-globulin (TGG) and stimulated with these antigens *in vitro*. The optimal time for addition of LPS to achieve suppression was 12 to 48 hr after initiation of culture, suggesting that this suppression is different from suppression of the response to SRBC, where optimal inhibition is achieved when LPS was added at the initiation of culture (Hoffmann *et al.*, 1975).

New evidence, in addition to that discussed in some detail in the next section, has recently been reported that suggests the existence of LPS-induced suppressor T cells. McGhee *et al.* (1980) have demonstrated that T cells purified from spleen or Peyer's patches of conventional mice can suppress the mitogenic and immune responses to TNP–LPS of B cells taken from germfree nude mice. Such suppressive activity was not recovered in T cells from germfree mice, but was present in conventionalized ex-germfree mice (McGhee *et al.*, 1981), suggesting that the presence of gut flora was responsible for the induction of naturally occurring suppressor T cells. Evidence linking LPS to the development of these suppressor T cells was found by Kiyono *et al.* (1982) in evaluating the tolerance induced by oral administration of antigen. LPS-responder mice fed SRBC by gastric intubation for 2 weeks did not produce antibody after parenteral challenge with the specific antigen. However, this regimen induced immunity rather than tolerance in LPS-nonresponder mice. Analysis of the responses of the F_1 and backcross generations demonstrated that tolerance was linked to the *Lps* gene, controlling the biological responses to LPS. That is, tolerance could be induced in animals that could respond mitogenically to LPS, but was not induced in individual animals that did not respond to LPS. In this system, tolerance was dependent on the presence of a T cell that suppressed the *in vitro* response of primed LPS-responder spleen cells, and such suppressor cells were not detected in orally immunized LPS-nonresponder mice (Kiyono *et al.*, 1982).

In addition to its immunomodulatory properties, LPS has many other properties affecting immunological and biological systems, and these are detailed in this volume and elsewhere (Morrison and Ulevitch, 1978; Morrison and Ryan, 1979). The potential therapeutic values of this material rest on a deeper understanding of its activities as they relate to behavior in these systems. The remainder of this chapter discusses work carried out in my laboratory during the past several years focusing on several aspects of LPS interaction with the immune system.

II. LPS and the Immune Response

A. Response of B Cells to LPS

1. Immunogenicity of TNP–LPS

Our early interest in LPS was based on the practical need for a T-independent antigen that was easily prepared, effective *in vitro*, and stimulated an easily measured response of limited specificity. At the time we began this work, LPS was known to be a potent T-independent (TI) antigen (Andersson and Blomgren), 1971; Möller and Michael, 1971). In addition, it had been shown to stimulate DNA synthesis (mitogenicity) and immunoglobulin synthesis (polyclonal activity) in murine B cells (Andersson *et al.*, 1972a,b; Gery *et al.*, 1972). On the basis of these observations, it was likely that chemical coupling of the TNP group to LPS would result in production of a molecule that would stimulate a TI anti-TNP response. The anti-TNP response in cultures of murine spleen cells was specifically stimulated by low doses of TNP–LPS and was dependent on the chemical coupling of the TNP moiety to LPS; simultaneous addition to cultures of TNP–SRBC and LPS did not stimulate an anti-TNP response, and LPS at these doses did not stimulate appreciable polyclonal responses (Jacobs and Morrison, 1975a). TNP–LPS was an effective TI antigen both *in vitro* and *in vivo*, stimulating a good PFC response in cultures of spleen cells from ATxBM mice, nude mice, or antithymocyte-treated normal spleen cells and in their *in vivo* counterparts (Jacobs and Morrison, 1975a; Jacobs, 1975). Since TNP–LPS is an immunogen in CBA/N mice, which possess an X-linked genetic deficiency in B-cell maturation, it has been characterized as a TI-1 antigen in contrast to many polysaccharide antigens, which are TI-2 (Mosier *et al.*, 1977).

TNP is covalently attached to LPS, probably to the amino groups of the phosphorylethanolamine of LPS; conjugation can be carried out with either purified or unpurified LPS preparations. Initially, we used a cacodylate buffer adjusted to pH 11.5, but 0.1% potassium carbonate or 0.1 M sodium borate, pH 10.5, are now routinely used. The alkaline pH is necessary, more for its ability to improve the accessibility of the amino groups to the reactant than for its effect on the chemical reaction itself, which can occur at neutral pH. LPS has also been conjugated with fluorescein isothiocyanate (FITC) (Scott *et al.*, 1979) and 4-hydroxy-3,5-dinitrophenol (NNP) (Coutinho *et al.*, 1974) and used to generate antihapten responses of these specificities.

2. Structural Requirements

As an immunogen, TNP–LPS possesses some of the same requirements for activity as LPS alone. The structure of LPS and the evidence for the importance of the lipid A moiety in the biological activity of the macromolecule have been presented in detail elsewhere (Galanos *et al.*, 1977). LPS from *S. minnesota* Re595, containing only lipid A and a small amount of carbohydrate, is also active as a B-cell mitogen and polyclonal activator (Andersson *et al.*, 1973). When conjugated with TNP, it induces an anti-TNP PFC response (Jacobs, 1975). Alkali treatment of LPS, which removes the ester-linked fatty acids of lipid A, results in a product with reduced toxicity (Neter *et al.*, 1958) and a loss of the ability to stimulate mitogenesis and polyclonal secretion (Andersson *et at.*, 1973). Such material conjugated in the same manner as native LPS no longer induces an anti-TNP response (Jacobs, 1975). Thus, the immunogenicity of TNP–LPS requires that the carrier portion have an intact, biologically active lipid A.

The structural requirements for immunogenicity of TNP–LPS reflected in the antihapten response appear to reflect the generalization that lipid A is the biologically relevant portion of LPS. However, it is not a direct reflection of the response to the LPS itself, which can be induced by the polysaccharide portion of LPS in the absence of active lipid A. Von Eschen and Rudbach (1974, 1976) induced a primary response to the O antigen using both native protoplasmic polysaccharide (NPP), the polysaccharide portion of LPS made in excess by some strains of *E. coli*, and alkali-treated LPS in which the lipid A has been rendered biologically inactive. Thus, the immunogenicity of LPS itself is not completely dependent on lipid A. Recently, Skelly *et al.* (1979)

measured the anti-TNP and anti-LPS PFC responses to TNP–LPS. Some strains of mice immunized with this antigen respond well to the TNP moiety but poorly to the O antigens of the carrier, but mice of the same strain respond well to the O antigens when immunized with native LPS.

3. Genetic Requirements

We have also used TNP–LPS to explore the genetic requirements for LPS immunological activity. Sultzer (1968) observed that the inflammatory response induced by LPS in C3H/HeJ mice was different from that induced in other strains, and this strain was resistant to LPS-induced toxicity. Furthermore, LPS did not induce mitogenicity in spleen cells from this strain, although such cells responded well to other T- and B-cell mitogens (Sultzer and Nilsson, 1972). The anti-TNP PFC produced in response to TNP–LPS is markedly reduced in this strain by comparison to other substrains of C3H as well as other strains with a variety of *H-2* haplotypes (Jacobs, 1975). Hybrid offspring of the nonresponder C3H/HeJ and other responder strains also responded to TNP–LPS, but the level of the responses was lower than that of the responder parent. These results suggested that LPS responsiveness is controlled by an autosomal codominant gene. Extensive studies by Watson *et al.* (1977, 1978) have demonstrated that the response to LPS is controlled by a single autosomal gene (*Lps*) on chromosome 4 of the mouse. Thus far, all the LPS biological activities examined have been found to be under the same genetic control. Extensive studies carried out by McGhee *et al.* (1979b) have confirmed our early impression that the expression of the gene is codominant.

4. Polymyxin B

The use of alkali-treated LPS as a technique for evaluating the requirements for lipid A in LPS responses has practical limitations. The conditions of treatment are critical and not always reproducible in different laboratories. The mitogenic response may increase rather than decrease after treatment, and responses tested *in vitro* may be affected by the presence of serum (Goodman and Sultzer, 1977; Betz and Morrison, 1977). Furthermore, it is wise to confirm that ester-linked fatty acids have been removed as predicted, and this may not be convenient for individual investigators. Finally, the molecular size of treated LPS is often changed, and this change alone may be responsible for observed differences in LPS activity. We therefore explored other methods to modify LPS activity. The cyclic peptide antibiotic polymyxin

B (PB), shown by Neter *et al.* (1958) to prevent LPS attachment to erythrocytes, also inhibits a number of LPS-mediated toxic responses (Rifkind, 1967; Rifkind and Palmer, 1966). We found that PB has profound effects on a number of LPS-induced activities in the immune system.

Addition of microgram quantities of PB to cultures of spleen cells stimulated with concanavalin A, phytohemagglutinin, LPS, and purified protein derivative resulted in selective inhibition of LPS-induced mitogenesis (Jacobs and Morrison, 1975b; Jacobs and Morrison, 1977). Inhibition was not due to the toxic effects of PB, as LPS–PB complexes, prepared by mixing the two materials and removing the free PB by dialysis, were far less mitogenic than untreated LPS. This complex also prevented the induction of mitogenicity by free LPS. Similarly, the LPS-induced polyclonal response is reduced by LPS (Jacobs, 1981).

In contrast, the immune response to TNP–LPS in our hands appeared initially to be enhanced (Jacobs and Morrison, 1975b). We interpreted these results to mean that, contrary to our earlier findings, mitogenicity of LPS was not a requirement for immunogenicity of its haptenated derivative. Similar experiments carried out by Smith and Hammarström (1976) were at variance with ours. They found that PB simply shifted the dose response of NNP–LPS and suggested that mitogenicity was indeed necessary. This point was relevant to the "one nonspecific signal" model of B-cell triggering proposed by Möller (1975) and Coutinho (1975) and has been somewhat clarified recently. We have now found that enhancement of the TNP–LPS response by PB occurs only when some batches of fetal calf serum are used and the cultures are carried for 4 days before assay (Jacobs, 1981). Most often, the result is suppression when standard amounts of TNP–LPS are used for stimulation and cells are assayed after 2½ days of culture. When the entire range of antigen doses is examined, the response to low doses of antigen is suppressed by PB; but the usual low responses achieved in the presence of high antigen doses are reversed and enhancement of the response occurs (Rosenspire and Jacobs, 1981). These results are consistent with our recently developed allosteric model of B-cell triggering, which integrates "one signal" and "two signal" models with the matrix model. The mathematical basis of this general model and the experimental verification are discussed elsewhere (Rosenspire *et al.*, 1981; Rosenspire and Jacobs, 1981).

To confirm that the effect of PB on LPS-induced responses was due to PB interaction with LPS, we examined the changes in the physical properties resulting from this interaction. We observed that there was an increase in the apparent molecular size and a decrease in the isopycnic density of the PB–LPS complex by comparison to the native material (Morrison and Jacobs, 1976). Since PB interacted with LPS Re595 as well as with LPS conventionally prepared from smooth organisms, we concluded that PB bound to the lipid A moiety of LPS. The interaction is stoichiometric, with one molecule of PB bound per monomer subunit of LPS. This is the same amount needed to experimentally inhibit LPS mitogenesis *in vitro* (Morrison and Jacobs, 1976; Jacobs and Morrison, 1977).

LPS prepared by the butanol extraction procedure (Bu-LPS) stimulates mitogenesis in cells from the LPS nonresponder C3H/HeJ (Skidmore *et al.*, 1975). We examined the effect of PB on the biological activities of this material in an effort to learn something about its structure that would explain why it behaved differently from LPS extracted with phenol, the more common method used. In fact, PB had no effect on the mitogenicity of Bu-LPS in either LPS-responder or -nonresponder mice. In addition, neither the molecular weight nor the density of Bu-LPS was altered in any way in the presence of excess PB (Morrison *et al.*, 1976).

These results could be explained by the presence of a contaminant in Bu-LPS that bound tightly to the lipid A and prevented PB binding to the same site. If the contaminant were mitogenic, it would account for the activity of Bu-LPS in C3H/HeJ mice. A material was recovered from a phenol extract of Bu-LPS that is mitogenic for both responder and nonresponder strains of mice. Since the PB data imply association with lipid A, this material has been referred to as *lipid A-associated protein* (LAP) (Morrison *et al.*, 1976). Material with similar activity has been identified in LPS prepared by extraction with trichloroacetic acid (the Boivin procedure), which has been called *endotoxin protein* (EP) by Sultzer and Goodman (1976). The presence of LAP, EP, and other contaminants in both commercial and laboratory preparations of LPS poses problems for investigators whose primary interest is in the mechanism of activation of lymphoid cells. Acceptance of the C3H/HeJ as a true LPS nonresponder was hindered in part due to the use of impure LPS preparations in attempts to confirm the initial observations. Whether or not such materials are important in the activity of LPS released from gram-negative organisms resident in the host remains to be determined.

B. Cellular Basis of LPS Immunomodulation

As discussed above, the results of investigations on stimulatory and suppressive effects of LPS have been inconsistent and often contradictory. Our studies on both aspects of immunomodulation indicate that LPS has multifocal effects on cells participating in antibody formation.

1. Suppression in Vivo

When we examined the *in vitro* activities of spleen cells and fractionated subpopulations of these cells taken from mice treated *in vivo* with LPS, we found that the primary anti-TNP PFC response to TNP–SRBC in C57BL/6 spleen cells was markedly depressed when donor mice had been injected with 10–50 μg LPS. This depression lasted for 5 days of *in vivo* treatment at which time recovery began, and maximum PFC responses were again achieved by 13 days after LPS administration. High doses of antigen added *in vitro* could not overcome the lowered response. Adjuvanticity was not observed under these conditions; PFC responses in cell cultures from LPS-treated mice never reached higher levels than responses in control cells from untreated mice.

The role of macrophages in this suppression was examined using the fact that the primary response in culture was macrophage dependent. Cells cultured after passage through Sephadex G-10 gave a lower response than unfractionated cells, and the response was restored to normal levels by addition of adherent spleen cells. Removal of macrophages from spleen cells of LPS-treated mice did not restore their capacity to respond, nor did nonadherent cells from the same donors recover activity in the presence of normal macrophages. Adherent cells from LPS-treated mice did, however, restore the activity of normal nonadherent spleen cells. Thus, macrophage dysfunction was not involved in the depressed response of spleen cells from LPS-treated mice (Uchiyama and Jacobs, 1978a).

We evaluated the ability of separate T- and B-cell populations from LPS-treated mice to cooperate with the corresponding normal cells *in vitro*. ATS-treated spleen cells (B cells plus macrophages), nylon wool-purified T cells, or irradiated spleen cells from SRBC-primed mice were cultured with TNP–SRBC for 4 days and the antihapten response then measured. In the presence of normal T cells, B cells from LPS-treated mice cooperated poorly. However, they gave responses identical to normal B cells when cultured with irradiated primed cells as the source of helper T cells. Nylon wool-passed T cells from LPS-treated mice were less effective in cooperating with normal B cells than were normal T cells. Thus, both B cells and T cells from LPS-treated mice had defective activities (Uchiyama and Jacobs, 1978a).

Deficient responses of a given cell population can be due to the absence of a cell, to deficient activity, or to the presence of a suppressive cell. We evaluated the possibility of suppressor cells by determining the effect of additions of various cell populations on the primary response of cultures of normal spleen cells. Unfractionated spleen cells from LPS-treated mice markedly suppressed the primary response of normal spleen cells. Suppressive activity was irradiation sensitive. Purified T cells from LPS-treated mice were able to exert suppressive effects on normal spleen cells. Further evidence that T cells are involved in suppression comes from the observation that spleen cells from LPS-treated animals consistently gave lower responses than T-depleted spleen cells from these same animals that were supplemented with an equal number of normal T cells. Normal B cells gave the same response as normal B cells supplemented with normal T cells. It appears that a suppressive effect was eliminated from LPS-treated spleen cells by treatment with anti-T-cell serum (Uchiyama and Jacobs, 1978a).

We concluded that the suppressive effect induced by LPS administration to mice in the absence of antigen resulted from interaction of LPS with both T cells and B cells. As a consequence, both lymphocyte populations become defective in cooperative activity, and suppressor cell activity in the spleen increases. We have not ruled out the possibility that administration of LPS *in vivo* induces only the redistribution of normal cell populations among the lymphoid organs so that the cells in the spleen no longer reflect the same proportions that they would in normal spleens. However, these results do indicate that LPS suppression of a primary response *in vivo* by previous administration of LPS is not likely to be due solely to redistribution of antigen.

2. Stimulation and Suppression in Vitro

Examination of the effect of LPS added to cultures of normal spleen cells provided an opportunity to study a system in which both enhancement and suppression of the antibody response could be achieved (Uchiyama and Jacobs, 1978b). Addition of 20 μg/ml LPS to normal spleen cell cultures markedly suppressed the antihapten response to TNP–SRBC. When carrier-primed cells were added to normal cells, the response increased, but this increase was much reduced in the presence of LPS, that is, helper cells did not overcome the suppression of LPS. However, when carrier-primed spleen cells were irradiated before addition to either normal spleen cells or normal B cells, the effect of LPS was different. In this case, the antihapten response achieved in the presence of LPS was higher than in its absence—LPS acted as an adjuvant. The suppressive activity detected when LPS was added to mixtures of carrier-primed and normal cells could be removed not only by irradiation but also by passage of the primed cells through nylon wool. Using cells prepared by this latter procedure, LPS stimulatory activity was still detectable. The elevated PFC response achieved by LPS in the presence of irradiated primed cells depended on a T cell in the added population. Carrier-primed cells treated with ATS and complement before irradiation had no helper activity when added to normal spleen cells in either the presence or the absence of LPS. The ability of LPS to suppress the response of mixtures of normal and carrier-primed cells is dependent on the genetic capacity of these cells to respond to LPS. In experiments carried out with cells from LPS-nonresponder mice, no suppressive effects of LPS were demonstrable (Jacobs, 1982). LPS stimulation of the antibody response is dependent on the presence of an irradiation-resistant T cell whose activity can be obscured by an irradiation-sensitive, nylon wool-adherent T cell. We concluded that LPS immunomodulation involves regulatory T cells with characteristics similar to those that normally regulate the immune response.

3. A Model for Adjuvanticity

During the performance of the experiments described above, we regularly found that LPS was unable to act as a T-cell replacing factor (TRF) when cultured with antigen and T-depleted spleen cells. Thus, our *in vitro* findings suggested that T cells were required for LPS to exert an adjuvant effect with a T-dependent antigen. This was consistent with the report of Parks *et al.* (1977), who found no adjuvant effect of LPS when tested *in vivo* in nude mice, but inconsistent with their results and those of others who could demonstrate adjuvanticity in nude cells *in vitro*. One explanation was that various lots of fetal calf serum used in cultures contained TRFs, thus enabling other investigators to demonstrate LPS adjuvanticity *in vitro* in the absence of T cells. If so, T-cell-dependent LPS adjuvanticity would not depend on LPS interaction with T cells but would reflect the B cell requirement for T cell interaction. Interaction of LPS with B cells alone could be seen as improving the response of these cells to helper T cells.

In order to examine this possibility, we carried out a series of experiments in which interaction of T cells with LPS was bypassed by replacing T cells with a cell culture supernatant containing TRF (Jacobs, 1979). When cells from nude mice and ATS-treated normal cells were cultured with TRF, LPS, and SRBC, higher levels of PFC were produced than would be expected from the additive effects of LPS and TRF. We have equated this synergistic effect with LPS adjuvanticity. The synergy obtained was dependent on the addition of antigen to cultures and was therefore not due simply to an enhanced polyclonal response. Furthermore, it could be obtained only with supernatants of allogeneic cell culture mixtures or normal spleen cells cultured with concanavalin A which contain TRF. Cultures of subpopulations of normal spleen cells did not produce supernatants with synergistic activity. We could find no evidence that macrophages were needed, as G-10-passed normal spleen cells served to demonstrate synergy between TRF and LPS as well as unfractionated spleen cells. Spleen cells from LPS-nonresponder mice did not respond synergistically to TRF and purified LPS, indicating that the response we measured was dependent on the genetic capacity of the B cells to respond to LPS (Jacobs, 1982). These results support a model of LPS-mediated adjuvanticity in which the T cell involvement is that of a cell cooperating in a T-dependent immune response, and the effect of LPS on B cells appears to increase the number of antigen-reactive precursor cells or their sensitivity to helper factors from T cells.

III. Future Work

The many mechanisms by which LPS modulates the activity of cells mediating the antibody response reflect the complexity of the immune system as much as they reveal the multiplicity of roles that LPS may play in immunomodulation. It is no longer sufficient to speak simply of LPS activation of B cells or T cells since it has become increasingly clear that each of the components involved is heterogeneous. The immune system is a network of effector and regulatory cells in a dynamic equilibrium that is influenced by LPS. LPS preparations commonly used, even when purified, consist of aggregates of different sizes, different structures, and differing ratios of lipid to carbohydrate. Investigations on the questions of immunomodulation must proceed to the use of homogeneous and well-characterized cell lines that reflect the naturally occurring populations and subpopulations of lymphocytes. Of equal importance is the need for preparations of LPS homogeneous in structure and composition. It is more than likely that the conflicting results between laboratories, if they do not reflect multiple LPS activities, are due to the fact that LPS preparations used by most investigators do not remotely meet increasingly stringent criteria of physicochemical purity. Increased attention to this point would continue the tradition already begun by investigators who have purified lipid A and the glycolipid from rough gram-negative organisms and made these materials available to their colleagues. Recent advances in receptor research have shown that ligand-receptor interactions and the resultant cell activation are easily modulated by small changes in ligand structure and the microenvironment of membrane receptors.

Use of homogeneous cell and LPS preparations will enable us to ask more direct questions on the nature of LPS interactions with target cells. It should be possible to distinguish between two possible modes of LPS–cell interactions: selective interaction with a specific membrane receptor for LPS or a nonspecific interaction of the hydrophobic portion of LPS with membrane lipids. Unpublished preliminary observations in my laboratory indicate that both types of interaction occur. We have detected selective binding of LPS to defined subpopulations of both T cells and B cells exposed to low concentrations of LPS, while nonselective binding to many lymphoid cells occurs at high concentrations of LPS. These two types of binding raise the possibility that two kinds of signals may be delivered by LPS, providing a basis for interpreting experimentally observed multifocal effects of LPS.

Regardless of the specificity of the interaction or the number of signals, we can ask questions about the biochemical events initiated by an LPS signal and compare these to the events that occur when the cell is activated by the physiologically relevant ligand. This approach should reveal whether LPS delivers a different kind of signal or simply bypasses the biological ligand in delivering the same signal. Such questions will eventually solve the dilemma of the LPS-nonresponder mouse, which in our hands has the same distribution of LPS-binding cells as the responder mouse.

Finally, I raise a question that may never be answered but that provides the impetus to develop an imaginative experimental approach. What is the biological role of responses of the cells of the immune system to LPS? Given that many complex organisms, including humans, are exposed periodically to small amounts of LPS shed from intestinal flora, and respond pathologically only to overwhelming infection, have we "learned to live with" LPS? Are the responses of B cells, for example, biologically "useful" to handle this common insult? Are these responses of evolutionary advantage? Or do they reflect biological functions unrelated to the immune response that provide a serendipitous tool with which the cell biologist can probe cell division, differentiation, and regulation?

IV. Conclusions

Studies on immunomodulation by LPS have demonstrated its capacity for both immunomodulation and immunosuppression. These effects appear to be achieved via a number of distinct pathways. While activation of B lymphocytes is most easily measured and is the basis for modulation in some cases, regulatory macrophage and T-cell activities have also been demonstrated to be affected by LPS. Our deeper understanding of the many characteristic responses of the immune system to

LPS depends on delineating the conditions under which they may be achieved, particularly in determining the cells that interact with LPS and characterizing the structure of the LPS that carries out these interactions. This information will clarify the possible role of gram-negative infections in inducing autoantibodies with pathological effects. Moreover, we will be in a better position to exploit the immunotherapeutic potential of LPS to modulate immune responses to foreign antigens, including tumor antigens. Finally, this approach will provide valuable information relevant to the efficacy of prophylactic immunization to the bacterial antigens carried on LPS for treatment of gram-negative septicemia.

V. References

Allison, A. C., and Davies, A. J. S., 1971, Requirement of thymus-dependent lymphocytes for potentiation by adjuvants of antibody formation, *Nature (London)* **233**:330.

Andersson, B., and Blomgren, H., 1971, Evidence for thymus-independent humoral antibody production in mice against polyvinylpyrrolidone and *E. coli* lipopolysaccharide, *Cell. Immuno.* **2**:411.

Andersson, J., Möller, G., and Sjöberg, O., 1972a, Selective induction of DNA synthesis in T and B lymphocytes, *Cell. Immunol.* **4**:381.

Andersson, J., Sjöberg, O., and Möller, G., 1972b, Induction of immunoglobulin and antibody synthesis *in vitro* by lipopolysaccharides, *Eur. J. Immunol.* **2**:349.

Andersson, J., Melchers, F., Galanos, C., and Lüderitz, O., 1973, The mitogenic effect of lipopolysaccharide on bone marrow derived mouse lymphocytes: Lipid A as the mitogenic part of the molecule, *J. Exp. Med.* **137**:943.

Armerding, D., and Katz, D. H., 1974, Activation of T and B lymphocytes *in vitro*. I. Regulatory influence of bacterial lipopolysaccharide (LPS) on specific T-cell helper function, *J. Exp. Med.* **139**:24.

Betz, S. J., and Morrison, D. C., 1977, Chemical and biological properties of a protein rich fraction of bacterial lipopolysaccharides. I. The *in vitro* murine lymphocyte response, *J. Immunol.* **119**:1475.

Brooke, M. S., 1965, Conversion of immunological paralysis to immunity by endotoxin, *Nature (London)* **206**:635.

Chiller, J. M., and Weigle, W. O., 1973, Termination of tolerance to human gamma globulin in mice by antigen and bacterial lipopolysaccharide (endotoxin), *J. Exp. Med.* **137**:740.

Claman, H. N., 1963, Tolerance to a protein antigen in adult mice and the effect of nonspecific factors, *J. Immunol.* **91**:833.

Coutinho, A., 1975, The theory of the 'one nonspecific signal' model for B cell activation, *Transplant. Rev.* **23**:49.

Coutinho, A., Gronowicz, E., Bullock, W. W., and Möller, G., 1974, Mechanism of thymus independent immunocyte triggering: Mitogenic activation of B cells results in specific immune responses, *J. Exp. Med.* **139**:74.

Dresser, D. W., 1961, Effectiveness of lipid and lipidophilic subtances as adjuvants, *Nature (London)* **191**:1169.

Franzl, R. E., and McMaster, P. D., 1968, The primary immune response in mice. I. The enhancement and suppression of hemolysin production by a bacterial endotoxin, *J. Exp. Med.* **127**:1087.

Galanos, C., Lüderitz, O., Rietschel, E. T., and Westphal, O., 1977, Newer aspects of the chemistry and biology of bacterial lipopolysaccharides, with special reference to their lipid A component, in: *Biochemistry of Lipids II* (T. W. Goodwin, ed.), *International Review of Biochemistry*, Vol. 14, pp. 239–335, University Park Press, Baltimore.

Gery, I., Kruger, J., and Spiesel, S. Z., 1972, Stimulation of B-lymphocytes by endotoxin: Reactions of thymus-deprived mice and karyotypic analysis of dividing cells in mice bearing T_6T_6 thymus grafts, *J. Immunol.* **108**:1088.

Golub, E. S., and Weigle, W. O., 1967, Studies on the induction of immunologic unresponsiveness. I. Effects of endotoxin and phytohemagglutinin, *J. Immunol.* **98**:1241.

Goodman, G. W., and Sultzer, B. M., 1977, Mild alkaline hydrolysis of lipopolysaccharide endotoxin enhances its mitogenicity for murine B cells, *Infect. Immun.* **17**:205.

Hamaoka, T., and Katz, D. H., 1973, Cellular site of action of various adjuvants in antibody responses to hapten-carrier conjugates, *J. Immunol.* **111**:1554.

Hoffmann, M. K., Weiss, O., Koenig, S., Hirst, J. A., and Oettgen, H. F., 1975, Suppression and enhancement of the T cell-dependent production of antibody to SRBC *in vitro* by bacterial lipopolysaccharide, *J. Immunol.* **114**:738.

Hoffmann, M. K., Galanos, C., Koenig, S., and Oettgen, H. F., 1977, B-Cell activation by lipopolysaccharide: Distinct pathways for induction of mitosis and antibody production, *J. Exp. Med.* **146**:1640.

Jacobs, D. M., 1975, Structural and genetic basis of the *in vivo* immune response to TNP-LPS, *J. Immunol.* **115**:988.

Jacobs, D. M., 1979, Synergy between T cell-replacing factor and bacterial lipopolysaccharide (LPS) in the primary antibody response *in vitro*: A model for lipopolysaccharide adjuvant action, *J. Immunol.* **122**:1421.

Jacobs, D. M., 1982, Lipopolysaccharide and the immune response, in: *Pharmacology of the Reticuloendothelial System* (D. Webb, ed.), pp. 231–251. Dekker, New York.

Jacobs, D. M., and Morrison, D. C., 1975a, Stimulation of T-independent primary antihapten response *in vitro* by TNP-lipopolysaccharide (TNP-LPS), *J. Immunol.* **114**:360.

Jacobs, D. M., and Morrison, D. C., 1975b, Dissociation between mitogenicity and immunogenicity of TNP-lipopolysaccharide (TNP-LPS), a T-independent antigen, *J. Exp. Med.* **141**:1453.

Jacobs, D. M., and Morrison, D. C., 1977, Inhibition of the mitogenic response to lipopolysaccharide (LPS) in mouse spleen cells by polymyxin B, *J. Immunol.* **118**:21.

Johnson, A. G., Gaines, S., and Landy, M., 1956, Studies on the O antigen of *Salmonella typhosa*. V. Enhancement of antibody response to protein antigens by the purified lipopolysaccharide, *J. Exp. Med.* **103**:225.

Kind, P., and Johnson, A. G., 1959, Studies on adjuvant action of bacterial toxins on antibody formation. I. Time limitation of enhancing effect and restoration of antibody formation in X-irradiated rabbits, *J. Immunol.* **82**:415.

Kiyono, H., McGhee, J. R., Wannemuehler, M. J., and Michalek, S. M., 1982, Lack of oral tolerance in C3H/HeJ mice, *J. Exp. Med.* **155**:605.

Louis, J. A., Chiller, J. M., and Weigle, W. O., 1973, The ability of bacterial lipopolysaccharide to modulate the induction of unresponsiveness to a state of immunity, *J. Exp. Med.* **138**:1481.

McGhee, J. R., Farrar, J. J., Michalek, S. M., Mergenhagen, S. E., and Rosenstreich, D. L., 1979a, Cellular requirements for lipopolysaccharide adjuvanticity: A role for both T lymphocytes and macrophages for *in vitro* responses to particulate antigens, *J. Exp. Med.* **149**:793.

McGhee, J. R., Michalek, S. M., Moore, R. N., Mergenhagen, S. E., and Rosenstreich, D.

L., 1979b, Genetic control of *in vivo* sensitivity to lipopolysaccharide: Evidence for condominant inheritance, *J. Immunol.* **122**:2052.

McGhee, J. R., Kiyono, H., Michalek, S. M., Babb, J. L., Rosenstreich, D. L., and Mergenhagen, S. E., 1980, Lipopolysaccharide (LPS) regulation of the immune response: T lymphocytes from normal mice suppress mitogenic and immunogenic responses to LPS, *J. Immunol.* **124**:1603.

McGhee, J. R., Michalek, S. M., Kiyono, H., Babb, J. L., Clark, M. P., and Mosteller, L. M., 1982, LPS regulation of the IgG immune response, in: *Mucosal Immunity* (W. Strober, L. A. Hanson, and K. Sell, eds.), pp. 57–72, Raven Press, New York.

Möller, G., 1975, One nonspecific signal triggers lymphocytes, *Transplant. Rev.* **23**:126.

Möller, G., and Michael, G., 1971, Frequency of antigen-sensitive cells to thymus-independent antigens, *Cell. Immunol.* **2**:309.

Morrison, D. C., and Jacobs, D. M., 1976, Binding of polymyxin B to lipid A portion of bacterial lipopolysaccharides *Immunochemistry* **13**:813.

Morrison, D. C., and Ryan, J. L., 1979, Bacterial endotoxins and host immune responses, *Adv. Immunol.* **28**:293.

Morrison, D. C., and Ulevitch, R. J., 1978, The effects of bacterial endotoxins on host mediation systems, *Am. J. Pathol.* **93**:527.

Morrison, D. C., Betz, S. J., and Jacobs, D. M., 1976, Isolation of a lipid A bound polypeptide responsible for "LPS initiated" mitogenesis of C3H/HeJ spleen cells, *J. Exp. Med.* **144**:840.

Mosier, D. E., Mond, J. J., and Goldings, E. A., 1977, The ontogeny of thymic independent antibody responses *in vitro* in normal mice and mice with an X-linked B cell defect, *J. Immunol.* **119**:1874.

Neter, E., Westphal, O., Lüderitz, O., Gorzynski, E. A., and Eichenberger, E., 1955, Studies of enterobacterial lipopolysaccharides: Effects of heat and chemicals on erythrocyte-modifying, antigenic, toxic and pyrogenic properties, *J. Immunol.* **76**:377.

Neter, E., Gorzynski, E. A., Westphal, O., and Lüderitz, O., 1958, The effects of antibiotics on enterobacterial lipopolysaccharides (endotoxins), hemagglutination and hemolysis, *J. Immunol.* **80**:66.

Newburger, P. E., Hamaoka, T., and Katz, D. H., 1974, Potentiation of helper T cell function in IgE antibody responses by bacterial lipopolysaccharide (LPS), *J. Immunol.* **113**:824.

Parks, D. E., Doyle, M. V., and Weigle, W. O., 1977, Effect of lipopolysaccharide on immunogenicity and tolerogenicity of HGG in C57BL/6J nude mice: Evidence for a possible B cell deficiency, *J. Immunol.* **119**:1923.

Persson, U., 1977, Lipopolysaccharide-induced suppression of the primary immune response to a thymus-dependent antigen, *J. Immunol.* **118**:789.

Rifkind, D., 1967, Prevention by polymyxin B of endotoxin lethality in mice, *J. Bacteriol.* **93**:1463.

Rifkind, D., and Palmer, J. D., 1966, Neutralization of endotoxin toxicity in chicken embryos by antibiotics, *J. Bacteriol.* **92**:815.

Rosenspire, A. J., and Jacobs, D. M., 1981, A general interactive model for B cell activation. II. Experimental verification, *Cell Biophys.* **3**:89.

Rosenspire, A. J., Rosen, R., and Jacobs, D. M., 1981, A general interactive model for B cell activation. I. The theory, *Cell Biophys.* **3**:71.

Rosenstreich, D. L., and Mizel, S. B., 1978, The participation of macrophages and macrophage cell lines in the activation of T lymphocytes by mitogens, *Immunol. Rev.* **40**:102.

Schmidtke, J. R., and Dixon, F. J., 1972, Immune response to hapten coupled to a nonimmunogenic carrier: Influence of lipopolysaccharide, *J. Exp. Med.* **136**:392.

Scott, D. W., Venkataraman, W., and Jandinski, J. J., 1979, Multiple pathways of B lymphocyte tolerance, *Immunol. Rev.* **43:**241.

Sjöberg, O., Andersson, J., and Möller, G., 1972, Lipopolysaccharide can substitute for helper cells in the antibody response *in vitro, Eur. J. Immunol.* **2:**326.

Skelly, R. R., Munkenbeck, P., and Morrison, D. C., 1979, Stimulation of T-independent antibody responses by hapten-lipopolysaccharides without repeating polymeric structure, *Infect. Immun.* **23:**287.

Skidmore, B. J., Morrison, D. C., Chiller, J. M., and Weigle, W. O., 1975, Immunological properties of bacterial lipopolysaccharide (LPS). II. The unresponsiveness of C3H/HeJ mouse spleen cells to LPS-induced mitogenesis is dependent on the method used to extract LPS, *J. Exp. Med.* **142:**1488.

Smith, E., and Hammarström, L., 1976, Inhibition of the mitogenicity of the carrier molecule results in loss of immunogenicity of a hapten–LPS conjugate, *Acta Path. Microbiol. Scand. Sect. C* **84:**495.

Spitznagel, J. K., and Allison, A. C., 1970, Mode of action of adjuvants: Effects on antibody responses to macrophage associated bovine serum albumin, *J. Immunol.* **104:**128.

Sultzer, B. M., 1968, Genetic control of leucocyte responses to endotoxin, *Nature (London)* **219:**1253.

Sultzer, B. M., and Goodman, G. W., 1976, Endotoxin protein: A B-cell mitogen and polyclonal activator of C3H/HeJ lymphocytes, *J. Exp. Med.* **144:**821.

Sultzer, B. M., and Nilsson, B. S., 1972, PPD-tuberculin—A B cell mitogen, *Nature New Biol.* **240:**198.

Uchiyama, T., and Jacobs, D. M., 1978a, Modulation of immune response by bacterial lipopolysaccharide (LPS): Multifocal effects of LPS-induced suppression of the primary antibody response to a T-dependent antigen, *J. Immunol.* **121:**2340.

Uchiyama, T., and Jacobs, D. M., 1978b, Modulation of immune response by bacterial lipopolysaccharide (LPS): Cellular basis of stimulatory and inhibitory effects of LPS on the *in vitro* IgM antibody response to a T-dependent antigen, *J. Immunol.* **121:**2347.

Unanue, E. R., Askonas, B. A., and Allison, A. C., 1969, A role for macrophages in the stimulation of immune responses by adjuvants, *J. Immunol.* **103:**71.

Von Eschen, K. B., and Rudbach, J. A., 1974, Immunological responses of mice to native protoplasmic polysaccharide and lipopolysaccharide: Functional separation of the two signals required to stimulate a secondary antibody response, *J. Exp. Med.* **140:**1604.

Von Eschen, K. B., and Rudbach, J. A., 1976, Antibody responses of mice to alkaline detoxified lipopolysaccharide, *J. Immunol.* **116:**8.

Walker, S. M., and Weigle, W. O., 1978, Effect of bacterial lipopolysaccharide on the *in vitro* secondary antibody response in mice. I. Description of the suppressive capacity of lipopolysaccharide, *Cell. Immunol.* **36:**170.

Watson, J., Riblet, R., and Taylor, B. A., 1977, The response of recombinant inbred strains of mice to bacterial lipopolysaccharide, *J. Immunol.* **118:**2088.

Watson, J., Kelley, K., Largen, M., and Taylor, B. A., 1978, The genetic mapping of a defective LPS response gene in C3H/HeJ mice, *J. Immunol.* **120:**422.

17

Lipopolysaccharide Regulation of the Immune Response

Beneficial Effects on Lymphoreticular Cells and a Model for Their Activation by LPS

Katherine A. Gollahon, Suzanne M. Michalek,
Michael J. Wannemuehler, and Jerry R. McGhee

I. Introduction

Bacterial endotoxin, which comprises a significant portion of the outer membrane of gram-negative bacteria, causes a large number of biological effects on a susceptible host such as man or the useful experimental animal, the mouse. When the classical phenol–water extraction method is employed to isolate endotoxin from the outer membrane of the gram-negative family Enterobacteriaceae, purified lipopolysaccharide (LPS) is obtained and this high-molecular-weight molecule exhibits most of the endotoxin-associated effects in the host (Morrison and Ryan, 1979). LPS is a potent immunomodulator and acts on lymphoreticular cells, e.g., B cells, macrophages ($M\phi$), and T cells (Table 1). There is a growing body of evidence that suggests that LPS may exert all host biological effects, including toxic manifestations, via an interaction with lymphoreticular cells (Michalek et al., 1980b; Vogel and Mergenhagen, 1982). In this chapter, we will limit ourselves to LPS effects on lymphoreticular cells that would be beneficial and also occur in our natural host–parasite environment. We encounter LPS in signifi-

Katherine A. Gollahon, Suzanne M. Michalek, Michael J. Wannemuehler, and Jerry R. McGhee • Department of Microbiology, Comprehensive Cancer Center and Institute of Dental Research, The University of Alabama in Birmingham, Birmingham, Alabama 35294.

**Table 1. Important Examples of Bacterial
Endotoxin Effects on Lymphoreticular Cells in
Susceptible Hosts**

Stimulation of the immune system
  B-Cell mitogenicity
  Polyclonal antibody induction
  Adjuvancy
  Macrophage activation
  Immunogenicity
  Activation of complement
Stimulation of mediator production
  Interleukin 1
  Colony-stimulating factor
  Interferon
  Glucocorticoid-antagonizing factor
  Tumor-necrotizing factor
Suppression–induction of tolerance
Pyrogenicity
Abortion
Tumor necrosis
Lethality

cant amounts from our gram-negative gastrointestinal flora, and there-fore, the LPS–lymphoreticular cell interaction probably first, and continually, occurs in gut-associated lymphoid tissue (GALT), e.g., Peyer's patches. We will briefly develop in this chapter how this interaction affects the host's immune response and offer mechanisms whereby LPS regulates both secretory and systemic immunity to mucosally encountered antigens. Furthermore, we now envision this LPS–lymphoreticular cell interaction as not only beneficial to the host, but, in the model presented below, as the major environmental stimulus that is involved in maturation of the host's immune system.

II. Environmental LPS Regulation of Polyclonal B-Cell Responses

During the past 10 years, numerous investigations have described several LPS effects on the host and these studies have been facilitated by the availability of unique mouse strains. The C3H/HeJ mouse is resistant to many known lipid A-induced effects (for review see Sultzer, 1968; Morrison and Ryan, 1979; Vogel *et al.*, 1982) (Table 1). Studies comparing this mouse strain with syngeneic LPS-responder C3H/HeN mice and recombinant inbred mice have led to the mapping of the LPS defect (Watson *et al.*, 1977). The Lps^d (d = defective) gene is present on chromosome 4 (Watson *et al.*, 1978a,b). This locus is closely linked to the gene coding for the major urinary protein (MUP-1) and the B-cell differentiation antigens Lyb 4 and Lyb 6 (see Section IV below). F_1 hybrid (Lps^n/Lps^d genotype) mice from matings of LPS-responder C3H/HeN (Lps^n/Lps^n) and C3H/HeJ (Lps^d/Lps^d) mice exhibit an intermediate ability to respond to LPS. The C3H/HeJ mouse has now achieved an important position in cellular immunology studies that investigate lymphoid cell responses in the absence of lipid A effects. We have used the C3H/HeJ mouse and LPS-responder strains to investigate environmental effects of endotoxin on GALT and the IgA response, and as summarized below (see Section III), we have found that gut bacterial LPS profoundly influences the IgA immune response.

In order to establish the influence of LPS from the indigenous gram-negative gut microflora, we have also used a second murine system. The germ-free (GF) mouse, which by definition does not possess endogenous LPS-bearing gram-negative bacteria, has been available for more than 30 years. One cannot preclude that some LPS exposure occurs via pyrogens present in diet; however, the significant difference in toxic manifestations elicited between GF and conventional (Conv) mice of the same strain (GF mice are more resistant to endotoxemia; Jensen *et al.*, 1963) would argue that environmental exposure of GF mice to LPS is minimal. The GF mouse thus offers an important model to evaluate the effects of LPS on maturation and function of lymphoreticular cells.

When lymphoid cell responses to either T-cell lectins (Con A and PHA) or B-cell mitogens (e.g., LPS) are assessed in GF mice, it is immediately apparent that both T and B cells from GF mice are fully responsive (McGhee *et al.*, 1980). In fact, when B-cell responses to various LPS preparations (including pure lipid A) are compared in spleen cell cultures from GF and Conv mice, significantly higher mitogenic responses occur in GF mice (Table 2, part A). GF mice also respond to LPS *in vivo* and when compared with Conv mice, the GF response occurs at lower LPS concentrations, follows different kinetics, and reaches higher levels of anti-LPS PFC and antibody levels (Kiyono *et al.*, 1980a). Thus, the GF mouse is exquisitely sensitive to LPS both *in vivo* and *in vitro*. The elevated LPS response pattern in GF mice, when compared with Conv animals, suggests that GF mice lack regulatory mechanisms present in Conv mice that are induced by the indigenous gram-negative microflora. We hypothesized that the simplest way to probe whether GF mouse B cells are responsive to both immunogen and LPS (without a requirement for thymus-derived helper cells) would be to employ the T-independent (TI) antigen trinitrophenyl (TNP)

Table 2. Comparison of *in Vitro* Responses to LPS of Spleen Cells from Germ-Free (GF) and Conventional (Conv) BALB/c Mice

A. LPS mitogenic response[a]

Mouse group	LPS(Ph)		Lipid A	
	cpm/culture	E/C	cpm/culture	E/C
GF	39,185 ± 2091	46.1	8352 ± 127	9.6
Conv	13,659 ± 966	15.7	2697 ± 68	3.1

[a] Data expressed as the mean cpm [^{3}H]-TdR incorporated/culture ± S.E.M. and as the mean of the stimulation ratio where E/C = cpm [^{3}H]-TdR incorporated with stimulant/cpm [^{3}H]-TdR incorporated in medium.

B. TNP–LPS immune response[a]

Anti-TNP PFC/culture

Mouse group	TNP–LPS concentration (μg/culture)		
	0.001	0.01	0.1
GF	1186 ± 227	1465 ± 158	1855 ± 185
Conv	105 ± 12	488 ± 39	1011 ± 88

[a] Data expressed as the mean PFC ± S.E.M. of cultures on day 4 of incubation.

C. T-Cell suppression of TNP–LPS responses[a]

Anti-TNP PFC/nude spleen cell culture

Mouse group	Source of T cells (1 × 10^6/culture)		
	None	Spleen	Peyer's patches
GF	1455 ± 41	1510 ± 105	1390 ± 55
Conv	1455 ± 41	512 ± 66	421 ± 54

[a] Data expressed as the mean PFC ± S.E.M. of cultures on day 4 of incubation.

haptenated LPS (TNP–LPS). As shown in Table 2 (part B), GF spleen cell cultures elicited higher anti-TNP PFC responses than Conv mouse spleen cells and responded at concentrations of TNP–LPS below the threshold of responses obtained with Conv spleen cell cultures. These results suggested that LPS (as a carrier) was sufficient to elicit TNP responses following immunization with TNP–LPS and indicated that Conv mice were less responsive to immunogen, perhaps via a negative regulatory mechanism.

To more directly determine if Conv mice harbor a naturally occurring lymphoreticular cell population that negatively regulates B-cell responses to LPS, we performed a number of studies with purified cell populations from either GF or Conv mice (McGhee *et al.*, 1980). Our results indicated that T cells from Conv mice diminished TNP–LPS responses in either nude mouse spleen or purified splenic B-cell cultures. Since the gut microflora could represent the major inducer of these suppressor T cells, we have also studied the effect of purified T cells from GALT (i.e., Peyer's patches) for this regulatory effect. From Table 2 (part C), it is clear that T cells from either Peyer's patches or spleen of Conv mice suppressed *in vitro* TNP–LPS responses. We therefore postulated that LPS from the gram-negative flora induced these T cells in GALT and it is their subsequent redistribution and homing to peripheral tissue that provides the host with a mechanism to control continued LPS-induced polyclonal B-cell activation in lymphoid tissues.

It is likely that this diminution of B-cell responses is broad range since ex-GF mice receiving a conventional gut flora challenge exhibit lower (Conv-like) B-cell mitogenic responses within 3 weeks of conventionalization (Fig. 1). These studies provide convincing evidence that lymphoid cells of GF mice are responsive to lectins, and that their B cells are quite sensitive to LPS. Although the T-cell regulation of LPS polyclonal activation *in vivo* may only be a primitive form of host response to bacterial endotoxin, we propose that LPS effects on GALT can determine the outcome of specific immune responses to orally-en-

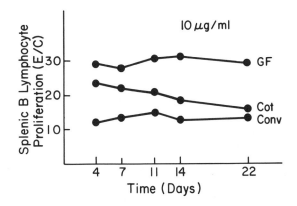

Figure 1. Comparison of splenic mitogenic responses to LPS in germ-free (GF). conventional (Conv), and ex-germ-free, fecally contaminated (Cot) BALB/c mice.

countered environmental antigens. In the model presented below, it is postulated that LPS may induce B-cell differention and Mφ activation, in addition to the observed T-cell effects.

III. LPS Regulation of IgA Responses and Oral Tolerance

It is well known that GALT, specifically Peyer's patches, contain T cells and precursor IgA B cells that, following antigenic stimulation, migrate to distant mucosal sites and differentiate into IgA-producing plasma cells with specificity to the gut-encountered antigen (Cebra *et al.*, 1976; Mestecky *et al.*, 1980; Roux *et al.*, 1977). Thus, murine Peyer's patches contain antigen-sensitive lymphocytes, normal numbers of functional Mφ, and an actively pinocytotic epithelium that readily transports environmental antigens (e.g., gut bacterial components and dietary antigens) to underlying lymphoreticular cells. Although the precise cellular events that lead to induction of IgA responses is not known, a number of studies suggest that the indigenous microflora influences the inductive events that follow GALT antigen exposure (Cebra *et al.*, 1980; McGhee *et al.*, 1981).

A major objective of our laboratory has been directed towards a cellular and molecular understanding of induction of the IgA immune response to orally administered antigens. We have shown that oral administration of certain bacteria, as well as other antigen forms, leads to the appearance of secretory IgA (sIgA) antibodies in various external secretions including saliva, colostrum, milk, and tears (Michalek *et al.*, 1976; Mestecky *et al.*, 1978). In some instances, the appearance of these antibodies correlated with protection against a particular bacterial disease (McGhee and Michalek, 1981). In order to fully understand the precise inductive events that occur in GALT, it will be necessary to evaluate the environmental influence of gut components on this lymphoid tissue. It has become clear in recent years that oral administration of antigen not only leads to induction of sIgA antibodies in external secretions but also to a state of host hyporesponsiveness to systemic encounters with the orally administered antigen. This inability of the host to respond to subsequent, peripheral administration of antigen after its prolonged oral administration has been termed *oral tolerance* (Tomasi, 1980). Since bacterial LPS may be associated with the induction of suppressor cells for polyclonal responses, we postulated that the LPS molecule from the gut environment may also be of central importance in the induction of immune responses or systemic unresponsiveness to antigens encountered by the natural oral route. Since oral administration of antigen can lead to a dual response pattern, e.g.,

IgA responses at mucosal surfaces and systemic unresponsiveness (Challacombe and Tomasi, 1980), we have begun to explore the potential role of the LPS molecule in this response pattern. It is well known that LPS exhibits immunomodulating properties (Morrison and Ryan, 1979). As discussed briefly above, LPS is a good polyclonal B-cell activator and immunogen (Table 1) and when administered with antigen, is a potent adjuvant. However, when LPS is administered significantly prior to antigen, instead of enhanced immune responses, the host becomes unresponsive to antigen (tolerance). Furthermore, when LPS is administered several times at spaced intervals, the host becomes tolerant to any additional LPS effect (endotoxin tolerance). Thus, the LPS molecule is capable of modulating the host response to endotoxin or to other antigens as well.

One way to evaluate the contribution of LPS to host responses to orally administered antigens would be to determine the kinetics of immune responses to thymus-dependent (TD) antigens given by the oral route in normal LPS-responder mice and in animals that are hyporesponsive to LPS, i.e., C3H/HeJ mice. In studies of this nature, we have administered sheep erythrocytes (SRBC) (and other similar antigens) by the oral route and have measured splenic isotype anti-SRBC PFC responses and antibody responses in both serum and secretions (Babb and McGhee, 1980). It was immediately apparent that C3H/HeJ mice exhibited a heightened immune response pattern to orally administered antigen when compared with LPS-responder mice such as the C3H/HeN strain. Although IgM and IgG isotype responses were seen, the IgA response was usually the most pronounced and was three- to fivefold higher in C3H/HeJ mice. To ensure that this immune response pattern was not simply due to strain differences in C3H mice, several additional LPS-responder mice and one additional LPS-nonresponder strain were investigated. In these experiments, mice were carrier-primed with SRBC by gastric intubation for two consecutive days followed by oral administration with TNP–SRBC 1 week later. Maximum splenic anti-TNP PFC responses occurred 7 days following this immunization (Kiyono et al., 1980b). As can be seen in Fig. 2A, all of the LPS-sensitive animals exhibited significant but somewhat low IgA anti-TNP PFC responses. On the other hand, the two LPS-nonresponder strains (C3H/HeJ and C57BL/10ScN) gave a much higher IgA response pattern suggesting a relationship between the inability of B cells to respond to lipid A with a heightened IgA response to the orally administered TD antigen. This same response pattern has been observed with other antigens including horse and chicken erythrocytes and bacteria. It would be difficult to explain these results by suggesting that lipid A stimulates IgA precursor B cells in GALT since

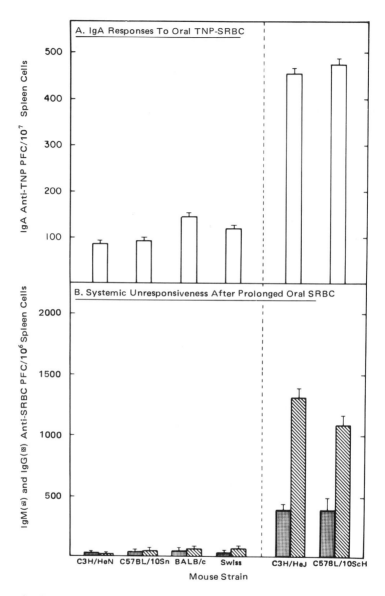

Figure 2. Comparison of responses to particulate antigen (erythrocytes) in LPS-responder and nonresponder mice. (A) IgA anti-TNP PFC responses in spleens of mice carrier-primed (SRBC) and immunized with TNP–SRBC by gastric intubation (GI). (B) Splenic anti-SRBC PFC responses following GI of SRBC for 14 consecutive days and challenged (i.p.) with SRBC 1 week later.

the high IgA responses were obtained in mice whose B cells have been shown to be nonresponsive to lipid A. A more likely explanation would be that LPS in the gut normally affects other cell types including perhaps the Mϕ and T cell, which are of importance in regulation of the IgA response. However, an equally plausible alternative to this would be that B cells in GALT of C3H/HeJ mice, which are not sensitive to polyclonal responses induced by lipid A, would remain sensitive to other polyclonal B-cell activators present in the gut. One could envision that precursor IgA B cells in GALT of these mice would also be sensitive to T cell signals required for their differentiation.

Our next studies were directed to assess whether differences in antigen-specific T cells occurred in Peyer's patches or spleen of LPS-responder and -nonresponder C3H mice. Greater numbers of T_{SRBC} helper cells for *in vitro* immune responses were noted in C3H/HeJ mice given SRBC by the oral route. The T_{SRBC} helper cell activity could be enhanced with submitogenic concentrations of Con A, a treatment that is known to polyclonally activate T-helper cells. Further, we found that Con A administered orally with SRBC to C3H/HeJ mice greatly enhanced their splenic IgA PFC response (Kiyono *et al.*, 1980b). It is well established that induction of T-helper cells is dependent upon antigen processing and presentation by Mϕ (accessory cells). Therefore, it is likely that this process is fully intact in Peyer's patches of C3H/HeJ mice. It remains to be explained why this process is exaggerated in GALT of LPS-nonresponder mice and we are currently investigating this question.

Continual oral administration of soluble or particulate antigens will often lead to an inability of these mice to respond to the same antigen subsequently given by a systemic route. The mechanism of oral tolerance induction is currently being studied and several mechanisms including T-suppressor cells, immune complexes, and anti-idiotypic antibodies have been proposed (Tomasi, 1980). We have postulated that gut LPS may induce precursor suppressor cells that may ultimately mediate oral tolerance. We have already discussed above our evidence that the indigenous gut microflora induces T cells that suppress polyclonal B-cell responses. It is thus conceivable that this T-cell population also contains precursors for antigen-specific suppression. To directly address this possibility, we performed a series of experiments where groups of LPS-responder or -nonresponder mice were continually given SRBC by the oral route. Subsequently, these mice were challenged with SRBC by i.p. injection and splenic anti-SRBC PFC assessed. LPS-responder mice manifest tolerance to SRBC (Fig. 2B); however, LPS-nonresponder mice gave significant immune responses. Although the data are not shown, LPS-nonresponder mice exhibit

IgG1, IgG2, and IgA isotype responses, suggesting that instead of induction of oral tolerance, these mice have been primed for anamnestic responses. Both mouse strains exhibited similar primary and secondary responses to SRBC given i.p. (data not shown). Therefore, these results suggest that LPS-nonresponder mice exhibit only positive effector responses to orally administered antigens and lack a suppressor pathway. In experiments not shown, we have found that these mice exhibit a preponderance of T-helper cells while LPS-responder strains given SRBC orally for prolonged periods exhibit antigen-specific suppressor cells in both Peyer's patches and spleen. It is tempting to speculate that bacterial LPS is responsible for induction of these precursor T-suppressor cells and thus provides the host with an important mechanism for regulation of systemic responses to orally administered antigen. A number of questions are thus raised from these experiments. (1) Are T cells for IgA responses affected by LPS? (2) Are different T-cell subsets responsible for systemic unresponsiveness and are these subsets more sensitive to LPS? (3) Since the host must respond to gut-encountered antigens with synthesis of IgA in external secretions, are these cells more resistant to LPS effects? (4) Since murine Peyer's patches contain significant numbers of Mϕ, does LPS modulate the immune response pattern to orally administered antigen by a direct effect on this cell type?

Of obvious importance to these experiments would be the demonstration that the heightened immune response pattern observed in LPS-nonresponder C3H/HeJ mice to orally administered antigen is linked to expression of the *Lps^d* gene. Genetic crosses between LPS-responder C3H/HeN and C3H/HeJ mice were performed and individual mice were assessed for both their ability to elicit IgA responses to orally administered antigen and splenic B-cell mitogenic responses to LPS. As shown in Fig. 3, C3H/HeN mice were fully responsive to LPS and exhibited low IgA anti-TNP PFC responses, while C3H/HeJ mice exhibited the opposite pattern, i.e., were unresponsive to LPS but gave high IgA anti-TNP PFC responses to orally administered antigen. F_1 mice exhibited an intermediate response pattern. F_2 mice segregated into a pattern where approximately one-fourth resembled the C3H/HeN, another one-fourth were C3H/HeJ-like, while approximately one-half exhibited an intermediate response pattern (Fig. 3). It should be noted that this would be expected if *Lps^d* expression and heightened IgA responses to oral antigen were both controlled by a single gene locus. A backcross linkage analysis proved that this was the case since one-half of progeny from $F_1 \times$ C3H/HeJ matings gave intermediate responses while the other backcross mice (approximately 50%) were nonresponsive to LPS and exhibited high IgA anti-TNP PFC re-

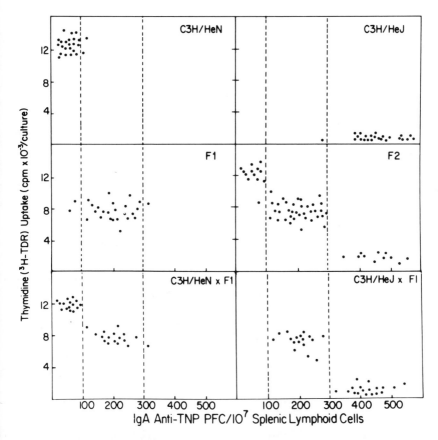

Figure 3. Genetic linkage of LPS nonresponsiveness (Lps^d/Lps^d) with elevated IgA immune responses. Mice were primed and immunized as described in Fig. 2A.

sponses. Backcross mice from $F_1 \times$ C3H/HeN matings further substantiated this relationship since approximately one-half gave intermediate responses while the other half were fully responsive to LPS and low IgA responders (Michalek *et al.*, 1980a). These results clearly established that the heightened immune response pattern to TD antigens in LPS-nonresponder mice is another manifestation of Lps^d expression.

The results summarized above clearly point out that the host's encounter with environmental LPS has far-reaching effects that are not only beneficial but of central importance for maturation and expression of the immune system. These include a T-cell regulation of B-cell polyclonal responses, effects on induction of the IgA response, and stimulation of precursor cells that, when subsequently sensitized by antigen, migrate to peripheral tissue and regulate subsequent host encounters with that particular antigen.

IV. Molecular Model for LPS-Induced Effects

From the above discussion, it is clearly evident that LPS is capable of inducing a broad range of effects, including endotoxemia and lymphoreticular cell effects such as adjuvanticity and polyclonal expansion of a significant percentage of the murine B-cell population (Table 1). A recent review (Morrison and Rudbach, 1981) has described several models for LPS activation of lymphocytes. Two current models include (1) destabilization of cellular membranes by the lipid A and (2) interaction of lipid A with a cell surface LPS receptor (Coutinho *et al.*, 1978). Although Coutinho *et al.* (1978) have provided supportive evidence for the existence of a putative LPS receptor, others have not been able to confirm this (Watson *et al.*, 1980). Additional support for the lack of a specific LPS receptor is provided by the finding that lipid A binds equally to B cells of LPS-responder or -nonresponder (C3H/HeJ) mice (Kabir and Rosenstreich, 1977; Gregory *et al.*, 1980). This controversy has led us to propose a model for LPS activation of lymphoreticular cells that does not involve a specific membrane receptor for LPS.

A detailed model proposed by Decker and Marchalonis (1978) describes the early molecular events in T-lymphocyte activation by lectins. In their model, ligand binds to a cell surface receptor, which triggers either cyclic nucleotide production (cAMP-cGMP) and/or alters membrane phospholipid content, such that an influx of Ca^{2+} ions occurs. The ions either interact with the cytoskeleton or with phosphorylated histone or nonhistone chromosomal proteins. The DNA-binding capacity of these phosphorylated proteins is altered so that mRNA and subsequent protein synthesis is stimulated (or inhibited) by the lectins. Although Decker and Marchalonis (1978) provide a molecular basis for the events that lead to T-cell activation by lectins, this may not fully explain LPS activation of B lymphocytes. It has been shown that LPS–membrane interaction followed by internalization are both necessary for B-cell mitogenic responses (Adler *et al.*, 1972; Smith, 1972).

We propose here that LPS activates the B cell by combination with an internal receptor in the cytoplasm that then interacts with the DNA. The concept of a cytoplasmic receptor is not unique; in fact, the model for steroid hormone regulation of eukaryotic cells has served as our prototype for a model of LPS B-cell activation. The steroid system has been described elsewhere (O'Malley and Means, 1976; Chan and O'Malley, 1978; Baxter and Funder, 1979); however, a brief review is presented so that the parallels between the two systems, LPS activation of B cells and steroid regulation of eukaryotic cells, can be appreciated.

Estrogen and progesterone are two of the most widely studied steroid hormones. They are derived from cholesterol, as are all steroids, and as such, are easily integrated into cellular membranes. This integration alone does not result in any detectable cellular effects. Following internalization of the hormone, the specific regulatory effects of the hormone are seen. In steroid-sensitive cells, the hormone binds to a cytoplasmic receptor, while cells that contain no internal receptor are not affected by the hormone. In sensitive cells, the binding of the hormone to the receptor causes a conformational change in the receptor. The altered receptor–hormone complex is then translocated into the cell nucleus where the complex either binds directly to the DNA or binds to a second receptor that in turn binds to DNA. This DNA-binding protein allows, or "activates," transcription of specific segments of DNA that are linked to functional or maturational processes. For instance, in the chick oviduct, progesterone is directly responsible for transcription of the message for avidin (O'Malley and Means, 1976). It is possible that other segments of DNA are also transcribed as a direct result of hormone-receptor-DNA interaction; however, these processes have not yet been clearly defined.

Our model for LPS activation is somewhat analogous to the steroid hormone system. Both lipid A (the mitogenic portion of LPS) and steroids are capable of integrating into cellular membranes without regard to the membrane type, i.e., both demonstrate "nonspecificity" in binding (Davies *et al.*, 1978; Morrison and Rudbach, 1981). Both molecules have defined actions in specific cells and apparently no action in other cells. From this it follows that binding to, or integration into, the cellular membrane is not sufficient for "activation" of the cell. Both steroids and LPS induce differentiation or maturation, in sharp contrast to peptide hormones, which initiate simple switch on, switch off signals. Furthermore, peptide hormones act on cell surface receptors, not cytoplasmic receptors, and exhibit immediate but transient effects. Steroids and LPS have been shown to cause phosphorylation of nonhistone chromosomal proteins (for review see Decker and Marchalonis, 1978). Phosphorylated proteins are the active form, allowing specific segments of DNA to be transcribed. The synthesis of new proteins in B lymphocytes following LPS stimulation was reported by Stott and Williamson (1978a,b), who described that phosphorylation of nuclear proteins occurs within 2 hr following stimulation. Furthermore, specific nonhistone chromosomal proteins are maximally synthesized by 24 hr. The finding that phosphorylation continues for at least 24 hr suggests that the phosphorylated form of the nuclear protein is necessary for increased synthesis of nonhistone chromosomal proteins. Phosphorylation of nonhistone chromosomal protein has also been demonstrated in B lymphocytes activated with anti-immunoglobulin (Nishizawa *et al.*, 1977). Thus, lipid A and steroid hormones induce similar intracellular events suggesting common molecular pathways of stimulation.

These analogies have led us to propose the following model of LPS activation of B cells. We suggest that LPS, like steroid hormones, binds to an internal cytoplasmic receptor that is a product of the *Lps* gene. The *Lps* gene protein is, for simplicity, termed the *lipid A receptor*. The binding of LPS to the lipid A receptor results in a conformational change that allows transcription of specific genes. It is postulated that among these genes are nonhistone chromosomal proteins that activate transcription of specific maturational markers. A report by Connett and Fleischman (1981) has shown that specific nonhistone chromosomal proteins are linked to the expression of gamma, alpha, and kappa immunoglobulin chains. We further suggest that LPS may mimic some naturally occurring signal, perhaps arachidonic acid or prostaglandin E_2, which would normally be triggered by antigen or antigen–antibody complexes binding to the B lymphocyte. For instance, it has been shown that the binding of antigen to IgG anchored in the Fc receptor of human B cells will trigger the activation of phospholipase A_2 and causes an increase in arachidonic acid, a precursor for prostaglandins (Suzuki *et al.*, 1980, 1981). This phospholipase A_2 activation has also been shown to be true for Con A stimulation of T lymphocytes (Hirata *et al.*, 1980). This mechanism may also account for Fc-mediated adjuvanticity (Morgan *et al.*, 1980). We postulate the LPS stimulation of all three lymphoreticular cell types, e.g., T and B lymphocytes and macrophages, occurs following lipid A–lipid A receptor binding in the cell cytoplasm.

A. LPS Clonal Expansion of B Lymphocytes

The *Lps* gene maps near the genes for Lyb 2, Lyb 4, and Lyb 6 on chromosome 4 in most mouse strains; however, in the C3H/HeJ strain the Lyb 4 locus fails to map between Lyb 2 and Lyb 6 (Howe *et al.*, 1979). The proximity of the *Lps* gene to these specific B-lymphocyte cell surface maturational markers and the mapping differences in the C3H/HeJ are suggestive of a linkage between the *Lps* gene product and the C3H/HeJ dysfunction (Watson *et al.*, 1980). Additional chromosomal irregularities have been described in C3H mice, since Horton and Hetherington (1980) have shown that the T-cell antigen Ly 6 does not map Ly 6–Thy 1–Trf on chromosome 9. Thus, several rearrangements (or mutations) have occurred in the C3H/HeJ mouse and these have provided a unique model for examination of the *Lps* gene defect, as well as for studies of B-cell activation.

In our model, we hypothesize that resting B cells characterized by the markers Fc$^+$, Ia$^+$, Lyb 2$^+$, sIgM$^+$, sIgD$^+$, and C3 receptor$^+$ (Ahmed *et al.*, 1978; McKenzie and Potter, 1979) would produce low levels of cytoplasmic lipid A receptor. It is proposed that the *Lps* gene product, in its unbound state, is self-regulating in the same manner as is seen with the arabinose operon (Englesberg *et al.*, 1969; Lee, 1978); a single protein can act both as a promoter and as an inactivator. This has been shown to occur with nonhistone chromosomal proteins in eukaryotic systems (Farber *et al.*, 1972; Stein *et al.*, 1974). The lipid A receptor may bind weakly to the *Lps* gene, blocking transcription but allowing low levels of the *Lps* gene product to be synthesized. Thus, the gene product is always present in the cell cytoplasm and nucleus, but only in threshold amounts. There are several interpretations of the regulatory events which can take place following lipid A internalization into the cell and three of these are discussed below. In the first instance, lipid A binds to the lipid A receptor in B lymphocytes from LPS responsive mice and causes a conformational change in the lipid A receptor. The LPS-receptor complex binds alternate sites on the DNA and can no longer bind the *Lps* gene which results in an increased synthesis of the lipid A receptor. The LPS-lipid A receptor complex acts as an inducer of transcription, triggers synthesis of new proteins which results in cellular maturation and differentiation. The lipid A receptor in cells of C3H/HeJ mice either fails to bind LPS, or binds but does not result in the appropriate conformational change and no increase in the synthesis of the *Lps* gene product occurs. In the second case, the C3H/HeN lipid A receptor behaves as described above; whereas, the C3H/HeJ lipid A receptor binds lipid A which results in a conformational change of the receptor, such that the *Lps* gene is bound tightly by the complex and no further *Lps* gene product (lipid A receptor) is synthesized. This renders the C3H/HeJ mouse refractory to the effects of LPS.

A third possibility, is that lipid A receptor is not self-regulating and may bind multiple sites on the DNA, such that many maturational events are regulated. When lipid A enters the cell, lipid A competitively inhibits the binding of the lipid A receptor to the DNA, so that if the receptor is complexed with lipid A, it cannot bind to any DNA. The DNA of the C3H/HeN mouse would have many "unblocked" sites and can initiate transcription of new mRNA's involved in the LPS driven differentiation. While in the C3H/HeJ mouse, the lipid A receptor–LPS interaction fails to occur, promoter sites remain complexed with the receptor and LPS driven differentiation is blocked.

In this model, the lipid A–lipid A receptor complex would directly affect differentiation of the B cell. This complex allows transcription of cell surface markers such as Lyb 4 and Lyb 6 as well as other maturational proteins (e.g., Lyb 3). The expression of these proteins may be either directly or indirectly regulated. Direct regulation would involve binding of the lipid A–lipid A receptor complex to a gene, for example the Lyb 4 gene, and this would allow transcription of the message (Fig. 4B). The indirect regulation would involve synthesis of additional nonhistone chromosomal proteins that in turn regulate expression of maturational markers (Fig. 4C). This would lead to a cascade or amplification process, not only in terms of the amount of a single protein produced, but also in the number of other proteins that can be triggered by a single stimulus (e.g., LPS). Of the two types of regulation, direct and indirect, the indirect mechanism seems more likely. In an indirect system, the induced nonhistone chromosomal protein may be self-regulating in a manner different from that suggested for self-regulation of the *Lps* gene. The induced nonhistone chromosomal protein, or its transcription product, may act as self inducer, in addition to acting as promoters for maturational markers; thus eliminating the need for further LPS stimulation.

Several lines of evidence lend support for this model. First, it has been shown that LPS must be internalized to activate the B cell (Adler *et al.*, 1972; Smith, 1972). Treatment of cells with neuraminidase prevents internalization of LPS and subsequent mitogenic responses in these cells. Second, the C3H/HeJ mouse binds LPS as readily as lipid A-responder strains (Kabir and Rosenstreich, 1977; Gregory *et al.*, 1980), further demonstrating that cell surface binding of LPS is not sufficient for mitogenic responses. Third, Watanabe and Ohara (1981) have used cell fusion techniques to show that C3H/HeN cytoplasts containing C3H/HeJ karyoplasts are stimulated by LPS in a "normal" manner. The authors proposed that this was due to a cell surface determinant on the C3H/HeN cytoplast. An alternate explanation would be that this was due to a cytoplasmic receptor difference. The C3H/HeN cytoplasmic receptor would bind lipid A and allow interaction with the C3H/HeJ DNA whereas the C3H/HeJ cytoplasmic receptor would not undergo conformational changes and would not bind to the C3H/HeN DNA.

Recent work from this laboratory has shown that germfree (GF) mice, presumably with more immature, less differentiated B cells, exhibit a more pronounced response to LPS than GF mice treated with LPS *in vivo*. As was seen in Table 2A, LPS induced elevated mitogenic responses in GF spleen cell cultures when compared with cultures from conventional mice. Gastric intubation of GF mice with a single dose of LPS, as low as 5 μg, gave a response pattern similar to that observed

A. Resting B Cell

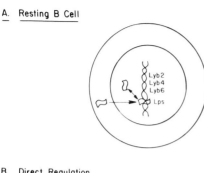

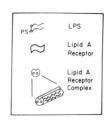

B. Direct Regulation

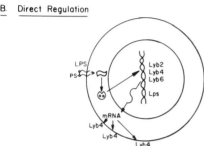

C. Indirect Regulation

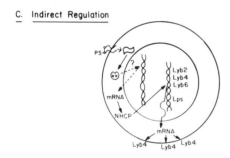

Figure 4. Lipid A activation of the B lymphocyte. (A) Resting B cell. In the absence of lipid A, the lipid A receptor binds to the *Lps* gene, blocking its transcription. (B) Direct regulation. LPS binds the lipid A receptor, changing the conformation of the receptor. The new form binds new segments of DNA, stimulates specific mRNA synthesis and B-cell maturation and proliferation. (C) Indirect regulation. Lipid A binds the lipid A receptor, changing the conformation of the receptor. The receptor binds new segments of DNA, stimulating synthesis of nonhistone chromosomal proteins. These proteins bind DNA and regulate division of B lymphocytes.

with Conv mice (Table 3). One possible explanation of the findings is that LPS triggers GF B cells to become more mature, while an alternative explanation that cannot be overlooked is the role of T cells, since

Table 3. Mitogenic Responses to LPS of Spleen Cells from Germ-Free BALB/c Mice Gastrically Intubated with Endotoxin[a]

Group	Pretreatment	Mitogenic response (E/C)		
		LPS	PHA	Con A
Germ-free	None	36.6	53.2	99.3
	5 μg	6.2	34.2	39.7
	50 μg	9.3	38.0	37.4
	500 μg	8.9	40.4	42.6
Monoassociated (*E. coli* K235)	None	10.9	38.9	70.8
Conventional	None	11.1	32.2	42.1

[a] Germ-free mice were assessed for mitogenic responses 5 days following gastric administration of LPS.

LPS may induce T cells that negatively regulate the B-cell response to LPS.

Additional support for our model comes from studies performed with F_1 hybrid mice (Fig. 3). F_1 mice (from C3H/HeN × C3H/HeJ crosses) gave intermediate mitogenic responses to LPS suggesting that *Lps* gene expression occurs either by allelic exclusion (Watson *et al.*, 1980) or by codominant expression. In allelic exclusion, each B cell in F_1 progeny expresses the lipid A receptor of either the C3H/HeJ (Lps^d) or the C3H/HeN (Lps^n) genotype. If codominance is manifested, then both receptor phenotypes would be equally expressed in a single cell. Both concepts are consistent with the observation that LPS responses of F_1 mice are approximately one-half of those seen in C3H/HeN animals.

Finally, Wolf and Merler (1979a,b) have shown that the lipid moiety of two mediators, soluble inhibitory factor (SIF) and lymphocyte mitogenic factor (LMF), are responsible for regulatory effects on B lymphocytes. This is in line with evidence that essential fatty acids can act as mediators of the immune response (for review see Mertin and Stackpoole, 1981). Thus, lipid A may mimic this lipid-mediated activation process.

In summary, we propose that LPS binds to a cytoplasmic receptor, and causes a conformational change. This receptor complex binds to DNA and stimulates transcription of mRNA and subsequent translation of differentiational and maturational proteins. It is conceivable that the endotoxin complex present in the outer membrane of our indigenous normal flora, serves as the differentiation signal. This simple model, although theoretical in nature, has enough circumstantial support to lead one to believe that many aspects will be proven to be correct.

B. Model for LPS Mediation of Macrophage Events

The consequences of LPS interactions with the macrophage ($M\phi$) have been widely studied in numerous laboratories (see Vogel *et al.*, 1981, for review). The lipid A portion of LPS induces the $M\phi$ to increase production of prostaglandins (Kurland and Bockman, 1978), to produce mediators, including interleukin 1 (IL 1; Hoffmann and Gilbert, 1980; Oppenheim *et al.*, 1980), and to enhance tumoricidal capacity of macrophages (Hibbs *et al.*, 1977). The macrophage of the C3H/HeJ mouse also expresses the *Lps* gene aberration; however, the defect remains to be defined at the molecular level (Rosenstreich and Vogel, 1980; Vogel *et al.*, 1981). As with B lymphocytes, we propose a mechanism for activating the macrophage, via LPS, that involves a cytoplasmic lipid A receptor. Lipid A enters the macrophage and binds with the lipid A receptor in the cytoplasm. The receptor–lipid A complex can now act as a DNA-binding ligand, which stimulates transcription of previously untranscribed regions of the macrophage DNA (Fig. 5). Many of the lipid A-induced functions involve the production of mediators, such as prostaglandins and IL 1. In the case of prostaglandin synthesis, the LPS-induced DNA transcription may code for enzymes that process arachidonic acid into the various prostaglandins or prostaglandin-like molecules. As was previously suggested, this may be directly or indirectly regulated; the processing enzymes are either directly transcribed, or nonhistone chromosomal regulatory proteins are produced that control transcription of the enzymes. Vogel *et al.* (1979) have demonstrated that the C3H/HeJ macrophage fails to increase production of prostaglandin (PGE_2) when stimulated with LPS. This can be interpreted as a failure of lipid A receptor to bind to the C3H/HeJ DNA, leading to a depression of mediator protein systhesis. Thus, LPS-stimulated prostaglandin production is regulated by a lipid A receptor that is a DNA-binding protein.

In this model, the production of IL 1 following LPS stimulation would also be regulated by the lipid A receptor. This monokine represents an early signal for T-cell activation and is likely controlled by an intricate set of mechanisms. Its synthesis could be directed by the lipid A receptor complex, which would bind directly to the $M\phi$ DNA and control the transcription of IL 1 (direct regulation). Alternatively, nonhistone chromosomal protein(s) triggered by the lipid A receptor com-

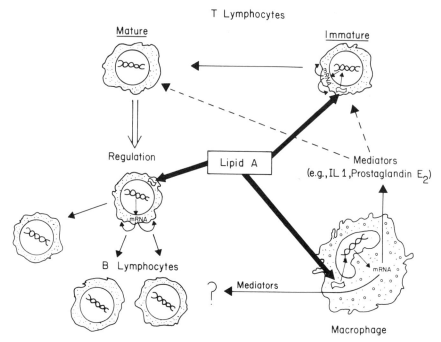

Figure 5. Lipid A activation of lymphoreticular cells.

plex may regulate the transcription of IL 1 (indirect regulation). Again, in the C3H/HeJ macrophage, it has been shown that LPS-stimulated IL 1 production is suppressed (Rosenstreich *et al.*, 1978; Vogel *et al.*, 1981). This supports the view that lipid A fails to cause a conformational change in the C3H/HeJ lipid A receptor, thus preventing its binding to the appropriate sequences of DNA and in turn blocking RNA and protein systhesis.

As in the case of the B lymphocyte, the Mφ cytoplasmic lipid A receptor is a DNA-binding protein. During the differentiation process, the Mφ and B cell undergo alternate DNA processing that limits their lipid A receptor-binding sites. Thus, the same receptor mediates completely different events in these cells. In the B lymphocyte, division and maturation occurs, while in the Mφ, mediator production is initiated

(Fig. 5). The basic sequence of events—lipid A–lipid A receptor interaction, conformational change of the lipid A receptor, binding of the
receptor to DNA, and triggering of protein synthesis—is the same in
both cells. Thus, a single locus regulates, in completely different ways,
these two lymphoreticular cell types.

C. LPS Interactions with T Lymphocytes

Although LPS has marked effects on B lymphocytes and Mϕ that
may ultimately lead to their activation, direct effects of LPS on T cells
is much more difficult to demonstrate. Several studies have documented that LPS will not directly activate thymocytes or mature T cells
(Andersson et al., 1972; Möller et al., 1972a; Peavy et al., 1974) or induce specific T-effector cells (Möller et al., 1972a,b); however, LPS
clearly influences the activation of thymocytes by T-cell lectins. Specifically, LPS enhances murine thymocyte responses to both Con A
(Forbes et al., 1975; Ozato et al., 1975) and PHA (Mizuguchi et al.,
1980). Enhancement of T-cell mitogenic responses was seen only when
LPS was added prior to the addition of the T-cell lectin, suggesting
that LPS in some fashion preconditioned the cell membrane for the
subsequent lectin interaction. It is not yet clear whether these LPS–T-
cell effects are lipid A mediated, since Dumont (1978) has presented
evidence that LPS enhances Con A thymocyte activity in C3H/HeJ
mice. However, the purity of the LPS preparation used in his experiments was not reported.

There is also documentation that LPS is capable of inducing T-cell
markers associated with maturation (Hämmerling et al., 1975; Winchurch et al., 1981) since LPS induces expression of Thy 1 and TL antigens in precursor T cells in vitro. Watson (1977) clearly showed that
lipid A induced Thy 1.2 expression in purified bone marrow cells of
LPS-responder mice; however, no effect was seen in cultures from
C3H/HeJ mice. He presented additional evidence that agents that increase intracellular levels of cAMP induce the expression of Thy 1.2
and TL antigens in precursor T cells (Watson, 1977) and he postulated
that LPS acts on both T- and B-lymphocyte subpopulations in this
manner. The above experiments are certainly compatible with our
model for LPS activation (Fig. 5) and would support our previous results (see Sections II and III above).

We thus propose that T cells have cytoplasmic lipid A receptors as
do both B cells and Mϕ and that lipid A interaction with this receptor
results in a maturational signal for this cell. This signal does not result
in cell division but instead in an ability of this cell to respond to antigen along an effector pathway (Fig. 5). The lipid A–lipid A receptor

binds to specific sites on the DNA of T cells and results in transcription of proteins that predispose the cell to a particular effector pathway (suppressor, helper, cytotoxic, etc.).

It should be reemphasized that lipid A probably mimics a naturally occurring cell product that regulates the cell. Lipid A expansion of B cells and regulatory T cells, however, may give the host a survival advantage. In this regard, it is well established that differentiation of B cells for IgA expression is a T-independent process; however, regulation of the IgA responses to antigen is T cell dependent. The regulated polyclonal expansion of B cells at mucosal surfaces, the most common route of antigen presentation, and the maturation of regulatory T cells at these sites would endow the host with a mature lymphoid system to deal with a range of viral and microbial agents that would constantly bombard the GALT immune system. This maturation is accomplished by mimicry of host factors (prostaglandins, monokines, or lymphokines) that stimulate T and B cells. Some of these host factors will interact with cytoplasmic receptors that can then bind DNA and stimulate synthesis of new proteins, thus allowing the host to respond more rapidly and effectively to virgin antigens.

V. Summary

In this chapter we have reviewed recent work concerning mechanisms of LPS regulation of the immune response. The host possesses a natural T-cell population (endotoxin induced?) that controls LPS B-cell responses, and this regulation prevents endotoxin in our environment from overstimulating our B-lymphocyte system. It is also clear that environmental endotoxins in the gastrointestinal tract profoundly affect host immune responses to orally encountered antigens. One manifestation of this is an association of LPS with the induction of IgA responses in GALT and the stimulation of T-suppressor cells that subsequently regulate systemic responses to antigen. We have presented a simple model that may explain, at the molecular level, how lymphoreticular cells are activated by LPS. We propose that the *Lps* gene product is a cytoplasmic protein that binds to internalized lipid A. The lipid A–lipid A receptor complex acts as a DNA-binding ligand, and this complex activates transcription of specific segments of DNA that are responsible for maturational events. Thus, the receptor complex induces transcription of specific mRNA that codes for translation of nonhistone chromosomal proteins. These proteins induce transcription of other DNA segments responsible for: Lyb surface antigens (B cells), mediators (Mϕ), or maturation into antigen-sensitive effectors (T cells).

ACKNOWLEDGMENTS

The authors greatly appreciate the critical assessment of this chapter and the proposed model by Drs. David Briles, Fiona Hunter, Gayle Knapp, and Stefanie Vogel. K.A.G. and M.J.W. are NIH trainees AI 07051.

Note Added to Proof. Since the submission of this manuscript three reports have been published which bear particular relevance to our model. In their earlier studies, Watanabe and Ohara (1981) considered only the cytoplasmic membrane in activation; however in their latest work (T. Watanabe, Y. Eda, and J. Ohara, 1982, *J. Exp. Med.* **152**:312), LPS activation via cytoplasmic-nuclear interactions was proposed. A recent report (H. J. Rahmsdorf, U. Mallick, H. Ponta, and P. Herrlick, 1982, *Cell* **29**:459) describes a B lymphocyte specific protein which disappears by 72 hours after stimulation with LPS. It is a basic glycoprotein with a molecular weight of 35,000 and is thought to be associated with the nuclear membrane. Finally, a report by Tarakoli and Moon (*Pathophysiological Effects of Endotoxins at the Cellular Level*, 1981, pp. 33–43, Liss, New York) examines the interactions of endotoxin with chromatin, DNA, and steroid receptors. They suggest that endotoxin may interact with the proteins associated with DNA rather than DNA itself.

VI. References

Adler, W. H., Osunkoya, B. O., Takiguchi, T., and Smith, R. T., 1972, The interactions of mitogens with lymphoid cells and the effect of neuraminidase on the cells' responsiveness to stimulation, *Cell. Immunol.* **3**:590.
Ahmed, A., Smith, A. H., Kessler, S. W., and Scher, I., 1978, Ontogeny of murine B cell surface antigens and functions, in: *Animal Models of Comparative and Developmental Aspects of Immunity and Disease* (E. Gershwin and E. L. Cooper, eds.), pp. 175–200, Pergamon Press, New York.
Andersson, J., Möller, G., and Sjöberg, O., 1972, Selective induction of DNA synthesis in B and T lymphocytes, *Cell. Immunol.* **4**:381.
Babb, J. L., and McGhee, J. R., 1980, Mice refractory to LPS manifest high IgA responses to orally administered antigen, *Infect. Immun.* **29**:322.
Baxter, J. D., and Funder, J. W., 1979, Hormone receptors, *N. Engl. J. Med.* **301**:1149.
Cebra, J. J., Gearhart, P. J., Kamat, R., Robertson, S. M., and Tseng, J., 1976, Origin and differentiation of lymphocytes involved in the secretory IgA response, *Cold Spring Harbor Symp. Quant. Biol.* **41**:201.
Cebra, J. J., Gearhart, P. J., Halsey, J. F., Hurwitz, J. L., and Shahin, R. D., 1980, Role of environmental antigens in the ontogeny of the secretory immune response, *J. Reticuloendothelial. Soc.* **28**:61s.
Challacombe, S. J., and Tomasi, T. B., Jr., 1980, Systemic tolerance and secretory immunity after oral immunization, *J. Exp. Med.* **152**:1459.

Chan, L., and O'Malley, B. W., 1978, Steroid hormone action: Recent advances, *Ann. Intern. Med.* **89**:694.

Connett, J., and Fleischman, J. B., 1981, Non-histone chromosomal proteins from immunoglobulin-producing mouse plasmacytoma cell, *Mol. Immunol.* **18**:573.

Coutinho, A., Forni, L., and Watanabe, T., 1978, Genetic and functional characterization of an antiserum to the lipid A-specific triggering receptor on murine B lymphocytes, *Eur. J. Immunol.* **8**:63.

Davies, M., Stewart-Tull, D. E. S., and Jackson, D. M., 1978, The binding of lipopolysaccharide from *Escherichia coli* to mammalian cell membranes and its effect on liposomes, *Biochim. Biophys. Acta* **508**:260.

Decker, J. M., and Marchalonis, J. J., 1978, Molecular events in lymphocyte activation: Role of non-histone chromosomal proteins in regulating gene expression, *Contemp. Top. Mol. Immunol.* **7**:365.

Dumont, F., 1978, Bacterial lipopolysaccharide (LPS) enhances concanavalin A reactivity of thymocytes from the low-responder mouse strain C3H/HeJ, *Experientia* **34**:125.

Englesberg, E., Squires, C., and Moronk, F., 1969, The L-arabinose operon in *E. coli* B/r: A genetic demonstration of two functional states of the product of a regulatory gene, *Proc. Natl. Acad. Sci. USA* **62**:1100.

Farber, J., Stein, A., and Baserga, R., 1972, Regulation of RNA synthesis during mitosis, *Biochem. Biophys. Res. Commun.* **47**:790.

Forbes, J. T., Nakao, Y., and Smith, R. T., 1975, T mitogens trigger LPS responsiveness in mouse thymus cells, *J. Immunol.* **114**:1004.

Gregory, S. H., Zimmerman, D. H., and Kern, M., 1980, The lipid A moiety of lipopolysaccharide is specifically bound to B cell subpopulations of responder and nonresponder animals, *J. Immunol.* **125**:102.

Hämmerling, U., Chin, A. F., Abbott, J., and Scheid, M. P., 1975, The ontogeny of murine B lymphocytes. I. Induction of phenotypic conversion of Ia$^-$ to Ia$^+$ lymphocytes, *J. Immunol.* **115**:1425.

Hibbs, J. B., Jr., Taintor, R. R., Chapman, H. A., Jr., and Weinberg, J. B., 1977, Macrophage tumor killing: Influence of local environment, *Science* **197**:279.

Hirata, F., Toyoshima, S., Axelrod, J., and Waxdal, M. J., 1980, Phospholipid methylation: A biochemical signal modulating lymphocyte mitogenesis, *Proc. Natl. Acad. Sci. USA* **77**:862.

Hoffmann, M. K., and Gilbert, K., 1980, Immune functions controlled by lipopolysaccharide-induced macrophage factor (IL-1), in: *Microbiology 1980* (D. Schlessinger, ed.), pp. 55–60, American Society for Microbiology, Washington, D.C.

Horton, M. A., and Hetherington, C. M., 1980, Genetic linkage of Ly-6 and Thy-1 loci in the mouse, *Immunogenetics* **11**:521.

Howe, R. C., Ahmed, A., Faldetta, T. J., Byrnes, J. E., Rogan, K. M., Dorf, M. E., Taylor, B. A., and Humphreys, R. E., 1979, Mapping of the Lyb-4 gene to different chromosomes in the DBA/2J and C3H/HeJ mice, *Immunogenetics* **9**:221.

Jensen, S. B., Mergenhagen, S. E., Fitzgerald, R. J., and Jordan, H. V., 1963, Susceptibility of conventional and germfree mice to lethal effects of endotoxin, *Proc. Soc. Exp. Biol. Med.* **113**:710.

Kabir, S., and Rosenstreich, D. L., 1977, Binding of bacterial endotoxin to murine spleen lymphocytes, *Infect. Immun.* **15**:156.

Kiyono, H., McGhee, J. R., and Michalek, S. M., 1980a, Lipopolysaccharide regulation of the immune response: Comparison of responses to LPS in germfree, *Escherichia coli*-monoassociated and conventional mice, *J. Immunol.* **124**:36.

Kiyono, H., Babb, J. L., Michalek, S. M., and McGhee, J. R., 1980b, Cellular basis for elevated IgA responses in C3H/HeJ mice, *J. Immunol.* **125**:732.

Kurland, J. I., and Bockman, R., 1978, Prostaglandin E production by human blood monocytes and mouse peritoneal macrophages, *J. Exp. Med.* **147**:952.

Lee, N., 1978, Molecular aspects of *ara* regulation, in: *The Operon* (J. H. Miller and W. S. Riznekoff, eds.), pp. 389–410, Cold Spring Harbor Laboratory, Cold Spring Harbor, N.Y.

McGhee, J. R., and Michalek, S. M., 1981, Immunobiology of dental caries: Microbial aspects and local immunity, *Annu. Rev. Microbiol.* **35**:595.

McGhee, J. R., Kiyono, H., Michalek, S. M., Babb, J. L., Rosenstreich, D. L., and Mergenhagen, S. E., 1980, Lipopolysaccharide (LPS) regulation of the immune response: T lymphocytes from normal mice suppress mitogenic and immunogenic responses to LPS, *J. Immunol.* **124**:1603.

McGhee, J. R., Michalek, S. M., Kiyono, H., Babb, J. L., Clark, M. P., and Mosteller, L. M., 1981, LPS regulation of the IgA immune response, in: *Mechanisms in Mucosal Immunity* (L. A. Hanson and W. Strober, eds.), pp. 55–72, Raven Press, New York.

McKenzie, I. F. C., and Potter, T., 1979, Murine lymphocyte surface antigens, *Adv. Immunol.* **27**:181.

Mertin, J., and Stackpoole, A., 1981, Prostaglandin precursors and the cell-mediated immune response, *Cell. Immunol.* **62**:293.

Mestecky, J., McGhee, J. R., Arnold, R. R., Michalek, S. M., Prince, S. J., and Babb, J. L., 1978, Selective induction of an immune response in human external secretions by ingestion of bacterial antigen, *J. Clin. Invest.* **61**:731.

Mestecky, J., McGhee, J. R., Crago, S. S., Jackson, S., Kilian, M., Kiyono, H., Babb, J. L., and Michalek, S. M., 1980, Molecular–cellular interactions in the secretory IgA response, *J. Reticuloendothelial. Soc.* **28**:45s.

Michalek, S. M., McGhee, J. R., Mestecky, J., Arnold, R. R., and Bozzo, L., 1976, Ingestion of *Streptococcus mutans* induces secretory IgA and caries immunity, *Science* **192**:1238.

Michalek, S. M., Kiyono, H., Babb, J. L., and McGhee, J. R., 1980a, Inheritance of LPS nonresponsiveness and elevated splenic IgA immune responses in mice orally immunized with heterologous erythrocytes, *J. Immunol.* **125**:2220.

Michalek, S. M., Moore, R. N., McGhee, J. R., Rosenstreich, D. L., and Mergenhagen, S. E., 1980b, The primary role of lymphoreticular cells in the mediation of host responses to bacterial endotoxin, *J. Infect. Dis.* **141**:55.

Mizuguchi, J., Kakiuchi, T., Nariuchi, H., and Matuhasi, T., 1980, Effect of lipopolysaccharide on the thymocyte response to PHA: Strain difference, *Immunology* **41**:393.

Möller, G., Andersson, J., and Sjöberg, O., 1972a, Lipopolysaccharides can convert heterologous red cells into thymus-independent antigens, *Cell. Immunol.* **4**:416.

Möller, G., Sjöberg, O., and Andersson, J., 1972b, Mitogen induced lymphocyte-mediated cytotoxicity *in vitro*: Effect of mitogens selectively activating T or B cells, *Eur. J. Immunol.* **2**:586.

Morgan, E. L., Walker, S. M., Thomas, M. L., and Weigle, W. O., 1980, Regulation of the immune response. I. The potentiation of *in vivo* and *in vitro* immune responses by Fc fragments, *J. Exp. Med.* **152**:113.

Morrison, D. C., and Rudbach, J. A., 1981, Endotoxin–cell-membrane interactions leading to transmembrane signalling, *Contemp. Top. Mol. Immunol.* **8**:187.

Morrison, D. C., and Ryan, J. L., 1979, Bacterial endotoxins and host immune responses, *Adv. Immunol.* **28**:293.

Nishizawa, Y., Kishimoto, T., Kikutani, H., and Yamamura, Y., 1977, Induction and properties of cytoplasmic factor(s) which enhance nuclear nonhistone protein phosphorylation in lymphocytes stimulated by anti Ig, *J. Exp. Med.* **146**:653.

O'Malley, B. W., and Means, A. R., 1976, The mechanism of steroid hormone regulation of transcription of specific eukaryotic genes, *Prog. Nucleic Acid Res. Mol. Biol.* **19**:403.

Oppenheim, J. J., Northoff, H., Greenhill, A., Mathieson, B. J., Smith, K., and Gillis, S., 1980, Properties of human monocyte derived lymphocyte activating factor (LAF) and lymphocyte-derived mitogenic factor (LMF), in: *Biochemical Characterization of Lymphokines* (A. L. deWeck, F. Kristensen, and M. Landy, eds.), pp. 399–404, Academic Press, New York.

Ozato, K., Adler, W. H., and Ebert, J. D., 1975, Synergism of bacterial lipopolysaccharides and concanavalin A in the activation of thymic lymphocytes, *Cell. Immunol.* **17**: 532.

Peavy, D. L., Adler, W. H., Shands, J. W., and Smith, R. T., 1974, Selective effects of mitogens on subpopulations of mouse lymphoid cells, *Cell. Immunol.* **11**:86.

Rosenstreich, D. L., and Vogel, S. N., 1980, Central role of macrophages in host responses to endotoxin, in: *Microbiology 1980* (D. Schlessinger, ed.), pp. 11–15, American Society for Microbiology, Washington, D.C.

Rosenstreich, D. L., Vogel, S. N., Jacques, A. R., Wahl, L. M., and Oppenheim, J. J., 1978, Macrophage sensitivity to endotoxin: Genetic control by a single codominant gene, *J. Immunol.* **121**:1664.

Roux, M. E., McWilliams, M., Phillips-Quagliata, J. M., Weiss-Carrington, P., and Lamm, M. E., 1977, Origin of IgA-secreting plasma cells in the mammary gland, *J. Exp. Med.* **146**:1311.

Smith, R. T., 1972, Specific recognition reactions at the cellular level in mouse lymphoreticular cell subpopulations, *Transplant. Rev.* **11**:178.

Stein, G., Hunter, G., and Lavie, L., 1974, Non-histone chromosomal proteins: Evidence for their role in mediating the binding of histones to deoxyribonucleic acid during the cell cycle, *Biochem. J.* **139**:71.

Stott, D. I., and Williamson, A. R., 1978b, Non-histone chromatin proteins of B lymphocytes stimulated by lipopolysaccharide. I. Synthesis, *Biochim. Biophys. Acta* **521**:726.

Stott, D. I., and Williamson, A. R., 1978a, Non-histone chromatin proteins of B lymphocytes stimulated by lipopolysaccharide. II. Phosphorylation, *Biochim. Biophys. Acta* **521**: 739.

Sultzer, B. M., 1968, Genetic control of leucocyte responses to endotoxin, *Nature (London)* **219**:1253.

Suzuki, T., Taki, T., Hashimine, K., and Sadasevan, R., 1981, Biochemical properties of biologically active Fc receptors of human B lymphocytes, *Molec. Immunol.* **18**:55.

Suzuki, T., Sadasevan, R., Saito-Taki, T., Stechschulte, D. J., Ballentine, L., and Helmkamp, G. M., 1980, Studies of Fc receptors of human B lymphocytes: Phospholipase A$_2$ activity of Fc receptor, *Biochemistry* **19**:6037.

Tomasi, T. B., Jr., 1980, Oral tolerance, *Transplantation* **29**:353.

Vogel, S., and Mergenhagen, S. E., 1982, The cellular basis of endotoxin susceptibility, in: Host-Parasite Interactions in Periodontal Diseases (R. J. Genco and S. E. Mergenhagen, eds.), pp. 160–168, American Society for Microbiology, Washington, D.C.

Vogel, S. N., Marshall, S. T., and Rosenstreich, D. L., 1979, Analysis of the effects of lipopolysaccharide on macrophages' differential phagocytic responses of C3H/HeN and C3H/HeJ macrophages *in vitro*, *Infect. Immun.* **25**:328.

Vogel, S. N., Weinblatt, A. C., and Rosenstreich, D. L., 1981, Inherent macrophage defects in mice, in: *Immunologic Defects of Laboratory Animals* (C. M. E. Gershwin and B. Merchant, eds.), pp. 327–357, Plenum Press, New York.

Watanabe, T., and Ohara, J., 1981, Functional nuclei of LPS-nonresponder C3H/HeJ mice after transfer into LPS-responder C3H/HeN cells by cell fusion, *Nature (London)* **290**:58.

Watson, J., 1977, Differentiation of B lymphocytes in C3H/HeJ mice: The induction of Ia antigens by lipopolysaccharide, *J. Immunol.* **118**:1103.

Watson, J., Riblet, R., and Taylor, B. A., 1977, The response of recombinant inbred strains of mice to bacterial lipopolysaccharides, *J. Immunol.* **118**:2088.

Watson, J., Largen, M., and McAdam, K. P. W. J., 1978a, Genetic control of endotoxin responses in mice, *J. Exp. Med.* **147**:39.

Watson, J., Kelly, K., Largen, M., and Taylor, B. A., 1978b, The genetic mapping of a defective LPS response gene in C3H/HeJ mice, *J. Immunol.* **120**:422.

Watson, J., Kelly, K., and Whitlock, C., 1980, Genetic control of endotoxin sensitivity, in: *Microbiology 1980* (D. Schlessinger, ed.), pp. 4–10, American Society for Microbiology, Washington, D.C.

Winchurch, R. A., Hilberg, C., Birmingham, W., and Munster, A. M., 1981, Bacterial lipopolysaccharides activate immune suppressor cells in newborn mice, *Cell. Immunol.* **58**:458.

Wolf, R. L., and Merler, E., 1979a, Role of lipids in the immune response. I. Localization to a lipid-containing fraction of the active moiety of an inhibitor (SIF) of lymphocyte proliferation, *J. Immunol.* **123**:1167.

Wolf, R. L., and Merler, E., 1979b, Role of lipids in the immune response. II. An enhancer of antibody production (APS) that is a lipid closely associated with lymphocyte mitogenic factor (LMF), *J. Immunol.* **123**:1175.

18

The Role of Humoral Factors in Endotoxin-Induced Reactions

L. Joe Berry and Jeanne E. Gaska

I. Introduction

Within the last half-decade it has become abundantly clear that many of the biological effects of bacterial endotoxin are mediated by what is believed to be a number of different protein and glycoprotein molecules (Berry, 1977; Moore and Berry, 1982). These molecules are derived from cells of the reticuloendothelial system (RES), especially macrophages but from other cells as well. It is intriguing to realize that a diffuse tissue such as the RES qualifies functionally as part of the endocrine system and the mediators as hormones. There is no anatomical or physiological reason why this should not be true since chromaffin tissue, a source of epinephrine and norephinephrine, is also diffusely distributed.

The question of whether mediators are preformed by the cells of origin to be released by a triggering mechanism or are synthesized in response to endotoxin (or other inducers) has not been fully resolved. The best evidence supports the latter concept since peak values appear in blood about 2 hr after an injection of endotoxin (Moore *et al.*, 1978c; Goodrum and Berry, 1978; Carswell *et al.*, 1975) and the appearance of some but not all can be blocked by inhibitors of protein synthesis (Wahl *et al.*, 1974; Kaiser and Wood, 1962; Bodel, 1970). Synthesis after the stimulus of endotoxin is also supported by the observation that RES activators such as BCG, zymosan, or *Corynebacterium*

L. Joe Berry and Jeanne E. Gaska • Department of Microbiology, University of Texas, Austin, Texas 78712.

parvum augment the amount of mediator released, possibly because of the RES proliferation these substances cause, but not the time when peak amounts are present (Goodrum and Berry, 1979). It should also be recognized that endotoxin is not the only substance capable of inducing the production of mediators since poly rI: poly rC (Cox and Rafter, 1971; Moore and Berry, 1976; Siegert *et al.*, 1976), sindbis virus and other viral infections (Gresser *et al.*, 1975; Moore *et al.*, 1978a; Siegert *et al.*, 1976), infection with gram-negative or, for some mediators, gram-positive organisms (Bodel and Miller, 1976; Kampschmidt *et al.*, 1973) are also active. This group of elicitors does not imitate the full breadth of endotoxin action nor does each mediator necessarily have the capability of the others but the inducers have certain effects in common; i.e., they cause fever, presumably as a result of endogenous pyrogen production. The demonstrable presence of one mediator may not mean that others are present as well.

Inasmuch as this volume focuses on the beneficial effects of endotoxin, it is relevant to point out that the distinction between "good" and "bad" is not always apparent and may have to be based on the frame of reference being used. The oft-asked question whether fever is or is not desirable is a case in point. Some pathogens grow poorly at 40°C (the gonococcus, for example) and body defenses such as phagocytosis may be enhanced up to this temperature (Harmon *et al.*, 1946). This would be a fortunate combination of events for the host but other pathogens such as *Campylobacter fetus* ssp. *jejuni* grow best at 42°C (Skirrow and Benjamin, 1980), a temperature that is inimical for man but one that is also intolerable for some bacteria. In *Campylobacter* enteritis in man, fever is the rule (Butzler and Skirrow, 1979) and provides no apparent advantage to the victim.

In the sections that follow, specific mediators will be described, especially those that have demonstrable benefit to the host. The effect of each mediator will be summarized, its chemical nature, to the extent known, will be presented and its cellular origin given.

II. Glucocorticoid-Antagonizing Factor

This mediator provides equivocal benefits to the host in some situations. It is a protein of macrophage origin (Moore *et al.*, 1978c; Goodrum and Berry, 1979) and has effects that have not yet been completely elucidated. It has been partially purified and the serum-borne form of the factor has an apparent molecular weight estimated to be about 150,000 (Moore *et al.*, 1978c; Goodrum and Berry, 1979). It is produced in barely detectable amounts by tolerant mice (Goodrum and

Berry, 1978) and by the nonresponder C3H/HeJ mice (Moore *et al.*, 1978c). It blocks at least some host responses to hydrocortisone or related hormones (Berry, 1977; Berry *et al.*, 1980). It has been shown experimentally that under manipulated experimental conditions, GAF functionally adrenalectomizes an animal inasmuch as the animal withstands stress inadequately and appears to have the handicaps that are associated with loss of adrenal cortical function (Moore *et al.*, 1978b; Couch *et al.*, 1979). On the other hand, hormones from the adrenal cortex are known to have immunosuppressive effects and were these counteracted by GAF, the animal would be benefited. A balance between GAF and glucocorticoid may well exist in the day to day life of healthy animals since both substances are present in measurable amounts under quiet resting conditions (Goodrum and Berry, 1979). When either is present in excess, a condition less than optimum may ensue. The checks and balances that characterize the endocrine system are exemplified by this relationship between GAF and the glucocorticoids.

Glucocorticoid hormones have a variety of effects (Martin, 1976). In the liver, the synthesis of several enzymes, such as tryptophan oxygenase, phosphoenolpyruvate carboxykinase, and tyrosine aminotransferase, is increased in response to hormone action because it causes a greater enzyme mRNA formation (Wicks *et al.*, 1974; Tilghman *et al.*, 1976). Some of the induced enzymes promote the conversion of amino acids into carbohydrates (gluconeogenesis and glyconeogenesis), a process essential for the animal's well-being during periods of fasting, such as overnight (Tilghman *et al.*, 1976). The mechanism whereby proteins are induced by steroid hormones is under active investigation (Koblinsky *et al.*, 1972; Rousseau *et al.*, 1973; Yamamoto and Alberts, 1976; Higgins and Gehring, 1978) and rapidly progress is being made. The general scheme of events has been elucidated. (1) After diffusion through the cytoplasmic membrane, the hormone complexes with a specific cytoplasmic receptor protein (Fig. 1, step 1), which undergoes a conformational change allowing the complex to enter the nucleus and bind to sites on the chromatin (Fig. 1, step 2), probably on DNA. (2) Binding results in stimulation of transcription of the appropriate genes by an unknown mechanism. Endotoxin blocks this inductive effect of glucocorticoids on all enzymes studied except tyrosine aminotransferase. The site where endotoxin interrupts the sequence of events that results in enzyme induction has not been identified. GAF or endotoxin, indirectly, does not alter the entry or binding of the hormone (Berry *et al.*, 1980) into the cytoplasm of hepatocytes nor does it appear to diminish the binding of the complex to DNA (unpublished observations). GAF then may inhibit induction at the level of transcription, transla-

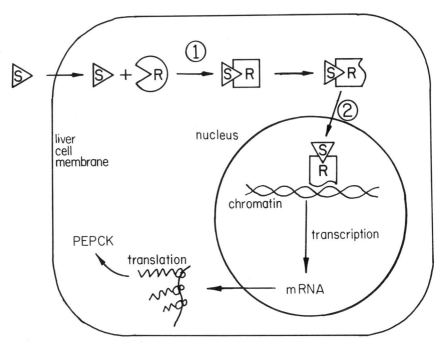

Figure 1. Mechanism of glucocorticoid hormone induction of enzymes in mammalian cells.

tion, or steps in between although transcription is the most attractive possibility since steroid induction is believed to act at this level only.

Macrophages have been identified as the source of GAF in a variety of ways. Antimacrophage serum raised in rabbits against mouse cells greatly diminishes those effects of endotoxin known to be mediated by GAF (Moore *et al.*, 1976). As an example, the ability of endotoxin to block the glucocorticoid induction of the hepatic enzyme phosphoenolpyruvate carboxykinase (PEPCK) is lost in mice pretreated with the antimacrophage serum. Endotoxin does not block hormonal induction of PEPCK in C3H/HeJ mice but if a mixture of endotoxin and macrophages derived from conventional animals is injected intraperitoneally into the C3H/HeJ mice, the induction of the enzyme is blocked (Moore *et al.*, 1978c). Neither macrophages nor endotoxin alone, when given intraperitoneally, have this effect. Finally, the supernatant fluid from an overnight culture of macrophages to which endotoxin was added will inhibit PEPCK induction by hydrocortisone in Reuber H35 hepatoma cells in culture. Endotoxin added directly to the hepatoma cells is inactive (Goodrum and Berry, 1979).

A strong argument can be made for the harmful effect of GAF since it is the cause of or contributes to the depletion of body reserves of carbohydrates (Couch *et al.*, 1979; Moore *et al.*, 1980a). This occurs when GAF is present in excess as it has been found to be following the administration of endotoxin (Goodrum and Berry, 1978). However, elevated amounts of GAF in febrile patients would have a protein-sparing effect that might help to counterbalance the harm that accrues from low levels of carbohydrates. The small amount of circulating GAF that can be measured (Goodrum and Berry, 1979) under normal conditions may well serve to prevent excessive protein catabolism for gluconeogenic purposes and excessive enzyme induction as well. The stimulus (stimuli) for the production of control amounts of GAF is (are) unknown. If there is a steady leakage of enteric microorganisms into the bloodstream of normal subjects, as many investigators believe (Berry and Smythe, 1959), then the entrapment of these bacteria by macrophages would serve to stimulate GAF formation. In fact, no stimulus may be necessary for GAF production under "resting" conditions and hence gram-negative organisms might not be necessary.

The thymus gland is another glucocorticoid target (Martin, 1976). Marked involution of the thymus occurs within 24 hr of glucocorticoid administration and by 2 days 95% of the gland's weight is lost. Exposure to stress, such as heat, cold, infection, trauma, hemorrhage, etc., has a similar effect on the thymus because these stimuli cause endogenous release of glucocorticoid hormones. When the stress is discontinued, the thymus rapidly returns to normal size. Steroid hormones inhibit glucose uptake by thymocytes and block RNA and protein synthesis. These changes are supposedly responsible for the involution. Whether GAF protects the thymus against elevated levels of glucocorticoid has not been determined nor is there a ready prediction as to the consequences of such an effect. Since recovery from thymus involution is prompt once steroid hormones return to normal levels in blood, a presumed survival value for thymus depletion during stress must be considered possible.

A large dose of glucocorticoid (1) promotes degradation of fats and release of fatty acids into the blood plasma, (2) causes loss of mass and weakening of skeletal muscles, (3) retards wound healing, thins the skin, and promotes bone demineralization, and (4) helps to maintain circulating blood volume by influencing transfer of water from cells to extracellular fluids (Martin, 1976). Both the fatty acids and the amino acids derived from muscle protein are used in gluconeogenesis. The underlying basis for these effects is not well understood nor have we acquired evidence to indicate whether endotoxin via GAF has an inhibitory action on any of these phenomena. Endotoxin was shown years ago, however, to cause no reduction in the amount of nitrogenous compounds excreted in urine of mice injected with glucocorticoid (Woodruff *et al.*, 1973) nor did it appear to alter most of the lipids in blood of appropriately treated animals (Sakaguchi and Sakaguchi, 1980). To this extent GAF does not appear to antagonize all of the metabolic alterations elicited by glucocorticoids.

Large doses of glucocorticoids alleviate the symptoms associated with inflammation. This is the treatment of choice for rheumatoid arthritis, delayed hypersensitivity reactions of the skin, such as poison ivy and similar conditions, and undesirable vascular reactions in infectious conditions of the eye (Martin, 1976). The capillary dilation that characterizes an inflammatory response is diminished by steroids and these compounds are believed to stabilize the lysosomal membranes in cells that accumulate in the inflamed area. Endotoxin was reported some years ago to labilize lysosomes (Janoff *et al.*, 1962). This observation provides a possible basis for antagonism at this level. More insight is needed.

The importance of adrenocortical function to survival of stressed animals has been recognized for decades. It does not seem to matter what type of stress is involved since heat, cold, hemorrhage, trauma, hypoxia, extreme physical exertion, and toxines are all poorly tolerated in adrenalectomized animals. The underlying protective role of the glucocorticoids is not well understood under these conditions but there is evidence to show that GAF (or endotoxin) neutralizes this role at least in cold-exposed mice (Couch *et al.*, 1979). It is difficult to hypothesize how or why this might be of value to the host.

Two additional mediators that may be related to GAF have been identified. Filkins (1980) has studied a macrophage factor that has what he calls insulin-like activity. He ascribes to it the changes in carbohydrate and fat metabolism that endotoxin is known to elicit. There are similarities between host responses to endotoxin and insulin, but sufficient differences between the action of the two have been identified to make it apparent that GAF is not insulin (Goodrum and Berry, 1978, 1979). Filkins' factor may well be the same as GAF.

Mishell and his associates have identified a macrophage factor that protects the *in vitro* immune response from the immunosuppressive action of glucocorticoids (Mishell *et al.*, 1977). The substance responsible has been named *glucocorticoid response-modifying factor* (GRMF). The primary target of GRMF appears to be T-helper cells. Another macrophage factor (see below) that has a similar activity, interleukin 1 (Oppenheim *et al.*, 1980), coelutes during gel filtration chromatography with GRMF (Mishell *et al.*, 1980) so they may be the same substance. GRMF is not, however, identical to GAF. Purified material sent to the authors by Mishell did not have GAF activity in rat hepatoma cells (unpublished observations).

III. Interferon

According to contemporary expectations, interferon may prove to be the most useful of all mediators produced in response to endotoxin. It was originally identified as an antiviral agent produced in response to a wide array of viruses (Isaacs and Lindenmann, 1957). Other inducers are known to yield interferon such as antigens, T-cell mitogens, poly rI: poly rC, and a number of modifications in either the poly rI or poly rC homopolymer. A variety of cell types serve as its source such as B and T lymphocytes, fibroblasts, epithelial cells, and macrophages. Three molecular types of interferon are produced, α, β, and γ (Johnson, 1982). The first two are formed in response to viruses and other inducers, whereas γ interferon is formed by T lymphocytes in response to antigens and T-cell mitogens. Endotoxin stimulates interferon production by B lymphocytes and macrophages, hence both α and β interferons. The three types differ in molecular weight. Both α and β interferons (INFα and INFβ, respectively) have been purified essentially to homogeneity. INFα has a molecular weight range between 18,500 and 23,000 while INFβ varies from 20,000 to 30,000 (cited by Johnson, 1982). INFα has not been purified as extensively as the other two but in humans its molecular weight is approximately 50,000 while in mice it is about 40,000. The range of values obtained for INFα and INFβ may be due to the fact that they are associated with small molecules. For example, ACTH is believed to be linked to INFα in the 23,000-dalton form but is dissociated from it in the 18,500 form (Blalock and Smith, 1980).

In addition to the antiviral properties of interferon, it is known to inhibit the hexose monophosphate shunt pathway in mouse spleen cells (Johnson, 1982) possibly by binding to cell membrane sulfhydryl groups, which may in turn inhibit membrane-bound glucose-6-phosphate dehydrogenase activity. The α and β interferons are able to inhibit colony formation by certain cell lines grown *in vitro* (see Johnson, 1982, for references) and they manifest immunoregulatory effects as well. For example, they may well serve as mediators of some forms of suppressor cell activity as indicated by a marked decrease under *in vitro* conditions in the number of plaque-forming spleen cells against the antigenic stimulus of sheep erythrocytes. Because the addition of anti-interferon immune serum to the plaque assay system blocks the suppression, interferon must be responsible for the effect. Moreover, anti-interferon immune serum administered to virus-infected mice results in a significantly more severe illness with a dramatic increase in number of animal deaths (Fauconnier, 1971). The antitumor activity of interferon may depend upon its ability to enhance the specific cytotoxicity of T lymphocytes, augment natural killer cell cytotoxicity, and increase antibody-dependent cell-mediated cytotoxicity.

The antitumor and antiviral properties of interferon serve at present as the primary impetus for interferon research. If an inexpensive source of interferon were to be found, it might prove to be the answer to the control of certain types of cancer. It would also provide a broad-spectrum therapeutic agent for viral diseases. Costs and limited supply have restricted efforts to test its therapeutic efficacy. Moreover, interferon is essentially species specific even though some types show a broader spectrum than others (Gordon and Minks, 1981). Should the bacterial cloning of the human genes encoding for interferon prove successful so that a fully active product is obtained, the therapeutic potential of interferon may be realized. It is too early to know how soon this time will come.

IV. Tumor Necrosis Factor

It was observed nearly four decades ago that an injection of endotoxin into mice bearing certain types of tumors resulted in hemorrhagic necrosis of the tumor (Shear and Turner, 1943). The necrosis started in the center of the tumor and spread outward but left unaltered malignant cells around the periphery. Not all tumors are subject to this effect but those that are become refractory to it after several treatments. The term *tumor necrosis factor* (TNF) was introduced by Carswell *et al.* (1975), who recognized that an active component was present in serum derived from mice that had been infected about 2 weeks previously with BCG and were then bled 2 hr after the i.v. administration of 10 μg of endotoxin. The BCG treatment resulted in proliferation of cells of the RES and hence a greater yield of TNF. This is further established by studies in which *C. parvum* was used to stimulate RES proliferation and a type of tumor immunity that is dependent upon T cells (North and Berendt, 1980). It was only in mice treated with *C. parvum* that endotoxin was able to necrotize syngeneic tumors. Neither *C. parvum* nor endotoxin given alone resulted in tumor regression. Both were required. While the investigators who reported these experiments did not suggest it, it may be that a critical blood level of TNF is reached only in mice primed with *C. parvum* and then injected with endotoxin. The mode of action of TNF is not understood but it is likely that it depends on an impairment in the microcirculation and a resulting anoxia. Hyperthermia inside the tumor may also be involved (Hämmerling *et al.*, 1980). TNF has been purified to near homogeneity and can kill tumor cells *in vitro* (Carswell *et al.*, 1975; Green *et al.*, 1975; Männel *et al.*, 1980). It is a larger molecule than interferon. Its molecular weight is 60,000 according to Männel *et al.* (1980) or 65,000 to Hämmerling *et al.* (1980). Carswell *et al.* (1975) originally estimated its size at about 150,000. This may represent a polymeric form of TNF. It is generally agreed, however, that it is a glycoprotein of macrophage origin (Green *et al.*, 1975; Männel *et al.*, 1980). An antiserum against TNF neutralizes its cytotoxic activity and it is antigenic when inoculated into a heterologous species (Männel *et al.*, 1980). Hämmerling *et al.* (1980) assayed their preparation for a variety of contaminants but it lacked, among other things, interferon and prostaglandin E.

It was recognized years ago that the tumor necrotic activity seen in response to an injection of endotoxin diminishes with successive treatments since the tumors become progressively less affected (Zahl *et al.*, 1943; Shear and Turner, 1943). This may be due to the appearance of resistant tumor cells, as suggested above, to the development of tolerance to endotoxin, or to both. Even though there is still poor understanding of the tolerant state, there is evidence to show that the production of endogenous pyrogen (Greisman and DuBuy, 1975), GAF, and TNF is greatly curtailed in these animals (Carswell *et al.*, 1975; Goodrum and Berry, 1979). This lack of mediator formation has not been explained and probably will not be until there is a better understanding of how endotoxin initiates mediator synthesis.

The biological effects of TNF extend well beyond its antitumor properties. It can replace T-helper cells in an *in vitro* immune response system, induce B-lymphocyte differentiation, induce thymocyte maturation, and induce thymocyte proliferation (Hämmerling *et al.*, 1980). The activities of TNF that involve lymphocyte regulation are similar to those of interleukin 1 (see below) and may, therefore, be dependent upon the presence of this substance as a contaminating mediator. A fraction of appropriate molecular size (15,000 to 20,000 daltons) has been separated from partially purified TNF preparations (Hämmerling *et al.*, 1980).

Since TNF is a macrophage product, it may be a major contributor to the antitumor immunity for which these cells are recognized. This may not be true for all types of tumors since TNF is not uniformly effective. This suggests that more than one mediator is responsible for this important defense mechanism.

V. Protection against X-Irradiation

It has been known for nearly three decades that small amounts of endotoxin administered about 24 hr prior to X-irradiation greatly increase the survival of X-rayed laboratory animals (Mefferd *et al.*, 1953;

Smith *et al.*, 1957). Prior injection is necessary for this effect since no protection is seen when the endotoxin is given at either the same time or after irradiation (Smith *et al.*, 1957). The reason why pretreatment is necessary has never been explained other than to show that the protected animals recover more rapidly than controls from the severe leukopenia that follows X-irradiation (Smith *et al.*, 1958). In mice subjected to about an LD_{80} dose of X-irradiation, the first deaths occur after 10 days to 2 weeks and deaths continue to occur until the end of the 30-day observation period (Addison and Berry, 1981). It is necessary to terminate experiments at this time because secondary causes of death (such as infections of endogenous origin) may occur subsequently (Taliaferro *et al.*, 1964).

Early efforts by Smith and associates to passively transfer irradiation protection with serum were unsuccessful (Smith *et al.*, 1957). This was attempted by bleeding mice a day or two after endotoxin was injected and then administering the serum i.v. to normal recipients to see if their survival was enhanced following X-irradiation. Behling *et al.* (1980) were able to protect mice with serum collected 2 hr after an injection of 50 μg endotoxin; 0.5 ml of the serum was injected i.p. 24 hr before irradiation and 80% of the mice survived, compared to 40% of untreated controls. Addison and Berry (1981) obtained no protection under similar conditions except that they injected only 0.3 ml of the postendotoxin serum and it was given i.v. 24 hr prior to irradiation. The reason for the difference in results cannot be explained except, possibly, by the strain of mice employed or the volume and route of administration of the serum used in the pretreatment.

Recent studies in the authors' laboratory (Moore *et al.*, 1978c; Goodrum and Berry, 1979) help explain why Smith *et al.* (1957) were unable to passively protect their mice. In the first place, for best results the serum must be collected from mice that have been primed with several i.v. injections of zymosan. (Infection with BCG 10 days to 2 weeks earlier or after several injections of *C. parvum* serve as well.) This treatment results in RES proliferation and an augmented yield of endotoxin-elicited mediators. Secondarily, the primed animals must be bled 2 hr after an i.v. injection of a small dose (about 50 μg) of endotoxin. As stated above, this is the interval when a maximum titer of GAF, TNF, and presumably of other mediators as well, is found in the blood.

Serum collected under these conditions can be given i.v. (0.3 ml) to mice made tolerant to endotoxin or to nonresponder C3H/HeJ mice about 24 hr prior to an LD_{80} irradiation and only about 20% of them die (Addison and Berry, 1981) (Table 1). Tolerant mice or C3H/HeJ mice are useful in these experiments since neither can be protected against X-irradiation by endotoxin. It is known that these mice fail to

Table 1. Percent Survival[a] of Tolerant CF-1 Mice and of C3H/HeJ
Mice following 650 Rads of X-Irradiation[b]

Treatment 24 hr prior to irradiation	Tolerant mice	C3H/HeJ mice
None (controls)	22%	36%
0.3 ml ES[c]	26%	34%
0.3 ml ZES[d]	80%	82%
10 μg endotoxin	22%	36%

[a] Survival 30 days after irradiation.
[b] Modified from Addison and Berry (1981) with permission.
[c] ES is serum collected from CF-1 mice 2 hr after 50 μg endotoxin was given i.v.
[d] ZES is the same as ES except that the mice were pretreated with three injections of zymosan.

produce GAF and colony-stimulating factor (CSF) after endotoxin (Goodrum and Berry, 1979; Urbaschek et al., 1977). Presumably that is why they cannot be protected against irradiation. They cannot form the necessary mediator(s) apparently because their macrophages are refractory. Increased survivorship in these animals cannot be attributed, therefore, to residual endotoxin that might remain in the serum.

The serum factor responsible for the protection has not been identified but partial fractionation of the serum has shown that it is present in the same fraction as GAF (Addison and Berry, 1981). This fraction does not appear to have colony-stimulating activity, whereas the whole serum does. Nevertheless, the white cell count and bone marrow count recover in the serum-protected animals (unpublished observations).

The nature of the passive protection is obscure. GAF peaks in serum after an endotoxin injection at 2 hr and it is back to control levels after 4 hr. It is assumed that the passively administered protective factor (or factors) has a short half-life in the circulation. It is, therefore, remarkable that increased survival of the pretreated irradiated mice extends over a period of 30 days. It is assumed that the factor binds to target cells and sets in motion the chain of events that is responsible for the larger number of survivors. The delay between the initial events and the final outcome is not understood and must await further study. It should also be noted, however, that pretreatment with endotoxin permits accelerated recovery from the immunosuppression that follows X-irradiation (Behling and Nowotny, 1978). More recently, Behling et al. (1980) have found that the polysaccharide and glycolipid moieties of the lipopolysaccharide molecule protect against the immunosuppression caused by X-irradiation. This could play a role in helping irradiated animals survive in larger numbers after pretreatment with endotoxin or one of its components.

Of the clearly established beneficial effects of endotoxin and its mediators, none may be of greater value than the ability to protect against ionizing irradiation. With the certain need for future expansion in nuclear power generation to meet the world's energy requirements and because of the potential hazard associated with such plants, radiation protection assumes a critical significance today. Moreover, because of the proliferation of nuclear weapons by the superpowers and by an ever-expanding number of developing nations, the threat of an atomic holocaust becomes all the more real. The need for a better understanding of radiation protection and ways of achieving it are, therefore, all the more critical. Observations of the type just described open the way for a variety of new approaches to the development of radioprotective measures.

VI. Colony-Stimulating Factor

Numerous publications during the last decade have been devoted to studies of a factor that stimulates the *in vitro* proliferation of myeloid precursor cells from bone marrow. Following an injection of endotoxin in mice, there occurs an initial granulocytopenia followed by a gradual increase in the number of granulocytes with a concomitant increase in serum levels of CSF. Production of this factor is also stimulated by X-irradiation (Onoda *et al.*, 1980), various antigens, and polynucleotides (Metcalf, 1972), and release of CSF from monocytes is stimulated by dextran sulfate (Gronowicz *et al.*, 1976).

CSF is represented by a family of proteins that have as a common function promotion of the proliferation and differentiation of granulocyte-macrophage progenitor cells but that are biochemically heterogeneous depending on the tissue and species of origin (Metcalf, 1972). When murine hemopoietic cells are grown in agar culture, CSF appears to be necessary for both the initiation and the continuation of growth and differentiation (Stanley *et al.*, 1975). CSF has been isolated from a number of species including mouse and man but most notably in human urine, murine L cells, serum, and lungs.

Several CSFs from different sources have been at least partially purified and characterized. Since most CSFs adhere to ConA and are positive for PAS staining, it appears that most CSFs are glycoproteins. As stated previously, CSF is in fact a heterogeneous class of proteins, and more than one type can often be separated from a single source. These may differ in molecular weight, the nature of the glycosidic group, and protein structure. Mouse lung conditioned medium (Burgess *et al.*, 1977) yielded a CSF of 29,000 molecular weight, whereas highly purified CSF from human urine (Stanley *et al.*, 1975) has been shown to be a glycoprotein of about 50,000 molecular weight. CSF derived from L-cell conditioned medium (Waheed and Shadduck, 1979) and pancreatic carcinoma cells (Wu *et al.*, 1979) have been purified to apparent homogeneity. The L-cell protein has a molecular weight of about 70,000 and consists of two subunits of equal size. The CSF from pancreatic carcinoma cells has a molecular weight of 50,000. Antibody against these purified proteins (Shadduck *et al.*, 1980; Yunis *et al.*, 1980) has been prepared and should be of great value both in elucidating the function of CSF and in procedures for its purification.

Roles other than the regulation of granulopoiesis have been suggested for CSF. Increased RNA synthesis in several myeloid cell types has been observed (Burgess and Metcalf, 1977). Moore *et al.*, (1980a) have suggested that CSF can stimulate macrophages to produce lymphocyte-activating factor (interleukin 1) as well as enhance the production of interferon in the presence of endotoxin (Moore *et al.*, 1980b). Other examples of macrophage activation by CSF include the killing of *Leishmania tropica* (Handman and Burgess, 1979) in infected mouse peritoneal macrophages and the induction of prostaglandin E synthesis in both normal and neoplastic macrophages (Kurland *et al.*, 1979). CSF can thus be seen to play a multifunctional role in the normal activities of certain cells and in serving as a regulator or cofactor in the mobilization of host defenses.

VII. Endorphins and Enkephalins

The role of these endogenously produced morphine-like compounds in an animal's response to endotoxin has not been fully established but recent observations provide indirect evidence that they are involved. The evidence is based on the ability of naloxone, an antagonist of endorphins, to neutralize some of the biological effects of endotoxin (Holaday and Faden, 1978; Faden and Holaday, 1980a,b; Reynolds *et al.*, 1980). An example of this, which is of considerable fundamental importance in attempting to understand the underlying mechanism of killing by endotoxin, is the ability of naloxone to correct the hypotension and poor vascular perfusion in appropriately treated rats and dogs without necessarily increasing their survival. Most endotoxinologists, including the authors, had believed previously that circulatory collapse was the ultimate physiologic change that was responsible for death in endotoxin poisoning. In light of the new findings, this no longer appears to be a tenable concept. Naloxone at a dose level somewhat larger than the one used in the study just described can protect endotoxemic rats against lethality but without any indication of further improvement in their circulation. In some experiments, naloxone prolongs life but does not affect survival (Raymond *et al.*, 1981). It appears, therefore, that the cardiovascular changes produced by endotoxin are linked to the increased production of endorphins and yet by reversing their effects on the circulatory system, greater survival does not inevitably ensue. Damage to other unidentified essential functions must, under these circumstances, contribute to death of the poisoned host.

Even though it may be unrelated directly to endotoxin, it is nevertheless of interest to note that naloxone not only is capable of reversing endotoxin shock, but can also correct hemorrhagic shock (Vargish *et al.*, 1980), surgical shock (Finck, 1977), and septic shock (Dirksen *et al.*, 1980; Peters *et al.*, 1981) as well. This suggests a linkage between endorphins and all three forms of shock, the endorphins providing the common denominator that has been suspected for years as underlying all types of shock.

If, however, one assumes that the action of naloxone depends on its ability to neutralize or displace endorphins from receptor sites on cells, then an inconsistency emerges when the circulatory data (referenced above) are compared to studies of its effect on food and water intake. Endotoxin in minute amounts has been known for years (Dubos and Schaedler, 1961) to inhibit food and water intake by laboratory animals, especially mice. This is due to a paralysis of gastric emptying (Turner and Berry, 1963) that lasts for at least 24 hr even after doses of a few micrograms. The intestinal tract under the same conditions becomes hypermotile, i.e., the animals have diarrhea. There is no published evidence that has come to the attention of the authors to suggest that endorphins play a role in these phenomena but there are several recent papers that show that naloxone impairs food and water intake in cats and rats (Brown and Holtzman, 1979, 1981; Foster *et al.*, 1981). Rather than reversing an effect of endotoxin, naloxone, in this situation, results in a similar response. It has also been shown that naloxone reverses the ability of the stress of an elevated intraluminal pressure to suppress peristalsis in the isolated ileum of guinea pigs (Kadlec and Horáček, 1980). It should be mentioned, however, that the guinea pig does not respond normally to endotoxin (Berry, 1964) and may not be the preferred animal for studies of this type.

A possible explanation of how endorphins act in initiating shock due to the various types of stress, mentioned above, has been offered by Dirksen *et al.* (1980). Figure 2 summarizes their hypothesis. Stress

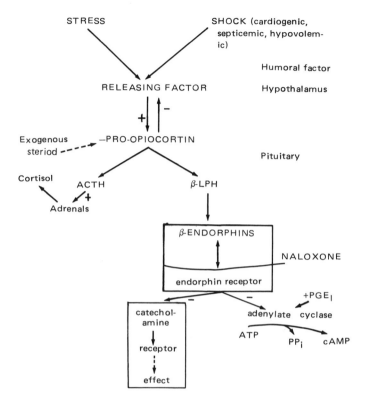

Figure 2. The neurophysiological mechanisms of endorphins in shock, as suggested by Dirksen *et al.* (1981). −, inhibition; +, stimulation. (Reproduced with permission.)

causes a hypothalamic releasing factor to be released and to stimulate the secretion of both β-endorphin and ACTH. A prohormone, pro-opiocortin, has a molecular weight of 31,000 and is a precursor of both ACTH and β-endorphin. β-Endorphin is released from the anterior lobe of the pituitary. β-Lipotropin is also released and is a precursor of smaller endorphins. Corticosteroid treatment serves as a negative feedback for the release of both ACTH and β-endorphins and hence serves as an effective treatment for shock. Naloxone, as indicated in Fig. 2, blocks the receptor site of β-endorphin.

VIII. Endogenous Pyrogen

There have been numerous studies of fever and the internally derived pyrogenic substance(s) responsible. An excellent review was published by Atkins (1960), which summarizes much of the early work, and an updated review by Atkins and Bodel was published in 1972. More recently, a history of pyrogen research appeared (Westphal *et al.*, 1977).

Endogenous pyrogen (EP) was originally obtained by Beeson (1948) as a saline extract of rabbit peritoneal exudate cells, primarily polymorphonuclear (PMN) leukocytes. Macrophages also produce an EP (Atkins *et al.*, 1967; Hahn *et al.*, 1967) that is apparently the same as that derived from PMN leukocytes (Dinarello *et al.*, 1977).

EP from rabbit cells was first purified from PMN leukocytes by Rafter *et al.*, (1966) and was estimated to have a molecular weight of 13,000. Using this as the size of the molecule, 30 ng was found to be sufficient to cause an easily detectable fever in homologous animals. Subsequently, other investigators purified EP but did not assign a molecular weight to their preparation (Murphy *et al.*, 1974). More recently, a size of 15,000 has been suggested (Dinarello *et al.*, 1977), the latter from human leukocytes. Antibody against this preparation was raised in rabbits and used to develop a radioimmunoassay for EP.

The question of how EP causes a rise in body temperature has not been fully resolved despite recent studies that have focused on this subject. It is generally believed that the pyrogen and not endotoxin acts on the thermoregulatory center in the hypothalamus. The preoptic area of the hypothalamus has been identified as the principal site of action (Cooper *et al.*, 1967) even though the periaqueductal gray matter may serve as an additional site (Rosendorf and Mooney, 1971). Apparently, the temperature of the body at which regulation occurs is elevated by the pyrogen and fever then results. Whether the pyrogen acts directly on the neurons or causes the formation or release of additional humoral factors is now under investigation. Prostaglandin E (PGE) has received a great deal of attention as the most likely candidate (Milton and Wendlandt, 1971; Feldberg and Cupta, 1973; Skarnes and McCracken, 1980) that forms in response to EP. However, some serious difficulties have arisen when attempts were made to establish PGE as the mediator (Cranston, 1980). A rise in PGE in the cerebrospinal fluid of EP-treated rabbits does not occur if sodium salicylate is given prior to treatment even though a normal fever develops (Cranston *et al.*, 1975). Moreover, two different antagonists of PGE, designated as SC 19220 and HR 546, were each able, when injected into the cerebral ventricles of rabbits, to block the febrile response to intraventricularly administered PGE but not to EP (Cranston *et al.*, 1976). If PGE is not the mediator of EP, it may be some other metabolite of arachidonic acid, the prostaglandin precursor (Cranston, 1980). It may also be that EP has a direct action on the thermoregulatory center.

IX. Leukocytic Endogenous Mediator

Kampschmidt and associates assigned the term *leukocytic endogenous mediator* (LEM) to a factor obtained from rat peritoneal exudate cells, blood leukocytes, or Kupffer cells appropriately stimulated with shellfish glycogen, endotoxin, or heat-killed *Staphylococcus aureus* (Kampschmidt *et al.*, 1973). LEM is preferred over EP by these investigators in recognition of the broad spectrum of biological activities it induces. In addition to causing fever, LEM also alters the concentration of iron, copper, and zinc in serum (Kampschmidt, 1979a); increases acute-phase proteins in blood (Eddington *et al.*, 1972); results in an elevation of fibrinogen (Bornstein and Walsh, 1978), haptoglobin (Bornstein and Walsh, 1978), α_1-macrofetoprotein (Eddington *et al.*, 1972), α_2-macrofetoprotein (Eddington *et al.*, 1972), ceruloplasmin (Bornstein and Walsh, 1978), hemopexin (Kampschmidt, 1979b), and C-reactive protein (Merriman *et al.*, 1975); raises the circulating level of neutrophils within about 1 hr (Kampschmidt *et al.*, 1972); and, after an interval of 24 hr, increases the resistance of rats to infection with *Salmonella typhimurium* (Kampschmidt and Pulliam, 1975). The last effect is likely attributable to the low level of circulating iron that follows LEM administration and thereby results in an inhibition of bacterial multiplication.

LEM has been purified and estimated to have a molecular weight of 13,000 to 16,000 (Merriman *et al.*, 1977). This agrees well with the weight of Ep (Dinarello *et al.*, 1977). However, the isoelectric point of LEM is 7.5 while that of EP is 5.0 (Dinarello *et al.*, 1977). Little is known about the mode of action of LEM or how it causes such diverse effects. If antibody to LEM were available along with antiserum to EP, it should be possible to determine whether the two mediators are the same.

VIII. The Interleukins

In 1972, Gery and Waksman reported that a factor present in the supernatant fluid of endotoxin-treated macrophages enhanced the proliferation of murine thymocytes. This substance was named *lymphocyte-activating factor* (LAF) but, more recently, has been designed *interleukin 1* (IL 1). In addition to its mitogenic effect on T cells, it has been shown to augment the induction of T-cell-dependent antibody responses by mouse spleen cells *in vitro* (Koopman *et al.*, 1977) and to replace the need for macrophages during the initiation of cytotoxic T-cell responses *in vitro* (Farrar *et al.*, 1980). IL 1 is a protein with a molecular weight of about 15,000 and a charge heterogeneity during gel filtration or electrophoresis (Koopman *et al.*, 1977). Thus, the adjuvanticity of endotoxin may depend on its ability to cause macrophages to produce IL 1 which activates helper T cells.

Production of IL 1 has been reported to promote thymocyte proliferation in response to antigenic or lectin stimulation (Chen and DiSabato, 1976). The responsible factor was originally called *T-cell mitogen factor* (TMF) but is now identified as *interleukin 2* (IL 2) (Mizel and Farrar, 1979). Farrar *et al.* (1980) demonstrated that IL 2 is produced only when macrophages are present in the same culture with splenic T cells. IL 2 can be distinguished from IL 1 since the former is a large molecule, 30,000 to 35,000 molecular weight, and maintains viability and proliferation of cytotoxic T cells in long-term culture (Watson *et al.*, 1979). Apparently immature thymic cells respond to this factor.

IX. Prostaglandins

As mentioned in the section on endogenous pyrogen, PGE has been implicated as a possible secondary mediator in initiating the events in the central nervous system that lead to fever. There is evidence for and against this concept, but in other types of response to endotoxin, prostaglandins play a putative role as mediators of endotoxemic effects.

It has been known for years that endotoxin acts as a powerful abortifacient in laboratory mice (Parant and Chedid, 1964). Skarnes and Harper (1972) found that indomethacin, which blocks the enzyme (cyclooxygenase) that converts arachidonic acid to a precursor of PGE (and other prostaglandins), prevents abortion due to endotoxin. An injection of PGF_2 into the uterus was followed promptly by abortion. This prostaglandin was shown to rise sharply in the uterus within 1 hr after i.v. administration of endotoxin. These observations serve as strong evidence that prostaglandins are involved in endotoxin-induced abortion in mice.

The initial pulmonary hypertension and systemic hypotension that occurs in experimental animals only minutes after an injection of endotoxin is suppressed by inhibitors of prostaglandin synthesis (Flynn, 1978). Pretreatment with indomethacin is required for this effect as well as for prolonging the onset of the progressively severe cardiovascular shock of endotoxemia. Somewhat similar findings were obtained in the author's laboratory. Indomethacin was found to block or diminish the *in vivo* production of GAF and its *in vivo* production by macrophages that had been stimulated with endotoxin and injected via the i.p. route into tolerant mice (Goodrum *et al.*, 1978). In the same report, indomethacin was shown to prolong the survival time of lead acetate-sensitized mice poisoned with endotoxin but not that of poisoned conventional mice. Lead acetate not only lowers the LD_{50} of endotoxin to a small fraction of its original value, but also reduces survival time to less than one-half that of control mice. Thus, the inhibition of prostaglandin synthesis by the drug appeared to protect mice over a period of 8 to 10 hr but not for a longer time. Neither PGE nor arachidonic acid duplicated the effects we observed after administering endotoxin. This may have been due to a concentration effect or to the *in vivo* distribution of the mediators.

That macrophages are capable of synthesizing prostaglandins *in vitro* was shown by Rietschel *et al.* (1980). Prolonged incubation in the presence of endotoxin or lipid A alone was required. Indomethacin blocked the synthesis while the addition of arachidonic acid increased it.

Prostaglandin synthesis by macrophages stimulated with endotoxin appears to be necessary for the synthesis of collagenase by these cells (Wahl *et al.*, 1974). This was made apparent by the ability of indomethacin to block the production of collagenase. This enzyme plays the key role in collagen destruction and is likely to be a factor in inflammatory processes.

X. Conclusions

The decade of the 1980s should see a broader understanding of the chemistry, mode of production, and ways of modifying the action of mediators of the biological effects of endotoxin. For example, Moore *et al.* (1980a) have shown that IL 1 formation is activated by CSF, and that prostaglandin regulates the production of CSF (Moore *et al.*, 1979). These interrelationships should prove informative as they are revealed and, in time, much of the current confusion should disappear.

A few experiments in which antiserum against a purified mediator has been used, have helped to establish unambiguously the specific role for that mediator. When this approach is expanded, it should be possible to prove whether a single factor has multiple effects and equally important, whether several factors are a single entity. Antibody against a mediator should help diminish undesirable effects while an injection of purified mediator might be used to enhance a desirable response. Should this prove feasible, the current effort to clone human genes for interferon in bacteria will no doubt be expanded to other mediators. This seems to be the direction much future research will take.

ACKNOWLEDGMENT

This work was supported in part by U.S. Public Health Service Grant AI-10087 from the National Institute of Allergy and Infectious Diseases.

XI. References

Addison, P. D., and Berry, L. J., 1981, Passive protection against X-irradiation with serum from zymosan-primed and endotoxin injected mice, *J. Reticuloendothelial Soc.* **30:** 301.

Atkins, E., 1960, Pathogenesis of fever, *Physiol. Rev.* **40:**580.

Atkins, E., and Bodel, P., 1972, Fever, *N. Engl. J. Med.* **286:**27.

Atkins, E., Bodel, P., and Francis, L., 1967, Release of an endogenous pyrogen *in vitro* from rabbit mononuclear cells, *J. Exp. Med.* **126:**357.

Beeson, P. B., 1948, Temperature-elevating effect of a substance obtained from polymorphonuclear leukocytes, *J. Clin. Invest.* **27:**524.

Behling, U. H., and Nowotny, A., 1978, Long-term adjuvant effect of bacterial endotoxin in prevention and restoration of radiation-caused immunosuppression, *Proc. Soc. Exp. Biol. Med.* **151:**348.

Behling, U. H., Pham, P. H., Madani, F., and Nowotny, A., 1980, Components of lipopolysaccharide which induce colony stimulation, adjuvancy and radioprotection, in: *Microbiology 1980* (D. Schlessinger, ed.), pp. 103–107, *American Society for Microbiology*, Washington, D.C.

Berry, L. J., 1964, Effect of endotoxin on the level of selected enzymes and metabolites, in: *Bacterial Endotoxins* (M. Landy and W. Braun, eds.), pp. 151–159, Rutgers University Press, New Brunswick, N.J.

Berry, L. J., 1977, Bacterial toxins, *CRC Crit. Rev. Toxicol.* **4:**239.

Berry, L. J., and Smythe, D. S., 1959, Effects of bacterial endotoxin on metabolism. II. Protein–carbohydrate balance following cortisone: Inhibition of intestinal absorption and adrenal response to ACTH, *J. Exp. Med.* **110:**407.

Berry, L. J., Goodrum, K. J., Ford, C. W., Resnick, I. G., and Shackleford, G. M., 1980, Partial characterization of glucocorticoid antagonizing factor in hepatoma cells, in: *Microbiology 1980* (D. Schlessinger, ed.), pp. 77–81, American Society for Microbiology, Washington, D.C.

Blalock, J. E., and Smith, E. M., 1980, Human leukocyte interferon: Structural and biological relatedness to adrenocorticotrophic hormone and endorphins, *Proc. Natl. Acad. Sci. USA* **77:**5972.

Bodel, P., 1970, Investigations of new protein synthesis in stimulated human blood leukocytes, *Yale J. Biol. Med.* **43:**145.

Bodel, P., and Miller, H., 1976, Pyrogen from mouse macrophages causes fever in mice, *Proc. Soc. Exp. Biol. Med.* **151:**93.

Bornstein, D. L., and Walsh, E. C., 1978, Endogenous mediators of the acutephase reaction. I. Rabbit granulocytic pyrogen and its chromatographic subfractions, *J. Lab. Clin. Med.* **91:**236.

Brown, D. R., and Holtzman, S. G., 1979, Suppression of deprivation-induced food and water intake in rats and mice by naloxone, *Pharmacol. Biochem. Behav.* **11:**567.

Brown, D. R., and Holtzman, S. G., 1981, Suppression of drinking by naloxone in the rat: A further characterization, *Eur. J. Pharmacol.* **69:**331.

Burgess, A. W., and Metcalf, D., 1977, The effect of colony-stimulating factor on the synthesis of ribonucleic acid by mouse bone marrow cells *in vitro*, *J. Cell. Physiol.* **90**:471.

Burgess, A. W., Camakaris, J., and Metcalf, D., 1977, Purification and properties of colony-stimulating factor from mouse lung-conditioned medium, *J. Biol. Chem.* **252**:1998.

Butzler, J. P., and Skirrow, M. B., 1979, Campylobacter enteritis, *Clin. Gastroenterol.* **8**:737.

Carswell, E. A., Old, L. J., Kassel, R. L., Green, S., Fiore, N., and Williamson, B., 1975, An endotoxin-induced serum factor that causes necrosis of tumors, *Proc. Natl. Acad. Sci. USA* **72**:3666.

Chen, D. M., and DiSabato, G., 1976, Further studies on the thymocyte stimulating factor, *Cell. Immunol.* **22**:211.

Cooper, K. E., Cranston, W. I., and Honour, A. J., 1967, Observations on the site and mode of action of pyrogens in the rabbit brain, *J. Physiol. (London)* **191**:325.

Couch, R. E., Jr., Moore, R. N., and Berry, L. J., 1979, Sensitization of tolerant mice to cold with a serum factor induced by endotoxin, *J. Appl. Physiol.: Respir. Environ. Exercise Physiol.* **46**:14.

Cox, C. G., and Rafter, G. W., 1971, Pyrogen and enzyme release from rabbit blood leukocytes promoted by endotoxin and polyinosinic polycytidylic acid, *Biochem. Med.* **5**:227.

Cranston, W. I., 1980, Mechanisms of production of fever by endogenous pyrogens, in: *Microbiology 1980* (D. Schlessinger, ed.), pp. 147–149, American Society for Microbiology, Washington, D.C.

Cranston, W. I., Hellon, R. F., and Mitchell, D., 1975, A dissociation between fever and prostaglandin concentration in cerebrospinal fluid, *J. Physiol. (London)* **253**:583.

Cranston, W. I., Duff, G. W., Hellon, R. F., Mitchell, D., and Townsend, Y., 1976, Evidence that brain prostaglandin synthesis is not essential for fever, *J. Physiol. (London)* **259**:239.

Dinarello, C. A., Renfer, L., and Wolff, S. M., 1977, Human leukocytic pyrogen: Purification and development of a radioimmunoassay, *Proc. Natl. Acad. Sci. USA* **74**:4624.

Dirksen, R., Otten, M. H., Wood, G. J., Verbaan, C. J., Haalebos, M. M. P., Verdouw, P. V., and Nijhuis, G. M. M., 1980, Naloxone in shock, *Lancet* **2**:1360.

Dirksen, R., Wood, G. J., and Nijhuis, G. M. M., 1981, Mechanism of naloxone therapy in the treatment of shock: A hypothesis, *Lancet* **1**:607.

Dubos, R., and Schaedler, R. W., 1961. The effect of bacterial endotoxins on the water intake and body weight of mice, *J. Exp. Med.* **113**:921.

Eddington, C. L., Upchurch, H. F., and Kampschmidt, R. F., 1972, Quantitation of alpha-2-AP globulin before and after stimulation with leukocytic extracts, *Proc. Soc. Exp. Biol. Med.* **139**:565.

Faden, A. I., and Holaday, J. W., 1980a, Naloxone treatment of endotoxic shock: Stereospecificity of physiologic and pharmacologic effects in the rat, *J. Pharmacol. Exp. Ther.* **212**:441.

Faden, A. I., and Holaday, J. W., 1980b, Experimental endotoxin shock: The pathophysiologic function of endorphins and treatment with opiate antagonists, *J. Infect. Dis.* **142**:229.

Farrar, W. L., Mizel, S. B., and Farrar, J. J., 1980, Participation of lymphocyte activating factor (interleukin 1) in the induction of cytotoxic T cell responses, *J. Immunol.* **124**:1371.

Fauconnier, B., 1971, Effet du serum anti-interferon sur la pathogénicité virale experimentale *"in vivo,"* *Pathol. Biol.* **19**:575.

Feldberg, W., and Gupta, K. P., 1973, Pyrogen fever and prostaglandin-like activity in cerebrospinal fluid, *J. Physiol. (London)* **228**:41.

Filkins, J. P., 1980, Endotoxin-enhanced secretion of macrophage insulin-like activity, *J. Reticuloendothelial Soc.* **27**:507.

Finck, A. D., 1977, Alleviation of prolonged postoperative central nervous system depression after treatment with naloxone, *Anesthesiology* **47**:392.

Flynn, J. T., 1978, Endotoxin shock in the rabbit: The effects of prostaglandin and arachidonic acid administration, *J. Pharmacol. Exp. Ther.* **206**:555.

Foster, J. A., Morrison, M., Dean, S. J., Hill, M., and Frenk, H., 1981, Naloxone suppresses food/water consumption in the deprived cat, *Pharmacol. Biochem. Behav.* **14**:419.

Gery, I., and Waksman, B. H., 1972, Potentiation of the T-lymphocyte response to mitogens. II. The cellular source of potentiating mediator(s), *J. Exp. Med.* **136**:143.

Goodrum, K. J., and Berry, L. J., 1978, The effect of glucocorticoid antagonizing factor on hepatoma cells, *Proc. Soc. Exp. Biol. Med.* **159**:359.

Goodrum, K. J., and Berry, L. J., 1979. The use of Reuber hepatoma cells for the study of a lipopolysaccharide-induced macrophage factor, *Lab. Invest.* **41**:174.

Goodrum, K. J., Moore, R. N., and Berry, L. J., 1978, Effect of indomethacin on the response of mice to endotoxin, *J. Reticuloendothelial Soc.* **23**:213.

Gordon, J., and Minks, M. A., 1981, The interferon renaissance: Molecular aspects of induction and action, *Microbiol. Rev.* **45**:244.

Green, S., Dobrjanksy, A., Carswell, E. A., Kassel, R. L., Old, L. J., Fiore, N., and Schwartz, M. K., 1975, Partial purification of a serum factor that causes necrosis of tumors, *Proc. Natl. Acad. Sci. USA* **73**:381.

Greisman, S. E., and DuBuy, B., 1975, Mechanism of endotoxin tolerance. IX. Effect of exchange transfusion, *Proc. Soc. Exp. Biol. Med.* **148**:675.

Gresser, I., Tovey, M. G., Maury, C., and Chouroulinkov, I., 1975, Lethality of interferon preparations for newborn mice, *Nature (London)* **1975**:258.

Gronowicz, E., Biberfeld, P., Wahren, B., and Coutinho, A., 1976, Characterization of dextran sulfate sensitive cells, *Scand. J. Immunol.* **5**:573.

Hahn, H., Char, D. C., Postel, W. B., and Wood, W. B., Jr., 1967, Studies on the pathogenesis of fever. XV. The production of endogenous pyrogen by peritoneal macrophages, *J. Exp. Med.* **126**:385.

Hämmerling, U., Old, L. J., Carswell, E., Abbott, J., Oettgen, H. F., and Hoffmann, M. K., 1980, Multifaceted properties of tumor-necrotizing serum, in: *Microbiology 1980* (D. Schlessinger, ed.), pp. 135–140, American Society for Microbiology, Washington, D.C.

Handman, E., and Burgess, A. W., 1979, Stimulation by granulocyte-macrophage colony-stimulating factor of *Leishmania tropica* killing by macrophages, *J. Immunol.* **122**:1134.

Harmon, D. R., Zarafonetis, C., and Clark, P. F., 1946, Temperature relations in phagocytosis, *J. Bacteriol.* **52**:337.

Higgins, S. J. and Gehring, U., 1978, Molecular mechanisms of steroid hormone action, in: *Advances in Cancer Research*, Vol. 28 (G. Klein and S. Weinhouse, eds.), pp. 313–397, Academic Press, New York.

Holaday, J. W., and Faden, A. I., 1978, Naloxone reversal of endotoxin hypotension suggests role of endorphins in shock, *Nature (London)* **275**:450.

Isaacs, A., and Lindenmann, J., 1957, Virus interference. I. The interferon, *Proc. R. Soc. London Ser. B* **147**:258.

Janoff, A., Weissmann, G., Zweifach, B. W., and Thomas, L., 1962, Pathogenesis of experimental shock. IV. Studies on lysosomes in normal and tolerant animals subjected to lethal trauma and endotoxemia, *J. Exp. Med.* **116**:451.

Johnson, H. M., 1982, Interferon and host defense systems, in: *Host Defenses to Intracellular Pathogens* (T. K. Eisenstein and H. Friedman, eds.), Plenum Press, New York.

Kadlec, O., and Horáček, J., 1980, Inhibition of peristaltic activity in the guinea-pig ileum by specific stress stimulus: Its reversal by naloxone and indomethacin, *Life Sci.* **27**:1557.

Kaiser, K. H., and Wood, W. B., Jr., 1962, Studies on the pathogenesis of fever. IX. The production of endogenous pyrogen by polymorphonuclear leukocytes, *J. Exp. Med.* **115**:27.

Kampschmidt, R. F., 1979a, Role of RES and leukocytic endogenous mediator in iron, zinc and copper metabolism, in: *Macrophages and Lymphocytes: Nature, Functions and Interaction* (M. R. Escobar and H. Friedman, eds.), pp. 403–411, Plenum Press, New York.

Kampschmidt, R. F., 1979b, Metabolic alterations elicited by endogenous pyrogens, in: *Fever* (J. M. Lipton, ed.), pp. 49–56, Raven Press, New York.

Kampschmidt, R. F., and Pulliam, L. A., 1975, Stimulation of antimicrobial activity in the rat with leukocytic endogenous mediator, *J. Reticuloendothelial Soc.* **17**:162.

Kampschmidt, R. F., Long, R. D., and Upchurch, H. F., 1972, Neutrophil releasing activity in rats injected with endogenous pyrogen, *Proc. Soc. Exp. Biol. Med.* **139**:1224.

Kampschmidt, R. F., Pulliam, L. A., and Upchurch, H. F., 1973, Sources of leukocytic endogenous mediator in the rat, *Proc. Soc. Exp. Biol. Med.* **144**:882.

Koblinsky, M., Beato, M., Kalimi, M., and Feigelson, P., 1972, Glucocorticoid binding protein of rat liver cytosol. I. Physical characterization and properties of the binding proteins, *J. Biol. Chem.* **247**:7897.

Koopman, W. J., Farrar, J. J., Oppenheim, J. J., Fuller-Bovar, J., and Dougherty, S., 1977, Association of a low molecular weight helper factor(s) with thymocyte proliferative activity, *J. Immunol.* **119**:55.

Kurland, J. I., Pelus, L. M., Ralph, P., Bockman, R. S., and Moore, M. A. S., 1979, Induction of prostaglandin E synthesis in normal and neoplastic macrophages: Role for colony-stimulating factor(s) distinct from effects on myeloid progenitor cell proliferation, *Proc. Natl. Acad. Sci. USA* **76**:2326.

Männel, D. N., Moore, R. N., and Mergenhagen, S. E., 1980, Endotoxin induced tumor cytotoxic factor, in: *Microbiology 1980* (D. Schlessinger, ed.), pp. 141–143, American Society for Microbiology, Washington, D. C.

Martin, C. R., 1976, *Textbook of Endocrine Physiology*, Williams & Wilkins, Baltimore.

Mefferd, R. B., Hendel, D. T., and Loeffer, J. B., 1953, Effect of piromen on survival of irradiated mice, *Proc. Soc. Exp. Biol. Med.* **85**:53.

Merriman, C. R., Pulliam, L. A., and Kampschmidt, R. F., 1975, Effect of leukocytic endogenous mediator on C-reactive protein in rabbits, *Proc. Soc. Exp. Biol. Med.* **149**:782.

Merriman, C. R., Pulliam, L. A., and Kampschmidt, R. F., 1977, Comparison of leukocytic pyrogen and leukocytic endogenous mediator, *Proc. Soc. Exp. Biol. Med.* **154**:224.

Metcalf, D., 1972, The colony stimulating factor (CSF), *Aust. J. Exp. Biol. Med. Sci.* **50**:547.

Milton, A. S., and Wendlandt, S., 1971, Effects on body temperature of prostaglandins of the A, E and F series on injection into the third ventricle of unanesthetized cats and rabbits, *J. Physiol. (London)* **218**:325.

Mishell, R. I., Lucas, A., and Mishell, B. B., 1977, The role of activated accessory cells in preventing immunosuppression by hydrocortisone, *J. Immunol.* **119**:118.

Mishell, R. I., Bradley, L. M., Chen, Y. U., Grabstein, K. H., and Shiigi, S. M., 1980, Protection of helper T cells from glucocorticosteroids by mediators from adjuvant-activated monocytes, in: *Microbiology 1980* (D. Schlessinger, ed.), pp. 82–86, American Society for Microbiology, Washington, D.C.

Mizel, S. B., and Farrar, J. J., 1979, Revised nomenclature for antigen-nonspecific T cell proliferation and helper factor, *Cell. Immunol.* **48**:433.

Moore, R. N., and Berry, L. J., 1976, Inhibited hormonal induction of hepatic phosphoenolpyruvate carboxykinase in polyI:C treated mice, an endotoxin-like glucocorticoid antagonism, *Eperientia* **32**:1566.

Moore, R. N., and Berry, L. J., 1982, Endocrine-like activities of the RES, in: *The Reticulo-*

endothelial System: A Comprehensive Treatise, Vol. IV, *Physiology* (J. P. Filkins and S. M. Reichard, eds.), Plenum Press, New York, in press.

Moore, R. N., Goodrum, K. J., and Berry, L. J., 1976, Mediation of endotoxin effects by macrophages, *J. Reticuloendothelial Soc.* **19**:187.

Moore, R. N., Berry, L. J., Garry, R. F., and Waite, M. R. F., 1978a, Effect of sindbis virus infection on hydrocortisone-induced hepatic enzymes in mice, *Proc. Soc. Exp. Biol. Med.* **157**:125.

Moore, R. N., Goodrum, K. J., Couch, R. E., Jr., and Berry, L. J., 1978b, Factors affecting macrophage function: Glucocorticoid antagonizing factor, *J. Reticuloendothelial Soc.* **23**:321.

Moore, R. N., Goodrum, K. J., Couch, R. E., Jr., and Berry, L. J., 1978c, Elicitation of endotoxemic effects in C3H/HeJ mice with glucocorticoid antagonizing factor and partial characterization of the factor, *Infect. Immun.* **19**:79.

Moore, R. N., Urbaschek, R., Wahl, L. M., and Mergenhagen, S. E., 1979, Prostaglandin regulation of colony-stimulating factor production by lipopolysaccharide-stimulated murine leukocytes, *Infect. Immun.* **26**:408.

Moore, R. N., Oppenheim, J. J., Farrar, J. J., Carter, C. S., Jr., Waheed, A., and Shadduck, R. K., 1980a, Production of lymphocyte-activating factor (interleukin 1) by macrophages activated with colony-stimulating factors, *J. Immunol.* **125**:1302.

Moore, R. N., Vogel, S. N., Wahl, L. M., and Mergenhagen, S. E., 1980b, Factors influencing lipopolysaccharide-induced interferon production, in: *Microbiology 1980* (D. Schlessinger, ed.), pp. 131–143, American Society for Microbiology, Washington, D.C.

Murphy, P. A., Chesney, P. J., and Wood, W. B., Jr., 1974, Further purification of rabbit leukocyte pyrogen, *J. Lab. Clin. Med.* **83**:310.

North, R. J., and Berendt, M. J., 1980, Immunological basis of endotoxin-facilitated tumor regression, in: *Microbiology 1980* (D. Schlessinger, ed.), pp. 144–146, American Society for Microbiology, Washington, D.C.

Onoda, M., Shinoda, M., Tsuneoka, K., and Shikita, M., 1980, X-Ray-induced production of granulocyte-macrophage colony-stimulating factor (GM-CSF) by mouse spleen cells in culture, *J. Cell. Physiol.* **104**:11.

Oppenheim, J. J., Moore, R. N., Gmelig-Meyling, F., Togawa, A., Wahl, S., Mathieson, B. J., Dougherty, S., and Carter, C., 1980, Role of cytokine- and endotoxin-induced monokines in lymphocyte proliferation, differentiation and immunoglobulin production, in: *Macrophage Regulation of Immunity* (E. R. Unanue and A. S. Rosenthal, eds.), pp. 379–398, Academic Press, New York.

Parant, M., and Chedid, L., 1964, Protective effect of chlorpromazine against bacterial-induced abortion, *Proc. Soc. Exp. Biol. Med.* **116**:906.

Peters, W. P., Johnson, M. W., Friedman, P. A., and Mitch, W. E., 1981, Pressor effect of naloxone in septic shock, *Lancet* **1**:529.

Rafter, C. W., Cheuk, S. F., Krause, D. W., and Wood, W. B., Jr., 1966, Studies on the pathogenesis of fever. XIV. Further observations on the chemistry of leukocytic pyrogen, *J. Exp. Med.* **123**:433.

Raymond, R. M., Harkema, J. M., Stoffs, W. V., and Emerson, T. E., Jr., 1981, Effects of naloxone therapy on hemodynamics and metabolism following a superlethal dosage of *Escherichia coli* endotoxin in dogs, *Sur. Gynecol. Obstet.* **152**:159.

Reynolds, D. G., Guill, N. J., Vargish, T., Lechner, R., Faden, A. I., and Holaday, J. W., 1980, Blockade of opiate receptors with naloxone improves survival and cardiac performance in canine endotoxic shock, *Circ. Shock* **7**:39.

Rietschel, E., Schade U., Lüderitz, O., Fischer, H., and Peskar, B. A., 1980, Prostaglandins in endotoxicosis, in: *Microbiology 1980* (D. Schlessinger, ed.), pp. 66–70, American Society for Microbiology, Washington, D.C.

Rosendorf, C., and Mooney, J. J., 1971, Central nervous system sites of action of a purified leukocyte pyrogen, *Am. J. Physiol.* **220:**597.

Rousseau, G. G., Baxter, J. D., Higgins, S. J., and Tomkins, G. M., 1973, Steroid-induced nuclear binding of glucocorticoid receptors in intact hepatoma cells, *J. Mol. Biol.* **79:** 539.

Sakaguchi, S., and Sakaguchi, O., 1980, Changes in serum lipid in endotoxin-poisoned mice, *Microbiol. Immunol.* **24:**357.

Shadduck, R. K., Pigoli, G., Waheed, A., and Boezel, F., 1980, The role of colony-stimulating factor in granulopoiesis, *J. Supramol. Struct.* **14:**423.

Shear, M. J., and Turner, F. C., 1943, Chemical treatment of tumors: Isolation of the hemorrhage producing fraction from *Serratia marcescens* culture filtrate, *J. Natl. Cancer Inst.* **4:**81.

Siegert, R., Philipp-Dormston, W. K., Radsak, K., and Menel, H., 1976, Mechanism of fever induction in rabbits, *Infect. Immun.* **14:**1130.

Skarnes, R. C., and Harper, M. J. K., 1972, Relationship between endotoxin-induced abortion and the synthesis of prostaglandin F, *Prostaglandins* **1:**191.

Skarnes, R. C., and McCracken, J. A., 1980, Mediating role of prostaglandin E in endotoxin fever, in: *Microbiology 1980* (D. Schlessinger, ed.), pp. 162–165, American Society for Microbiology, Washington, D.C.

Skirrow, M. B., and Benjamin, J., 1980, '1001' campylobacters: Cultural characteristics of intestinal campylobacters from man and animals, *J. Hyg.* **85:**427.

Smith, W. W., Alderman, I. M., and Gillespie, R. E., 1957, Increased survival in irradiated animals treated with bacterial endotoxin, *Am. J. Physiol.* **191:**124.

Smith, W. W., Alderman, I. M., and Gillespie, R. E., 1958, Hemopoietic recovery induced by bacterial endotoxin in irradiated mice, *Am. J. Physiol.* **192:**549.

Stanley, E. R., Hansen, G., Woodcock, J., and Metcalf, D., 1975, Colony-stimulating factor and the regulation of granulopoiesis and macrophage production, *Fed. Proc.* **34:**2272.

Taliaferro, W. H., Taliaferro, L. G., and Jaroslow, B. N., 1964, *Radiation and Immune Mechanisms*, Academic Press, New York.

Tilghman, S. M., Hanson, R. W., and Ballard, F. J., 1976, Hormonal regulation of phosphoenolpyruvate carboxykinase (GTP) in mammalian tissues, in: *Gluconeogenesis: Its Regulation in Mammalian Tissues* (R. W. Hanson and M. A. Mehlman, eds.), pp. 47–91, Wiley, New York.

Turner, M. M., and Berry, L. J., 1963, Inhibition of gastric emptying in mice by bacterial endotoxins, *Am. J. Physiol.* **205:**1113.

Urbaschek, R., Mergenhagen, S. E., and Urbaschek, B., 1977, Failure of endotoxin to protect C3H/HeJ mice against lethal X-irradiation, *Infect. Immun.* **18:**860.

Vargish, T., Reynolds, D. G., Guill, N. J., Lechner, R. B., Holaday, J. W., and Faden, A. I., 1980, Naloxone reversal of hypovolemic shock in dogs, *Circ. Shock* **7:**31.

Waheed, A., and Shadduck, K., 1979, Purification and properties of L-cell-derived colony-stimulating factor, *J. Lab. Clin. Med.* **94:**180.

Wahl, L. M., Wahl, S. M., Mergenhagen, S. E., and Martin, G. R., 1974, Collagenase production by endotoxin activated macrophages, *Proc. Natl. Acad. Sci. USA* **71:**3598.

Watson, J., Gillis, S., Marbrook, J., Mochizuki, D., and Smith, K. A., 1979, Biochemical and biological characterization of lymphocyte regulatory molecules. I. Purification of a class of murine lymphokines, *J. Exp. Med.* **150:**849.

Westphal, O., Westphal, U., and Sommer, T., 1977, The history of pyrogen research, in: *Microbiology 1977* (D. Schlessinger, ed.), pp. 221–238, American Society for Microbiology, Washington, D.C.

Wicks, W. D., Barnett, C. A., and McKibbin, J. B., 1974, Interaction between hormones and cyclic AMP in regulating specific hepatic enzyme synthesis, *Fed. Proc.* **33:**1105.

Woodruff, P. W. H., O'Carroll, D. I., Koigumi, S., and Fine, J., 1973, Role of the intestinal flora in major trauma, *J. Infect. Dis.* **128**(Suppl.):290.

Wu, M.C., Cini, J. K., and Yunis, A. A., 1979, Purification of a colony-stimulating factor from cultured pancreatic carcinoma cells, *J. Biol. Chem.* **254**:6226.

Yamamoto, K. R., and Alberts, B. M., 1976, Steroid receptors: Elements for modulation of eukaryotic transcription, in: *Annual Review of Biochemistry*, Vol. 45 (E. E. Snell, P. D. Boyer, A. Meister, and C. C. Richardson, eds.), pp. 721–746, Annual Reviews, Inc., Palo Alto, Calif.

Yunis, A. A., Wu, M. C., Miller, A. M., Ingram, M., and Files, N., 1980, Antibody to purified colony-stimulating factor: Use in the identification and isolation of granulocyte macrophage progenitor cells, *Blood Cells* **6**:679.

Zahl, P. A., Hutner, S. H., and Cooper, F. S., 1943, Action of bacterial toxins on tumors. VI. Protection against tumor hemorrhage following heterologous immunization, *Proc. Soc. Exp. Biol. Med.* **54**:187.

19

Induction of Interferon by Endotoxin

Monto Ho

I. The Phenomenon

The ability of bacterial endotoxin to modify virus infections was well documented before it was found it could induce interferon. Endotoxin decreased both the toxicity and the replication of some viruses. Hook and Wagner (1959) found that as little as 0.008 μg of purified *E. coli* lipopolysaccharide (LPS) inoculated intracerebrally in mice 75 min to 3½ hr before an intracerebral challenge of 3–8 LD_{50} of WS influenza virus decreased the direct neurotoxic effects of this virus. Typhoid vaccine was similarly effective. This protective effect lasted for about 7 days. No effect on viral multiplication in the brain was found.

Wagner *et al.* (1959) also studied the effect of endotoxin on Eastern equine encephalomyelitis (EEE) virus infection in mice. One microgram of purified LPS prepared from *Salmonella abortus equi*, inoculated intracerebrally, decreased the intracerebral LD_{50} of EEE by 0.5 to 0.6 log if inoculated 24 hr before the virus. Other mice were treated with endotoxin intracerebrally, or intravenously, or by both routes, and were challenged with EEE intraperitoneally. The endotoxin treatment prolonged the survival of the mice or increased the numbers surviving. There was a small but significant decrease in viral replication in the brains of endotoxin-treated animals.

Gledhill (1959a) found that when 50 μg of partially purified endotoxin from *Chromobacterium prodigiosum*, or 0.2 ml of a formalinized filtrate of *Salmonella typhimurium*, was inoculated intraperitoneally into

Monto Ho ● Department of Microbiology, Graduate School of Public Health, and Department of Medicine, School of Medicine, University of Pittsburgh, Pittsburgh, Pennsylvania 15261.

mice 24 hr before 10^6 LD_{50} of ectromelia virus by the same route, the mice lived significantly longer than controls, 45% of whom were dead within 1½ days after virus inoculation.

Several workers examined the action of endotoxin on viral effects in cell cultures. Wagner (1955) attempted without success to demonstrate increased resistance to the toxic action of influenza virus on HeLa cells previously exposed to xerosin or *Cholera vibrio* filtrates. These endotoxin-containing materials had been found effective in animals. The inability of endotoxin to affect the resistance of chick cell cultures was also reported in the case of EEE virus (Wagner *et al.*, 1959), and VSV (Youngner and Stinebring, 1964). Endotoxin also did not increase the resistance of rabbit kidney cells to VSV (Ho, 1964b), or of monkey kidney cells to poliovirus (Ho, unpublished data).

In contrast to these negative findings, Likar *et al.* (1959) observed that crude filtrates of three out of four human strains of *E. coli* inhibited the replication of Type 1 (Mahoney) poliovirus, if the filtrates were incubated with monkey kidney cell cultures for 3 days prior to virus inoculation. Assuming that endotoxin was the protective agent, the authors used LPS from *S. typhimurium*, and this was even more effective in reducing viral replication than the bacterial filtrates. Unfortunately, it is impossible to deduce from their paper the purity, quantity, or biological potency of the preparations used. This work has not been confirmed.

It is clear from the above that bacterial endotoxins under certain circumstances inhibit viral infections at least in the intact host. Gledhill (1959b) then found that the serum of animals inoculated with endotoxin had a "sparing" or passive protective effect against viral infections. Serum (0.2 to 0.4 ml) obtained from mice 2 hr after intraperitoneal inoculation of a growth filtrate of *S. typhimurium*, which contained endotoxin, was given intraperitoneally to weanling mice. It significantly reduced the mortality caused by infection 3 hr later with ectromelia virus, or with the combination of *Eperythrozoon coccoides* and mouse hepatitis (MHV-1) virus. The sparing effect was as good, if not better, when the serum was inoculated several hours before the virus (Gledhill, 1964). This suggested to the author that "the obvious conclusion is that the sparing substance in the serum does not directly inactivate virus." Heating of the serum at 56°C for 30 min did not eliminate its protective effect. It was non dialyzable and resistant to pH 2, and it passed through a Gradocol membrane of average pore diameter of 130 nm. Sera obtained from mice 1, 2, 4, and 6 hr after intraperitoneal inoculation of endotoxin all had about the same sparing effect, but it was less in sera obtained 12 hr after inoculation. The author did not assay the inhibitor in an *in vitro* assay for interferon, or show that it was species specific. He suspected that this inhibitor was related to interferon (Gledhill, 1959c, 1964). In retrospect, it is clear from the properties of this inhibitor and the kinetics of its appearance that the inhibitor was probably interferon.

The next development was the finding by Stinebring and Youngner (1964) that "inhibitor activity appears in the serum (of chickens) at about 2 hours after *E. coli* endotoxin. . . . It was necessary to inject birds (chickens) intravenously with at least 40 mg. Inhibitor activity, which was barely detectable at 2 hours, had disappeared by 4–6 hours after injection. The small and transient amounts in the serum made it too difficult to investigate the properties of the inhibitor. . . ."

This brief statement cannot be considered the definitive report that endotoxin induced interferon, although it did represent progress because unlike Gledhill's material, the inhibitor was detected by an *in vitro* assay rather than by passive protection of an animal. In view of Gledhill's description of a presumed interferon induced by endotoxin, the next essential step clearly was to describe the properties of endotoxin-induced inhibitor and to show that it was indeed interferon, at least to the extent allowed by the state of the art at the time. The chicken material could not serve this purpose.

This step and these descriptions came independently and concurrently from the works of Ho (1964b) and Stinebring and Youngner (1964). Ho (1964b) found that *E. coli* O113 endotoxin in doses of 2 μg or greater induced a circulating interferon-like inhibitor in rabbits. The inhibitor appeared 1½ to 2 hr after inoculation of endotoxin and disappeared in 7 to 24 hr. It was species specific in being active only *in vitro* in rabbit cell cultures and not in chick cell cultures; it was active against more than one type of virus; its action increased with the length of incubation of the inhibitor with rabbit kidney cell cultures, which suggests that it acted intracellularly, as does interferon. It was shown that the factor was not residual endotoxin in the serum, since endotoxin alone did not inhibit VSV in rabbit cell cultures, nor was it a simple reaction product between serum and endotoxin, since it could not be produced by incubating endotoxin in serum. The inhibitor was nondialyzable, did not sediment at 105,000g, and was inactivated by trypsin. While these properties are those of an interferon, the inhibitor was less stable to heat or acid than rabbit interferon induced by viruses. That interferons of different physiochemical stability can be induced by endotoxin in rabbits has since been repeatedly confirmed (Maehara and Ho, 1977).

Stinebring and Youngner (1964) reported that in mice, 2 hr after an intravenous injection of 25 to 250 μg of endotoxin, interferon with titers as high as 1:768 was obtained from the plasma. This inhibitor was also shown to be nondialyzable, sensitive to trypsin, and resistant to pH 3 and heat. Later, Hallum *et al.* (1965) found by chromatography on Sephadex G-100 that non-virus-induced mouse interferons had much larger molecular weights than NDV-induced interferon. The molecular weights for NDV- and endotoxin-induced mouse interferons were respectively 25,000 and 89,000. Ke *et al.* (1966a) found that rabbit endotoxin-induced interferon was also larger than virus-induced interferon.

Recently, Sauter and Wolfensberger (1980) reported that after injection of 2 μg *E. coli* endotoxin intravenously in man, up to 160 units of interferon appeared in the serum between 2 and 5 hr after injection.

II. Mechanism of Induction of Interferon by Endotoxin

The discovery that substances other than viruses could induce interferon naturally raised the question whether they induced interferon in the same way. Because of the rapid appearance of interferon in the bloodstream of mice 2 hr following intravenous injections of endotoxin, Stinebring and Youngner (1964) suggested that interferon was already preformed in cells, and was not "induced" but merely released by this agent. Interferon was similarly rapidly produced when gram-negative endotoxin-containing bacteria such as *Serratia marcescens* or *Salmonella typhimurium* were injected. However, when either live *Brucella abortus*, which seek an intracellular site, or NDV was injected into mice, interferon appeared in the serum only after 6 hr.

The induction of interferon by *B. abortus* is complex and probably due to at least two substances. Endotoxin isolated from the organism by the Boivin method (Staub, 1965), or LPS extracted from the organism with the phenol–water method of Westphal (Westphal and Jann, 1965) produced a typical endotoxin-type response in mice with peak levels of interferon appearing 2 hr after injection (Keleti *et al.*, 1974). On the other hand, a nonviable aqueous ether-extracted pellet of *B. abortus* (Bru-pel) induced high-titered interferon in mice of a virus-induced" type with peak titers 6½ hr after injection. The interferon-inducing and antiviral properties of Bru-pel were thought to be a unique unextractable component of the organism plus a lipid, probably not endotoxin, which is present in Bru-pel and extractable by chloroform–methanol (Feingold *et al.*, 1976). Bru-pel represents better than endotoxin the mode of interferon induction by live *B. abortus*.

The problem of the mechanism of induction of interferon by endotoxin was studied further in experiments using metabolic inhibitors. Ho and Kono (1965a) found that endotoxin-induced interferon formation in rabbits was unaffected by pretreatment with actinomycin D. While the effect of the dose of actinomycin used (1 mg/kg) on RNA synthesis was not measured, it was sufficient to depress and delay interferon formation induced by sindbis virus.

Youngner et al. (1965) also found that endotoxin- and B. abortus-induced interferon production in mice was unaffected by pretreatment with a large dose of actinomycin. But virus-induced interferon was not affected either, even though RNA synthesis in the liver was reduced to 3% of the normal level. Perhaps the synthetic capabilities of the liver may not reflect what is occurring in the cells producing interferon.

The production of interferon induced by B. abortus was inhibited by pretreatment of mice with a dose of cycloheximide that completely inhibited protein synthesis in the mouse liver (Youngner et al., 1965). There was no inhibition of the formation of interferon induced by injection of endotoxin; on the contrary, pretreatment with cycloheximide produced an enhanced and prolonged interferon response. These authors noted a similar phenomenon when the animals were pretreated with puromycin, and they suggested that the antibiotic increased the release of a preformed interferon from the cells. In retrospect, the enhancement of interferon production is best explained by the inhibition of a control protein that restricts interferon production (Ho et al., 1973). Ke et al. (1966b) also observed that interferon production induced by injection of rabbits with endotoxin was unaffected by pretreatment of the animals with a dose of puromycin that completely inhibited protein synthesis in the liver.

Up to this point, results from animal studies subjected to antimetabolites were erroneously interpreted by all investigators that neither synthesis of DNA-dependent RNA nor synthesis of protein was required for the induction of interferon by endotoxin. Work with endotoxin in cell cultures had repeatedly cast doubt on the interpretation that endotoxin-induced interferon represented preformed interferon released from cells. Interferon produced by peritoneal macrophages after addition of endotoxin was reportedly inhibited by actinomycin and puromycin and hence presumably required synthesis of mRNA and protein (Smith and Wagner, 1967). Finkelstein et al. (1968) also found that production of interferon by endotoxin-stimulated macrophages was inhibited by actinomycin D and puromycin, although only in higher doses than those that inhibited virus-stimulated cells. Kobayashi et al. (1969) maintained that there was an actinomycin-sensitive (early) and an actinomycin-resistant (late) phase of interferon production under stimulation by endotoxin, while both phases were inhibited by puromycin.

A similar dichotomy of results obtained in vitro and in the animal pertained to polyriboinosinic:polyribocytidylic acid (poly I:C). Youngner and Hallum (1968) first reported that the production of interferon induced by poly I:C in mice was not inhibited by cycloheximide, but later work in cell cultures and tissue slices showed that production required synthesis of RNA as well as protein (Finkelstein *et al.*, 1968; Field *et al.*, 1968; Vilcek, 1970; Tan *et al.*, 1979; Ho and Ke, 1970).

Ho *et al.* (1973) reinvestigated the problem of endotoxin-induced interferon by measuring interferon production in spleen, liver, and lung tissue slices of 1-kg rabbits injected with endotoxin. If the animals received 1 mg of actinomycin/kg 1 hr before administration of endotoxin, interferon production in the spleen was increased to 1660 units in 100 mg of tissue compared to 230 units in control tissue from animals not treated with actinomycin D. However, if the dose was increased to 5 mg/kg, no detectable interferon was produced. RNA synthesis in the spleen was inhibited 70% by 1 mg/kg and 86% by 5 mg/kg of actinomycin D. Production of interferon in slices was also reversibly inhibited by cycloheximide or puromycin. These findings suggested that induction of interferon production by endotoxin, like induction by poly I:C or virus, required synthesis of mRNA and protein. The accentuation of interferon production by incomplete inhibition of RNA synthesis (1 mg/kg) was explained by inhibited synthesis of mRNA needed for a synthesis of a control protein that inhibits interferon synthesis. This protein was presumably induced during the course of interferon production. Such accentuation was also observed when induction of interferon by poly I:C and by viruses was partially inhibited by actinomycin D (Tan *et al.*, 1970). Previous findings in animals that showed lack of inhibition of endotoxin-induced interferon by actinomycin D and puromycin were probably due to difficulty of achieving adequate inhibition of RNA or protein synthesis in interferon-producing cells in the intact animal.

Viruses and endotoxin are now thought to induce the same type of interferon (α interferons), while immune or γ interferon is induced by sensitized lymphocytes in the presence of antigen, or by nonspecific mitogens. The two interferons are antigenically distinct (Youngner and Salvin, 1973).

III. Hyporeactivity to Repeated Induction

It has long been noted that the induction and production of interferon in cell cultures is a "one-shot affair" that does not continue despite the continued presence of the inducer (Ho, 1964a). Cantell and Paucker (1963) found that after interferon is induced in L cells inoculated with NDV, the cells are rendered refractory to further induction and interferon production. Apparently it is only the descendants of the refractory cells that regain their capacity to be induced.

Ho and Kono (1965b) and Ho *et al.* (1965) found that after one intravenous inoculation of sindbis virus into rabbits, a second inoculation

24 hr later failed to elicit a second crop of interferon production. The situation is similar with endotoxin-induced interferon, where a dose of endotoxin as small as 0.1 μg rendered the animal tolerant of a second inducing dose of endotoxin. The state of tolerance persisted for about 6 days, after which the animal was again sensitive. There was "crossed tolerance" in the sense that an injection of endotoxin made the animal tolerant to the effect of a subsequent injection of virus, but an injection of virus did not produce tolerance to endotoxin. Later, it was found in the rabbit that endotoxin produced hyporeactivity against poly I:C, but an injection of poly I:C actually accentuated the production of interferon by a subsequent injection of endotoxin (Ho et al., 1970). Hence, of the three substances tested, endotoxin was the most effective in producing homologous and heterologous hyporeactivity, although it is the poorest interferon inducer. The explanation for this phenomenon is still not entirely clear. It is reminiscent of the well-known development of tolerance to endotoxins and pyrogens in general (Atkins, 1960). In the case of interferon induction, a serum humoral factor was found in tolerant animals that could inactivate the interferon-inducing capacity of endotoxin (Ho et al., 1965). This factor was thought to be related to the endotoxin-detoxifying factor found in the normal serum of various animals (Ho and Kass, 1958; Skarnes et al., 1958). But it is unlikely that this serum factor explains the hyporeactivity phenomenon. Very likely, the explanation will come when lymphocytes and macrophages can be studied conveniently in their physiologic states in vitro for at least 2 weeks.

Youngner and Stinebring (1965) independently described hyporeactivity in mice. They reported complete crossing of hyporeactivity in mice. Animals made tolerant with NDV reacted poorly to endotoxin, and those made tolerant with endotoxin reacted poorly to NDV. They also showed that endotoxin produced hyporeactivity to the interferon-inducing capacity of B. abortus. Another interesting phenomenon that they described was that animals infected with the avirulent BCG strain of tubercle bacillus responded to an injection of endotoxin with greater than usual production of interferon. This "hyperreactivity" to endotoxin paralleled the known enhancing effect of infection with mycobacteria on the toxicity of endotoxin (Suter and Kirsanow, 1961).

IV. Cellular Source and Nature of Endotoxin-Induced Interferon

The fact that the biosynthetic requirements of endotoxin-induced interferon may be similar to virus-induced interferon should not obscure the differences between those two groups of interferons or their production.

In terms of the mode of production, interferon induced by NDV was increased by elevating the ambient temperature of rabbits to 35°C whereas cooling to 4°C decreased interferon production. In contrast, neither elevation nor lowering of body temperature affected endotoxin-induced interferon (Postic *et al.*, 1966). In macrophage cell cultures or in tissue slices, endotoxin-induced interferon is better produced at 23 or 25°C than 37°C (Laskovic *et al.*, 1967; Ho *et al.*, 1973). Mention has already been made of the more rapid production of endotoxin-induced interferon.

In 1-kg rabbits, cortisol did not inhibit virus-induced interferon unless 250 mg was administered, while 5 mg was sufficient to suppress endotoxin-induced interferon. Adrenalectomy enhanced the latter 10-fold, while it had no effect on virus-induced interferon (Postic *et al.*, 1967). These differences point either to different metabolic requirements for interferon production, or to responses of different target cells. What these might be is still unclear.

Jullien and De Maeyer (1966) found in mice that whole-body irradiation suppressed virus-induced interferon, while in the rat, Billiau (1969) found that such irradiation did not diminish, but actually enhanced endotoxin-induced interferon. Taken together, these results may be interpreted to mean that in response to endotoxin, interferon is produced by cells that are sensitive to cortisol, stimulated by higher temperatures, and are radioresistant. Perhaps such cells are certain types of lymphocytes and/or macrophages. The accentuation of interferon production by X-irradiation may be related to inhibition of a control protein as described above for antimetabolites, or due to increased endotoxemia and infection following irradiation.

Ito *et al.* (1973) suggested that both macrophages and lymphocytes may be required for production of interferon. Splenectomized mice did not produce interferon when injected with endotoxin. When such mice were given either mouse (syngeneic) or rat (xenogeneic) spleen cells, the capacity to produce interferon was restored. The responsible cell type was found to be the macrophage. When xenogeneic cells were transferred, the type of interferon induced by endotoxin was still specific for the mouse. Hence, the interferon-producing cells were not the transferred macrophages, although these were essential for production. The producer was assumed to be the lymphocyte (Ho *et al.*, 1973).

Maehara and Ho (1977) found that endotoxin induced interferons *in vitro* with different properties in mouse macrophages and B lymphocytes. Macrophage interferon was labile at 56°C and was neutralized by antiserum against mouse α/β interferon at a dilution of 1:6142. B-Cell interferon was more heat stable and was neutralized by the same antiserum only at a dilution of 1:276. Serum obtained early (1 hr) after an intravenous injection of 100 μg endotoxin resembled macrophage interferon, whereas serum obtained at later times resembled more and more B-cell interferon. The diverse cellular origin of endotoxin-induced interferon may explain the broad hyporeactivity produced by endotoxin in animals mentioned above, and some of its differences from virus-induced interferon.

Maehara *et al.* (1977) later showed that irrespective of the inducer, whether it be endotoxin, NDV, or poly I:C, interferon produced by T and B lymphocytes were relatively heat stable and of low antigenicity when reacted against antiserum against L-cell interferon (α/β). Macrophage interferon was more like L-cell interferon. These two types of interferons could also be distinguished by chromatography on CH-Sepharose 4B. Kawada *et al.* (1980) have recently reported that various viruses or poly I:C induced interferons in B or T lymphocytes, L cells (fibroblasts), or macrophages that consist of mixtures of F (fast) or α interferon and S (slow) or β interferon. These two basic forms of α and β interferons may be distinguished antigenically, by charge, and by carbohydrate concentration. Their molecular weights are respectively 22–24,000 and 32–40,000. Human and mouse α interferons shared antigenic sites. Endotoxin-induced interferon was not studied. It remains to be seen how different mixtures of α and β interferons can produce the different types of interferons described by Maehara *et al.* (1977), and why endotoxin-induced interferon is higher in molecular weight than virus-induced interferon (Hallum *et al.*, 1965; Ke *et al.*, 1966a).

V. Therapeutic Significance of Endotoxin as an Inducer

In general, there are two approaches to applying the interferon system therapeutically. There is now tremendous interest in manufacturing interferon in induced cell culture or from microorganisms by recombinant DNA techniques and using it in clinical trials against viral infections or against tumors. Another approach would be to administer an inducer to the patient to stimulate him to make his own interferon. By and large, one can produce more interferon endogenously than can be administered exogenously (Ho and Postic, 1967).

Endotoxin is not very useful as an inducer in most cell cultures because it is restricted to macrophages and lymphocytes. It would not be employed to make exogenous interferon. It can be used to induce interferon endogenously in the intact animal host. In terms of potency, it is a relatively poor inducer of interferon, much less effective than double-stranded RNA or viruses (Ho and Armstrong, 1975). Even the best inducers, such as the synthetic double-stranded RNAs, suffer from two major flaws: (1) they are usually too toxic and (2) they produce hyporeactivity. Unfortunately, no inducer demonstrates these two flaws as well as endotoxin (Ho, 1981).

From the biochemical point of view, it is unlikely that manipulation of the LPS molecule can separate toxicity from inducing ability. Youngner and Feingold (1967) found that the component of endotoxin essential for its interferon-inducing action is lipid A. Endotoxins obtained from bacterial mutants that lacked the O-specific side chains of heptose were fully active as inducers. Borecky (1968) also noted that a strain of *Salmonella typhi* devoid of O-antigen was an effective interferon inducer. Feingold *et al.* (1976) found that lipid could be freed from KDO (1-keto-3-deoxyoctonate) without loss of inducing activity. Unfortunately, the active fraction (Westphal's lipid A), as would be expected, was a potent lethal toxin for mice at low doses. These results do not augur well for the possibility of developing from endotoxin a nontoxic interferon-inducing component.

VI. Biological Significance of Endotoxin-Induced Interferon

The ultimate biological significance of induction of interferon by endotoxin, like many phenomena associated with endotoxin, is obscure. One might ask whether interferon induced by endotoxin occurs in nature and during natural disease. Michaels *et al.* (1965) studied 10 patients with gram-negative infections and demonstrated an inhibitor in the serum of two children who had *Hemophilus influenzae* bacteremia and meningitis. The inhibitor had all the properties of interferon. It appeared transiently at the height of the disease process, following the administration of antibiotics. It is possible that antibiotics may have contributed to endotoxemia by killing the bacteria. Possibly interferon is often induced during gram-negative infections, but it is infrequently found because interferon is cleared rapidly from the circulation. In prolonged infections of endotoxemia, hyporeactivity to induction is probably established in patients.

Is the small amount of interferon induced by endotoxin important biologically? There is no clear answer. Interferon should no longer be considered as merely an antiviral substance. It is a broad immunomodulator with both stimulatory and inhibitory activities on both humoral and cellular immunity (Stewart, 1979). Its stimulatory effect on natural killer cell activity, which may be an important component of host defenses against tumors and against viral infections, is particularly impressive (Trinchieri and Santoli, 1978). Interferon also suppresses cell multiplication, and it has even been suggested that interferon mediates the activity of suppressor T lymphocytes (Kadish *et al.*, 1980).

Another function of interferon may well be that it is a new endogenous pyrogen. The first endogenous pyrogen, well known to workers in the endotoxin field, was originally described by Beeson (1948). It is a natural product of polymorphonuclear leukocytes that is also stimulated by endotoxin. It is now known to be produced by all bone marrow-derived phagocytes, such as monocytes (Hahn et al., 1967), macrophages (Atkins et al., 1967), and eosinophils (Mickenberg et al., 1972). It is a smaller protein than interferon, and unlike interferon, does not appear to be species specific (Bodel and Atkins, 1966a). It has not been purified to the extent interferon has, but a radioimmunoassay for human endogenous pyrogen has been described (Dinarello et al., 1977). One of the most consistent side effects of interferon in man has been its pyrogenicity. For some time, it was unknown whether the pyrogenicity was due to impurities, since purified interferon has until recently not been available for human or animal trials. Recently, it has been shown that after purification of α interferon by monoclonal antibody affinity chromatography by more than 100×, about the same degree of pyrogenicity was produced by a comparable amount of interferon (Scott et al., 1981). The mechanism by which interferon produces fever, whether by stimulating the first endogenous pyrogen or the prostaglandin system, or by direct effect on the thermoregulatory center (Bernheim et al., 1979) is unknown. It is distinguished from endogenous pyrogen in one important respect: tolerance to interferon-induced fever develops readily (Stewart, 1979).

Fever accompanies many pathologic conditions that involve the production of interferon: viral infections, hypersensitivity reactions, and endotoxemia. Viruses induce interferon directly. Sensitized lymphocytes produce γ interferons upon contact with specific antigen. And endotoxemia may be a part of many gram-negative infections. In these situations, it is possible that interferon accounts for or at least contributes to the production of fever. In a study on the fever or hypersensitivity in rabbits. Atkins and Francis (1978) postulated that a lymphokine produced during a hypersensitivity reaction released endogenous pyrogen. This lymphokine was produced by sensitized lymphocytes in the presence of specific antigen. In all likelihood, γ interferon was produced in this reaction, although it was not looked for. The question is whether it was the responsible lymphokine. Clearly this interesting area of the interrelationships between endotoxin, endogenous pyrogen, and interferon has barely been explored.

Endotoxins, in the words of Bennett (1964), "seem to have been endowed by nature with virtues and vices in the exact and glamorous proportions needed to render them irresistible to any investigator who comes to know them." These identical words may be applied to the interferons. The study of the interrelationships between these two natural substances of abiding interest cannot but be equally seductive.

VII. Summary

Bacterial endotoxin induces moderate amounts of interferon in the intact animal, and in macrophages and lymphocytes *in vitro*. The animal is resistant to reinduction. Induction requires mRNA and protein synthesis and is subject to similar controls as induction by viruses and polyribonucleotides. The type of interferon induced is antigenically similar to that induced by viruses, but differences in molecular weight and physiochemical stability remain to be explained. It is unlikely that endotoxin or any component thereof will be therapeutically useful as an inducer. The significance of interferon induced by endotoxin during disease states, and the role of interferon in the genesis of fever remain to be explored.

VIII. References

Atkins, E., 1960, Pathogenesis of fever, *Physiol. Rev.* **40:**580.
Atkins, E., and Francis, L., 1978, Pathogenesis of fever in delayed hypersensitivity: Factors influencing release of pyrogen-inducing lymphokines, *Infect. Immun.* **21:**806.
Atkins, E., Bodel, P., and Francis, L., 1967, Release of an endogenous pyrogen *in vitro* from rabbit mononuclear cells, *J. Exp. Med.* **126:**357.
Beeson, P. B., 1948, Temperature-elevating effect of a substance obtained from polymorphonuclear leucocytes, *J. Clin. Invest.* **27:**524.
Bennett, I. L., 1964, Approaches to the mechanism of endotoxin action, in: *Bacterial Endotoxins* (M. Landy and W. Braun, eds.), pp. xiii–xiv, Rutgers University Press, New Brunswick, N.J.
Bernheim, H. A., Block, L. H., and Atkins, E., 1979, Fever: Pathogenesis, physiology and purpose, *Ann. Intern. Med.* **91:**261.
Billiau, A., 1969, Systemic induction of interferon, Thesis for degree of geaggregeerde bij bet Hoger Onderwijs, University of Leuven, Belgium.
Bodel, P. T., and Atkins, E., 1966, Human leucocyte pyrogen producing fever in rabbits, *Proc. Soc. Exp. Biol. Med.* **121:**943.
Borecky, L., 1968, Interferon induction by non-viral agents, in: *Medical and Applied Virology* (M. Sanders and E. H. Lennette, eds.), p. 181, Green, Missouri.
Cantell, K., and Paucker, K., 1963, Quantitative studies on viral interference in suspended L-cells. IV. Production and assay of interferon, *Virology* **21:**11.
Dinarello, C. A., Renfer, L., and Wolff, S. M., 1977, The production of antibody against human leucocytic pyrogen, *J. Clin. Invest.* **60:**456.
Feingold, D. S., Keleti, G., and Youngner, J. S., 1976, Antiviral activity of *Brucella abortus* preparations: Separation of active components, *Infect. Immun.* **13:**763.
Field, A. K., Tytell, A. A., Lampson, G. P., and Hilleman, M. R., 1968, Inducers of interferon and host resistance. V. *In vitro* studies, *Proc. Natl. Acad. Sci. USA* **61:**340.
Finkelstein, M. S., Bausek, G. H., and Merigan, T. C., 1968, Interferon inducers *in vitro*: Difference in sensitivity to inhibitors of RNA and protein synthesis, *Science* **161:**465.
Gledhill, A. W., 1959a, The effect of bacterial endotoxin on resistance of mice to ectromelia, *Br. J. Exp. Pathol.* **40:**195.
Gledhill, A. W., 1959b, Sparing effect of serum of mice treated with endotoxin upon certain murine virus diseases, *Nature (London)* **183:**185.

Gledhill, A. W., 1959c, The interference of mouse hepatitis virus with ectromelia in mice and a possible explanation of its mechanism, *Br. J. Exp. Pathol.* **40**:291.

Gledhill, A. W., 1964, Influence of endotoxin upon the pathogenesis of viral infections, in: *Bacterial Endotoxins* (M. Landy and W. Braun, eds.), p. 410, Rutgers University Press, New Brunswick, N.J.

Hahn, H. H., Char, D. C., Postell, W. B., and Wood, W. B., Jr., 1967, Studies on the pathogenesis of fever. XV. The production of endogenous pyrogen by peritoneal macrophages, *J. Exp. Med.* **126**:385.

Hallum, J. V., Youngner, J. S., and Stinebring, W. R., 1965, Interferon activity associated with high molecular weight proteins in the circulation of mice injected with endotoxin or bacteria, *Virology* **27**:429.

Ho, M., 1964a, Identification and induction of interferon, *Bacteriol. Rev.* **28**:367.

Ho, M., 1964b, Interferon-like viral inhibitor in rabbits after intravenous administration of endotoxin, *Science* **146**:1472.

Ho, M., 1981, Induction of the interferon protein in the animal: General considerations, in: *The Interferon System: A Review to 1982* (S. Baron, F. Dianzani, and J. Stanton, eds.), *Texas Reports on Biology and Medicine*, in press.

Ho, M., and Armstrong, J. A., 1975, Interferon, *Annu. Rev. Microbiol.* **29**:131.

Ho, M., and Kass, E. H., 1958, Protective effect of components of normal blood against the lethal action of endotoxin, *J. Clin. Med.* **51**:297.

Ho, M., and Ke, Y. H., 1970, The mechanisms of stimulation of interferon production by a complexed polyribonucleotide, *Virology* **40**:683.

Ho, M., and Kono, Y., 1965a, Effect of antinomycin D on virus and endotoxin-induced interferon-like inhibitors in rabbits, *Proc. Natl. Acad. Sci. USA* **53**:220.

Ho, M., and Kono, Y., 1965b, Tolerance to the induction of interferons by endotoxin and virus, *J. Clin. Invest.* **44**:1059.

Ho, M., and Postic, B., 1967, Prospects for applying interferon to man, PAHO/WHO Scientific Publication 147, pp. 632–649, Washington, D.C.

Ho, M., Kono, Y., and Breinig, M. K., 1965, Tolerance to the induction of interferon by endotoxin and virus: Role of a humoral factor, *Proc. Soc. Exp. Biol. Med.* **119**:1227.

Ho, M., Breinig, M. K., Postic, B., and Armstrong, J. A., 1970, Effect of preinjections on the stimulation of interferon by a complexed polynucleotide, endotoxin and virus, *Ann. N.Y. Acad. Sci.* **173**:680.

Ho, M., Ke, Y. H., and Armstrong, J. A., 1973, Mechanism of interferon induction by endotoxin, *J. Infect. Dis.* **128**(Suppl.):220.

Hook, E. W., and Wagner, R. R., 1959, The resistance-promoting activity of endotoxins and other microbial products. II. Protection against the neurotoxic action of influenza virus, *J. Immunol.* **83**:310.

Ito, Y., Nagata, I., and Kunii, A., 1973, Mechanism of endotoxin-type interferon production in mice, *Virology* **52**:439.

Jullien, P., and DeMaeyer, E., 1966, Interferon synthesis in X-irradiated animals. I. Depression of circulating interferon in C_3H mice after whole body irradiation, *Int. J. Radiat. Biol.* **11**:567.

Kadish, A. S., Tansey, F. A., Yu, G. S. M., Doyle, A. T., and Bloom, B. R., 1980, Interferon as a mediator of human lymphocyte suppression, *J. Exp. Med.* **151**:637.

Kawada, Y., Yamamoto, Y., Fujisawa, J., and Watanabe, Y., 1980, Antigenic relationships among various interferon species, *Ann. N.Y. Acad. Sci.* **350**:422.

Ke, Y. H., Ho, M., and Merigan, T. C., 1966a, Heterogeneity of rabbit serum interferons, *Nature (London)* **211**:541.

Ke, Y. H., Singer, S. H., Postic, B., and Ho, M., 1966b, Effect of puromycin on virus and endotoxin-induced interferonlike inhibitors in rabbits, *Proc. Soc. Exp. Biol. Med.* **121**:181.

Keleti, G., Feingold, D. S., and Youngner, J. S., 1974, Interferon induction in mice by lipopolysaccharide from *Brucella abortus, Infec. Immun.* **10**:282.

Kobayashi, S., Yasui, O., and Masuzumi, M., 1969, Studies on early-appearing interferon *in vitro*. I. Production of endotoxin-induced interferon by mouse spleen cells cultured *in vitro, Proc. Soc. Exp. Biol. Med.* **131**:487.

Laskovic, V., Borecky, L., Sike, D., Master, L., and Bauer, S., 1967, The temperature requirement for interferon production in cells stimulated by Newcastle disease virus or microbial agents *in vitro, Acta Virol. (Engl. Ed.)* **11**:500.

Likar, M., Bartley, E. O., and Wilson, D. C., 1959, Observations on the interaction of poliovirus and host cells *in vitro*. III. The effect of some bacterial metabolites and endotoxins, *Br. J. Exp. Pathol.* **40**:391.

Maehara, N., and Ho, M., 1977, Cellular origin of interferon induced by bacterial lipopolysaccharide, *Infect. Immun.* **15**:78.

Maehara, N., Ho, M., and Armstrong, J. A., 1977, Differences in mouse interferons according to cell source and mode of induction, *Infect. Immun.* **17**:572.

Michaels, R. H., Weinberger, M. M., and Ho, M., 1965, Circulating interferon-like viral inhibitor in patients with meningitis due to *Haemophilus influenzae, N. Engl. J. Med.* **272**:1148.

Mickenberg, I. D., Root, R. K., and Wolff, S. M., 1972, Bactericidal and metabolic properties of human eosinophils, *Blood* **39**:67.

Postic, B., DeAngelis, C., Breinig, M. K., and Ho, M., 1966, Effect of temperature on the induction of interferons by endotoxin and virus, *J. Bacteriol.* **91**:1277.

Postic, B., DeAngelis, C., Breinig, M. K., and Ho, M., 1967, Effects of cortisol and adrenalectomy on induction of interferon by endotoxin, *Proc. Soc. Exp. Biol. Med.* **125**:89.

Sauter, C., and Wolfensberger, C., 1980, Interferon in human serum after injection of endotoxin, *Lancet* **2**:852.

Scott, G. M., Secher, D. S., Flowers, D., Bate, J., Cantell, K., and Tyrrell, D. A. J., 1981, Toxicity of interferon, *Br. Med. J.* **282**:1345.

Skarnes, R. C., Rosen, F. S., Shear, M. J., and Landy, M., 1958, Inactivation of endotoxin by a humoral component. II. Interaction of endotoxin with serum and plasma. *J. Exp. Med.* **108**:685.

Smith, T. J., and Wagner, R. R., 1967, Rabbit macrophage interferon. I. Conditions for biosynthesis of virus-infected and uninfected cells, *J. Exp. Med.* **125**:559.

Staub, H. M., 1965, Bacterial lipido-proteino-polysaccharide ("O" somatic antigens) extracted with trichloroacetic acid, in: *Methods in Carbohydrate Chemistry* (R. L. Whistler, J. N. BeMiller, and M. L. Wolfram, eds.), Vol. 5, pp. 92–93, Academic Press, New York.

Stewart, W. E., II, 1979, *The Interferon System*, Springer-Verlag, Berlin.

Stinebring, W. R., and Youngner, J. S., 1964, Patterns of interferon appearance in mice injected with bacteria or bacterial endotoxin, *Nature (London)* **204**:712.

Suter, E., and Kirsanow, E. M., 1961, Hyperreactivity to endotoxin in mice infected with mycobacteria—Induction and elicitation of the reactions, *Immunology* **4**:354.

Tan, Y. H., Armstrong, J. A., Ke, Y. H., and Ho, M., 1970, Regulation of cellular interferon production: Enhancement by antimetabolites, *Proc. Natl. Acad. Sci. USA* **67**:464.

Trinchieri, G., and Santoli, D., 1978, Anti-viral activity induced by culturing lymphocytes with tumor derived or virus-transformed cells: Enhancement of human natural killer cell activity by interferon and antagonistic inhibition of susceptibility of target cells to lysis, *J. Exp. Med.* **147**:1314.

Vilcek, J., 1970, Metabolic determinants of the induction of interferon by a synthetic double-stranded polynucleotide in rabbit kidney cells, *Ann. N.Y. Acad. Sci.* **173**:390.

Wagner, R. R., 1955, Cytotoxic action of influenza virus: Failure to induce acquired resistance phenomenon in tissue culture, *Proc. Soc. Exp. Biol. Med.* **90**:214.

Wagner, R. R., Snyder, R. M., Hook, E. W., and Luttrell, C. N., 1959, Effect of bacterial endotoxin on resistance of mice to viral encephalitides, *J. Immunol.* **83**:87.

Westphal, O., and Jann, K., 1965, Bacterial lipopolysaccharides: Extraction with phenol-water and further application of the procedure, in: *Methods in Carbohydrate Chemistry* (R. L. Whistler, J. N. BeMiller, and M. L. Wolfram, eds.), Vol. 5, pp. 83–91, Academic Press, New York.

Youngner, J. S., and Feingold, D. S., 1967, Interferon production in mice by cell wall mutants of *Salmonella typhimurium, J. Virol.* **1**:1164.

Youngner, J. S., and Hallum, J. V., 1968, Interferon production in mice by double-stranded synthetic polynucleotides: Induction or release?, *Virology* **35**:177.

Youngner, J. S., and Salvin, S. B., 1973, Production and properties of migration inhibitory factor and interferon in the circulation of mice with delayed hypersensitivity, *J. Immunol.* **111**:1914.

Youngner, J. S., and Stinebring, W. R., 1964, Interferon production in chickens injected with *Brucella abortus, Science* **144**:1022.

Youngner, J. S., and Stinebring, W. R., 1965, Interferon appearance stimulated by endo-toxin, bacteria, or viruses in mice pretreated with *Escherichia coli* endotoxin or infected with *Mycobacterium tuberculosis, Nature (London)* **208**:456.

Youngner, J. S., Stinebring, W. R., and Taube, S. G., 1965, Influence of inhibitors of protein synthesis on interferon formation in mice, *Virology* **27**:541.

20

Endotoxin-Induced Release of Colony-Stimulating Factor

Dov H. Pluznik

I. Introduction

The introduction of an *in vitro* technique for cloning granulocyte and macrophage precursor cells by Pluznik and Sachs (1965) and Bradley and Metcalf (1966) provided a new approach to the evaluation of the regulation of granulopoiesis and macrophage formation. Committed precursor cells for granulocyte and macrophage differentiation can proliferate in soft agar cultures to form colonies of granulocytes and/or macrophages (Ichikawa *et al.*, 1966). Colonies arise from single cells termed *colony-forming cells* (CFC) (Pluznik and Sachs, 1966). Colony formation is wholly dependent upon the constant presence of a stimulatory glycoprotein substance named *colony-stimulating factor* (CSF).

This stimulating factor can be derived from a number of sources. A variety of cells either freshly isolated or in continuous lines can spontaneously release this factor (Pluznik and Sachs, 1966; Ichikawa *et al.*, 1966; Bradley and Metcalf, 1966; Bradley and Sumner, 1968; Metcalf, 1971). CSF can be obtained from body fluids such as serum and urine (Robinson *et al.*, 1967, 1969), and from organs by direct release or after extraction (Bradley *et al.*, 1971; Sheridan and Metcalf, 1973).

Release of CSF can be induced in lymphoid cell cultures by mixed lymphocyte culture, antigens, and mitogens (Parker and Metcalf, 1974a,b; Ruscetti and Chervenick, 1975). Among the most potent inducers of CSF generation are bacterial endotoxins (Metcalf, 1971; Quesenberry *et al.*, 1972; Apte and Pluznik, 1976a).

Dov H. Pluznik • Department of Life Sciences, Bar-Ilan University, Ramat-Gan, Israel; present address: Laboratory of Microbiology and Immunology, National Institute of Dental Research, NIH, Bethesda, Maryland 20205.

Bacterial endotoxins have a pronounced effect on the hemopoietic system, (for review see Kass and Wolff, 1973). They have a particular effect on the granulopoietic system (Metcalf, 1971; Quesenberry *et al.*, 1972, 1973). After an injection of endotoxin into mice, there is an initial early granulocytopemia, together with a striking increase in serum CSF. After the initial granulocytopemia, there is a release of marrow granulocytes into the peripheral blood and a subsequent wave of proliferation and differentiation along the granulocytic pathway in hemopoietic tissues, evidenced by a rise in marrow and splenic precursors for granulocyte-macrophage differentiation.

Bacterial endotoxins are lipopolysaccharides (LPS), products of gram-negative bacteria. They are macromolecules composed of three regions differing in their chemical and biological properties. The O-specific polysaccharide (region I) carries the main serological specificity of bacteria. It is linked to a core polysaccharide (region II) that is common to groups of bacteria. This polysaccharide core is linked through a 2-keto-3-deoxyoctonoate (KDO) trisaccharide to the lipid component (regions III), termed *lipid A* (Westphal, 1975).

The predominant assay for CSF is based on the CSF-dependent formation of colonies of granulocytes/macrophages by a fixed number of bone marrow cells in soft agar cultures (Pluznik and Sachs, 1966). The number of colonies developing in the soft agar supplemented with various dilutions of a CSF source represent the CSF activity (Apte and Pluznik, 1976b). To quantitate CSF activity in the serum, mice are injected i.v. with LPS. Six hours later, the mice are sacrificed and blood is collected for serum. Samples of sera at various dilutions are incorporated into the hard agar, to a final concentration of 10%. To quantitate CSF activity in supernatants of spleen cell cultures, usually 2×10^6 spleen cells/ml are incubated in the presence of LPS for 4 days. At the end of the incubation period, the culture fluids are collected, diluted, and incorporated into the hard agar at a final concentration of 15% (Apte *et al.*, 1977a). Bone marrow cells (10^5) in soft agar medium are seeded on top of the hard agar medium. After 7 days of incubation at 37°C in 5% CO_2, cell colonies developing in the agar are counted. The number of colonies represent CSF activity. In the present chapter the following questions will be elucidated: Which part of the LPS molecule is responsible for inducing generation of CSF? What is the genetic control that governs the generation of CSF by LPS? What are the cellular interactions involved in generation of LPS-induced CSF?

II. Identification of the Active Portion of LPS That Induces CSF

In order to examine which part of the LPS molecule is active in inducing generation of CSF, two experimental approaches have been used: The first approach was to separate between the lipid A moiety and the polysaccharide moiety by acid hydrolysis and test each portion separately. The second approach was to use defective lipopolysaccharides (i.e., glycolipids obtained from *Salmonella* R mutants), which lack the O-specific side chains and different parts of the core polysaccharide; however, they all contain the lipid region (Lüderitz *et al.*, 1966). Such glycolipids showed potent endotoxin activity, indicating that lipid A may be the active part of the LPS molecule (Kasai and Nowotny, 1967; Kim and Watson, 1967; Lüderitz *et al.*, 1966).

The effects of LPS, its breakdown products obtained by acid hydrolysis, and glycolipids from R mutants of *Salmonella minnesota* on CSF activity in the serum and in spleen cell cultures were tested. For the *in vivo* experiments, groups of mice were injected i.v. with increasing amounts of lipid A, PS, and the parent LPS (10–100 μg/mouse) to obtain serum CSF. An increase in CSF activity (Table 1) is shown in post-lipid A sera as the amount of the lipid A injected increased. Post-PS serum showed no significant CSF activity. Under the same experimental conditions, groups of mice were injected i.v. with mR595, mR7, or

Table 1. Effect of LPS, Its Breakdown Products, and Its R-Mutant Glycolipids on *in Vivo* and *in Vitro* Generation of CSF

	μg	CSF activity[a]	
		In vivo[b]	*In vitro*[c]
LPS	10	130	211
	50	186	234
	100	198	234
Breakdown products			
Lipid A	10	75	234
	50	135	248
	100	162	257
PS	10	0	57
	50	0	91
	100	12	111
Glycolipids			
R595	10	98	271
R7	10	110	260
R345	10	117	242
LPS	10	107	211

[a] Number of colonies/10^5 bone marrow cells represents CSF activity.

[b] Mice were injected i.v. with the various products. Six hours later the serum was obtained and assayed for CSF activity.

[c] Spleen cells (2×10^6/ml) were cultured with the various products. Four days later the supernatants were collected and assayed for CSF activity. (For details see Apte *et al.*, 1976, 1977a).

mR345 glycolipid (10 μg/mouse) to generate serum CSF. The results of these experiments are also shown in Table 1. All three preparations of glycolipids used and the complete LPS (from *Salmonella abortus equi*) induced a similar increase in serum CSF as measured by their ability to support growth of bone marrow colonies in soft agar cultures. In the *in vitro* experiments (Apte *et al.*, 1977a), lipid A, PS, and the glycolipids were incubated with spleen cell cultures for 4 days, after which the supernatants were assayed for CSF activity. Lipid A was shown to be very active, inducing high levels of CSF comparable to or even higer than those obtained by LPS. The PS was significantly less active and showed some enhancement of CSF generation only at high doses (Table 1). The colonies induced by PS were diffuse and contained a relatively small number of cells, whereas those induced by all concentrations of lipid A or LPS were compact and contained many more cells, indicating that more CSF or a better source of CSF was secreted in the latter case. All glycolipids were active to the same extent as complete LPS (Table 1).

It has been shown that lipid A induces biological activities as pyrogenicity, interferon release, tumor inhibition and necrosis, and enhancement of the dermal reactivity to epinephrine (Galanos *et al.*, 1972; Lüderitz *et al.*, 1973), as well as bone marrow-derived (B) lymphocyte mitogenesis and adjuvanticity (Andersson *et al.*, 1973; Chiller *et al.*, 1973). However, there are also reports that preparations of lipid A have not induced some of these activities (Mergenhagen *et al.*, 1963; Neter and Ribi, 1963; Ribi *et al.*, 1964). In view of these reports, it is interesting to mention that it was shown by Chang *et al.* (1974) that PS also has the ability to induce generation of CSF in the serum.

III. Genetic Control of Generation of Serum CSF

The biological reactions that are the basis of the endotoxic response became possible to analyze genetically when a mouse strain (C3H/HeJ) was discovered to be defective in LPS-induced responses (Sultzer, 1968).

These mice are resistant to the lethal toxic effects of LPS, and the number of mononuclear cells in the peritoneal cavity increases after an i.p. injection of LPS (Sultzer, 1968). The toxic response and the mobilization of leukocytes to the site of injection are under a polygenic inheritance control (Sultzer, 1972). In addition, it was shown that these mice are low responders to the mitogenic and immunogenic responses of LPS, and these two latter responses are monogenically controlled (Coutinho *et al.*, 1974; Watson and Riblet, 1974). Apte and Pluznik (1976a) have shown that the C3H/HeJ mice fail to induce granulopoietic responses after an injection of LPS, and no increase in serum CSF was observed. This unresponsiveness was not due to rapid clearance of endotoxin from the circulation, or to inactivation of the endotoxin by opsonizing antibodies, or to the presence of high concentrations of serum inhibitors extractable by chloroform (Apte and Pluznik, 1976a).

For studying the genetic control of host response in generation of LPS-induced CSF, two strains of mice have been used: C3H/eB, a high responder to LPS, and C3H/HeJ, a low responder to LPS. Reciprocal breeding was conducted to produce F_1 and F_2 progeny. In addition, F_1 mice were backcrossed to the C3H/HeJ parental type.

It was shown that C3H/eB mice responded to LPS injected by generating high levels of serum CSF (Apte and Pluznik, 1976b). As the amount of the LPS injected was increased, there was a corresponding increase in the levels of CSF, up to an optimum dose of 25 μg/mouse. Higher doses caused no additional stimulation and a plateau in the levels of CSF was observed. C3H/HeJ mice failed to show significant levels of serum CSF at the LPS concentrations used, and only a small rise in CSF level was observed at the highest concentration used (100 μg/mouse).

The genetic control of CSF generation was studied by injecting a single dose of LPS (10 μg/ml) into F_1, F_2, and the backcrossed generation of F_1 to the original low-responder parental type (C3H/HeJ). The CSF activity present in the sera of the various groups is shown in Fig. 1. As seen in Fig. 1, C3H/eB mice and the F_1 hybrid generated high levels of serum CSF, whereas C3H/HeJ mice did not generate any detectable levels of serum. Eighty percent of the mice from the F_2 population generated CSF activity similar to the high-responder C3H/eB parent strain, whereas the remaining 20% showed no response at all. In 45% of the mice from the backcross (F_1 × C3H/HeJ) population, elevated levels of serum CSF were observed, whereas 55% of the mice from this generation did not give any detectable amounts of CSF. Sex differences were not found to affect the generation of serum CSF in the different hybrids that were tested. These results suggest that the control of CSF production in response to LPS is governed by a single autosomal dominant gene.

It has been shown that lipid A is the active component of the LPS molecule in stimulating CSF generation in the serum (Apte et al., 1976). A series of experiments was performed, therefore, to examine whether the genetic control mechanisms governing the CSF response to LPS are affected by the complexity of the LPS molecule. We used a glycolipid from a rough (R) mutant of S. minnesota 595, which contains only the lipid A portion linked to a trisaccharide of KDO. Mice from the F_1 generation, as well as mice from each parental type, were injected with 10 μg of LPS-R595 and their serum CSF responses were evaluated as previously described. Results of these experiments (Fig. 2)

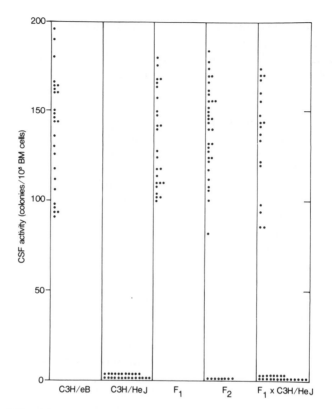

Figure 1. CSF activity in sera of the parental strains, the F_1 hybrid, the F_2 generation, and the backcross ($F_1 \times$ C3H/HeJ), after injection of LPS. Number of colonies/10^5 bone marrow cells represents CSF activity. Each point represents CSF activity of an individual mouse.

indicate that F_1 mice generated a high level of serum CSF similar to the response of C3H/eB mice. The results suggest that the genetic control mechanisms of LPS-R595-induced CSF are the same as those observed with complete LPS; generation of CSF is monogenically controlled.

The mechanisms affecting *in vivo* granulopoiesis and macrophage formation are unknown at present and the role of CSF as an *in vivo* "granulopoietin" has not been unequivocally demonstrated. An experimental approach to elucidate the relation between CSF activity in the serum and proliferation of CFC in the spleen was undertaken. Individual mice from the backcross ($F_1 \times$ C3H/HeJ) and F_2 generations were examined for their ability to generate serum CSF and to increase the splenic CFC after a single injection of 10 μg LPS. Each of the mice was bled, 6 hr after injection, to obtain serum for quantitating CSF activity.

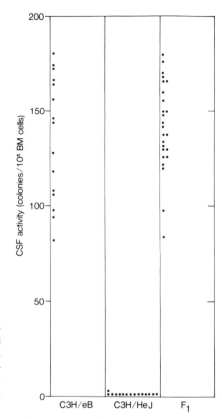

Figure 2. CSF activity in sera of the parental strains (C3H/eB and C3H/HeJ) and the F_1 hybrid, after injection of the glycolipid R595. Number of colonies/10^5 bone marrow cells represents CSF activity. Each point represents CSF activity of an individual mouse.

The same mice were allowed to remain alive six more days after which their spleen cells were cloned in soft agar cultures to determine the values of CFC. Results of such an experiment are shown in Table 2. Mice showing low CSF activity (0 colonies/10^5 bone marrow cells) also showed a low CFC response (83.3%). As for mice showing a high CSF activity (> 74 colonies/10^5 bone marrow cells), 58.3% showed a high CFC response, 33.3% showed an intermediate response, and only 8.3% showed a low response.

From the results of this last experiment in which the association between serum CSF and splenic CFC responses was tested, it seems that CSF may have a physiological regulative role in the process of granulopoiesis. This is evidenced by the fact that all mice that showed a high CFC response generated high levels of CSF, and most of the mice that did not react to the LPS (no increase in splenic CFC) showed no detectable levels of CSF. Mice that showed an intermediate response of splenic CFC usually generated high levels of CSF. This last

Table 2. Association between Generation of CSF in the Serum and Proliferation of CFC in the Spleen, in F_2 and Backcross ($F_1 \times$ C3H/HeJ) Mice Injected with LPS[a]

	Number of mice			
	CFC response			
CSF activity	High	Intermediate	Low	Total
High	14 (58.3%)	8 (33.3%)	2 (8.3%)	24
Low	0 (0%)	2 (16.6%)	10 (83.3%)	12

[a] High CFC response, > 280 CFC/5×10^6 spleen cells; intermediate, 150–280 CFC/5×10^6 spleen cells; low, < 150 CFC/5×10^6 spleen cells. High CSF activity, > 74 CFC/10^5 bone marrow cells; low, CFC/10^5 bone marrow cells. (For details see Apte and Pluznik, 1976a.)

fact may indicate that CSF activity can be interfered by other factors, leukopoietic or endocrine, and, thus, CSF may be an important regulator but not the sole factor regulating granulopoiesis.

IV. Cellular Interactions in the Generation of CSF

The ability of CFC from C3H/HeJ mice to proliferate in the presence of an external source of CSF obtained from post-LPS serum of C3H/eB mice (Apte and Pluznik, 1976a) shows that the unresponsiveness of these CFC is not due to a defect in the progenitor cells but rather to a defect in generation of serum CSF in the C3H/HeJ mice. To identify the origin of the cellular population responsible for LPS-induced serum CSF, low-responder C3H/HeJ mice were lethally irradiated and hemopoietically reconstituted with bone marrow cells from the high-responder strain C3H/eB and C3H/eB nude mice, as well as with C3H/HeJ bone marrow cells as controls. Only mice reconstituted with bone marrow cells from C3H/eB or C3H/eB nude mice restored the serum CSF response to LPS, demonstrating that cells generating CSF in response to LPS are of bone marrow origin (Apte and Pluznik, 1976c).

Bone marrow cells are the source of all hemopoietic cells in the adult mouse. These cells differentiate and mature to the various blood cells. The spleen is the organ that contains all these blood cells at various stages of differentiation and maturation. This organ was selected, therefore, for further experiments to identify the cells participating in generating CSF after LPS administration. Large amounts of CSF were generated by spleen cells from C3H/eB mice incubated with LPS, whereas only a weak activity of CSF was detected in supernatants of spleen cells from C3H/HeJ mice incubated with LPS (Apte *et al.*, 1977a). Experiments were performed, therefore, to identify the cell types that participate in generation of CSF. Since LPS is a B-cell mitogen, special emphasis was put on the lymphoid cells in the spleen.

The ability of spleen cells to secrete CSF in response to LPS decreased gradually with macrophage depletion (Apte *et al.*, 1980). Furthermore, when LPS-induced nonadherent lymphocyte cultures were reconstituted with increasing numbers of 10% proteose peptone-elicited peritoneal macrophages, large amounts of CSF were generated (Apte *et al.*, 1980). Optimal levels of CSF were elaborated in mixed cultures containing 2×10^6 nonadherent cells and 2×10^5 macrophages/ ml; higher concentrations of macrophages actually exerted an inhibitory effect on CSF generation. Macrophages, when cultured at the same cell concentration as that added to lymphocyte cultures, generated only minute amounts of CSF (Table 3). Thus, it seems that an interaction between macrophages and lymphocytes leads to secretion of large amounts of CSF in macrophage–lymphocyte mixed cultures.

In further experiments, the abilities of resident and elicited peritoneal macrophage populations to generate CSF were compared. Peritoneal resident macrophages, from unstimulated mice of the strains C3H/eB and C3H/HeJ, were cultured (2×10^5/ml) for 4 days with increasing concentrations of LPS, and the CSF activity was tested in the culture fluids collected. Elevated levels of CSF were detected in supernatants of macrophage cultures from C3H/eB mice, whereas only small

Table 3. Generation of LPS-Induced CSF in Spleen Cell Cultures of C3H/eB and C3H/HeJ Mice Using Various Combinations of Cells or Cell Products

Cell populations[a]	CSF activity[b]
Spleen (C3H/eB)	175
Nonadherent spleen (NA) (C3H/eB)	38
Macrophages (Mφ) (C3H/eB)	18
NA (C3H/eB) + Mφ (C3H/eB)	190
NA (C3H/eB) + Mφ supernatant (C3H/eB)	155
NA (C3H/eB) + Mφ supernatant (C3H/HeJ)	54
NA (C3H/HeJ) + Mφ supernatant (C3H/eB)	164
NA (C3H/HeJ) + Mφ supernatant (c3H/HeJ)	46

[a] Cell populations were incubated for 4 days with LPS, after which the supernatants were assayed for CSF activity. Macrophage (Mφ) supernatants were obtained from proteose peptone-elicited peritoneal macrophages incubated with LPS.

[b] Number of colonies/10^5 bone marrow cells represents CSF activity. (For details see Apte *et al.*, 1980.)

amounts of CSF were detected in C3H/HeJ macrophage cultures (Apte *et al.*, 1979).

Different results were obtained with elicited macrophages. Such macrophages were obtained from the peritoneal cavity of C3H/eB mice that had been injected 3 days before with proteose peptone (2 to 10%) or with thioglycolate broth (2.93%). Injection of mice with proteose peptone broth of different concentrations caused a suppression in secretion of CSF as the concentration of the proteose peptone solution injected increased. Peritoneal macrophages from such treated mice, when cultured with LPS (50 μg/ml), elaborated only a small amount of CSF (Apte *et al.*, 1979).

To elucidate the nature of the cellular interactions between proteose peptone-elicited macrophages and splenic lymphocytes in the process of generation of CSF, the following experiments were undertaken. Macrophage supernatants were prepared by incubating proteose peptone-elicited macrophages (2×10^6/ml) for 24 to 48 hr with or without LPS (50 μg/ml), and the culture fluids were thereafter collected and diluted 1 : 2 in culture medium. Splenic lymphocytes (2×10^6/ml) were suspended in such macrophage supernatants and were subsequently cultured for 4 days, either with or without LPS (50 μg/ml). The culture fluids of such lymphocyte suspensions were incorporated into the agar as a source of CSF at a final concentration of 15%. The results indicate that active macrophage supernatants were generated when LPS was added to cultures, and such supernatants could activate lymphocytes to secrete CSF, even when the lymphocyte cultures were not stimulated directly by the mitogen. Supernatants prepared from macrophage cultures not stimulated by LPS manifested only a weak lymphostimulatory activity, and the addition of LPS to such lymphocyte cultures improved only partially the secretion of CSF (Apte *et al.*, 1980).

The availability of LPS low- and high-responder strains of mice provided a useful tool for studying whether the pathway of the cellular interactions involved in generation of CSF in lymphoid cell cultures is unidirectional.

LPS-induced macrophage supernatants were prepared from proteose peptone-elicited macrophage cultures from both C3H/HeJ and C3H/eB mice. The activity of each type of macrophage culture fluid was cross-tested on lymphocytes from both strains of mice. The results of these experiments (Table 3) show that macrophages obtained from C3H/HeJ mice did not elaborate any significant lymphocyte helper activity. However, supernatants generated by macrophages from C3H/eB mice activated lymphocytes from C3H/HeJ mice to secrete large amounts of CSF. These data provide evidence that generation of CSF in mixed cultures of proteose peptone-elicited macrophages and splenic lymphocytes occurs via a unidirectional pathway; that is, proteose peptone-elicited macrophages obtained from a high-responder strain secrete a soluble factor(s) that induces lymphocytes from an LPS low-responder strain to generate CSF (Apte *et al.*, 1980).

In further experiments, it was found that the population of splenic lymphocytes generating high levels of CSF in response to LPS-induced macrophage supernatants most likely consists of B lymphocytes. This was evidenced by showing that nonadherent lymphocytes from congenitally athymic nude (*nu/nu*) mice responded to such macrophage helper factor(s), as did lymphocytes from conventional mice (Apte *et al.*, 1980). In addition, it was also observed that thymocytes cultured with or without LPS-induced macrophage culture fluids did not elaborate any detectable amounts of CSF.

It also appears that the B-cell subpopulation that interacts with such macrophage factors may be distinct from that that undergoes mitogenesis and blast transformation in response to LPS. In a previous study (Apte *et al.*, 1977a), we have shown that large amounts of CSF are present in the culture fluids of LPS-stimulated spleen cells from CBA/N mice. Such mice carry an X-linked inability to mount normal immune response to thymus-independent antigens and poor *in vitro* responsiveness to B-cell mitogens (Scher *et al.*, 1975). Associated with this defect is a reduced number of splenic B cells, low serum IgM relative to putative IgD, and an immature pattern of cell surface immunoglobulin density distribution (Finkelman *et al.*, 1975). In addition, modification of the LPS molecule with polymyxin B caused a marked reduction in mitogenic activity, although the ability to induce CSF was not significantly altered. DNA inhibitors did not affect CSF release, although they completely inhibited mitogenicity (Apte *et al.*, 1977a). Thus, the B-cell population participating in the process of LPS-induced CSF generation is a nondividing population without need for DNA synthesis.

V. Cellular Interactions Involved in Overcoming Tolerance to LPS-Induced CSF

As mentioned previously, injection of LPS into mice leads to a rapid elevation in serum CSF. Like other biological effects of LPS, repeated injections result in diminished generation of serum CSF (Quesenberry *et al.*, 1975), a state termed *endotoxin resistance* or *endotoxin tolerance*. This state could not be transferred with serum from tolerant mice to normal mice (Staber, 1980), thus excluding a humoral factor responsible for this tolerance.

To elucidate the cellular interactions involved in LPS induction of serum CSF tolerance, the following experiments were undertaken. ICR mice were injected with 25 μg LPS, and at 24-hr intervals were reinjected with the same dose of LPS. Six hours later, the mice were bled and sera assayed for CSF activity in soft agar cultures. It was found that an interval of 72–96 hr between the two injections of LPS induced the best state of tolerance, with almost no CSF activity detected in the sera. Intraperitoneal injection of 5×10^7 spleen cells from untreated ICR mice to tolerant mice, 24 hr prior to the second administration of LPS, was followed by generation of serum CSF in the tolerant mice. Such spleen cells could efficiently be substituted by injecting a mixture of 5×10^7 nonadherent spleen cells with 2.5×10^6 thioglycolate-elicited peritoneal macrophages into the tolerant mice. Injection of nonadherent cells or macrophages alone was not as efficient in enabling the tolerant mice to generate serum CSF. In the absence of exogenous spleen cells, generation of LPS-induced serum CSF is delayed for several days in the tolerant mice (Williams *et al.*, 1981).

These data support the previously described *in vitro* experiments in which a cooperation between macrophages and lymphocytes is required in order to generate CSF by LPS (Apte *et al.*, 1979). They may suggest that similar cellular interactions are taking place *in vivo* in generation of LPS-induced CSF.

The present experiments describing the cellular interactions involved in LPS-induced CSF generation can be summarized as follows (Fig. 3): LPS interacts first with macrophages. Resident macrophages will directly generate CSF. Elicited macrophages will release a substance that interacts with B lymphocytes, which in turn then release CSF. For resident macrophages, LPS is the only signal required for generation of CSF, whereas with elicited macrophages, LPS is the first signal, and the subsequent substance released by the macrophages is the second signal, which activates the B lymphocytes to release CSF. It remains to be tested whether the substance released by elicited macrophages is related to interleukin 1 (Mizel and Mizel, 1981). It also has to be tested whether the two CSFs, one that is directly released from the resident macrophages and the second that is released from the B lymphocytes, are identical in their biochemical properties.

VI. Association between LPS-Induced Serum CSF and LPS-Induced Serum Interferon (IFN)

Injection of LPS into mice causes the appearance of IFN in the bloodstream (Stinebring and Youngner, 1964; Hallum *et al.*, 1965). The time kinetics of appearance of IFN in the serum is very rapid and

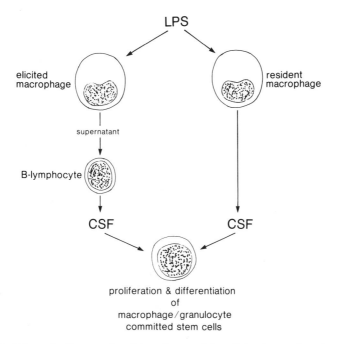

Figure 3. Schematic diagram describing the possible cellular interactions involved in LPS-induced generation of CSF.

reaches its peak values between 2 and 6 hr after injection of LPS (Apte *et al.*, 1977b). Such time course very much resembles the kinetics of LPS-induced serum CSF. The genetic control of generation of LPS-induced serum IFN was investigated by using two strains of mice: C3H/eB, a high responder to LPS, and C3H/HeJ, a low responder to LPS. Analysis of IFN titers in the sera of the F_1, F_2, and the backcross of F_1 with C3H/HeJ generations indicated that a single dominant autosomal gene controls LPS-induced IFN, similar to CSF (Apte *et al.*, 1977b).

Moreover, *in vitro* analysis of the cellular interactions involved in the generation of CSF (Apte *et al.*, 1980) and IFN (Ascher *et al.*, 1981) has indicated that LPS interacts with macrophages, which in turn activate lymphocytes to release CSF and IFN. Thus, it was tempting to test whether the production of both lymphokines is linked. To test this possibility, basically the same experimental approach was used as that to test the association between CSF and CFC. Individual mice from the backcross ($F_1 \times$ C3H/HeJ) and F_2 generations were injected with 100 μg/mouse LPS and 3 hr later bled for serum. Each serum sample was simultaneously tested for CSF activity in soft agar cultures and for IFN titer in the plaque reduction assay against vesicular stomatitis virus. Re-

Dov H. Pluznik

Table 4. Association between Generation of CSF and IFN
in the Serum, in F_2 and Backcross ($F_1 \times$ C3H/HeJ) Mice
Injected with LPS[a]

	Number of mice		
	IFN activity		
CSF activity	High	Low	Total
High	12 (75%)	4 (25%)	16
Low	3 (20%)	15 (80%)	18

[a] High CSF activity, > 90 CFC/10^5 bone marrow cells; low, 0–30 CFC/10^5 bone marrow cells. High IFN activity, > 30 units/ml; low, < 10 units/ml. (For details see Apte *et al.*, 1977b.)

sults of these experiments are seen in Table 4. It is clear from the data that a close association exists between LPS-induced CSF and IFN.

In addition to its antiviral activity, IFN has also been shown to inhibit CFC proliferation in soft agar cultures (McNeill and Fleming, 1971; McNeill and Gresser, 1973). This fact raises the question whether "contaminating" IFN present in LPS-induced CSF samples interferes with the CSF activity. Support for such a potential interference is obtained from experiments dealing with the interaction of CSF and its target cells. It is known that IFN interacts with its target cells via ganglioside receptors present on the cell membrane (Besancon and Ankel, 1974; Besancon *et al.*, 1976). Such interaction can be inhibited by preincubation of the target cells with cholera toxin (Friedman and Kohn, 1976), which also interacts with gangliosides (Cuatrecasas, 1978). Recent experiments (Lenz and Pluznik, 1980) have indicated that preincubation of bone marrow cells with cholera toxin can inhibit the stimulatory effect of CSF; no colonies develop in the soft agar cultures. Such preincubation of bone marrow cells with cholera toxin has no effect on the pluripotent stem cells (CFU-S). In addition, we demonstrated that preincubation of CSF with gangliosides also inhibits the colony stimulatory effect of CSF. These data suggest that CSF also interacts with its target cells via membrane ganglioside receptors. It is possible that IFN competes with CSF on these receptors. Experiments with purified CSF and IFN can probably elucidate such a possibility.

ACKNOWLEDGMENTS

I would like to acknowledge the collaboration of my students, R. N. Apte, O. Ascher, C. F. Hertogs, R. Lenz, and Z. Williams, and to thank them for their contributions to these studies. I would also like to thank N. S. Tare for helpful suggestions during the preparation of this chapter, and Peggy Pogue and Erin Spangelo for their secretarial assistance.

VII. References

Andersson, J., Melchers, F., Galanos, C., and Lüderitz, O., 1973, The mitogenic effect of lipopolysaccharide on bone marrow derived mouse lymphocytes: Lipid A as the mitogenic part of the molecule, *J. Exp. Med.* **137**:943.

Apte, R. N., and Pluznik, D. H., 1976a, Control mechanisms of endotoxin and particulate material stimulation of hemopoietic colony forming cell differentiation, *Exp. Hematol.* **4**:10.

Apte, R. N., and Pluznik, D. H., 1976b, Genetic control of lipopolysaccharide induced generation of serum colony stimulating factor and proliferation of splenic granulocyte/macrophage precursor cells, *J. Cell. Physiol.* **89**:313.

Apte, R. N., and Pluznik, D. H., 1976c, Regulation of murine granulopoiesis by bacterial endotoxins, in: *Progress in Differentiation Research* (N. Müller-Berat, ed.), North-Holland, Amsterdam.

Apte, R. N., Galanos, C., and Pluznik, D. H., 1976, Lipid A, the active part of bacterial endotoxins in inducing serum colony stimulating activity and proliferation of splenic granulocyte/macrophage progenitor cells, *J. Cell. Physiol.* **87**:71.

Apte, R. N., Hertogs, C. F., and Pluznik, D. H., 1977a, Regulation of lipopolysaccharide-induced granulopoiesis and macrophage formation by spleen cells. I. Relationship between colony stimulating factor release and lymphocyte activation *in vitro*, *J. Immunol.* **118**:1435.

Apte, R. N., Ascher, O., and Pluznik, D. H., 1977b, Genetic analysis of generation of serum interferon by bacterial lipopolysaccharide, *J. Immunol.* **119**:1898.

Apte, R. N., Hertogs, C. F., and Pluznik, D. H., 1979, Generation of colony stimulating factor by purified macrophages and lymphocytes, *J. Reticuloendothelial Soc.* **26**:491.

Apte, R. N., Hertogs, C. F., and Pluznik, D. H., 1980, Regulation of lipopolysaccharide-induced granulopoiesis and macrophage formation by spleen cells. II. Macrophage–lymphocyte interaction in the process of generation of colony-stimulating-factor, *J. Immunol.* **124**:1223.

Ascher, O., Apte, R. N., and Pluznik, D. H., 1981, Generation of lipopolysaccharide-induced interferon in spleen cell cultures. I. Genetic analysis and cellular requirements, *Immunogenetics* **12**:117.

Besancon, F., and Ankel, H., 1974, Binding of interferon to gangliosides, *Nature (London)* **252**:478.

Besancon, F., Ankel, H., and Basu, S., 1976, Specificity and reversibility of interferon ganglioside interaction, *Nature (London)* **259**:576.

Bradley, T. R., and Metcalf, D., 1966, The growth of mouse bone marrow cells *in vitro*, *Aust. J. Exp. Biol. Med. Sci.* **44**:287.

Bradley, T. R., and Sumner, M. A., 1968, Stimulation of mouse bone marrow colony growth *in vitro* by conditioned medium, *Aust. J. Exp. Biol. Med. Sci.* **46**:607.

Bradley, T. R., Stanley, E. R., and Sumner, M. A., 1971, Factors from mouse tissues stimulating colony growth of mouse bone marrow cells *in vitro*, *Aust. J. Exp. Biol. Med. Sci.* **49**:595.

Chang, H. L., Thompson, J. J., and Nowotny, A., 1974, Release of colony stimulating factor (CSF) by non-endotoxic breakdown product of bacterial lipopolysaccharides, *Immunol. Commun.* **3**:401.

Chiller, J. M., Skidmore, B. J., Morrison, D. C., and Weigle, W. O., 1973, Relationship of the structure of bacterial lipopolysaccharides to its function in mitogenesis and adjuvanticity, *Proc. Natl. Acad. Sci. USA* **70:**2129.

Coutinho, A., Gronowicz, E., and Sultzer, B. M., 1974, Genetical control of B cell responses. I. Selective unresponsiveness to lipopolysaccharide, *Scand. J. Immunol.* **4:**139.

Cuatrecasas, P., 1973, Gangliosides and membrane receptors for cholera toxin, *Biochemistry* **12:**3558.

Finkelman, F. D., Smith, A. H., Scher, I., and Paul, W. E., 1975, Abnormal ratio of membrane immunoglobulin classes in mice with an X-linked B-lymphocyte defect, *J. Exp. Med.* **142:**1316.

Friedman, R. M., and Kohn, L. D., 1976, Cholera toxin inhibits interferon action, *Biochem. Biophys. Res. Commun.* **70:**1078.

Galanos, C., Rietschel, E. T., Lüderitz, O., and Westphal, O., 1972, Biological activities of lipid A complexed with bovine serum albumin, *Eur. J. Biochem.* **31:**230.

Hallum, J. V., Youngner, J. S., and Stinebring, W. R., 1965, Interferon activity associated with high molecular weight proteins in the circulation of mice injected with endotoxin or bacteria, *Virology* **27:**429.

Ichikawa, Y., Pluznik, D. H., and Sachs, L., 1966, *In vitro* control of development of macrophage and granulocyte colonies, *Proc. Natl. Acad. Sci. USA* **56:**488.

Kasai, N., and Nowotny, A., 1967, Endotoxic glycolipid from heptoseless mutant of *Salmonella minnesota*, *J. Bacteriol.* **94:**1824.

Kass, E. H., and Wolff, S. M. (guest editors), 1973, *Bacterial Lipopolysaccharides: Chemistry, Biology and Clinical Significance of Endotoxins*, *J. Infect. Dis.* **128**(Suppl.).

Kim, Y. B., and Watson, D. W., 1967, Biologically active endotoxin from *Salmonella* mutants deficient in O and R-polysaccharides and heptose, *J. Bacteriol.* **94:**1320.

Lenz, R., and Pluznik, D. H., 1980, Membranal gangliosides present on committed hemopoietic stem cells (CFU-c): Possible receptors for colony-stimulating factor (CSF), in: *Experimental Hematology Today—1980* (S. J. Baum, G. D. Ladney, and D. W. Bekkum, eds.), pp. 13–18, Karger, Basel.

Lüderitz, O., Staub, A. M., and Westphal, O., 1966, Immunochemistry of O and R antigens of *Salmonella* and related Enterobacteriaceae, *Bacteriol. Rev.* **20:**192.

Lüderitz, O., Galanos, C., Lehmann, V., Munminen, M., Rietschel, E. T., Rosenfelder, G., Simon, M., and Westphal, O., 1973, Lipid A: Chemical structure and biological activity, *J. Infect. Dis.* **128**(Suppl.):577.

McNeill, T. A., and Fleming, W. A., 1971, The relation between serum interferon and inhibitor of mouse hemopoietic colonies *in vitro*, *Immunology* **21:**76.

McNeill, T. A., and Gresser, I., 1973, Inhibition of hemopoietic colony growth by interferon preparations from different sources, *Nature New Biol.* **244:**173.

Mergenhagen, S. E., Martin, G. R., and Schiffman, E., 1963, Studies on an endotoxin of a group C *Neisseria* meningitidis, *J. Immunol.* **90:**312.

Metcalf, D., 1971, Acute antigen-induced proliferation *in vitro* of bone marrow precursors of granulocytes and macrophages, *Immunology* **21:**427.

Mizel, S. B., and Mizel, D., 1981, Purification to apparent homogeneity of murine interleukin 1, *J. Immunol.* **126:**834.

Neter, E., and Ribi, E., 1963, Effects of *Salmonella enteritidis* endotoxin preparations and lipid A on dermal reactivity in rabbits to epinephrine, *Proc. Soc. Exp. Biol. Med.* **112:**269.

Parker, J. W., and Metcalf, D., 1974a, Production of colony stimulating factor in mixed leucocyte cultures, *Immunology* **26:**1039.

Parker, J. W., and Metcalf, D., 1974b, Production of colony-stimulating factor in mitogen stimulated lymphocyte cultures, *J. Immunol.* **112:**502.

Pluznik, D. H., and Sachs, L., 1965, The cloning of normal "mast" cells in tissue culture, *J. Cell. Physiol.* **66:**319.

Pluznik, D. H., and Sachs, L., 1966, The induction of colonies of normal "mast" cells by substance in conditioned medium, *Exp. Cell. Res.* **43:**553.

Quesenberry, P. J., Morley, A., Stohlman, F., Jr., Rickard, K., Howard, D., and Smith, M., 1972, Effect of endotoxin on granulopoiesis and colony-stimulating factor, *N. Engl. J. Med.* **286:**227.

Quesenberry, P. J., Morley, A., Miller, M., Rickard, K., Howard, D., and Stohlman, F., Jr., 1973, Effect of endotoxin on granulopoiesis and the *in vitro* colony forming cell, *Blood* **41:**391.

Quesenberry, P. J., Halperin, J., Ryan, M., and Stohlman, F., Jr., 1975, Tolerance to granulocyte-releasing and colony stimulating factor elevating effects of endotoxin, *Blood* **45:**789.

Ribi, E., Anacker, R. L., Fukushi, K., Haskins, W. T., Landy, M., and Milner, K. C., 1964, Relationship of chemical composition to biological activity, in: *Bacterial Endotoxins* (M. Landy and W. Braun, eds.), pp. 16–28, Rutgers University Press, New Brunswick, N.J.

Robinson, W. A., Metcalf, D., and Bradley, T. R., 1967, Stimulation by normal and leukemic mouse sera of colony formation *in vitro* by mouse bone marrow cells, *J. Cell. Physiol.* **69:**83.

Robinson, W. A., Stanley, E. R., and Metcalf, D., 1969, Stimulation of bone marrow colony growth *in vitro* by human urine, *Blood* **33:**396.

Ruscetti, F. W., and Chervenick, P. A., 1975, Regulation of the release of colony stimulating activity from mitogen stimulated lymphocytes, *J. Immunol.* **114:**1513.

Scher, I., Ahmed, A., Strong, D. M., Steinberg, A. D., and Paul, W. E., 1975, X-Linked B-lymphocyte immune defect in CBA/HN mice. I. Studies of the function and composition of spleen cells, *J. Exp. Med.* **141:**788.

Sheridan, J. W., and Metcalf, D., 1973, A low molecular weight factor in lung conditioned medium stimulating granulocyte and monocyte colony formation *in vitro*, *J. Cell. Physiol.* **81:**11.

Staber, F. G., 1980, Diminished response of granulocyte-macrophage colony-stimulating factor (GM-CSF) in mice after sensitization with bacterial cell wall components, *Exp. Hematol.* **8:**120.

Stinebring, W. R., and Youngner, J. S., 1964, Patterns of interferon appearance in mice injected with bacteria or bacterial endotoxin, *Nature (London)* **204:**712.

Sultzer, B. M., 1968, Genetic control of leucocyte responses to endotoxins, *Nature (London)* **219:**1253.

Sultzer, B. M., 1972, Genetic control of host responses to endotoxin, *Infect. Immun.* **5:**107.

Watson, J., and Riblet, R., 1974, Genetic control of responses to bacterial lipopolysaccharides in mice. I. Evidence for a single gene that influences mitogenic and immunogenic responses to lipopolysaccharide, *J. Exp. Med.* **140:**1147.

Westphal, O., 1975, Bacterial endotoxins, *Int. Arch. Allergy Appl. Immunol.* **49:**1.

Williams, Z., Hertogs, C. F., and Pluznik, D. H., 1981, Involvement of murine spleen cells in tolerance to lipopolysaccharide (LPS) induced serum colony-stimulating-factor (CSF), *Exp. Hematol.* **9**(Suppl.):9.

21

Colony-Stimulating Activity Induction in Patients with Acute Leukemia

Wolfgang Hinterberger, Wolfgang Graninger, and Walter R. Paukovits

The myeloid system is designed to combat bacterial invasion. Nonactivated myelopoiesis is rapidly capable of responding to onwarding infection. The mechanisms by which activating impulses are mediated are not completely understood. This survey aims to elucidate early myeloid reactions to gram-negative infection in normal and myeloid leukemic individuals. It is shown that this response involves consequences both for the proliferation and defense function of myelopoiesis as well.

I. Regulation of Immature Myelopoiesis and the Response to LPS

Myeloid reactions are better understood by dividing the system in two branches: Firstly, mature granulocytes, stored in the bone marrow and circulating in the blood, are the immediate responders to bacterial invasion, being rapidly consumed as a consequence of their defense function. Secondly, a pool of stem cells can maintain its size by self-replication, but can also give rise to a progeny of mature effector cells. Since bacterial infection leads to an increased turnover rate of granulocytes, the stem cell pool needs rapidly to cover the increased cellular

Wolfgang Hinterberger ● First Department of Medicine   Wolfgang Graninger ● Department of Chemotherapy   Walter R. Paukovits ● Institute for Cancer Research, University of Vienna, Vienna, Austria.

requirement. Therefore, immediate reactions of both branches can be expected following bacterial infection.

Myeloid committed stem cells are detectable by their capacity to form colonies of granulocytes and macrophages (Pluznik and Sachs, 1966; Bradley and Metcalf, 1966). Their growth is absolutely dependent on the presence of a group of proteins that are operationally termed *colony-stimulating factors* (CSF) (reviewed by Burgess and Metcalf, 1980). Such CSF act specifically on cells forming myeloid colonies, since no other hemopoietic cell line shows comparable proliferative responses.

A. Quantitative Assessment of Immature Myelopoiesis

The counting of myeloid committed stem cells (*colony-forming unit in culture*, CFU-C) along with the estimation of the activity of CSF (*colony-stimulating activity*, CSA) offers the possibility to monitor myeloid reactions in various states of infection and proliferative disorders. Attempts to purify human CSF have not as yet been successful. This is attributed by most investigators to low antigenicity of CSF molecules preventing antibody formation in immunized animals and also to low physiologic concentrations, which hinder the chemical enrichment and subsequent biological assay.

Most laboratories employ normal human or mouse bone marrow as the target system for the measure of CSA in body fluids. A sigmoid dose–response curve between CSA and the number of stem cells forming colonies within the target bone marrow forms the basis for this widely employed bioassay. The comparison of results obtained in human and mouse bone marrow assays has shown a nonequivalence, which was attributed to differences in specificity of CSF for either bone marrow (Lind *et al.*, 1974). Though human CSA seems more properly measured in human bone marrow cultures, human marrow contains CSA-releasing cells, which otherwise can mask the activity of CSF added for assay. In addition, human bone marrow contains cells that secrete prostaglandins when a certain concentration of CSA is reached in the culture medium (Kurland *et al.*, 1978). Among other biological effects, such prostaglandins are inhibitory for myeloid colony formation (Kurland *et al.*, 1978). The existence of a colony-"enhancing" activity has also been reported, which may alone or in concert with CSA alter the growth of the target bone marrow (Metcalf *et al.*, 1975). Though the use of mouse bone marrow for assaying human CSA is limited by differences in specificity, its use seems rather advantageous for small amounts of CSA, since mouse bone marrow completely lacks endogenous CSA production.

In summary, both assays share considerable drawbacks, which can make their parallel application helpful. No absolute levels of CSA can yet be given for biological fluids such as serum or urine, but there is wide agreement that CSA is detectable in the serum and urine of all normal individuals. The estimation of CSA awaits further improvement of the assays employed, but the techniques described and applied in these investigations roughly satisfy the requirements for the study of myeloid reactions following infection.

B. LPS-Induced Alterations of CSA in Experimental Animals and Man

Among the countless biological effects of LPS *in vivo*, studies in mice have shown a rapid increase of CSA in the serum and urine following LPS injection (Metcalf, 1971; Quesenberry *et al.*, 1972). Such studies were repeated in other experimental animals, including dogs (Hinterberger *et al.*, 1979b; MacVittie and Walker, 1978) and rats (Shadduck, 1974), showing an essentially similar response.

Not much is known about LPS and subsequent alterations of CSA in man. Experimental studies in volunteers are limited to the potential harmful effects of LPS, and infectious diseases are probably not a good model to study early alterations of CSA levels. A challenge with a low-dose preparation of LPS in a group of six healthy volunteers was initially performed by Golde and Cline (1975): Six adult males were injected intravenously with Piromen (0.1–0.15 μg/kg), a purified endotoxin from *Pseudomonas*, and body temperature, leukocyte count, and serum CSA levels were monitored. All individuals developed neutropenia with subsequent rebound leukocytosis. The CSA level showed a marked increase after LPS administration with maximum levels at 60 min and high levels persisting for 180 min. This response of the CSA level was comparable to that obtained in experimental animals.

In these studies, the dose of LPS was sufficiently high to cause profound neutropenia, rebound leukocytosis, and the clinical symptomatology included shaking chills and malaise. Therefore, the CSA response was not clearly attributable to LPS, since profound neutropenia itself might have triggered a CSA response by following a feedback mechanism.

Similar experiments were therefore carried out with a dose of LPS that in our experience did not significantly alter the neutrophil count. Pyrifer, a detoxified preparation of LPS from *E. coli*, was injected intravenously in six young male volunteers. (One milliliter of Pyrifer contained 50 units of endotoxin; 1 unit was prepared from approximately 10^6 apathogenic coli bacteria.) The volunteers and, as discussed in Section III, the patients with acute leukemia were exhaustively informed about the aim of this trial. Consent for the administration of LPS was obtained in all cases. Pyrifer is a medication traditionally given for non-specific "immunostimulation" in patients with recurrent episodes of minor infections.

Serum samples were obtained at intervals of 15 min and the levels of CSA were determined in parallel bone marrow cultures employing 75,000 mouse bone marrow cells (C57BL mice) and 100,000 human bone marrow cells obtained from patients with idiopathic thrombocytopenia prior to therapy. A detailed description of the technical procedure is given elsewhere (Hinterberger *et al.*, 1979c). Briefly, 0.1 ml serum was mixed with McCoy medium, fetal calf serum (20% v/v), together with the bone marrow cells in a 0.3% agar base. Colonies with more than 50 cells in mouse bone marrow and more than 20 cells in human bone marrow were counted on days 7 and 10, respectively.

A rise in the serum levels of CSA was detectable within 15 to 30 min in all individuals, with results comparable in both assays employed (Fig. 1). No clinical symptomatology was observed and febrile reactions were only mild. No alteration of the neutrophil count was observed in

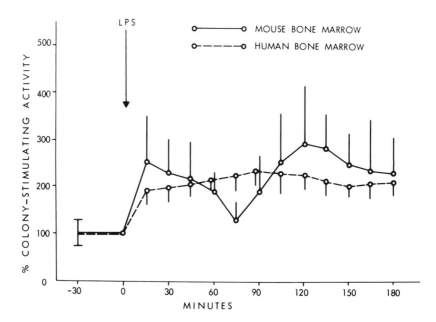

Figure 1. Colony-stimulating activity of LPS (Pyrifer) in six normal individuals. CSA is expressed as a percentage of the initial value (= 100%). Initial colony counts for mouse target bone marrow ranged between 6 and 22 CFU-C/75,000 bone marrow cells and for human bone marrow between 3 and 18 CFU-C/100,000 bone marrow cells.

four individuals, while in the remaining a slight decrease of the neutrophil count was shown, which was rapidly returning to the initial level not followed by rebound leukocytosis.

While the serum samples assayed with human bone marrow showed a plateau of the CSA levels following the initial rise, CSA measured with mouse bone marrow gave the impression of a curve with a second peak. These differences may come from an altered distribution of CSFs within the collectively measured CSA, differing also in their specificity for either bone marrow. These results demonstrate that neutropenia itself is not responsible for the increase of CSA following LPS administration.

Repeated administration of LPS into mice has led to a state of tolerance in which the CSA response may become completely suppressed (Metcalf, 1971, 1974b; Quesenberry *et al.*, 1975). This is only one part of a phenomenon summarized by most as "endotoxin tolerance." Repeated doses of LPS have never been given to human volunteers, but patients with chronic diseases of the liver may include individuals with chronic endotoxinemia. The situation in such patients may resemble the state of experimental animals that have been repeatedly challenged with LPS. We measured CSA levels in 55 normal individuals and 5 patients with cirrhosis of the liver in which endotoxinemia was detected by the *Limulus* assay. There was no difference of CSA levels in both groups, as determined with mouse bone marrow as the target system (Hinterberger *et al.*, 1979a). Thus, similar to repeated administration of LPS in experimental animals, chronic endotoxinemia in man is very likely not associated with increased levels of CSA.

C. Cellular and Humoral Requirements for the LPS-Induced CSA Increase

In contrast to earlier reports (Greenberg *et al.*, 1971), it is now widely accepted that granulocytes are incapable of releasing CSA. This is in accordance with the observation of Shadduck (1974), who found neutropenic rats responding equally with increased CSA levels following LPS. T lymphocytes represent a proven source of CSA, but a binding of several hours of a lectin appears necessary for a subsequent CSA release (Ruscetti and Chervenick, 1975). Therefore, granulocytes and T lymphocytes are unlikely cellular sources for the CSA released following LPS. Cells from the monocyte-macrophage pathway (Chervenick and LoBuglio, 1972; Cline *et al.*, 1974; Eaves and Bruce, 1974) and, more recently, vascular endothelial cells (Quesenberry and Gimbrone, 1980) are now commonly accepted sources of the LPS-induced increase of CSA.

Quesenberry and Gimbrone (1980) studied human vascular endothelial cells incubated with endotoxin from *Salmonella typhosa* at 1 μg/ml culture medium and the increase of CSA was monitored over a period from 6 to 72 hr. A progressive increase of CSA in the culture medium was observed; unfortunately, no data were reported about a CSA increase within the first hour of culture. Interestingly, the incubation of vascular endothelial cells with granulocyte lysate had a comparable effect: CSA production was markedly increased from 6 to 72 hr. This observation should be seen in connection with the well-known margination of granulocytes following LPS administration *in vivo*, which is probably caused by activation of the complement cascade, in particular, by cleavage of the component C5 (Fehr, 1977).

Human monocytes were found to release CSA spontaneously in culture (Chervenick and LoBuglio, 1972; Cline *et al.*, 1974), but incubation with LPS was not accompanied by a marked and rapid release of CSA within 2 to 48 hr (Cline *et al.*, 1974; Shah *et al.*, 1977). Monocytes cultured over a period of 14 days and differentiated into macrophages were, however, capable of increasing the production rate of CSA within 2 to 48 hr upon contact with LPS (Cline *et al.*, 1974).

The well-known effect of LPS on complement activation together with the expression of receptors for certain complement components on cells from the monocyte-macrophage pathway prompted us to study various culture conditions including decomplementation of human serum present in the culture medium. We had observed earlier that heat inactivation of autologous serum could abrogate an otherwise rapid increase of CSA by human adherent mononuclear leukocytes following LPS (Hinterberger *et al.*, 1980). In an extension of these studies, we tried to more closely analyze the role of complement activation as follows. Mononuclear leukocytes from normal human donors were obtained by separation on Ficoll–Hypaque and incubated at a concentration of 10^6 cells/ml in Falcon culture flasks. Nonadherent cells were removed after 1 hr incubation at 37°C with culture medium. The adherent cells were incubated in RPMI 1640 medium containing 5% heat-inactivated fetal calf serum and 20% human serum, which was (a) fresh or (b) heat-inactivated (56°C, 30 min) or (c) depleted of complement component C3 (Cappell, Switzerland) and (d,e) supplemented with native C3 (Cordis) in one of two concentrations. LPS from *Salmonella typhimurium* was added to all culture flasks giving a final concentration of 7 μg/ml. The supernatants were harvested after 60 min, centrifuged (1000*g*, 30 min), and subsequently assayed for CSA employing 10^5 human bone marrow cells/ml in agar–McCoy medium.

In the presence of C3-deficient serum, LPS induced an increase of CSA comparable to that obtained in cultures with heat-inactivated se-

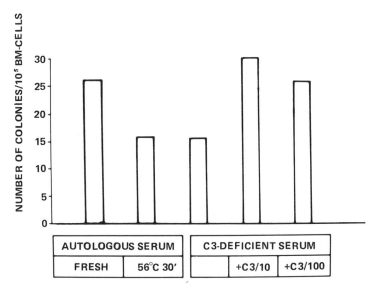

Figure 2. Colony-stimulating activity of supernatants harvested after 60 min from normal adherent mononuclear leukocytes. CSA is expressed in number of CFU-C/100,000 normal, adherent cell-depleted human bone marrow cells.

rum (Fig. 2). However, C3-deficient serum supplemented with native C3 allowed a rapid CSA release by adherent leukocytes upon incubation with LPS. In this experimental setting, bound C3b was demonstrated on the membrane of the cells of assays employing sera a, d, and e by means of immunofluorescence.

The result of this experiment suggests that human adherent mononuclear leukocytes, which mainly consist of monocytes, are equally capable of rapidly releasing CSA upon contact with LPS. However, together with the observations on endothelial cells and granulocyte lysates, complement activation seems critical for the LPS-induced CSA release *in vivo*.

II. Biological Significance of the LPS-Induced Increase of CSA

The understanding of the physiological significance of CSA in the regulation of myelopoiesis forms the basis for prospective therapeutic trials with purified material. The most likely function of newly released CSA is to activate immature myeloid cells in order to accelerate the production rate of mature cells. However, recently obtained experimental data also point to an effect of CSA on mature myeloid cells, which are not capable of further division: RNA synthesis was induced in mature myeloid cells of mouse bone marrow within 10 min upon incubation with mouse lung-derived purified CSF (Burgess and Metcalf, 1977). RNA synthesis was not observed in other hemopoietic cell lines other than myeloid, indicating the specific nature of this process. RNA synthesis in myeloid cells that are otherwise not capable of further division strongly implies activation of protein synthesis. Consequently, CSA may potentially activate circulating or stored granulocytes thereby mediating a certain "arming" effect. This possibility prompted us to investigate the effect of a commercially available CSF preparation on the phagocytosis of normal granulocytes by means of chemiluminescence (Stevens *et al.*, 1978), as follows. Neutrophil granulocytes obtained by dextran sedimentation and ammonium chloride lysis were used at a concentration of 10^6 cells/ml. As phagocytic particles, zymosan (Sigma) opsonized with fresh human serum was used (5×10^7 particles/ml). The neutrophils were incubated for 5, 10, and 30 min in RPMI medium enriched with commercially available CSF (GCT-CSF; DiPersio *et al.*, 1978), giving a final concentration of 50%, 10%, and 1% of CSF. Chemiluminescence associated with phagocytosis was measured after mixing 100 μl of the neutrophils and 100 μl of the zymosan suspension with 100 μl of luminol (10^{-3}M) to enhance luminescence intensity. Luminescence was measured in a Lumac biocounter at intervals of 60 sec. The results of this experiment showed no difference in luminescence intensity, measured either with cells incubated with various concentrations of CSF or with cells incubated without CSF.

The result from this pilot experiment certainly does not exclude activation of granulocyte functions by CSF other than luminescence-associated phagocytosis. Also, the subset of CSF chosen for these experiments need not necessarily enhance this particular granulocyte function. We still feel that the data reported by Burgess and Metcalf (1977) strongly implicate an influence of CSF molecules on granulocytes, so that further investigations should be done in spite of the negative result of our pilot study.

III. CSA in Myeloid Leukemic Hemopoiesis

A. Role of CSA on Leukemic Proliferation and Differentiation

The significance of CSA in the regulation of myeloid leukemic proliferation and differentiation is difficult to evaluate and the present knowledge is mainly derived from *in vitro* culture studies.

Though acute leukemic cell populations appear rather homogeneous by microscopy, the introduction of bone marrow cultures has shown functional heterogeneity. As in normal bone marrow, a minority of clonogenic cells responds to and proliferates in the presence of CSA. While in our series of 23 normal bone marrows an amount of 25 to 220 CFU-C/10^5 nucleated bone marrow cells was counted, the range of leukemic colonies in our series of 30 patients was between 0 and 600/10^5 bone marrow cells. However, without an exogenous source of CSA, proliferation was not demonstrable in normal or in leukemic bone marrow. This observation is in line with the experience of many laboratories, i.e., that the growth and occasional differentiation of myeloid leukemic cells will only occur in the presence of CSA (summarized by Metcalf, 1976). Blasts with monocytic differentiation generally retain the capacity of their normal counterparts to release CSA (Golde et al., 1974b; Goldman et al., 1976; Hinterberger et al., 1977, 1981). This leukemic CSA is in all cases capable of inducing clonal growth of autologous (leukemic) and heterologous (normal) bone marrow (Hinterberger et al., 1981). The latter observation is of some interest, since attempts of purification of leukemic CSA have brought evidence for molecular weights of leukemic CSF different from normal and also from each other (Hinterberger et al., 1981), and for the demonstration of only one subset of CSF released spontaneously by leukemic blasts (Price et al., 1978). The biological significance of CSA released by leukemic blasts is at present completely unknown. Though monoblastic leukemic cells generate enormous amounts of CSA in vitro, the prognosis of this type of leukemia is not really different from other types of acute myeloid leukemia, at least under the therapeutic regimens currently applied.

While clonal growth of leukemic cells in vitro is generally dependent on the presence of CSA, maturation of leukemic blasts under such stimulation is only rarely observed. CSA is preferably acting as a promoter of leukemic proliferation, but differentiation up to a stage of functional maturity rarely occurs. Few leukemic cell lines are known in which CSA reproducibly induced differentiation (Gallo et al., 1978; Koeffler and Golde, 1978). We have observed a continuous cell line derived from a patient with promyelocytic leukemia, whose cells required the continuous presence of CSA for proliferation as well as for survival. Removal of CSA rapidly caused cell death. The interpretation of such observations is difficult, and the currently available knowledge simply does not allow prediction of the response of an acute leukemic cell population to a given subset of CSF.

Numerous experiments have been performed to prove or disprove a different sensitivity of leukemic cells for CSA in comparison with normal, residual stem cells (Metcalf, 1974a; Francis *et al,* 1979; Brennan *et al.,* 1979), but as yet, no convincing data for either possibility have been reported. Such a concept had gained some attraction from experiments with leukemic blasts in liquid culture (Brennan *et al.,* 1979), in which clonal growth and self-replication from one patient were extended with CSA, while the remission marrow of the same patient was not. It has been attempted to supress leukemic myelopoiesis *in vitro* by serial recloning in the presence of CSA, in order to provide a continuous pressure for differentiation and for a subsequent loss of proliferative capacity ("clonal extinction"). While no reliable data are available for human leukemic cell populations, studies in the mouse myelomonocytic leukemia WEHI 3B have failed to achieve clonal extinction (Metcalf, 1979).

Though the majority of blast cells in patients with acute myeloid leukemia are incapable of releasing CSA, alveolar macrophages in these patients appear to retain this particular capacity (Golde *et al.,* 1974a). Also, some adherent bone marrow cells from patients with acute myeloid leukemia may be capable of a CSA release, which is only subnormal (Greenberg *et al.,* 1978). The release of CSA by adherent cells was positively correlated with the chance of obtaining remission. In addition, the longitudinal assessment of adherent cell-derived CSA predicted the duration of remission: Patients with short remission had decreased marrow CSA, while long-term remission was associated with increased marrow CSA. Similar results were reported by Alsabati and Saleh (1979): The presence of CSA-releasing cells in the peripheral blood of patients with acute myeloid leukemia was predictive for obtaining subsequent remission. In complete accordance, Hornsten *et al.* (1977) reported the presence of CSA-releasing cells in the peripheral blood of patients with acute nonlymphocytic leukemia similarly predictive for obtaining remission. However, since only a minority of circulating cells were capable of releasing CSA, the question remains as to whether coexisting normal cells provided such CSA or whether production of CSA was associated with some differentiation of leukemic cells. Further, it is unknown whether such CSA provides better conditions for the growth of normal stem cells, or if it simply reflects a useful parameter to monitor microenvironmental influences.

B. LPS-Induced Alterations of CSA in Patients with Acute Myeloid Leukemia

In acute myeloid leukemia, normal hemopoiesis is replaced by a population of blasts that in their majority are incapable of releasing CSA. Conversely, tissue macrophages and a variable number of marrow-derived adherent cells seem still capable of generating CSA. This complex situation makes it very difficult to predict a normal vs. abnormal CSA response in infection. In such patients a number of clinical situations may themselves evoke alterations of the CSA level: neutropenia, infection, renal dysfunction, and even the application of cytostatic drugs (Myers and Robinson, 1975). The assessment of instantaneous serum CSA levels probably does not suffice to enlighten possible abnormal regulatory conditions. In an extensive survey, Metcalf et al. (1971) investigated 251 sera from 33 patients with acute myeloid and myelomonocytic leukemia: 30% of the samples tested showed an increased CSA level, and patients in relapse tended to respond inadequately in infectious complications with high levels of serum CSA. In addition, the occurrence of inhibitors for colony formation was described in approximately half of the sera tested. The concurrent presence of inhibitors of colony formation makes the assessment of CSA levels even more complex, since ignorance of the chemical nature of blast-derived inhibitors (Chiyoda et al., 1976; Hinterberger et al., 1977) makes their separate evaluation impossible. Studies from Mintz and Sachs (1973) have shown that serum from patients with untreated acute myeloid leukemia had about the same average CSA levels as normal sera, but in accordance with Metcalf et al. (1971), some sera of these patients contained inhibitory material.

In order to get more precise information on the CSA response in patients acutely ill from acute myeloid leukemia, but free of infectious complications, we challenged four voluntarily participating patients with the same low-dose LPS preparation as has been used for the normal volunteers. Patients with a high cell count and those with monocytic leukemia were excluded because of the risk of developing consumption coagulopathy.

The patients were either newly diagnosed or in a first relapse following complete remission (Table 1). They had low leukocyte counts, but their bone marrows were packed with blast cells. Diagnosis was obtained by evaluation of a bone marrow aspirate and cytochemistry, which included nonspecific esterase, PAS, and peroxidase.

A mild febrile reaction was seen in all patients, though it did not exceed the temperature elevations of the normal individuals. The leukocyte count remained unaltered, again reflecting the low dose of LPS applied.

The changes of the serum CSA levels, determined in a mouse and human bone marrow culture assay, are shown in Fig. 3. In sharp contrast to the group of normal individuals, none of the four patients showed an increase in the serum CSA level. This result was equally demonstrable in the human and mouse bone marrow assay.

Table 1. Clinical and Hematological Data of Four Patients

Patient	Diagnosis	WBC $\times$ 10^9	Differential count
K.H., 16 years, m	Acute myeloid leukemia		
	Relapse	3.4	92 Bl, 5 Ly, 3 Mo
	Remission	4.5	44 Gr, 8 Mo, 48 Ly
E.L., 29 years, m	Acute myeloid leukemia	1.9	54 Bl, 2 Gr, 5 Mo, 39 Ly
K.E., 48 years, m	Acute myeloid leukemia, first relapse	1.0	90 Bl, 10 Ly
O.H., 26 years, f	Acute myeloid leukemia, first relapse	4.2	93 Bl, 1 Gr, 1 Mo, 5 Ly

Patient K.H. obtained complete remission and was challenged a second time with the same dose of LPS (Fig. 4). The CSA response in remission was comparable to normal. The onset of fever was after 7 hr in relapse and after 10 hr in remission.

These results denote an abnormality of regulatory conditions under which the proliferation and differentiation of leukemic and residual, normal cells occurs. Similar results were also obtained in patients with blast crisis of chronic myelocytic leukemia, which, in contrast to pa-

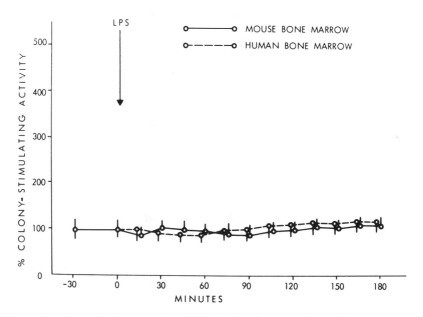

Figure 3. Colony-stimulating activity of LPS (Pyrifer) in four patients with acute myeloid leukemia. Sera were assayed with the same target bone marrows as shown in Fig. 1.

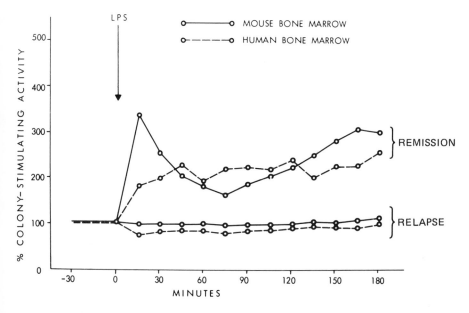

Figure 4. CSA induction in patient K.H. in remission and in relapse.

tients with "chronic phase" myeloid leukemia, gave an essentially normal response of CSA following LPS (Hinterberger *et al.*, 1979). Thus, replacement of normal myelopoiesis by an acute leukemic cell population profoundly alters basic regulatory conditions.

The complex interplay between the different CSFs, their target cells, and the existence of yet poorly defined inhibitors makes the analysis of such experiments difficult: The injection of LPS may cause a parallel increase of an inhibitor, which could mask an otherwise normal CSA response. Further, no data are yet available describing the capacity of vascular endothelial cells in acute leukemia to release CSA. Mature myeloid cells in acute leukemia grossly fail to express receptors for the complement component C3b, which implies a potentially disturbed interaction between LPS, complement activation, and granulocyte sequestration (Hofmeister *et al.*, 1981): The expression of C3b receptors is again restored in remission.

A number of still unresolved functional disturbances may be the underlying cause for the basic regulatory abnormality observed in our experiments. The analysis of such disturbances by *in vitro* techniques is of practical interest for the understanding of leukemic growth regulation.

IV. Possible Beneficial Effects of LPS in Patients with Acute Leukemia

The evidence summarized in this review points to abnormal regulatory conditions in acute leukemia under which the growth of cells and the defense against infection occurs. The experimental results find their confirmation by the clinical observation that patients with acute leukemia are more susceptible to major infections than a group of non-leukemic patients otherwise matched for the number of neutrophils (Bodey *et al.*, 1966). Some experienced hematologists claim that infectious complications have occasionally improved the natural course in patients with acute leukemia not suitable for chemotherapy. However, such anecdotal reports are not confirmed by larger number of patient histories (Reiffers *et al.*, 1981): Early major infections were indicative of a bad prognosis in a group of 88 patients. Certainly, systemic infection accompanying chemotherapy will never be appreciated, but the application of CSA without parallel infection may become a useful tool. Recent experiments with ubiquinone derivatives in mice have shown that only ubiquinone Q7, which *in vitro* and *in vivo* stimulates the release of CSA by macrophages, has a protective effect in artificial infection, in contrast to ubiquinone derivatives that do not induce a CSA release (Mayer *et al.*, 1980).

Many clinical centers have obtained disappointing results from nonspecific immunotherapy with BCG and levamisole for patients with acute leukemia in remission. No attempts have yet been reported with regard to repeated stimulation of myelopoiesis with detoxified LPS. If one assumes a critical role of CSA for remission maintenance, repeated administration of increasing doses of LPS for CSA induction might be beneficial. We feel that with respect to the experimental evidence so far available and also based on theoretical considerations, such a clinical trial would be justified.

Repeated administration of LPS in remission might also increase nonspecific resistance against endotoxin, thereby ameliorating the severe consequences of gram-negative infection in relapse. Repeated administration of LPS in mice has led to an increased tolerance to lethal doses of endotoxin and has also demonstrated a radioprotective effect (Kinosita *et al.*, 1963). No such protective effects have yet been studied in man, but total-body irradiation has become more widely applied for the treatment of lymphomas and also in the course of bone marrow transplantation for leukemias.

Once human CSF becomes available in purified preparations, it may be used to profoundly alter growth conditions of normal and leukemic clonogenic cells. While LPS may be used to induce CSA during remission, defined CSF molecules might be applied in relapse: Normal, suppressed myelopoiesis may be forced to generate a sufficient number of mature myeloid cells, while leukemic blasts may otherwise be triggered into differentiation thereby losing the proliferative potential. Though our current knowledge of CSF molecules does not allow to distinguish between factors more likely to support proliferation or differentiation, subsets with a defined capacity may find their use according to the clinical situation: Hypoplastic normal bone marrow ablated from the leukemic population may benefit from CSF more capable of promoting proliferation, while leukemic blasts resistant to chemotherapy may alternatively benefit from a differentiation-inducing subset of CSF. The spectrum of CSF application may also include the blockage of CSF receptors in blast populations highly sensitive for CSF.

The finding of RNA synthesis induced by CSF in mature myeloid cells points to an action of CSF similar to that of erythropoietin, which induces growth of immature red cells and hemoglobin synthesis in more mature cells. Though our initial experiments measuring phagocytosis of opsonized zymosan particles did not support such a possibility, further experiments will be necessary to elicit an action of CSF on mature myeloid cells. Investigation of the biological significance of CSA in human infection has surprisingly been neglected, but the recent data will hopefully stimulate further investigations.

ACKNOWLEDGMENTS

This work was supported by the Gertrud Keller Stiftung, Faculty of Medicine, University of Vienna, and by Grant 4100, Österreichischer Forschungsrat.

V. References

Alsabati, E. A., and Saleh, K., 1979, Colony stimulating and colony forming cells in peripheral blood as prognostic markers in acute myeloid leukemia, *Oncology* **34**:180.
Bodey, G. P., Buckley, M., Sathe, Y. S., and Freireich, E. J., 1966, Quantitative relationship between circulating leucocytes and infection in patients with acute leukemia, *Ann. Intern. Med.* **64**:328.
Bradley, T. R., and Metcalf, D., 1966, The growth of mouse bone marrow cells *in vitro*, *Aust. J. Exp. Biol. Med. Sci.* **44**:287.
Brennan, J. K., DiPersio, J. F., Abboud, C. N., and Lichtman, M. A., 1979, The exceptional responsiveness of certain human myeloid leukemia cells to colony stimulating activity, *Blood* **54**:1230.
Burgess, A. W., and Metcalf, D., 1977, The effect of colony stimulating factor on the synthesis of ribonucleic acid by mouse bone marrow cells *in vitro*, *J. Cell. Physiol.* **90**:471.

Burgess, A. W., and Metcalf, D., 1980, The nature and action of granulocyte-macrophage colony stimulating factors, *Blood* **56**:947.

Chervenick, P. A., and LoBuglio, A. F., 1972, Human blood monocytes: Stimulators of granulocyte and mononuclear colony formation *in vitro*, *Science* **178**:164.

Chiyoda, S., Mizoguchi, H., Asano, S., Takaku, F., and Miura, Y., 1976, Influence of leukemic cells on the colony formation of human bone marrow cells *in vitro*. II. Suppressive effect of leukemic cells extracts, *Br. J. Cancer* **33**:379.

Cline, M. J., Rothman, B., and Golde, D. W., 1974, Effect of endotoxin on the production of colony stimulating factor by human monocytes and macrophages, *J. Cell. Physiol.* **84**:193.

DiPersio, J. F., Brennan, J. K., Lichtman, M. A., and Speiser, B. L., 1978, Human cell lines that elaborate colony stimulating activity for the marrow cells of man and other species, *Blood* **51**:507.

Eaves, A. C., and Bruce, W. R., 1974, *In vitro* production of colony stimulating activity. I. Exposure of mouse peritoneal cells to endotoxin, *Cell Tissue Kinet.* **7**:19.

Fehr, J., 1977, Complement as a mediator of granulocyte adherence and margination: Studies based on the acute neutropenia of filtration leukapheresis, in: *The Granulocyte: Function and Clinical Utilization* (I. J. Greenwood and G. A. Jamieson, eds.), Liss, New York.

Francis, G. E., Berney, J. J., Chipping, P. M., and Hoffbrand, A. V., 1979, Stimulation of human haemopoietic cells by colony stimulating factors—Sensitivity of leukemic cells, *Br. J. Haematol.* **41**:545.

Gallo, R., Ruscetti, F., Collins, S., and Gallagher, R., 1978, Human myeloid leukemia cells: Studies on oncornaviral related information and *in vitro* growth and differentiation, in: *Hematopoietic Cell Differentiation* (D. W. Golde, M. J. Cline, D. Metcalf, and C. Fox, eds.), pp. 335–355, Academic Press, New York.

Golde, D. W., and Cline, M. J., 1975, Endotoxin induced release of colony stimulating activity in man, *Proc. Soc. Exp. Biol. Med.* **149**:845.

Golde, D. W., Finley, T. N., and Cline, M. J., 1974a, The pulmonary macrophage in leukemia, *N. Engl. J. Med.* **290**:875.

Golde, D. W., Rothman, B., and Cline, M. J., 1974b, Production of colony stimulating factor by malignant leucocytes, *Blood* **43**:749.

Goldman, J. M., Th'ng, K. H., Catovsky, D., and Galton, D. A. G., 1976, Production of colony stimulating factor by leukemia leukocytes, *Blood* **47**:381.

Greenberg, P. L., Nichols, W. C., and Schrier, S. L., 1971, Granulopoiesis in acute myeloid leukemia and preleukemia, *N. Engl. J. Med.* **284**:1225.

Greenberg, P. L., Mara, B., and Heller, P., 1978, Marrow adherent cell colony stimulating activity production in acute myeloid leukemia, *Blood* **52**:362.

Hinterberger, W., Frischauf, H., Kletter, K., Paukovits, W. R., and Bais, P., 1977, Humoral function of acute leukemic blasts, *Scand. J. Haematol.* **19**:121.

Hinterberger, W., Fridrich, P., Ferenci, P., and Paukovits, W. R., 1979a, Colony stimulating activity in acute and chronic endotoxinemia in man, *Experientia* **35**:1398.

Hinterberger, W., Kinast, H., Paukovits, W. R., Keiler, A., and Moschl, P., 1979b, Endotoxin induced myeloid reactions in dogs, *Exp. Pathol.* **17**:113.

Hinterberger, W., Paukovits, W. R., Mittermayer, K., and Wallner, U., 1979c, Endotoxin induced colony stimulating activity in normal and myeloid leukaemic subjects, *Scand. J. Haematol.* **22**:280.

Hinterberger, W., Mittermayer, K., Paukovits, W. R., and Singer, J., 1980, Different types of endotoxin induced release of colony stimulating factors by adherent leucocytes in the presence of fresh and heat-inactivated autologous serum, *Scand. J. Haematol.* **25**:221.

Hinterberger, W., Neumann, E., Resch, F., Schwarzmeier, J., Paukovits, W. R., and

Lechner, K., 1981, The release of granulocyte-macrophage colony stimulating activity (CSA): A property of blasts restricted to monocytic differentiation, in: *Leukemia Markers* (W. Knapp, ed.), Academic Press, New York.

Hofmeister, B. G., Carrera, C. J., and Barrett, S. G., 1981, Neutrophil surface markers in patients with acute leukemia, *Blood* 57:372.

Hornsten, P., Granstrom, M., Wahren, B., and Gahrton, G., 1977, Prognostic value of colony stimulating and colony forming cells in peripheral blood in acute nonlymphocytic leukemia, *Acta Med. Scand.* 201:405.

Kinosita, R., Nowotny, A., and Shikata, T., 1963, Effects of endotoxin on bone marrow, *Blood* 21:779.

Koeffler, P. H., and Golde, D. W., 1978, Acute myelogenous leukemia: A human cell line responsive to colony stimulating activity, *Science* 200:1153.

Kurland, J. L., Broxmeyer, H., Pelus, L. M., Bockman, R. S., and Moore, M. A. S., 1978, Limitation of excessive myelopoiesis by the intrinsic modulation of macrophage derived prostaglandin E, *Science* 199:552.

Lind, E. D., Bradley, M. L., Gunz, F. W., and Vincent, P. C., 1974, The non-equivalence of mouse and human marrow culture in the assay of granulopoietic stimulatory factors, *J. Cell. Physiol.* 83:35.

MacVittie, T. J., and Walker, R. I., 1978, Endotoxin induced alterations in canine granulopoiesis: Colony stimulating factor, colony forming cells in culture, and growth of cells in diffusion chambers, *Exp. Haematol.* 6:613.

Mayer, P., Hamberger, H., and Drews, J., 1980, Differential effects of ubiquinone Q_7 and ubiquinone analogues on macrophage activation and experimental infections in granulocytopenic mice, *Infection* 8:256.

Metcalf, D., 1971, Acute antigen induced elevation of serum colony stimulating factor, *Immunology* 21:427.

Metcalf, D., 1974a, Regulation by colony stimulating factor of granulocyte and macrophage colony formation *in vitro* by normal leukemic cells, in: *Control of Proliferation in Animal Cells* (R. Baserga, ed.), pp. 887–905, Cold Spring Harbor Laboratory, Cold Spring Harbor, N.Y.

Metcalf, D., 1974b, Depressed responses of the granulocyte macrophage system to bacterial antigens following preimmunization, *Immunology* 26:1115.

Metcalf, D., 1976, Colony formation by myeloid leukemic cells, in: *Hemopoietic Colonies* Springer-Verlag, Berlin.

Metcalf, D., 1979, Clonal analysis of the action of granulocyte macrophage colony stimulating factor on the proliferation and differentiation of myelomonocytic leukemic cells, *Int. J. Cancer* 24:616.

Metcalf, D., Chan, S. H., Gunz, F. W., Vincent, P., and Ravich, R. B., 1971, Colony stimulating factor and inhibitor levels in acute granulocytic leukemia, *Blood* 38:143.

Metcalf, D., MacDonald, H. R., and Chester, H. M., 1975, Serum potentiation of granulocytic and macrophage colony formation *in vitro*, *Exp. Hematol.* 3:261.

Mintz, U., and Sachs, L., 1973, Differences in inducing activity for human bone marrow colonies in normal serum and serum from patients with leukemia, *Blood* 42:331.

Myers, A. M., and Robinson, W. A., 1975, Colony stimulating factor levels in human serum and urine following chemotherapy, *Proc. Soc. Exp. Biol. Med.* 148:694.

Pluznik, D., and Sachs, L., 1966, The induction of colonies of normal "mast" cells by substance in conditioned medium, *Exp. Cell Res.* 43:553.

Price, G. B., Krogsrud, R., Stewart, S., and Senn, J. S., 1978, Heterogeneity and biochemistry of colony stimulating activities, in: *Haematopoietic Cell Differentiation* (D. W. Golde, M. J. Cline, D. Metcalf, and C. F. Fox, eds.), pp. 417–432, Academic Press, New York.

Quesenberry, P. J., and Gimbrone, M. A., 1980, Vascular endothelium as a regulator of

granulopoiesis: Production of colony stimulating activity by cultured human endothelial cells, *Blood* **56:**1060.

Quesenberry, P. J., Morley, A., Stohlman, F., Jr., Rickard, K., Howard, D., and Smith, M.,
1972, Effect of endotoxin on granulopoiesis and colony-stimulating factor, *N. Engl. J.
Med.* **286:**227.

Quesenberry, P. J., Halperin, J., Ryan, M., and Stohlman, F., Jr., 1975, Tolerance to the
granulocyte releasing and colony stimulating factor elevating effects of endotoxin,
Blood **45:**789.

Reiffers, G., Vezon, G., and Broustet, A., 1981, Acute myeloblastic leukemia: Prognositic
value of infectious complications observed during induction treatment, *Sem. Hop.* **57:**
133 (in French).

Ruscetti, F. W., and Chervenick, P. A., 1975, Regulation of the release of colony stimulating activity from mitogen-stimulated lymphocytes, *J. Immunol.* **114:**1513.

Shadduck, R. K., 1974, Colony stimulating factor response to endotoxin in normal and
leukopenic recipients, *Exp. Hematol.* **2:**147.

Shah, R. G., Caporale, L. H., and Moore, M. A. S., 1977, Characterization of colony stimulating activity produced by human monocytes and phytohaemagglutinin-stimulated
lymphocytes, *Blood* **50:**811.

Stevens, P., Winston, D. J., and van Dyke, K., 1978, *In vitro* evaluation of opsonic and cellular granulocyte function by luminol dependent chemiluminescence: Utility in patients with severe neutropenia and cellular deficiency states, *Infect. Immun.* **22:**41.

22

Effect of Endotoxin on Graft-versus-Host Reactions

P. Liacopoulos

I. Introduction

The graft-versus-host (GVH) reaction, in the same way as other reactions of humoral or cellular immunity, is regulated by a set of cellular interactions produced either by direct cell-to-cell contact or through the mediation of soluble factors released by the activated cells. It would therefore not be surprising if the potent immunoregulatory properties of bacterial lipopolysaccharides (LPS) were also observed in the GVH reaction. This reaction is initiated by a "graft" of immunologically competent lymphocytes introduced into a "host" that confronts the graft with a major histocompatibility (MHC) difference, without being itself able to mount a similar immunological attack against the donor lymphoid cells. Indeed, if the host is immunologically competent, the host-versus-graft (HVG) reaction precedes by a few days the GVH reaction and eliminates or inactivates allogeneic cells. Therefore, *in vivo* GVH reaction can only develop (1) in immunologically immature animals confronted with competent lymphoid cells from a histoincompatible adult, (2) in adult animals protected from the lethal effects of irradiation by allogeneic hemopoietic cell inoculum containing mature lymphoid cells, and (3) in adult F_1 hybrids derived from two inbred strains of different MHC antigens given lymphoid cells of the one parental strain that recognize the antigens inherited from the other parental strain and react against them.

P. Liacopoulos ● Institut d'Immunobiologie (INSERM U.20, CNRS, L.A. 143), Hôpital Broussais, 75674 Paris Cedex 14, France.

II. Immunoregulatory Processes during GVH Reactions

The antihost reaction of grafted cells is underpinned by intricate cellular interactions resulting in several distinct clinical disorders and eventually in the death of the host. In GVH reactions as in other cell-mediated immunological responses, the effector cells are T cells. In fact, several subpopulations of T cells synergistically interact for optimal development of the antihost immune response. The T_1 cells (or Lyt 123+ cells), short-lived and located in the thymus and spleen, are activated through stimulation by host antigens. The T_2 cells (or Lyt 1+ cells), peripheral recirculating and long-lived cells, after activation by the same antigens amplify the reactivity and promote differentiation of Lyt 123+ cells towards Lyt 23+ cells that behave either as killer cells that destroy the target cells bearing the stimulating antigens, or as suppressor cells that inhibit activity of T_2/Lyt 1+ cells (Cantor and Gershon, 1979). Alternatively, these killer and suppressor activities may be properties of two different Lyt 23+ subsets since the suppressor cells express an *I-J* subregion antigen whereas killer cells do not bear this antigen (Murphy, 1978). It seems that T_1/Lyt 123+ cells are stimulated by *H-2K* and *D* region encoded antigens and T_2/Lyt 1+ cells are stimulated by *I* region encoded antigens (Wolters and Benner, 1980).

These interactions between T-cell subpopulations also involve macrophages, which bear cell membrane structures binding T-helper as well as T-suppressive factors. When the macrophage carries such factors, it conveys information signals from one T-cell set to another. This transfer of information is an energy-requiring process since inactivated macrophages lose the ability to present signals in an active form to the appropriate target cell (Ptak *et al.*, 1978). The role of macrophages in GVH reactions is illustrated by work of Yamashita *et al.* (1980). Addition of macrophages to an inoculum of parental lymph node cells depleted of macrophages, greatly potentiates the GVH reaction induced in F_1 hybrids depleted of macrophages and lymphocytes. Conversely, adherent cells from spleens of mice experiencing a GVH reaction in an advanced phase exert a strong nonspecific suppressor activity (Villeneuve *et al.*, 1980).

In addition to immunoregulatory cellular circuits, humoral factors such as enhancing antibody, antigen–antibody complexes, anti-idiotypic and anti-cell receptor antibody may all contribute to the regulation of GVH reaction. Out of this variety of control relationships some are specific, i.e., regulating the response to a single antigen, whereas others are nonspecific, generally modifying immunoreactivity to all antigens. Due to this modulation, a GVH reaction, once evoked, does not invariably evolve until the host is consumed. Not only do local GVH reactions inevitably subside after reaching their height, but in many cases animals afflicted with GVH disease may survive the acute phase of the disease and present an apparent clinical recovery, whilst donor cells still persist in the host (Fox, 1966). Animals that have recovered from GVH disease become particularly resistant to a second GVH reaction induced either by cells syngeneic to the first donor or by third-party cells (Grebe and Streilein, 1976). This evolution of the GVH reaction to resolution and to resistance of recovered animals to further challenges, strongly suggests that GVH reactions are subject to regulatory processes that may be either endogenous or elicited by exogenous treatments.

III. Modulation of GVH Reactions by LPS

Since GVH reaction follows the immunological attack of donor cells against the recipient cells through their membrane histocompatibility antigens, various experimental maneuvers at different levels are conceivable in order to modify the reactivity of donor cells: on the donor, prior to the removal of cells for injection; on the donor cells themselves *in vitro* before their injection into the recipient; or they may even be applied to the host while the reaction is developing. Several kinds of experimental manipulations are capable of bringing GVH reactions under control (cf. Grebe and Streilein, 1976). Among them can be mentioned those acting on the nonspecific regulation of immunological reactions, e.g., treatment with bacterial LPS. The immunoregulatory properties of LPS have been studied for many years, and these studies have clearly established that according to the timing of the treatment, LPS may either increase or decrease various immunological reactions.

A. Effect of Donor Treatment

The immunoregulatory activity of endotoxins on antibody production and on cell-mediated immunity elicited by various antigens is well documented (Morrison and Ryan, 1979). In general, when endotoxin is given simultaneously or after the antigen, it provokes an increase of the response, whereas when given before the antigen, it inhibits it (Franzl and McMaster, 1968; McMaster and Franzl, 1968). Indeed, injection of 10 to 50 μg of E. coli LPS to mice 1 to 3 days before their spleen cells were cultured with TNP–SRBC, decreased by almost 20 times their responsivity (Uchiyama and Jacobs, 1978a). In a more systematic study, Liacopoulos et al. (1980) studied the effect of E. coli LPS (EC-LPS) pretreatment of C57BL/6 mice on five *in vitro* immunological and mito-genic reactions of the spleen cells of these mice (Fig. 1A). The anti-SRBC PFC response was decreased by 74%, the MLC and CML responses to BALB/c cells were decreased by 36 and 28%, respectively, and the response to PHA was decreased by 27%; the response to LPS was not affected. This inhibition could be attributed to the development of a suppressive cell population since, when cells from pretreated animals were cocultured in 1 to 1 proportion with cells from normal syngeneic animals, they consistently lowered the reactivity of the latter (Fig. 1B).

The broad inhibition of several cell reactivities and especially MLC and CML responses to histocompatibility antigens by LPS would account for results of earlier studies that had shown that donor pretreatment with various bacterial endotoxins could inhibit GVH activity of their cells (Liacopoulos *et al.*, 1967a). Two bacterial endotoxins were used: *Salmonella enteritidis* LPS (SE-LPS) and *Serratia marcescens* LPS (SM-LPS), the latter either in its original form or modified by treatment

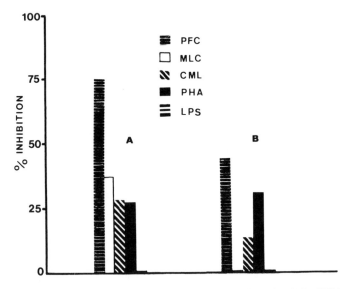

Figure 1. Suppressive effects of EC-LPS on immunological (PFC, MLC, CML) or mitogenic (PHA, LPS) *in vitro* responses. (A) Responses of spleen cells from C57BL/6 mice given an i.p. injection of 5 μg EC-LPS. Each column represents the percentage inhibition of the corresponding response of cells from treated mice as compared to the response of cells from normal mice. (B) Spleen cells from treated mice cocultured in 1:1 proportion with cells from normal mice. Each column represents the percentage inhibition of the half value of the sum of the responses of separate cultures of cells from normal and treated mice. (From Liacopoulos *et al.*, 1980.)

with CH_3OK (SM-LPS/CH_3OK) in order to reduce its toxicity (Nowotny, 1963). Spleen cells from untreated C57BL/6 ($H2^b$) transferred in midlethally irradiated (500 R) adult C3H/HeN ($H2^k$) mice regularly produced mortality of recipients given more than 10×10^6 donor cells. Within the range of 10×10^6 to 22.5×10^6 donor cells, the mortality rate was a linear function of the number of transferred cells (Fig. 2). This well-defined correlation between normal donor cell dosage and ensuing mortality in recipients, provided a standard dose–response reference plot against which the quantitative effectiveness of spleen cells from endotoxin-treated donors could be matched and compared.

It has previously been found that various protein antigens could inhibit GVH reactions when injected in high doses to cell donors (Liacopoulos *et al.*, 1967b). As bacterial LPS possess an important toxicity, it was necessary to precisely delimit the doses that could be supported by the animals. These preliminary studies showed that amounts of 30 μg of SE-LPS, 100 μg of SM-LPS, and 200 μg of SM-LPS/CH_3OK per day, given by i.p. injections for 6 to 8 days, were well tolerated. At the end of these periods of treatment, measured numbers of harvested donor spleen cells were injected i.v. into irradiated (500 R) C3H/HeN recipients. It will be seen in Table 1 that, in all experimental groups, a significant proportion of the hosts survived up to the 15th

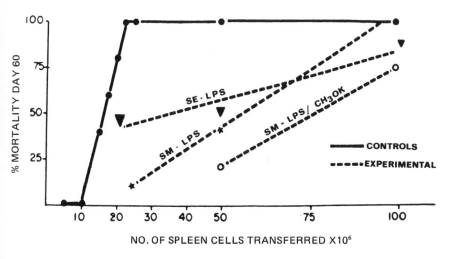

Figure 2. Final mortality levels in midlethally irradiated (500 R) C3H/HeN mice receiving measured numbers of normal (——) or LPS-pretreated (- - - -) donor C57BL/6 spleen cells. Data for experimental curves are reported in Table 1 (SM-LPS/CH_3OK 200 μg/day for 6 days). (From Liacopoulos *et al.*, 1967b.)

day as compared with 30, 0, and 0% survival in the corresponding control groups. Most of the hosts in these experimental groups were still
alive on the 60th day, especially those that received cells from SM-
LPS/CH₃OK-treated donors. Survival at 60 days was considered to be
evidence that transferred cells were incapable of initiating a fatal form
of GVH disease. The efficiency of SM-LPS/CH$_3$OK in inhibiting donor
cell GVH reactivity motivated further study of this antigen involving
other modes of donor treatment, e.g., a 4-day donor treatment with 30
μg of SM-LPS/CH$_3$OK per day and single immunizing injections of 10
μg either 7 days or 4 days prior to spleen cell transfer. The results of
these experiments indicate that even these modest donor treatments
resulted in significantly increased survival of recipients in most experimental groups. However, inhibition of donor cell GVH activity following these modes of treatment was not as great as that following
treatment with 200 μg per day for 6 days (Table 1).

Recipients of spleen cells from LPS-treated donors also displayed a
significant delay in the time at which they succumbed to GHV disease.
Increasing the number of transferred normal spleen cells from 15 $\times$
10^6 up to 100 $\times$ 10^6 cells did not substantially shorten the fatality interval in control groups (12.9 to 10.3 days, respectively; 11.3 days on the
average). Among hosts given cells from LPS-treated donors, besides a
high proportion of definite survivors, the fatality interval of dead animals was about twice as long (SE-LPS, 19.4 days; SM-LPS, 25.3 days;
and SM-LPS/CH$_3$OK, 25.3 days on the average). This increase of the
fatality interval in experimental groups clearly indicates that GVH effector donor spleen cell dilution as a result of the spleen hyperplasia
provoked by the LPS treatment (Table 1) cannot alone explain the observed inhibition of GVH activity. Furthermore, pretreatment with SE-
LPS, which provoked an 8-fold increase of the spleen cellularity, inhibited GVH activity of spleen cells much less than the treatment
with SM/CH$_3$OK, which increased the spleen cellularity by only
2.5-fold (Table 1). In Fig. 2 are reported the mortality rates of recipients given measured numbers of either control or pretreated spleen
cells. It can be seen that the slopes of the curves for control vs. experimental groups are very different. The reduced slopes of the experimental dose–response curves (Fig. 2) suggest a decreased GVH
effectiveness for increasing doses of transferred donor cells. For example, 100 $\times$ 10^6 cells from SE-LPS-treated donors are as effective in
provoking host mortality as 21 $\times$ 10^6 normal cells, whereas 20 $\times$ 10^6
treated donor cells are as active as 15 $\times$ 10^6 normal cells. Thus, two
properties characterize the GVH activity of cells from treated donors:
(1) increase of fatality interval (from 11.3 days in control to 25 days in
experimental animals) and (2) decrease of GVH effectiveness of cells

Table 1. Effect of Donor C57BL/6 Mice Pretreatment with SE-LPS, SM-LPS, or SM-LPS/CH₃OK on Competence of Donor Spleen Cells to Produce GVH Mortality in Adult C3H/HeN Recipients Irradiated with 500 R[a]

Donor treatment		Donor spleen cellularity ($\times 10^6$ cells/spleen)	No. of donor cells transferred ($\times 10^6$)	No. of recipients	Percent survival at		Final survival (>60 days)
LPS	Amount, days				Day 15	Day 30	
None (controls)	—	135	20	20	30	20	20
			50	20	0	0	0
			100	15	0	0	0
SE	30 µg/day for 7 days	1090	20	28	57	54	54
			50	12	75	67	50
			100	8	25	12	12
SM	100 µg/day for 8 days	606	20	10	100	90	90
			50	10	100	80	60
			100	10	100	0	0
	10 µg on day 7	170	20	6	83	83	83
			50	6	33	33	33
			100	5	0	0	0
SM/CH₃OK	10 µg on day 4	265	20	8	87	75	75
			50	8	25	25	25
			100	8	0	0	0
	30 µg/day for 4 days	446	20	8	100	100	100
			50	8	87	87	87
			100	8	100	37	25
	200 µg/day for 6 days	340	20	8	100	100	100
			50	8	100	100	100
			100	11	91	73	64

[a] From Liacopoulos and Merchant (1969).

from treated animals as the number of transferred cells increase. Both are suggestive of an inhibitory process produced by the endotoxin treatment of donor animals.

Recipients surviving GVH reaction were grafted with donor skin 30 to 40 days after transfer of treated donor cells in order to disclose whether a transplantation tolerance of donor tissues was induced in these animals, as the result of a possible establishment of donor cells in the recipients. All of these animals rejected the graft in a somewhat accelerated manner (8 to 9 days). It thus appeared that irradiation with 500 R would not sufficiently reduce the number of immunocompetent cells of the host in order to allow permanent establishment of donor cells.

In another series of experiments (Liacopoulos and Merchant, 1969), C3H/HeN recipients were lethally irradiated with 700 R and given 25×10^6, 50×10^6, or 100×10^6 spleen cells from C57BL/6 donors either normal or pretreated with SM-LPS/CH$_3$OK (200 μg/day for 8 days). As shown in Fig. 3, increase of the irradiation dose did not modify the rate of mortality in control recipients (8 to 15 days' survival, 11 days on the average). Experimental recipients given cells from treated donors displayed a longer survival: those given 25×10^6 cells survived 26.8 days and those given 50 or 100×10^6 cells survived 18.25 days on the average. These results again show that pretreatment of donors with endotoxin substantially reduces the GVH reactivity of their spleen cells, but this reduction is temporary and the donor cells recover their reactivity and kill the host.

However, endotoxin treatment tested in a milder histocompatibility barrier combination (CBA and A/J mice, which share the K through E loci of MHC but differ at C through D) not only inhibited GVH reaction but also promoted establishment of a long-lasting tolerance of donor skin graft in the recipient (Liacopoulos et al., 1968). CBA donors were treated with EC-LPS for 10 days (50 μg/day for 3 days, 75 μg/day for 3 days, and 100 μg/day for 4 days) and their spleen cells were transferred in A/J recipient irradiated with 700 R (Table 2). In this combination, mortality rate in control recipients is milder and resulted in a final survival of 22% of recipients. Among recipients of spleen cells from LPS-treated donors, the mortality was markedly reduced and 72% of recipients survived. All survivors were grafted with donor skin and this was tolerated (> 100 days) in 18.8% of control survivors, whereas in experimental survivors the graft was tolerated in 62% of recipients. In a more recent study, Damais et al. (1972) treated adult C57BL/6 mice with SE-LPS for 7 days (30 μg/day) and then transferred their spleen cells to (CBA $\times$ AKR)F$_1$ hybrids or to Swiss neonatal recipient mice. This pretreatment of donor animals drastically reduced the GVH mor-

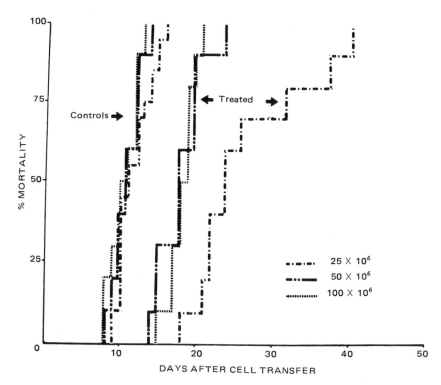

Figure 3. Mortality in lethally irradiated (700 R) C3H/HeN mice receiving indicated numbers of spleen cells from donor C57BL/6 mice either control or pretreated with SM-LPS/CH₃OK (200 μg/day for 8 days). (From Liacopoulos and Merchant, 1969.)

tality of recipients to nearly 0% as compared to a mortality rate of about 80% in control recipients transfused with cells from nontreated donors.

B. Effect of *in Vitro* Treatment of Donor Cells

The definite inhibitory effect of donor pretreatment with various LPS on the GVH activity of donor cells led Chedid (1973) to treat the donor cells *in vitro* before their transfer into the recipient. Adult C57BL/6 spleen cells were incubated *in vitro* (2×10^8 cells with 30 μg of radiolabeled SE-LPS in 1 ml) for 1 hr at 37°C, in parallel to samples with no added LPS. After washing, the cells were transferred into neonatal Swiss mice (10^7/mouse). Control cells produced a mortality of 64% of recipients, whereas among recipients of SE-LPS-treated cells mortality was only 27.3% ($p < 0.01$, Table 3). Since the SE-LPS used was radiolabeled, it was possible to calculate that no more than 0.07 μg

Table 2. Inhibition of GVH Reaction and Induction of Tolerance of Donor Skin Graft by Treatment of Donor CBA Mice with EC-LPS before Transfer of Their Spleen Cells into Adult A/J Recipients Irradiated with 700 R

Donor treatment	No. of donor cells transferred ($\times 10^6$)	No. of recipients	Survival at 60 days		Tolerance of skin graft	
			No. of mice	%	No. of grafted animals	Graft survival at 100 days
None (controls)	25	18	3	16.6		
	50	20	5	25		
	100	16	4	25		
	200	15	3	20		
Total		69	15	22	15	13/69 (18.8)
EC-LPS (775 µg in 10 days)	25	8	8	100		
	50	18	12	66		
	100	14	9	64		
	200	18	13	72		
Total		58	42	72	42	36/58 (62)

[a] Tolerant/total (percentage).

of LPS was injected with treated cells. Furthermore, as a control for investigating whether the observed effect on GVH was due to the *in vitro* incubation or to an action *in vivo* of LPS that still contaminated the cells, fresh cells were injected together with 0.5 μg of SE-LPS. In this group, the mortality was 95.6%, i.e., greater than in the control group (Table 3).

Another method for *in vitro* treatment of donor cells was used by Thomson *et al.* (1978). They separated adherent spleen cells of CBA mice treated daily for 7 days with 60 μg EC-LPS/day. Then, on these adherent cells, they overlayed normal spleen cells and incubated them together for 1 to 2 hr. After incubation for 2 hr, nonadherent cells were washed out and transferred into newborn BALB/c mice. The ensuing mortality up to the 25th day after transfer was only 8% as compared with an 86.6% mortality provoked by spleen cells incubated with normal adherent cells. It must be stressed, however, that the number of nonadherent cells recovered after 2 hr incubation of 20×10^6 total spleen cells with adherent cells from LPS-treated mice was only $5–7 \times 10^6$ because of an intimate association of normal lymphocytes with LPS-activated macrophages adherent to plastic dishes. Thus, to experimental recipients $5–7 \times 10^6$ donor cells were given, whereas the control recipients received approximately 18×10^6 donor cells recovered after incubation of donor spleen cells with adherent cells from normal mice where no association of lymphocytes with macrophages occurred.

In spite of the large number of survivors after *in vitro* treatment of donor cells, no information was provided by these authors concerning

Table 3. Inhibition of GVH Reaction by Prior Incubation of Normal C57 BL Spleen Cells with SE-LPS (30 μg/ml for 1 hr at 37°C) before Their Injection into Newborn Swiss Recipients[a]

Treatment of donor cells	No. of donor cells transferred	Mortality of recipients at	
		Day 15	Day 30
Normal cells	10^7	38/86 (44.2)[b]	55/86 (64)
Normal cells incubated with SE-LPS (30 μg/ml)	10^7	8/55 (14.5)	15/55 (27.3)[c]
Normal cells + 0.5 μg SE-LPS	10^7	26/46 (56.5)	44/46 (95.6)

[a] From Chedid (1973) with permission of the publishers.
[b] Dead/total (percentage).
[c] $p < 0.01$.

establishment of tolerance of donor tissues in the recipients that survived GVH disease.

C. Effect of the Host Treatment

The effects of host treatment with EC-LPS were systematically
studied by Skopińska (1972) in C57BL/6 × A/J)F₁ adult hybrid recipients of parental A/J mouse spleen cells. These effects were evaluated
by assessing both the splenomegaly in nonirradiated recipients 21 days
after cell transfer and the mortality rate in recipients irradiated with
500 R. The EC-LPS was given i.p. in a single dose of 20 µg/mouse at
various times either before or after donor cell transfer. The injection of
EC-LPS provoked a different effect on GVH splenomegaly according to
the time of this injection with regard to the injection of donor cells. As
shown in Fig. 4, EC-LPS given 4 days before cell transfer significantly

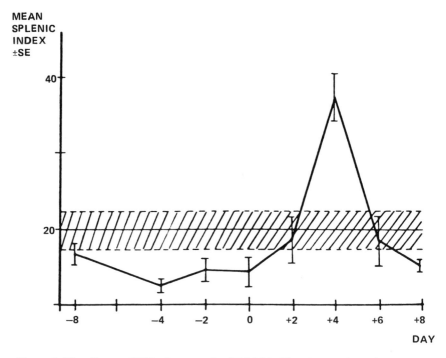

Figure 4. The effect on GVH splenomegaly of EC-LPS (20 µg/mouse, i.p.) given on various days with respect to the day of donor spleen cell injection. Day 0 refers to the day
of A strain spleen cell injection into B6AF₁ recipients. Splenomegaly was evaluated 21
days after the injection of cells. The cross-hatched zone refers to the mean splenic index
± S.E. seen in animals injected with donor spleen cells alone. (From Skopińska, 1972,
with permission of the publishers.)

decreased the splenic index in the recipients of donor cells alone. In contrast, when EC-LPS was given 4 days after cell transfer, the splenic index was increased by almost twice as much. The mortality rate in irradiated recipients was also increased by the injection of EC-LPS 4 days after cell transfer. In this donor–recipient combination, the mortality rate was 50% on the 50th day after cell transfer. In the group of recipients given EC-LPS, all of the recipients died by the 34th day. In a control group where similarly irradiated recipients were given syngeneic spleen cells, injection of EC-LPS did not provoke any appreciable effect.

IV. Mechanism of Immunoregulatory Effect of LPS

The preceding analysis of the effects of LPS on GVH reactions clearly shows that when LPS is administered to the donor animal before removal of cells or even to the recipient before cell transfer, it inhibits the subsequent GVH splenomegaly or mortality. Given to the recipient with or after donor cell transfer, it considerably enhances both these manifestations of GVH disease.

Previous *in vitro* experiments on antibody production or cellular immunity reactions (Persson, 1977; Liacopoulos *et al.*, 1980) have evidenced that the effects of LPS can be imposed on the activity of normal cells by cells from LPS-treated animals. Therefore, enhancement or suppression should follow the development of cell populations modulating directly or through soluble mediators, the emergence or the activity of effector cells.

Of the three types of cells cooperating in immunological responses, namely, macrophages, B cells, and T cells, the two former are directly sensitive to LPS, whereas the latter seem insensitive. Macrophages are extremely sensitive to LPS in concentrations of a few nanograms per milliliter. Among the various consequences of this contact (cf. Morrison and Ryan, 1979), perhaps the most significant in immunoregulation is the release of monokines from LPS-activated macrophages. One of these monokines, the lymphocyte-activating factor (LAF) first described by Gery *et al.* (1971), markedly enhances the proliferative response of T lymphocytes to PHA and also provokes a direct proliferative response of T cells, without producing detectable effects on B cells (Gery *et al.*, 1972a). Generation of LAF, after stimulation of macrophage tumor line $P388D_1$ by LPS (Meltzer and Oppenheim, 1977), as well as absence of LAF activity in supernatants of peritoneal exudate cells from the endotoxin-unresponsive C3H/HeJ mice (Mizel *et al.*, 1978) clearly show that LAF is released by macrophages directly stimulated by LPS.

B cells in the presence of microgram concentrations of LPS undergo a series of alterations including blast formation, initiation of RNA and DNA synthesis ultimately leading to cell division (Gery et al., 1972b). In addition, B cells stimulated by LPS undergo a differentiation process resulting in increased number of PFC against unrelated antigens, i.e., present a polyclonal activation (Andersson et al., 1972). LPS can also circumvent the requirement of many antigens for T-helper cells in the initiation of antibody response by B cells (Möller et al., 1972). Not all B cells are sensitive to mitogenic stimulation by LPS. Gronowicz and Coutinho (1975) showed that the B-cell subset responsive to dextran sulfate is different from that responsive to LPS and appears earlier in ontogeny or in bone marrow than the latter.

In contrast to the above-described effects of LPS on macrophages and B cells, T cells do not appear to be directly influenced by LPS in spite of the finding that they bind LPS almost equally well on their surface (Andersson et al., 1972). However, mitogenic T-cell responses to Con A could be enhanced 2- to 10-fold by prior or simultaneous addition of LPS (Ozato et al., 1975). This effect has also been observed in thymocytes from LPS-unresponsive CH3/HeJ mice (Dumont, 1978) and shows that this must be a direct effect on activated T cells since neither macrophages nor B cells from these mice are triggered by LPS. Furthermore, Haas et al. (1978) showed that although adjuvant effect after simultaneous administration of LPS and antigen (SRBC) is absent in C3H/HeJ, pretreatment with LPS of such mice or of their responder counterparts C3Heb/FeJ mice results in the suppression of their subsequent PFC response to SRBC, and this suppressive effect can be transferred in cultures of normal spleen cells. This interesting observation gives rise to the possibility of preferential generation of suppressor cells by direct action of LPS on quiescent T cells.

The direct effects of LPS on macrophages, B and T lymphocytes take place whether or not an antigen is present. It could therefore be assumed that the nonspecific triggering of lymphoid tissues by LPS would be beneficial to an antigen-specific reaction if the antigen is present at the same time, whereas if the antigen is given after that period of stimulation, it will encounter lymphoid tissues during the homeostatic reversion of cellular activation and the response to it will be inhibited.

In GVH reactions, donor cells confront antigen when they are transfused to unresponsive allogeneic recipients. A proliferative reaction ensues rapidly after recognition of host antigens, which culminates on about the 10th to 20th day after transfer (Simonsen, 1962; Skopińska, 1972). If LPS is present at that time (i.e., given with or a few days after cell transfer), it would potentiate this proliferative reaction in two ways: (1) T cells activated after contact with allogeneic antigens could be further stimulated by LPS in the manner of mitogenic T-cell responses elicited by Con A, which are markedly enhanced by LPS (Ozato *et al.*, 1975); and (2) through LAF release by LPS-activated macrophages, which greatly potentiates the proliferative response of T cells responding to PHA (Gery *et al.*, 1971) and induces thymocytes to respond to histocompatibility antigens (Beller *et al.*, 1978).

As a consequence of the proliferation of donor T_1 and T_2 cells triggered by host histocompatibility antigens, increasing numbers of killer cells appear (maturing T_1/Lyt 123$^+$ and T_2/Lyt 23$^+$) resulting from the reaction of donor cells to *H2K* and *D* region encoded recipient antigens (Wolters and Benner, 1980). Moreover, LPS treatment increases the killer and the helper activity of thymocytes by selecting T_2 cells (Lyt 1$^+$ and Lyt 23$^+$ cells) (Baroni *et al.*, 1976). This enhancement of T-cell responses results in increase of both proliferative response and mortality rate in LPS-treated recipients.

Thus, when antigen is present, the reaction to it is markedly strengthened by LPS and lasts longer than the cellular mitogenic and polyclonal reaction produced by LPS alone *in vitro* (Sjöberg *et al.*, 1972) or *in vivo* (Gronowicz and Coutinho, 1974). These reactions to LPS alone decline after the second or third day following LPS administration. The inhibition of immunological responses to different antigens occurs precisely when the antigen is given within that period of decline of the cellular reaction to LPS. Since the inhibition determined by prior injection of LPS can be imposed on normal cells by LPS-treated cells, both *in vitro* and *in vivo* (Liacopoulos *et al.*, 1980; Thomson *et al.*, 1978), it can be deduced that this inhibition is an active process brought about by suppressor cells in charge of the homeostatic reestablishment of the normal state.

Concerning the identity of the suppressor cells modulating the GVH activity of normal spleen cells, Thomson and Jutila (1974) clearly showed that these cells belong to the population of adherent cells, since injection to neonatal mice of a lethal dose of normal spleen cells mixed with adherent spleen cells from endotoxin-treated mice greatly reduced the GVH mortality of recipients. These authors further showed: (1) that *in vitro* incubation for 2 hr of normal spleen cells with adherent cells from LPS-treated mice, also suppressed the GVH activity of the former cells, and (2) that although the supernatant of *in vitro* incubated adherent cells did not impair the activity of normal cells, serum from LPS-treated mice also suppressed the GVH reaction (Thomson *et al.*, 1978). Yet, the adherent cell population is a mixture of cells comprising macrophages, B cells, and T cells. The responsibility of macrophages in the suppressive effect can be excluded since LPS-treated cells depleted of macrophages either by filtration through a G-10 column (Uchiyama and Jacobs, 1978a) or by iron powder treatment (Persson, 1977), still retained their suppressive activity. In a more systematic study, Uchiyama and Jacobs (1978b) examined the peculiar defects of each of the three types of cells from LPS-treated mice. They found that splenic macrophages from normal mice and LPS-treated mice were equally efficient in supporting development of PFC by macrophage-depleted normal spleen cells, whereas T cells from LPS-treated mice have a reduced capacity to interact with normal B cells. They further showed (Uchiyama and Jacobs, 1978b) that increase of antibody response by LPS is dependent on a helper cell that is radiation resistant and nonadherent to nylon wool. Inhibition of this response is due to a suppressor cell that is radiation sensitive and adherent to nylon wool. Both cell types are sensitive to anti-T-cell serum.

V. Conclusions

GVH reactions can markedly be modulated by bacterial LPS. As in other immunological responses, the outcome of this regulation, enhancement or inhibition, appears to depend on the time of LPS administration with regard to antigenic stimulation. When LPS is given to the donor animals or to the donor cells *in vitro* before confronting the host antigens, the GVH reaction is consistently inhibited and in suitable donor–host combinations a long-lasting transplantation tolerance may ensue. When LPS is given with or after cell transfer, the GVH reaction is greatly enhanced.

The cooperation between ancillary and effector cells could explain these opposite effects of LPS administration. Of the three cell types cooperating in immunological responses, macrophages, B cells, and T cells, the two former types are well-known targets for LPS. B cells do not seem to play any significant role in GVH reactions. Macrophage activation by LPS results in release of factors that promote differentiation of T_1 to T_2 cells, especially when these cells are simultaneously stimulated by host antigens. This increased maturation of T cells would explain why, when LPS is given with or after cell transfer, both proliferative reaction and mortality rate of recipients are greatly potentiated. In the absence of host antigens, i.e., when LPS is given to the donor before cell transfer, these different processes of differentiation also take place. They are rapidly followed, however, by the appearance of the suppressor cell variety of T_2 cells that accumulate, in significant numbers, when donor cells encounter host antigens. An inhibition of antigen-specific helper and killer cell triggering will ensue resulting in attenuation of GVH disease.

VI. References

Andersson, J., Sjöberg, O., and Möller, G., 1972, Mitogens as probes for immunocyte activation and cellular cooperation, *Transplant. Rev.* **11**:131.

Baroni, C. D., Ruco, L., Soravito De Franceschi, S., Uccini, S., Adorini, L., and Doria, G., 1976, Biological effects of *Escherichia coli* lipopolysaccharide (LPS) *in vivo*. I. Selection in the mouse thymus of killer and helper cells, *Immunology* **31**:217.

Beller, D. I., Farr, A. G., and Unanue, E. R., 1978, Regulation of lymphocyte proliferation and differentiation by macrophages, *Fed. Proc.* **37**:91.

Cantor, H., and Gershon, R. K., 1979, Immunological circuits: Cellular composition, *Fed. Proc.* **38**:2058.

Chedid, L., 1973, Possible role of endotoxemia during immunologic imbalance, in: *Bacterial Lipopolysaccharides* (E. H. Kass and S. M. Wolff, eds.), pp. 104–109, University of Chicago Press, Chicago.

Damais, C., Lamensans, A., and Chedid, L., 1972, Endotoxins bacteriennes et maladie homologue du nouveau-né, *C. R. Acad. Sci.* **274**:1113.

Dumont, F., 1978, Bacterial lipopolysaccharide (LPS) enhances concanavalin A reactivity of thymocytes from the low-LPS-responder mouse strain C3H/HeJ, *Experientia* **34**: 125.

Fox, M., 1966, Lymphoid repopulation by donor cells in the graft-versus-host reaction, *Transplantation* **4**:11.

Franzl, R. E., and McMaster, P. D., 1968, The primary immune response in mice. I. The enhancement and suppression of hemolysin production by a bacterial endotoxin, *J. Exp. Med.* **127**:1087.

Gery, I., Gershon, R. K., and Waksman, B. H., 1971, Potentiation of cultured mouse thymocyte responses by factors released by peripheral leucocytes, *J. Immunol.* **107**:1778.

Gery, I., Gerson, R. K., and Waksman, B. H., 1972a, Potentiation of the T-lymphocyte response to mitogens. I. The responding cell, *J. Exp. Med.* **136**:128.

Gery, I., Kruger, J., and Spiesel, S. Z., 1972b, Stimulation of B-lymphocytes by endotoxin: Reactions of thymus-deprived mice and karyotypic analysis of dividing cells in mice bearing T$_6$T$_6$ thymus grafts, *J. Immunol.* **108**:1088.

Grebe, S. C., and Streilein, J. W., 1976, Graft-versus-host reactions: A review, *Adv. Immunol.* **22**:119.

Gronowicz, E., and Coutinho, A., 1974, Selective triggering of B cell subpopulations by mitogens, *Eur. J. Immunol.* **4**:771.

Gronowicz, E., and Coutinho, A., 1975, Functional analysis of B cell heterogeneity, *Transplant. Rev.* **24**:3.

Haas, G. P., Johnson, A. G., and Nowotny, A., 1978, Suppression of the immune response in C3H/HeJ mice by protein-free lipopolysaccharides, *J. Exp. Med.* **148**:1081.

Liacopoulos, M., Lambert, F., and Liacopoulos, P., 1980, Nonspecific inhibitory processes of immunological and mitogenic cellular responses. I. Comparative effect of four suppressive agents, *Immunology* **41**:143.

Liacopoulos, P., and Merchant, B., 1969, Effet du traitement des donneurs avec des endotoxines sur la maladie homologue des receveurs adultes irradiés, in: *La structure et les effets biologiques des produits bactériens provenants de germes Gram-négatifs* (L. Chedid, ed.), pp. 341–356, Colloq. Int. sur les Endotoxines, CNRS No. 174, Paris.

Liacopoulos, P., Merchant, B., and Harrel, B. E., 1967a, Effect of donor immunization with somatic polysaccharides on the graft-versus-host reactivity of transferred donor splenocytes, *Proc. Soc. Exp. Biol. Med.* **125**:958.

Liacopoulos, P., Merchant, B., and Harrel, B. E., 1967b, Inhibition of the graft-versus-host reaction by pretreatment of donors with various antigens, *Transplantation* **5**:1423.

Liacopoulos, P., Herlem, G., and Perramant, M. F., 1968, Suppression de la maladie homologue et induction de tolerance dans une combinaison de souris adultes de locus H$_2$ different, in: *Advances in Transplantation* (J. Dausset, J. Hamburger, and G. Mathe, eds.), p. 183–187, Munksgaard, Copenhagen.

McMaster, P. D., and Franzl, R. E., 1968, The primary immune response in mice. II. Cellular responses of lymphoid tissue accompanying the enhancement or complete suppression of antibody formation by a bacterial endotoxin, *J. Exp. Med.* **127:**1109.

Meltzer, M., and Oppenheim, J. J., 1977, Bidirectional amplification of macrophage–lymphocyte interactions: Enhanced lymphocyte activation factor production by activated adherent mouse peritoneal cells, *J. Immunol.* **118:**77.

Mizel, S. B., Oppenheim, J. J., and Rosenstreich, D. L., 1978, Characterization of lymphocyte-activating factor (LAF) produced by a macrophage cell line, P388D$_1$, *J. Immunol.* **120:**1504.

Möller, G., Andersson, J., and Sjöberg, O., 1972, Lipopolysaccharide can convert heterologous erythrocytes into thymus-independent antigens, *Cell. Immunol.* **4:**416.

Morrison, D. C., and Ryan, J. L., 1979, Bacterial endotoxins and host immune responses, *Adv. Immunol.* **28:**293.

Murphy, D. B., 1978, The I. J. subregion of the murine H$_2$ gene complex, *Springer Semin. Immunopathol.* **1:**111.

Nowotny, A., 1963, Endotoxoid preparations, *Nature (London)***197:**721.

Ozato, K., Adler, W. H., and Ebert, J. D., 1975, Synergism of bacterial lipopolysaccharides and concanavalin A in the activation of thymic lymphocytes, *Cell. Immunol.* **17:**532.

Persson, U., 1977, Lipopolysaccharide-induced suppression of the primary immune response to a thymus-dependent antigen, *J. Immunol.* **118:**789.

Ptak, W., Zembala, M., and Gershon, R. K., 1978, Intermediary role of macrophages in the passage of suppressor signals between T-cell subsets, *J. Exp. Med.* **148:**424.

Simonsen, M., 1962, Graft versus host reactions: Their natural history and applicability as tools of research, *Prog. Allergy* **6:**349.

Sjöberg, O., Andersson, J., and Möller, G., 1972, Lipopolysaccharide can substitute for helper cells in the antibody response *in vitro*, *Eur. J. Immunol.* **2:**326.

Skopińska, E., 1972, Some effects of *Escherichia coli* endotoxin on the graft-versus-host reaction in mice, *Transplantation* **14:**432.

Thomson, P. D., and Jutila, J. W., 1974, The suppression of graft-versus-host disease in mice by endotoxin-treated adherent spleen cells, *J. Reticuloendothelial Soc.* **16:**327.

Thomson, P. D., Rampy, P. A., and Jutila, J. W., 1978, A mechanism for the suppression of graft-versus-host disease with endotoxin, *J. Immunol.* **120:**1340.

Uchiyama, T., and Jacobs, D. M., 1978a, Modulation of immune response by bacterial lipopolysaccharide (LPS): Multifocal effects of LPS induced suppression on the primary antibody response to a T-dependent antigen, *J. Immunol.* **121:**2340.

Uchiyama, T., and Jacobs, D. M., 1978b, Modulation of immune response by bacterial lipopolysaccharide (LPS): Cellular basis of stimulatory and inhibitory effects of LPS on the *in vitro* IgM antibody response to a T-dependent antigen, *J. Immunol.* **121:**2347.

Villeneuve, L., Brousseau, P., Chaput, J., and Elie, R., 1980 Role of the adherent cells in graft-versus-host induced suppression of the humoral immune response, *Scand. J. Immunol.* **12:**321.

Wolters, E. A. J., and Benner, R., 1980, Different target antigens for T-cell subsets acting synergistically *in vivo*, *Nature, (London)* **286:**895.

Yamashita, A., Hattori, Y., and Fukumoto, T., 1980, Augmenting effect of exogenous macrophages on the rat graft-versus-host reaction in F$_1$ hybrids depleted of macrophages, *Transplantation* **30:**122.

23

The Effect of Endotoxins on Skin and Kidney Transplantation Reactions

Ewa Skopińska-Różewska

I. Introduction

Transplantation immunity involves the development of cellular-type immunity to the donor histocompatibility antigens, and the production of humoral antibodies stimulated by these antigens. It is reasonable to expect that in the near future the classical nonspecific immunosuppressive therapy used to prevent or to reduce immune reactions caused by transplantation will be replaced by more "specific" treatment that will not interfere with the immunological response of the host to other antigens besides those of the donor. Such treatment in the future will be based upon reinforcing the reactions leading to the production of protective factors and suppressing reactions that lead to the damage of the transplant. Before commencing such a regimen of immunosuppression (or, rather, immunomodulation), the influence of various incidental factors should be established. Among these, the bacterial endotoxins appear to have explicit importance.

Endotoxins are known to act upon immunological reaction. Small and intermediate doses cause a strong adjuvant effect upon antibody production and stimulate the activity of the RES, whereas large doses act suppressively. They exert toxic effects such as fever and shock, as well as alterations in the blood coagulation system, leading to features of disseminated intravascular coagulation. Furthermore, administration of an endotoxin mixed with homologous antibody will not elicit toxic

Ewa Skopińska-Różewska • Division of Experimental Immunosuppression, Transplantation Institute, Warsaw Medical School, 02-006 Warsaw, Poland.

reactions but will induce nonspecific resistance enhancement (Nowotny *et al.*, 1965).

I felt that there were at least two reasons for the initiation of complex studies on the influence of endotoxin on transplantation immunity. First, as gram-negative microbes are constant components of the intestinal flora in the experimental animal and in man, infections caused by these microbes may constitute potential complications of tissue and organ transplantation. Second, gram-negative bacterial lipopolysaccharides (LPS) cause changes in the reactivity of cells that belong to the hemopoietic, lymphatic, and reticuloendothelial systems, that is, those systems that are directly involved with the process of transplant rejection. It should also be noted that endotoxin may increase the toxic effects of immunosuppressive drugs (Marecki and Bradley, 1973).

This chapter is not a complete review of my work done between 1972 and 1978. During this period I studied the direct and indirect effect of LPS on the graft-versus-host reaction in mice, including the role of anti-LPS immunoglobulins on these effects. Most of these data were published (Skopińska, 1972, 1974a,b, 1976). In this review I will survey our findings obtained during the investigations of LPS-induced modification of transplantation reaction in mice and rabbits elicited by skin or kidney grafts.

II. Brief Review of the Literature

Experiments devoted to this question are few in number. The results obtained from the literature are presented in Table 1.

Al-Askari *et al.* (1964) studied in rabbits the influence of endotoxins injected i.v. or locally in the area of the allogeneic skin graft on the survival time of this skin graft and the type of its rejection. They reported that a dose of endotoxin in the range of 0.01 to 20 μg/kg markedly shortens the survival time of the allogeneic graft, with no influence upon the autogeneic skin graft. In animals "tolerant" to endotoxins, this phenomenon did not appear. In a discussion of their results, they proposed that the reaction is a type of local Shwartzman reaction, where the immunological process occurring in the allogeneic graft replaces the sensitizing LPS injection. The authors compared this effect to the experiments of Stetson (1955), who found that an i.v. injection of endotoxin caused a necrotic hemorrhagic reaction in the areas of positive tuberculin reactions in sensitized rabbits.

Klassen and Milgrom (1970, 1971) studied the phenomenon of the generalized Shwartzman reaction in rabbits with autogeneic or alloge-

Table 1. Experimental Studies of the Effect of Endotoxins on Transplantation Reactions

Animal	Graft	Kind of effect	Reference
Rabbit	Skin	Accelerated rejection (treatment after grafting)	Al-Askari et al. (1964)
Mouse	Skin	Prolongation of survival (pretreatment)	Terino et al. (1964)
Guinea pig	Skin	Prolongation of survival (pretreatment)	Chase and Rapaport (1965)
Fowl	Lymphocytes	30% suppression of normal lymphocyte transfer reaction (treatment after grafting)	Floersheim (1969)
Rabbit	Kidney	Suppression of Shwartzman reaction in de-nervated kidney	Klassen and Milgrom (1970, 1971)
Mouse	Skin	Prolongation of survival (pre- and posttreat-ment)	Kirpatovski and Stani-slavski (1971)
Mouse	Skin	Treatment after grafting: prolongation of survival, most intense in mice adoptively immunized against LPS	Skopińska and Oluwasanmi (1972)
Dog	Small bowel	Accelerated death of unprepared graft. The preparation of the small bowel with anti-biotics led to prolongation of survival	Toledo-Pereyra et al. (1974)
Rabbit, mouse	Kidney, skin	Administration of LPS 2 days after grafting caused suppression of LyMIF production in the first, and enhancement in the sec-ond week after grafting	Skopińska et al. (1974)
Rabbit	Kidney	Administration 2 days after grafting caused suppression of DNA synthesis in donor–recipient MLC	Skopińska et al. (1977)
Mouse	Skin	Enhancement of LyMIF production 2 weeks after grafting, present only if LPS was giv-en 2 days after transplantation. Adminis-tration of LPS 2 or 4 days before grafting suppressed production of LyMIF	Skopińska (1978)

neic renal transplants. They administered *E. coli* endotoxin (400 μg/ rabbit) via an i.v. infusion, lasting 4 to 6 hr. Renal cortex necrosis that appeared in these animals was more intense in the animal's own kidney, which was allowed to remain *in situ*, than in the transplanted kidney regardless of whether it was allogeneic or autogeneic. In the opinion of the authors, such results were due to the fact that the transplanted kidney was not innervated. The adrenergic nervous system appears to have a major role in the pathogenesis of the Shwartzman reaction.

In mice, all investigators obtained a prolongation of the skin graft survival time of various degrees, using high doses of endotoxin. In the experiments of Kirpatovski and Stanislawski (1971), the higher the degree of tissue incompatibility, the more marked was the influence of LPS. With skin grafting from C57BL/6J to BALB/c mice, the survival time was prolonged by 144%, but when the skin grafting was performed from C3H to CBA mice, graft survival time was prolonged by 47%. When the skin was grafted from the male to the female of the same strain (C57BL), survival time was increased by only 40%.

Different results were obtained by Terino *et al.* (1964), who administered killed gram-negative bacteria to mice, and obtained the induction of tolerance to skin homografts across the weak male–female Y chromosome incompatibility barrier and showed only a prolongation of the graft survival in *H-2* incompatibility.

Toledo-Pereyra *et al.* (1974) reported accelerated death of unprepared small-bowel grafts in dogs. The preparation of the small bowel with antibiotics led to prolongation of survival.

There also exist some reports from human renal transplant clinics where the authors attempt to explain several cases of acute and hyperacute episodes of renal rejection (especially those that appeared in recipients compatible with the donors in HLA antigens) as caused by Shwartzman-type reaction. The source of the endotoxin could be the dialyzing fluid in the dialyzing machine or, perhaps, an infection from the kidney itself. They also describe three cases where a nephrectomy of the autogeneic pyelonephritic kidney was performed some time after the grafting of an allogeneic kidney, and in all three cases there appeared a strong febrile reaction and a rapid disintegration of function of the transplanted kidney, which in one case led to the removal of the graft (Husberg *et al.*, 1971; Morris *et al.*, 1969; Starzl *et al.*, 1968).

In our own experiments, we administered low doses of LPS once, on various days in relation to the grafting of allogeneic skin (mouse, rabbit) or kidney (rabbit). We evaluated the influence of LPS on the skin graft survival time and also on the cellular and humoral immunity developing after transplantation.

III. Results

A. Differences in the Response to LPS between Various Inbred Strains of Mice

Figure 1 presents the results of B10D2 skin graft survival in five inbred mouse strains. *E. coli* LPS was administered once in a dose of 20 μg on the second day after grafting, causing a statistically significant prolongation of the skin graft survival time in only one genetic combination (B10D2 to B10AF$_1$).

Tests were conducted on the effect of LPS on the production of the factor that inhibits spleen lymphocyte migration (LyMIF), in three

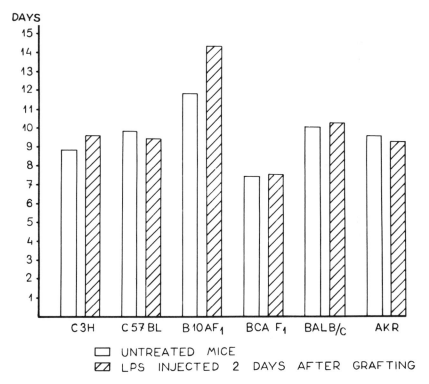

DAYS

□ UNTREATED MICE
▨ LPS INJECTED 2 DAYS AFTER GRAFTING

Figure 1. The effect of 20 μg of LPS given s.c. to groups of mice, 2 days after skin transplantation from B10D2 donors, on the survival time of skin grafts. The following inbred strains of mice and their F$_1$ hybrids were used as skin graft recipients: C3H/A, C57BL 10 ScSn, BALB/c, AKR, F$_1$ hybrids (BALB/c × C3H/A further called BCAF$_1$, F$_1$ hybrids (C57BL 10 ScSn × A/ Praha) further called B10AF$_1$. Each bar represents value of mean skin graft survival time obtained for the group of 20 mice. The values for B10AF$_1$ were: 11.8 ± 0.71 (S.E.) for untreated and 14.3 ± 0.65 for LPS-treated groups ($p < 0.01$).

of five genetic systems. The results are presented in Figs. 2, 3, and 4. Spleen cells from mice that were injected with a single dose of LPS on the second day after skin grafting, demonstrated 7 days after grafting, in tissue culture, decreased migration activity in relation to analogous cells derived from animals that were not treated with LPS. This effect was most marked in AKR mice and weakest in B10AF$_1$ mice.

Under the influence of B10D2 antigen, AKR mice produced the migration inhibitory factor, acting equally on their own lymphocytes as well as on the splenic lymphocytes derived from the untreated syngeneic mice and placed in the same chamber in a distinct capillary tube. Some mice produced the migration stimulatory factors. In the group treated with LPS, there was an increase in the number of animals whose cultured lymphocytes produced the inhibitory factor.

The BCAF$_1$ mice system behaved differently. In these mice, LPS caused an increase in the number of cultures producing the stimulatory

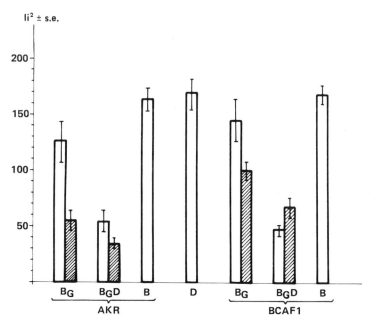

Figure 2. *In vitro* migration of spleen cells from AKR and BCAF$_1$ mice. B, cells from nongrafted animals; B$_G$, cells from mice that were grafted 7 days earlier with B10D2 skin; B$_G$D, cells as above, mixed in a 1:1 ratio with B10D2 spleen cells; D, B10D2 cells alone. Open bars, untreated mice; hatched bars, mice that received 20 μg of LPS s.c. 2 days after grafting. The results of cell migration from capillary tubes were scored after 18 hr incubation of triplicate cultures at 37°C (direct migration test). They are presented as the mean area of migration ± S.E. Each bar represents a mean result from a group of 10 animals.

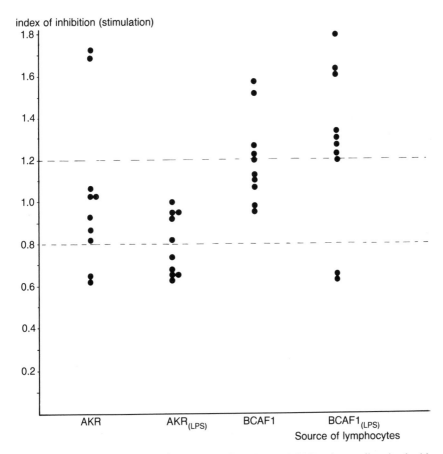

Figure 3. Production of LyMIF and LyMSF by AKR or BCAF$_1$ spleen cells mixed with B10D2 splenocytes and cultured in a capillary tube. Target cells were spleen cells from untreated syngeneic mice, placed in the same culture chamber in a distinct capillary tube (one-stage indirect test); for further details see Skopińska and Nowaczyk (1979). The factors released from the "productive" tube into the culture medium influence the migration of target cells. All mice were grafted with B10D2 skin 7 days before. Half of the mice received 20 μg of LPS 2 days after grafting.

factor. However, both in AKR and BCAF$_1$ mice, on the 16th day after grafting no production of factors influencing splenic cell migration was observed, independent of whether the mice obtained LPS or not.

In B10AF$_1$ mice, a decreased production of LyMIF and an intensification of LyMSF production appeared on the 7th day after the graft and was followed by an intensification of LyMIF production thereafter (Skopińska et al., 1974).

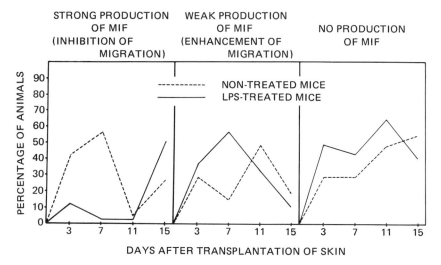

PRODUCTION OF MIF BY SPLENIC LYMPHOCYTES OF
F1 B10A ♀ MICE GRAFTED WITH B10D2 SKIN

------- 33 MICE, GRAFT ONLY

―――― 31 MICE INJECTED WITH E. COLI ENDOTOXIN TWO DAYS AFTER GRAFTING

PERCENTAGE OF ANIMALS EXHIBITING:

STRONG PRODUCTION   WEAK PRODUCTION
OF MIF   OF MIF   NO PRODUCTION
(INHIBITION OF   (ENHANCEMENT OF   OF MIF
MIGRATION)   MIGRATION)

----------- NON-TREATED MICE
―――― LPS-TREATED MICE

PERCENTAGE OF ANIMALS

DAYS AFTER TRANSPLANTATION OF SKIN

Figure 4. The effect of a single dose of 20 μg of LPS, administered s.c. to B10AF$_1$ mice 2 days after grafting of B10D2 skin, on the inhibition or stimulation of *in vitro* spleen cell migration caused by B10D2 antigen (Skopińska *et al.*, 1974). Results are expressed as a percentage of the animals exhibiting strong production of LyMIF, weak production of LyMIF (or, perhaps, an increased production of LyMSF), and no production of factors (or, perhaps, an equivalent production of both LyMIF and LyMSF). For technical details see Ostrowski *et al.* (1969).

B. Dependence on the Time of LPS Administration

A definitive temporal relationship has been found for the influence of LPS on parameters of transplantation immunity in B6AF$_1$ and B10AF$_1$ mice such as LyMIF production under the influence of donor antigen, skin graft survival, cytotoxic and cytophilic antibody production, and production of serum factors blocking the regional lymph node blastic response to B10D2 lymphatic cells (Figs. 5–9, Table 2).

Statistically significant (or on the border of statistical significance) was the prolongation of graft survival obtained upon administration of LPS on the second or fourth day after grafting. Administration of LPS on these days also caused an increase in the cytotoxic and cytophilic antibody production.

However, serum factors blocking regional lymph node reaction, and intensification of LyMIF production appeared only in the group

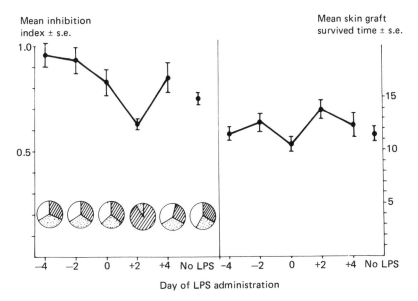

Figure 5. Time dependence of the effect of LPS on B10D2 skin graft survival in B10AF₁ mice and on the *in vitro* inhibition of spleen cell migration caused by donor B10D2 antigen. Cultures of spleen fragments were established 16 days after skin grafting. For further technical details see Ostrowski *et al.* (1969) and Skopińska (1978). Areas inside the circles represent the percentage of animals exhibiting inhibition of migration (hatched zone), enhancement of migration (dotted zone), or no effect (white zone).

treated 2 days after grafting (Table 2). Administration of LPS 2 or 4 days before grafting caused an inhibition of LyMIF production (Skopiń ska, 1978).

In rabbits treated with one dose of LPS, administered on the second, fourth, or fifth day after an allogeneic skin graft, some animals demonstrated prolonged graft survival time (Fig. 10). A majority of the rabbits, however, rejected their skin grafts in the same pattern as the untreated group.

C. The Influence of ATS and the Adoptive Immunity to LPS with Respect to LPS Effect on Skin Graft Survival

In B6AF₁ mice treated with ATS, the administration of LPS after skin grafting further prolonged the time of the graft survival (Fig. 6). The longest mean skin graft survival time was observed in the group that, besides ATS (anti-thymocytic serum) and LPS, also received an i.p. inoculation of spleen cells derived from syngeneic mice treated with multiple injections of LPS (serum anti-LPS antibody titer 1:128). Cells were transplanted on the fifth day after the skin graft, and 20 μg of

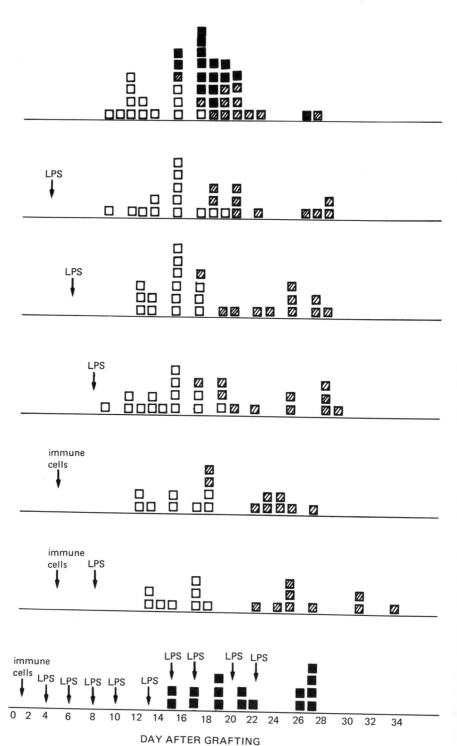

DAY AFTER GRAFTING

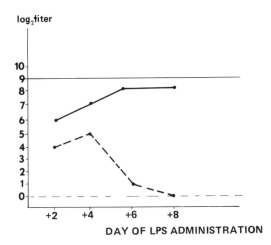

Figure 7. The effect of LPS given on various days after grafting into B6AF₁ mice on cytotoxic antidonor B10D2 antibody response. Broken curve represents the results obtained from pools of six mice, bled 20 days after first skin grafting. Solid curve represents the results obtained from pools of 7–13 mice, bled 14 days after the second skin grafting. Horizontal broken line represents primary, and horizontal solid line secondary antibody response of LPS-untreated mice.

LPS was injected 3 days later. The mean graft survival time in this group was 27.1 days and it differed highly significantly from the survival time of the group treated with ATS alone, that is, 19.7 days (Skopińska and Oluwasanmi, 1972). The question arises as to whether this effect connected with the presence of some suppressive population in the spleens of the LPS-immunized mice, or whether it is the result of the anti-LPS antibody production by the transferred spleen cells. In subsequent experiments, where B10A mice received ATS, a skin graft, and six doses of mouse antiendotoxin serum (MAES) between days 7 and 19 after the skin graft (Fig. 11), in the group treated with MAES, a sta-

←⎯⎯⎯⎯⎯⎯⎯⎯⎯⎯⎯⎯⎯⎯⎯⎯⎯⎯⎯⎯⎯⎯⎯⎯⎯⎯⎯⎯⎯⎯⎯

Figure 6. The effect of LPS given on various days after skin grafting on the survival of B10D2 skin in B6AF₁ mice. Each square represents one graft rejected. Open squares correspond to animals not treated with ATS. Hatched squares represent animals treated with ATS (0.25 ml 1 day before grafting, 0.15 ml on the day of grafting, and 0.10 ml 2 days after grating). Solid squares represent animals treated with a lower dose of ATS (0.25 ml 1 day before grafting). LPS was administered s.c., in a single dose of 20 μg, or in multiple 10-μg doses. In the last three groups, mice received "immune cells," that is, i.p. inoculation of 6 × 10⁷ living spleen and mesenteric lymph node cells from B6AF₁ mice that had been immunized against LPS and had an antibody titer of 1:128 as determined by passive hemagglutination. For further technical details see Skopińska and Oluwasanmi (1972).

The longest mean skin graft survival time was obtained in the group treated with three doses of ATS and immune cells, followed by a single dose of LPS (27.1 ± 1.32 days).

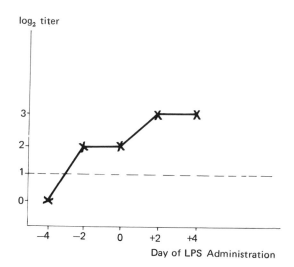

Figure 8. The effect of LPS given at various days before or after B10D2 skin grafting on antidonor cytophilic antibody production by B10AF$_1$ mice. Groups of five mice were bled 7 days after grafting, sera pooled. Broken line represents log$_2$ titer of serum pool derived from five grafted but LPS-untreated mice. Determination of the cytophilic antibody was done according to the passive direct technique, described by Mazzolli and Barrera (1974), using mouse peritoneal macrophages. Cells from lymph nodes of B10D2 mice were used as antigen.

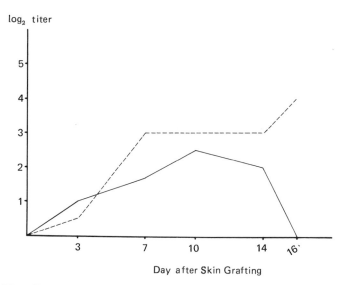

Figure 9. The effect of LPS given to B10AF$_1$ mice 2 days after B10D2 skin grafting on the appearance of antidonor cytophilic antibodies on various days after grafting. Groups of five mice were bled 3, 7, 10, 14, and 16 days after grafting, sera pooled and assayed for cytophilic activity by passive direct technique (Mazzolli and Barrera, 1974). Solid line represents mean log$_2$ titers of two serum pools from untreated and broken line from LPS-treated group of mice.

Table 2. Time Dependence of the LPS Effect on the Production of Serum Factors Blocking the Regional Lymph Node Reaction[a,b]

Group No.	Day of LPS administration into donors of serum with respect to the day of skin grafting	Skin grafting of serum donors	Mean ‰ of blastic cells ± S.E.	Difference from group 1
1	No LPS	No	4.86 ± 0.74	—
2	No LPS	Yes	5.09 ± 1.04	NS
3	−4	Yes	3.56 ± 0.45	NS
4	−2	Yes	4.89 ± 0.20	NS
5	+2	Yes	1.98 ± 0.65	$p < 0.05$
6	+4	Yes	2.79 ± 0.86	NS
7	LPS given 5 days before serum harvesting	No	5.84 ± 0.47	NS

[a] Sera were harvested from groups of B6AF$_1$ mice, 7 days after B10D2 sking grafting. Mice were treated with 20 μg of LPS on various days with respect to the day of skin grafting.

[b] The regional lymph node reaction was evoked by injecting spleen cells (2.5×10^6) from B10D2 mice into both hind foot pads of B10AF$_1$ (groups of 6 mice). One day later the mice received 0.1 ml of tested serum s.c. into the hind legs. Three days after mice were sacrificed, smears from the popliteal lymph nodes were prepared, stained with MGG, and the percentage of blastic cells was evaluated by counting 2000 cells from each slide. For technical details see Ostrowski et al. (1968).

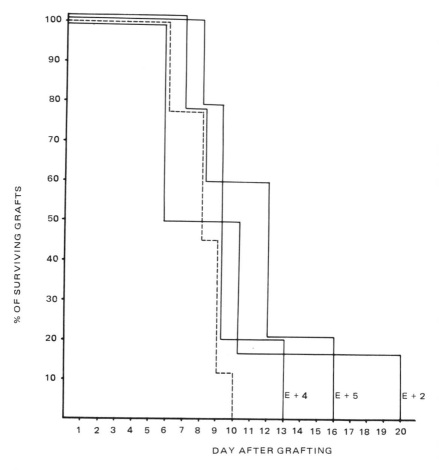

Figure 10. The effect on graft survival time of a single dose of LPS (2.5 μg/kg body wt) given i.v. to a rabbit 2, 4, or 5 days after allogeneic skin grafting. Broken line represents the results obtained in untreated rabbits.

tistically significant prolongation of graft survival time appeared. Hence, the anti-LPS antibody may work synergistically with ATS in the prolongation of the skin graft survival.

D. The Effect of LPS on Kidney Transplantation Immunity in Rabbits

The results of experiments conducted on rabbits after renal grafts are presented in Figs. 12 and 13. Just as in the experiments conducted on B10AF$_1$ mice, the administration of LPS on the second day after the graft inhibited the production of LyMIF in the first week and intensified it in the second week after grafting (Skopińska et al., 1974). Similar

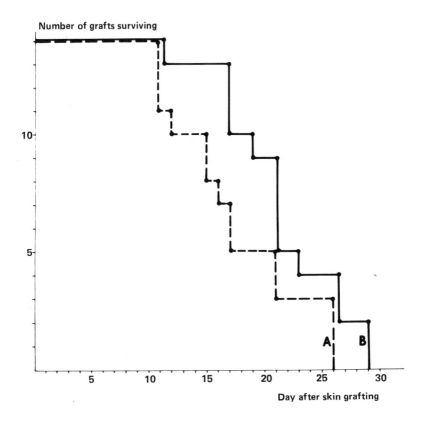

Figure 11. Time course of rejection of B10D2 skin grafts by B10AF₁ mice treated with ATS (A) or ATS and MAES (B). ATS (0.2 ml) was administered s.c., twice, 2 days before and on the day of grafting. MAES (0.1 ml) was injected six times, s.c., on days +7, +11, +13, +15, +17, and +19, where day 0 refers to the day of skin grafting procedure.

inhibition in the first week was observed in the synthesis of DNA in the mixed cultures of donor and recipient lymphocytes. This effect occurred only in rabbits treated on the second day after grafting. Rabbits treated the day before grafting did not differ from untreated rabbits (Skopińska *et al.*, 1977).

Cytotoxic antibodies were not present in the first week, but in the second week they appeared in 30% of the recipients that had not been treated, and in 80% of the recipients that had been treated with LPS on the second day after the renal graft.

The histological picture of the kidney, 6 days after the graft procedure, was similar in both treated groups. It was found that there was less intensification of the acute rejection symptoms than in the rabbits that were not treated, but a characteristic of both of these LPS-treated

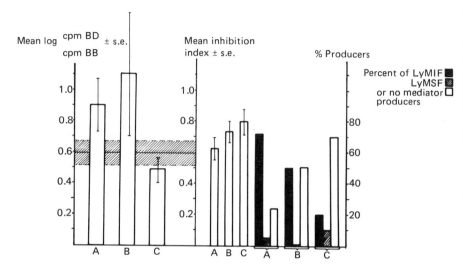

Figure 12. The effect of a single dose of LPS (2.5 μg/kg body wt) given i.v. to a rabbit 1 day before or 2 days after kidney grafting on the appearance of LyMIF or LyMSF and DNA synthesis in the mixed donor + recipient lymphocyte cultures, established 6 days after kidney transplantation. A, untreated rabbits; B, LPS 1 day before grafting; C, LPS 2 days after grafting. Transplantation of the kidney was performed orthotopically, by "cuff" technique (Skopińska *et al.*, 1971, 1977). Target cells used for evaluating LyMIF activity of MLC supernatants were murine spleen cells migrating from capillary tubes. For further technical details see Skopińska *et al.* (1977) and Skopińska and Nowaczyk (1979).

groups was the neutrophilic and eosinophilic infiltration. These infiltrations, although of a reduced degree, also appeared in the autogeneic kidney of the recipient of the allograft, as well as in its liver, and they were accompanied by such changes as focal hepatocellular necrosis and passive hyperemia.

IV. Discussion

The influence of gram-negative bacterial LPS on the phenomena of transplantation immunity must be taken into close consideration.

In the case of the GVH reaction, the influence may be dual in nature: first, that LPS inhibits the immunological activity of the thymus-dependent lymphocytes of the donor, regardless of whether it is administered to the recipient before or after the cellular inoculation. This inhibiting effect is not found in thymectomized animals, which allows us to speculate that LPS may activate some sort of thymus-dependent nonspecific suppressive population, recipient in origin, which limits the GVH activity of donor T cells. Rowlands *et al.* (1965) described the

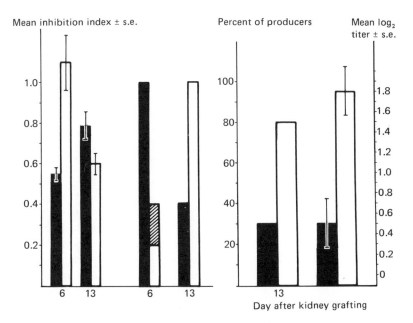

Figure 13. The effect of LPS on LyMIF production and cytotoxic antidonor lymphocyte antibody response in rabbits. LPS was injected i.v. 2 days after kidney grafting. Solid bars, untreated rabbits; open bars, LPS-treated rabbits; hatched bar, percentage of LyMSF producers in LPS-treated group. In untreated rabbits, no production of LyMSF was observed. Cytotoxic antibody was not detectable 6 days after grafting in both LPS-treated and untreated groups.

changes that occur in the mouse thymus after the administration of LPS, that is, a rapid loss of thymic lymphocytes, associated with an apparent increase in cells with a pyroninophilic cytoplasm and an increase in the RNA/DNA ratio.

Mouse thymus cells are essentially unresponsive to LPS. However, the thymus cell population probably contains a very small B-cell subset, which could be directly triggered by LPS, under circumstances favorable to T-cell stimulation (Forbes *et al.*, 1975). Thymocytes from LPS-treated mice were found to be more responsive than normal thymocytes to PHA and Con A (Adorini *et al.*, 1976) and displayed increased GVH and helper activities in the irradiated recipients (Baroni *et al.*, 1976). Persson (1977) presented some evidence of the fact that B lymphocytes activated by LPS may have a suppressive activity on reactions where T lymphocytes are involved.

Second, LPS may intensify the GVH reaction, and this effect, in direct contrast to the previous effect, is markedly dependent on the time of LPS administration. Because an equally exact dependence has been found for the stimulation of antidonor antibodies by LPS, we may hypothesize that the intensification of the GVH disease may be caused by the intensification of antibody production by the parental B lymphocytes and by the increased activity of the recipient's "killer" lymphocytes that destroys its own tissue coated with antibodies.

The intensification in the production of the antirecipient antibodies probably is not due to the direct activity of LPS on the B lymphocyte, but rather to an enhanced function of helper T lymphocytes of the donor provided by the stimulation of the macrophage system or the inactivation of the suppressor T lymphocytes (Armerding and Katz, 1974; Elie and Lapp, 1977; Kagnoff et al., 1974; Pickel and Hoffmann, 1977).

Intensification of the GVH reaction in animals treated with MAES and LPS may be explained by the coating of cells with LPS–MAES complexes and an intensification of the ADCC reaction. Extensive granulocytic infiltrations present in the organs of these mice may suggest the involvement of these cells in the observed autoaggressive lesions. It was suggested by Allison et al. (1973) that some of the in vivo effects of the endotoxins might be due to the attachment of the endotoxin and the antibody onto the host cells and their sensitization for interaction with killer cells.

I obtained the inhibition of all of these autoaggressive phenomena by treating the mice with ATS before inoculating with parental cells. In the available literature, two reports were found on the subject of decreased GVH reactions after treating the recipients with ALS (anti-lymphocytic serum) before the inoculation of parental cells. In one report, the authors attempt to explain this effect by the remaining ALS activity in the organism of the recipient, which acts on the cells of the donor (Haskovcova and Nouza, 1970). From my experiments, this cannot be, as demonstrated by the experiments with the repopulation with syngencic cells. The second report involves the local GVH reaction under the renal capsule that does not occur in animals treated previously with ALS. The author explains this fact as the elimination by ALS of the recipient's proliferative component (Elkins, 1966).

In my experiments, I have found a difference in the ATS-sensitivity of the host cellular population that probably suppresses the donor T-cell activity (unsusceptible to ATS) and the tissue-damaging host cell population (very sensitive to ATS). It should be noted that ALS apparently does not influence the host's reactivity to endotoxins (Abdelnoor et al., 1972; Veit and Michael, 1972).

In the skin grafting system, probably the fact that the ATS does not act upon the mouse response to the thymus-independent antigen (which is LPS), allowed to obtain a synergistic activity in the prolongation of the graft survival time. It appears that in B6AF$_1$ mice, antibodies acting against the endotoxin could fulfill an "enhancing" role. One should also clarify whether there is any cross-reaction between antigen 31 of the mouse *H-2* histocompatibility system and LPS B from *E. coli* strain O26:B6. It was demonstrated that certain LPS from gram-negative bacteria cross-react with some human HLA specificities (Hirata *et al.*, 1973).

One may suggest that the transient suppression of cellular immune response and prolongation of skin allograft survival may be connected in my experiments with the phenomenon of antigenic competition. This cannot be true, as it has been documented that LPS does not initiate antigenic competition (Veit and Michael, 1972).

There is another possibility that could explain the observed prolongation of skin graft survival time in B10AF$_1$ mice, which is the appearance of serum factors that interfere with the afferent sensitization arc. We found these serum factors to be present in B10AF$_1$ mice, but absent in AKR and BALB/c mice, which correlated well with the results of the skin grafting. In mice, administration of LPS caused in two genetic systems the intensification of LyMSF factor production, 1 week after the transplant.

We have reported the appearance of factors stimulatory for lymphocyte migration in mixed lymphocyte cultures of rabbits in the first week after the kidney transplant (Skopińska *et al.*, 1978a,b). According to Fox and Rajaraman (1978), the production of another mediator, the macrophage migration inhibitory factor, appears to be under the control of the suppressor cells, i.e., T lymphocytes. In all of these situations where migration stimulatory activity is found, this is thought to be due to the increased suppressor cell activity.

In experiments carried out on rabbits, we found that both peripheral blood T and B lymphocytes can produce *in vitro*, under the influence of Con A, the inhibitory and stimulatory factors affecting the *in vitro* migration of T and B lymphocytes. The production of factors acting upon the migration of B lymphocytes was independent of monocytes, whereas the production of factors affecting T lymphocytes was dependent upon the presence of monocytes (Skopińska and Łukasik, 1979).

On the subject of the cellular source of LyMIF and LyMSF in mice, nothing is known. However, it is known that the production of another mediator, the macrophage inhibitory factor in mice, may be mediated by both T and B lymphocytes (Adelman *et al.*, 1980). Nothing is known on the subject of the direct relation between LyMIF and mediators influencing the migration of other types of cells. Our preliminary experiments, conducted on human lymphocytes activated by PHA, indicate

that LyMIF is a protein of a molecular weight 10,000 to 15,000 (Skopińska *et al.*, 1983), and therefore smaller than human MIF, which has a
molecular weight of 23,000. The functional significance of the
lymphokines acting upon the migration of the lymphocytes is unknown;
perhaps they fulfill some role in the processes of various lymphocytic
interactions. Their appearance in transplantological systems may be involved with the rejection processes (Eidemiller and Bell, 1972; Skopiń
ska *at al.*, 1977), but the inhibition of their appearance does not necessarily lead to a prolongation of the graft survival time (Sankowski *et al.*,
1969; Skopińska *et al.*, 1969, Skopińska, 1978).

In rabbits treated with LPS before or after the renal transplant, the
intensity of the acute rejection reaction (especially lymphocyte infiltration) was decreased in both groups, in spite of the fact that the LyMIF
production and DNA synthesis were disturbed only in the group treated 2 days after the graft procedure.

In summary, administering LPS to rabbits prolonged the skin graft
survival time in some of them and decreased the acute rejection reaction of the grafted kidney. On the other hand, in rabbits treated with
LPS, it was observed that granulocytic infiltrations were present in the
grafted kidney, and also to a lesser degree in the autogeneic kidney.
This fact may be due to the known effect of LPS on granulopoiesis and
colony-stimulating factor (CSF) production (Quesenberry *et al.*, 1972).
Hence, the presence of granulocytes may intensify the rejection process
in the transplanted kidney.

ACKNOWLEDGMENTS

I wish to thank Professor Paul S. Russell, my mentor, for the fruit-
ful inspiration instilled originally, years ago. The author's research reported herein was supported by Grants AM-07055 and AI-06320 from
the U.S. Public Health Service and Grants 0932/III/9/d and
10.5.08.2.4 from the Polish Academy of Sciences.

V. References

Abdelnoor, A., Chang, C. M., and Nowotny, A., 1972, Effect of antilymphocyte serum on
   endotoxin reactions, *Bacteriol. Proc.* **1972:**92.
Adelman, N. E., Ksiazek, J., Yoshida, T., and Cohen, S., 1980, Lymphoid sources of murine migration inhibition factor, *J. Immunol.* **124:**825.
Adorini, L., Ruco, L., Uccini, S., Soravito De Franceschi, G., Baroni, C. D., and Doria,
   G., 1976, Biological effects of *Escherichia coli* lipopolysaccharide (LPS) *in vivo*. II. Selection in the mouse thymus of PHA- and Con A-responsive cells, *Immunology* **31:**225.

Al-Askari, S., Zweiman, B., Lawrence, H. S., and Thomas, L., 1964, The effect of endo-toxin on skin homografts in rabbits, *J. Immunol.* **93**:742.

Allison, A. C., Davies, P., and Page, R. C., 1973, Effects of endotoxin on macrophages and other lymphoreticular cells, *J. Infect. Dis.* **128**(Suppl.):212.

Armerding, D., and Katz, D. H., 1974, Activation of T and B lymphocytes *in vitro*. I. Reg-ulatory influence of bacterial lipopolysaccharide (LPS) on specific T-cell helper func-tion, *J. Exp. Med.* **139**:24.

Baroni, C. D., Ruco, L., Soravito De Franceschi, G., Uccini, S., Adorini, L., and Doria, G., 1976, Biological effects of *Escherichia col.* lipopolysaccharide (LPS) *in vivo*. I. Selec-tion in the mouse thymus of killer and helper cells, *Immunology* **31**:217.

Chase, R. M., and Rapaport, F. T., 1965, The bacterial induction of homograft sensitivi-ty. I. Effects of sensitization with group A streptococci, *J. Exp. Med.* **122**:721.

Eidemiller, L. R., and Bell, P. R. F., 1972, Migration inhibition and homograft rejection in rats, *Transplantation* **13**:5.

Elie, R., and Lapp, W. F., 1977, Graft versus host-induced immunosuppression: Mecha-nism of depressed T-cell helper function *in vitro*, *Cell. Immunol.* **34**:38.

Elkins, W. L., 1966, The interaction of donor and host lymphoid cells in pathogenesis of renal cortical destruction induced by a local graft versus host reaction, *J. Exp. Med.* **123**:103.

Floersheim, G. L., 1969, Suppression of cellular immunity by gram-negative bacteria, *Antibiot. Chemother.* **15**:407.

Forbes, J. T., Nakao, Y., and Smith, R. T., 1975, T mitogens trigger LPS responsiveness in mouse thymus cells, *J. Immunol.* **114**:1004.

Fox, R. A., and Rajaraman, K., 1978, The role of suppressor cells in the production of macrophage migration inhibition factor, *Immunol. Commun.* **7**:311.

Haskovcova, H., and Nouza, K., 1970, Treatment and pretreatment of graft-versus-host reaction by antilymphocyte serum, *Folia Biol. (Prague)* **16**:397.

Hirata, A. A., McIntire, F. C., Terasaki, P. I., and Mittal, K. K., 1973, Cross reactions be-tween human transplantation antigens and bacterial lipopolysaccharides, *Transplanta-tion* **15**:441.

Husberg, B., Nilsson, T., and Fritz, H., 1971, Acute rejection of allogeneic transplanted kidneys following nephrectomy of autogeneic pyelonephritic kidneys, *Transplantation* **11**:1971.

Kagnoff, M. F., Billings, P., and Cohn, M., 1974, Functional characteristics of Peyer's patch lymphoid cells. II. LPS is thymus dependent, *J. Exp. Med.* **139**:407.

Kirpatovski, I. D., and Stanislawski, E. S., 1971, Immunosuppressive effect of cell-free ex-tracts from *Escherichia coli*, *Transplant. Proc.* **3**:831.

Klassen, J., and Milgrom, F., 1970, The generalized Shwartzman phenomenon in rabbits with denervated kidneys, *Proc. Soc. Exp. Biol. Med.* **134**:980.

Klassen, J., and Milgrom, F., 1971, Studies on cortical necrosis in rabbit renal homograft, *Transplantation* **11**:35.

Marecki, N. M., and Bradley, S. G., 1973, Enhanced toxicity for mice of combinations of bacterial endotoxin with antitumor drugs, *Antimicrob. Agents Chemother.* **3**:599.

Mazzolli, A. B., and Barrera, C., 1974, A method for detecting cytophilic activity in a ho-mologous system, *J. Immunol. Methods* **4**:31.

Morris, P. J., Kincaid, S. P., and McKenzie, I. F. C., 1969, Leukocyte antigens in renal transplantation: Immediate renal allograft rejection with a negative cross-match, *Med. J. Aust.* **56**:379.

Nowotny, A., Radvany, R., and Neale, N. E., 1965, Neutralization of toxic bacterial O-an-tigens with O-antibodies while maintaining their stimulus on non-specific resistance, *Life Sci.* **4**:1107.

Ostrowski, K., Skopińska, E., Sankowski, A., Pienkowski, M., and Priegnitz, A., 1968, Antigen-induced changes in regional lymph nodes of mice. I. Allogeneic cells, *Bull. Acad. Polon. Sci. Sér. Biol.* **16**:217.

Ostrowski, K., Skopińska, E., Lazarewicz, J., Gorski, A., Zaleska-Rutczynska, Z., and Sankowski, A., 1969, The range and sensitivity of migration inhibition test used for estimation of antigenicity of splenic microsomal fraction, *Folia Biol. (Prague)* **15**:146.

Persson, U., 1977, Lipopolysaccharide-induced suppression of the primary immune response to a thymus-dependent antigen, *J. Immunol.* **118**:789.

Pickel, K., and Hoffmann, M. K., 1977, Suppressor T cells arising in mice undergoing a graft-vs.-host response, *J. Immunol.* **118**:653.

Quesenberry, P., Morley, A., Stohlman, F., Jr., Rickard, K., Howard, D., and Smith, M., 1972, Effect of endotoxin on granulopoiesis and colony-stimulating factor, *N. Engl. J. Med.* **286**:227.

Rowlands, D. T., Claman, H. N., and Kind, P. D., 1965, The effect of endotoxin on the thymus of young mice, *Am. J. Pathol.* **46**:165.

Sankowski, A., Skopińska, E., Lazarewicz, J., Gorski, A., and Ostrowski, K., 1969, Effect of azathioprine on allograft sensitivity: Study of spleen cell migration, *Bull. Acad. Polon. Sér. Sci. Biol.* **17**:223.

Skopińska, E., 1972, Some effects of *Escherichia coli* endotoxin on the graft-versus-host reaction in mice, *Transplantation* **14**:432.

Skopińska, E., 1974a, The effect of *E. coli* endotoxin on the GVH reaction in mice. II. Non H-2 systems, *Arch. Immunol. Ther. Exp. (Engl. Ed.)* **22**:301.

Skopińska, E., 1974b, The effect of *E. coli* endotoxin on the GVH reaction in mice. III. Runt disease in newborns, *Arch Immunol. Ther. Exp. (Engl. Ed.)* **22**:305.

Skopińska, E., 1976, The effect of *E. coli* lipopolysaccharide (endotoxin) on the GVH reaction in mice. IV, *Ann. Med. Sect. Pol. Acad. Sci.* **21**:185.

Skopińska, E., 1978, Effect of lipopolysaccharide on the development of cell-mediated immunity to transplantation antigens, *Transplantation* **26**:420.

Skopińska, E., and Lukasik, R., 1979, Production of lymphocyte migration inhibitory (LyMIF) and stimulatory (LyMSF) factors by T and B lymphocytes from rabbit peripheral blood stimulated with Con A, *Bull. Acad. Polon. Sér. Sci. Biol.* **27**:701.

Skopińska, E., and Nowaczyk, M., 1979, A one-stage test for simultaneous study of migration inhibition factor of peripheral leukocytes (LIF) and migration inhibition factor of mouse spleen cells (MSIF), *Arch. Immunol. Ther. Exp. (Warsaw)* **27**:171.

Skopińska, E., and Oluwasanmi, J. O., 1972, Effect of *Escherichia coli* endotoxin on skin allografts in mice, *Folia Biol. (Prague)* **18**:270.

Skopińska, E., Sankowski, A., and Nouza, K., 1969, Effect of azathioprine on the development of allograft sensitivity in mice, *Bull. Acad. Polon. Sér. Sci. Biol.* **17**:85.

Skopińska, E., Wasik, M., and Wojtulewicz-Kurkus, J., 1971, Indirect migration inhibition test in xenogeneic systems. III. Cellular immunity to kidney grafts, *Arch. Immunol. Ther. Exp. (Engl. Ed.)* **19**:377.

Skopińska, E., Nowaczyk, M., Gorski, A., Wasik, M. Krzywicka, E., Wojtulewicz-Kurkus, J., and Orlowski, T., 1974, The influence of antithymocytic serum, α-globulins, and *E. coli* endotoxin on the development of cellular immunity in experimental skin and kidney transplantation, Abstracts, International Conference of Allergologists and Clinical Immunologists of Socialist Countries, Prague, p. 43.

Skopińska, E., Nowaczyk, M., Gorski, A., and Glyda, J., 1977, Studies on MIF production following transplantation of an allogeneic kidney in the rabbit, *Immunol. Pol.* **2**:153.

Skopińska, E., Ziembikiewicz, A., Lukasik, R., and Misztal, B., 1978a, The appearance of factors influencing migration of lymphocytes (LyMIF and LyMSF) in rabbit mixed lymphocyte cultures, *Arch. Immunol. Ther. Exp. (Engl. Ed.)* **26**:389.

Skopińska, E., Wasik, M., Moscicka-Wesolowska, M., Wojtulewicz-Kurkus, J., and Lukasik, R., 1978b, The effect of ATG treatment on LyMIF and LyMSF production in MLC of rabbits after renal allotransplantation, *Arch. Immunol. Ther. Exp. (Engl. Ed.)* **26**:987.

Skopińska, E., Madalinski, W., Lukasik, R., 1983, Some characteristics of human lymphocyte migration inhibitory factor and the effect of ampicillin on its appearance in PHA-stimulated cultures, *Acta Physiol. Pol.* (in press).

Starzl, T. E., Lorner, R. A., Dixon, F. J., Groth, C. G., Brettschneider, L., and Terasaki, P. I., 1968, Shwartzman reaction after human renal homotransplantation, *N. Engl. J. Med.* **278**:642.

Stetson, C. A., 1955, Studies on the mechanism of the Shwartzman phenomenon: Similarities between reactions to endotoxin and certain reactions of bacterial allergy, *J. Exp. Med.* **101**:421.

Terino, E. O., Miller, J., and Glenn, W. W. L., 1964, Tolerance induction and skin graft prolongation by competing antigens, *Surgery* **56**:256.

Toledo-Pereyra, L. H., Raij, L., Simmons, R. L., and Najarian, J. S., 1974, Role of endotoxin and bacteria in long-term survival of preserved small-bowel allografts, *Surgery* **76**:474.

Veit, B. C., and Michael, J. G., 1972, The lack of thymic influence in regulating the immune response to *E. coli* O127 endotoxin, *J. Immunol.* **109**:547.

24

Adoptive Immunotherapy of Leukemia by Endotoxin-Induced Modulation of the Graft-versus-Host Reaction

Robert L. Truitt

I. Introduction

Allogeneic bone marrow transplantation is used for the treatment of a variety of hematological dyscrasias including aplastic anemia, severe combined immunodeficiency disease, and leukemia (Thomas *et al.*, 1975; Bortin and Rimm, 1977; Storb *et al.*, 1978; Bortin and Rimm, 1978, 1981). Two of the major problems associated with allogeneic marrow transplantation are the immunologic attack of grafted cells or their progeny against histocompatibility antigens on host cells, i.e., graft-versus-host (GVH) reactivity, and susceptibility of the immuno-suppressed host to endogenous and exogenous sources of infection (Thomas *et al.*, 1975). Attempts to resolve these problems have consumed a great deal of experimental and clinical effort with moderate, but by no means complete success.

Following allogeneic spleen or bone marrow cell transplantation, an unusual syndrome of diarrhea, wasting, and skin lesions was observed in transplanted animals (van Bekkum and de Vries, 1967) as well as man (Glucksberg *et al.*, 1974). This syndrome was given the name *secondary disease* to distinguish it from radiation sickness or primary disease

Robert L. Truitt • May and Sigmund Winter Research Laboratory, Mount Sinai Medical Center, Milwaukee, Wisconsin 53233.

(van Bekkum and de Vries, 1967). Now there is little doubt that secondary disease is primarily the consequence of a GVH reaction; however,
factors other than GVH reactivity influence the severity and incidence
of secondary disease. In particular, a role for bacteria in the etiology of
secondary disease was suggested by several lines of evidence. van
Bekkum and Vos (1961) demonstrated that the addition of aureomycin
to the diet markedly suppressed the symptoms and mortality of secondary disease in F_1 hybrid mice given immunocompetent parental cells.
Connell and Wilson (1965) reported a substantial increase in the proportion of mice surviving 30 days after lethal irradiation and allogeneic
marrow transplantation if the hosts were germfree. Subsequent work in
several laboratories using germfree animals or animals whose intestinal
microflora had been eliminated by oral antibiotic treatment substantiated a major role for the microbial component of the host in secondary
disease mortality (Jones *et al.*, 1971; van Bekkum *et al.*, 1974; Heit *et al.*,
1977; Truitt, 1978b).

Reed and Jutila (1965; Jutila, 1973) had postulated that secondary
disease and other wasting syndromes shared an infectious or toxemic
process as a consequence of immunoincompetence, and animals undergoing a GVH reaction were found to be acutely sensitive to the lethal
effects of bacterial endotoxins (Keast, 1973; Walker *et al.*, 1975; Walker, 1978). This led several investigators to consider the possibility that
secondary disease (the consequences of a GVH reaction and hereafter
referred to simply as GVH disease) could be modulated directly or indirectly by rendering the donor or host tolerant or hyporeactive to endotoxin. The purpose of this chapter is to briefly review some of the
experimental approaches to the prevention of GVH disease by the
adoptive transfer of hemopoietic or immunocompetent cells from endotoxin-treated donors. In addition, I will describe attempts to use this
approach to prevent GVH disease associated with allogeneic spleen cell
transplantation for adoptive immunotherapy of murine leukemia.

II. Experimental Approaches to the Mitigation of GVH Disease

Liacopoulos *et al.* (1967; Liacopoulos and Merchant, 1969) first
showed that the pretreatment of allogeneic, *H-2*-incompatible mice
with endotoxin inhibited the ability of their spleen cells to cause GVH
disease in sublethally irradiated adult recipient mice. They found that
modulation of GVH disease was linked to the total quantity of antigen
used for pretreatment of the donors and attributed the results to prior
immunologic commitment which precluded GVH reactivity. Subse-

quently, Damais *et al.* (1972) reproduced these results in a GVH assay using mortality in unirradiated newborn mice as an end point. Outbred Swiss mice and F_1 hybrids of inbred mice were given spleen cells from *H-2*-incompatible C57BL/6 mice treated with 30 µg of bacterial endotoxin daily for 7 days. When spleen cells from endotoxin-tolerant adult donors were transplanted, the mortality rate at 40 days was 4% (2/45) as compared to 80% (36/45) with spleen cells from untreated donors. Mice receiving cells from endotoxin-treated donors did not show the typical stunted appearance of mice given spleen cells from untreated donors.

Chedid (1973) subsequently reported that GVH-associated mortality could be inhibited by incubation of the donor's spleen cells with endotoxin *in vitro*. In his experiments, 2×10^8 spleen cells of C57BL adult mice were incubated with 30 µg of endotoxin for 1 hr at 37°C; viability of treated cells was approximately 90%. Control splenocytes were incubated under the same conditions without endotoxin prior to injection into newborn Swiss mice. Preincubation of the splenocytes with endotoxin greatly reduced GVH disease; mortality was 64% (55/86) among control mice injected with untreated cells as compared to 27% (15/55) in animals receiving endotoxin-incubated spleen cells 30 days earlier. The decreased GVH reaction could not be attributed to transfer of endotoxin since the amount of endotoxin contamination was minimal as measured with radiolabeled endotoxin.

Thomson and Jutila (1974) confirmed the effect of endotoxin pretreatment on GVH reactivity using mortality in neonatal BALB/c mice given spleen cells from CBA/J adult donors. Mortality on day 24 for neonates given normal CBA spleen cells was 77% (24/31) as compared to 15% (4/27) for mice given spleen cells from CBA donors pretreated with seven daily i.p. injections of 60 µg of *E. coli* endotoxin. These authors also separated the spleen cells into surface-adherent and nonadherent subpopulations and demonstrated that the GVH reactivity resided in the nonadherent spleen cell population of the endotoxin-treated donor animals. In addition, they noted a marked reduction in mortality from GVH disease when adherent spleen cells from endotoxin-treated mice were administered together with unseparated, normal spleen cells. When normal adherent or endotoxin-treated nonadherent spleen cells were added to normal spleen cells, they failed to protect recipient mice from GVH disease. The authors interpreted these results to indicate that a cell capable of suppressing the GVH reaction was present in the adherent cell population of spleens from mice treated with endotoxin and that this cell interfered with activities of nonadherent cell populations in promoting GVH reactivity. In a subsequent study, these investigators found that serum from endotoxin-treated mice impaired GVH reactivity of normal CBA spleen cells (Thomson *et al.*, 1978). Mortality in BALB/c neonates given whole spleen cells in combination with normal serum was 70% (19/27) as compared to 17% (6/35) when serum from endotoxin-treated mice was used. When the antiendotoxin activity was removed from the immune serum by absorption with endotoxin before injection, the GVH reactivity returned to the level seen in the normal spleen cell controls. The predominant antibody class in the serum of endotoxin-treated mice was IgM. The authors hypothesized that macrophages from endotoxin-treated CBA mice were capable of suppressing the GVH reaction by some unique lymphocyte–macrophage encounter promoted by IgM antibody, but the precise mechanism by which macrophages from endotoxin-treated animals decreased the GVH reactivity (as well as delayed rejection of allogeneic BALB/c skin grafts) was unclear.

   Rose *et al.* (1976) also demonstrated endotoxin-induced modulation of GVH disease in a model system that mimicked clinical bone marrow transplantation. In this system, *H-2*-incompatible DBA/2 donors were rendered hyporeactive to *S. typhosa* endotoxin by a series of seven injections of increasing doses over a 13-day period. Recipient adult AKR mice were immunosuppressed with 5 Gy of total-body radiation (TBR) and 185 mg cyclophosphamide (CY)/kg body wt. Transplantation of spleen and lymph node cells from normal DBA/2 mice into immunosuppressed AKR recipients resulted in acute GVH disease with a median survival time (MST) of 14 days; only 3 of 16 mice survived 30 days after transplant. In contrast, acute GVH disease was prevented when donor mice were treated with endotoxin prior to transplantation; the MST was 52 days, and all 16 mice survived the first 30 days. Endotoxin treatment of the AKR recipient mice had no significant effect on the acute GVH disease. Of particular note in this study was that while the mortality attributable to acute GVH disease (death within 30 days) was significantly decreased following endotoxin treatment of the donor, mortality from chronic GVH disease (death occurring more than 30 days posttransplant) was not abrogated. Four mice from these experiments were tested for lymphoid cell chimerism at 200 days posttransplant using unidirectional mixed leukocyte culture (MLC) assays. Reciprocal nonstimulation was found when cells from the experimental mice were cultured with DBA/2 cells, whereas reciprocal stimulation was observed when the cells were cultured with cells from AKR mice. Thus, the surviving animals were considered to be permanent lymphoid chimeras. The ability of spleen cells from endotoxin-treated mice to respond to host antigens in MLC assays was not affected by endotoxin treatment. In addition, spleen cells from endotoxin-treated DBA/2 mice failed to suppress the responses of normal DBA/2 and AKR spleen cells to each other in third-party MLC tests.

Galley *et al.* (1975) grafted spleen cells from CBA mice made tolerant to ferric chloride-treated endotoxin into irradiated, antibiotic-decontaminated (C57BL/6 × CBA)F$_1$ mice to ascertain whether nontoxic endotoxin could modulate the death rate from GVH disease. All mice grafted with spleen cells from donors tolerant to either untreated or ferric chloride-treated endotoxin survived for 30 days; antibiotic-decontaminated mice grafted with spleen cells from normal CBA mice died 10–12 days after injection.

Rampy and Jutila (1979) examined the influence of endotoxin on GVH reactivity in neonatal BALB/c mice inoculated with spleen cells from *H-2*-incompatible donors that were responsive (CBA/J) or unresponsive (C3H/HeJ) to endotoxin at the B-lymphocyte level. Prior treatment of CBA adult mice with endotoxin decreased the ability of their spleen cells to cause fatal GVH disease in BALB/c neonates; in contrast, there was no difference in the GVH reactivity of spleen cells from normal or endotoxin-treated C3H/HeJ mice. It is noteworthy that for CBA donors the method for endotoxin preparation affected the results; butanol-extracted endotoxin failed to modulate GVH reactivity, whereas phenol-extracted endotoxin did. In contrast to Rose *et al.* (1976), Rampy and Jutila (1979) found that the reactivity of donor CBA spleen cells to host BALB/c cells in MLC assays was decreased almost twofold by pretreatment with *E. coli* endotoxin; however, polyclonal B-cell activation apparently was not responsible for the suppression of allogeneic recognition. Decreased GVH reactivity following endotoxin treatment correlated with stimulation of the reticuloendothelial system as measured in a carbon clearance assay. The authors confirmed an earlier report (Thomson *et al.*, 1978) that serum from endotoxin-treated CBA mice was capable of abrogating a GVH reaction by normal CBA spleen cells. In addition, they demonstrated that the humoral factor involved in the suppression was strain specific, suggesting that antiendotoxin antibody alone was not responsible for reduced GVH mortality.

III. Adoptive Immunotherapy of Leukemia Using Endotoxin-Treated Donors

Allogeneic bone marrow transplantation in combination with high-dose chemoradiotherapy is being applied with increasing frequency for patients with refractory or poor-prognosis acute leukemia (Thomas *et al.*, 1977; Bortin and Rimm, 1978; Gale *et al.*, 1979; Powles *et al.*, 1980). As presently employed, allogeneic marrow transplantation is used for hemotpoietic and lymphoid reconstitution permitting the use of supralethal doses of chemoradiotherapy, which hopefully result in a higher proportion of leukemia cells being killed. Although this approach has met with some success, recurrent leukemia has been a major cause of failure (Thomas *et al.*, 1977; Bortin and Rimm, 1978; Gale *et al.*, 1979). A graft-versus-leukemia (GVL) reaction has been reported to occur in man following high-dose chemoradiotherapy and transplantation of HLA-compatible allogeneic bone marrow (Weiden *et al.*, 1979, 1981; OKunewick and Meredith, 1981); however, a GVL effect was detected only in the presence of concurrent acute and/or chronic GVH disease of moderate to severe intensity. Recurrent leukemia was a major complication if GVH disease was absent or mild, particularly in patients with acute lymphoblastic leukemia transplanted in second or subsequent remission or in relapse (Weiden *et al.*, 1979, 1981; McIntyre and Gale, 1981).

In view of these findings, intentional modification of the transplant patient regimen so as to increase the incidence of acute and chronic GVH disease in patients who are at high risk of leukemia recurrence has been proposed (Weiden *et al.*, 1979, 1981). Such modification might include administration of donor buffy coat cells with the marrow inoculum or reduced doses or abbreviated schedules of drugs used to prevent GVH disease. Obviously, this is a dangerous course to pursue because of the potentially lethal consequences of GVH disease.

We reported that cells from *H-2*-incompatible donors exerted a significant GVL effect in experimental animal systems; however, as one would anticipate, use of *H-2*-incompatible donors magnified the already formidable problem of GVH disease (LeFeber *et al.*, 1977; Bortin *et al.*, 1978). A method to mitigate GVH reactivity without abolishing the GVL reactivity of adoptively transferred, allogeneic immunocompetent cells was needed. Based on the reports described earlier in this chapter that mortality from GVH disease in mice could be mitigated if the donor animals were treated with bacterial endotoxin prior to transplantation, we undertook a series of studies to determine whether transplantation of immunocompetent cells from endotoxin-treated *H-2* -incompatible donors permitted a successful GVL reaction to occur without the lethal consequenes of GVH disease.

Long-passage (BW5147) or spontaneous (AKR-L) acute T-cell lymphoblastic leukemias originating in AKR/J (*H-2^k*) mice were used in these studies. AKR leukemia resembles several human malignancies but most closely simulates human T-cell leukemias (Bortin and Truitt, 1977). BW5147 was originally obtained from The Jackson Laboratory, Bar Harbor, Maine, and was maintained in our laboratory as an acute lymphoblastic leukemia in young AKR mice by weekly i.v. passage (Bortin, 1974). Spontaneous leukemia develops in AKR mice at a median age of 8 months, and more than 90% of the mice develop and die of the malignancy by 12 months of age (Bortin and Truitt, 1977). AKR-L is characterized by a thymoma and is easily diagnosed on the basis of spleen and lymph node enlargement. AKR mice bearing advanced leukemia were obtained from a colony of 6- to 12-month-old female mice and used in the experiments on adoptive immunotherapy of AKR-L. Antileukemic reactivity of *H-2*-incompatible donor immunocompetent cells was measured either in a GVL bioassay system (Bortin *et al.*, 1973; Bortin, 1974; Truitt *et al.*, 1978) or by survival of leukemic mice treated with adoptively transferred cells following chemoradiotherapy (Truitt *et al.*, 1978; Truitt, 1979).

In all experiments, donor DBA/2 ($H-2^d$) mice were pretreated with seven injections of *Salmonella typhosa* O901W lipopolysaccharide (endotoxin) over a 13-day period (Truitt *et al.*, 1978). Their cells were then tested *in vivo* for reactivity against BW5147 carried in AKR ($H-2^k$) mice as follows. On day 0, normal young AKR mice were given 2.5×10^4 BW5147 leukemic blast cells, and the leukemia was allowed to multiply for 6 days. If no further treatment was given, the mice died of leukemia 2 days later (MST = 8.4 days). Based on cytokinetic studies, the tumor burden at the time of treatment was estimated to exceed 1×10^8 leukemic cells, which is less than one $\log_{10}$ below the lethal number (Bortin, 1974). On day 6, the leukemic mice were given CY plus TBR for tumor cytoreduction and immunosuppression. The chemoradiotherapy regimen induced a remission but was not curative by itself. Four to six hours later, the immunosuppressed, leukemic recipients were administered donor spleen cells i.v. with or without admixed lymph node cells. Endotoxin-treated DBA/2 mice were used as cell donors for the experimental groups; control groups of AKR mice were prepared in an identical manner, but were given no cells or transplanted with cells from syngeneic or untreated DBA/2 donors.

Shown in Fig. 1 are cumulative mortality curves for experimental and control groups that received no cells (Panel A) or immunocompetent cells from untreated syngeneic (Panel B), untreated DBA/2 (Panel C), or endotoxin-treated DBA/2 mice (Panel D). The MSTs of control groups ranged from 18 to 20 days, and all mice were dead in 34 days. In control groups given no cells or syngeneic cells, all deaths were attributed to leukemia. Mice that received cells from untreated DBA/2 mice showed no gross evidence of leukemia, and their deaths were attributed to acute GVH disease. In contrast, leukemic mice that received chemoradiotherapy plus a transplant of 10 to 40×10^6 spleen plus 2×10^6 lymph node cells from endotoxin-treated DBA/2 mice survived significantly longer (MST range = 38 to 68 days; $p < 0.01$).

Thorough necropsy examinations failed to reveal leukemia in the organs of AKR mice transplanted with cells from endotoxin-treated

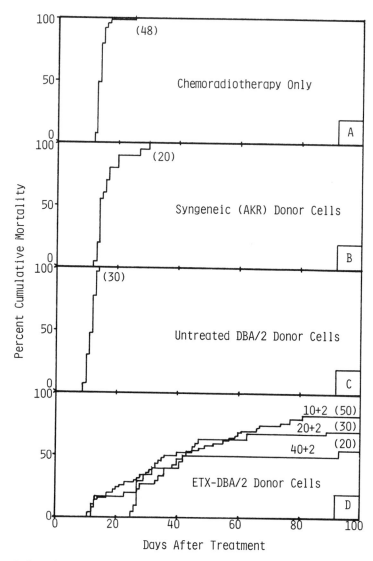

Figure 1. Cumulative mortality of BW5147 leukemia-bearing AKR mice treated with 150 mg/kg CY and 4.25-Gy TBR alone (Panel A) or in combination with spleen and lymph node cells from syngeneic (AKR) donors (Panel B), untreated DBA/2 donors (Panel C), or endotoxin-treated DBA/2 donors (ETX-DBA/2) (Panel D). Mice in Panels B and C received 10×10^6 spleen cells admixed with 2×10^6 lymph node cells. Mice in Panel D were given 10, 20, or 40×10^6 spleen and 2×10^6 lymph node cells as indicated. The number of mice transplanted are shown in parentheses.

DBA/2 donors. The absence of leukemia was confirmed by the results of GVL bioassays in which we found that as few as 5×10^6 splenocytes admixed with 2×10^6 lymph node cells from endotoxin-treated DBA/2 mice could eliminate residual leukemia from 77% (23/30) of the AKR mice; transplantation of 12×10^6 spleen or 10×10^6 spleen plus 2×10^6 lymph node cells eliminated leukemia from 90% (27/30) to 100% (30/30) of the mice within 6 days (Truitt et al., 1978). All deaths in the experimental groups (Panel D) were attributed to GVH disease or its complications. It is important to note that GVH disease was not completely eliminated when cells from endotoxin-treated donors were used for adoptive immunotherapy; however, the pattern of GVH disease was changed from the acute type (most deaths within 30 days) as found when normal DBA/2 cells were used to the delayed type (most deaths beyond 30 days).

Similar results were obtained when spleen cells from endotoxin-treated DBA/2 donors were used for adoptive immunotherapy of AKR mice bearing spontaneous AKR-L in a clinically relevant treatment model (unpublished data). In these experiments, AKR mice diagnosed as bearing advanced, spontaneous AKR-L were randomized into control and experimental groups (Table 1). One control group was left untreated (Group 1); all other groups were given chemoradiotherapy for remission-induction and pretransplant immunosuppression. All

Table 1. Adoptive Immunotherapy of Spontaneous Acute T-Cell Lymphoblastic Leukemia in AKR Mice Using Spleen Cells from Untreated or Endotoxin-Treated DBA/2 Donors

Group	Chemoradio-therapy[a]	Donor cells[b]	No. leukemic AKR mice[c]	MST^c (days)	Percent survival on day[c]			
					30	60	120	240
1	−	None	77	17	12	0	0	0
2	+	None	12	14	8	8	8	—
3	+	DBA/2	9	8	0	0	0	0
4	+	ETX-DBA/2[d]	10	38	50[e]	30	30[f]	20

[a] 100 mg/kg CY plus 10 mg/kg methyl cyclohexylnitrosourea on days −22, −15; 10 mg/kg palmotyl cytosine arabinoside on days −7, −5, −3; and 150 mg/kg CY and 7-Gy TBR on day −1.
[b] 20×10^6 spleen and 2×10^6 lymph node cells injected i.v. on day 0.
[c] Number of AKR mice bearing advanced, spontaneous leukemia. Median survival time (MST) and percent survival for Groups 2–4 are based on mice completing treatment and calculated from the last day of treatment. All mice were given high-dose antibiotics and housed in a laminar airflow protective environment as described elsewhere (Truitt, 1978a).
[d] Endotoxin treatment of DBA/2 (ETX-DBA/2) donors; 0.4 mg/kg i.p. on days −14, −13, −12; 1 mg/kg on day −9; 2.0 mg/kg on day −7; and 4.0 mg/kg on days −5 and −2. Mice used as donors on day 0.
[e] $p < 0.001$ vs. Group 1; $p < 0.05$ vs. Groups 2 and 3.
[f] $p < 0.001$ vs. Group 1; $p = 0.19$ vs. Group 2; $p = 0.07$ vs. Group 3.

mice were housed in sterile cages in a laminar airflow protective environment and given high dose antibiotics for intestinal decontamination as described elsewhere (Truitt, 1978a) in order to diminish infectious complications. For adoptive immunotherapy, Groups 3 and 4 were given 20×10^6 spleen cells plus 2×10^6 lymph node cells from normal or endotoxin-treated DBA/2 donors, respectively. The survival results are presented in Table 1.

All untreated AKR-L mice died with leukemia; their MST was 17 days after diagnosis (Group 1, Table 1). The chemoradiotherapy regimen employed for remission-induction and immunosuppression was lethal in the absence of hemopoietic support (Group 2). Mice given spleen and lymph node cells from untreated DBA/2 donors died from acute GVH disease within 12 days of transplant (Group 3). In contrast, adoptive transfer of cells from endotoxin-treated DBA/2 donor mice resulted in higher survival rates at 30, 60, 120, and 240 days posttransplant.

Endotoxin treatment of experimental animals has been reported to increase their antitumor reactivity (Grohsman and Nowotny, 1972; Parr *et al.*, 1973; Yang and Nowotny, 1974). Therefore, it was of interest to determine whether endotoxin treatment increased the reactivity of DBA/2 cells against AKR-L. We quantified the antileukemic reactivity of cells from untreated and endotoxin-treated DBA/2 donors using a GVL bioassay described in detail elsewhere (Bortin *et al.*, 1973). Lethally irradiated, young AKR mice were given graded doses (1×10^5, 1×10^6, and 1×10^7) of AKR-L cells. The following day, groups of 20 AKR mice at each leukemia dose were given no cells or injected (i.v.) with 40×10^6 spleen cells from untreated DBA/2 or endotoxin-treated DBA/2 donors. The GVL reaction was allowed to proceed for 6 days; then, all spleen cells from individual AKR primary recipients were transferred i.p. to individual AKR secondary recipients to determine whether any leukemia remained. The procedure of Reed and Muench (1938) was used to calculate the amount of leukemia that could be eliminated from 50% of the AKR mice after treatment with 40×10^6 allogeneic spleen cells (TD_{50} = tumor dose for 50%). The TD_{50} for mice receiving untreated DBA/2 cells was 6.1×10^6. In contrast, the TD_{50} for mice given endotoxin-treated DBA/2 spleen cells was 1.1×10^6 (data not shown). Thus, rather than enhancement of antitumor activity, endotoxin treatment resulted in an 82% reduction in antileukemic reactivity of the donor cells. Nevertheless, as seen in Table 1 and Fig. 1, cells from endotoxin-treated DBA/2 donors retained sufficient GVL reactivity to be therapeutically effective against spontaneous and long-passage AKR leukemias.

IV. GVH-Reactive Cell Populations in Endotoxin-Treated Donors

Nonlymphoid cellular elements in the spleens of mice are known to increase severalfold following treatment with endotoxin (Boggs et al., 1973). In our experiments, we found that the number of spleen cells had increased an average of 2.4-fold above untreated spleens by 2 days after completion of the endotoxin regimen (Truitt et al., 1978; unpublished data). An increase in nonlymphoid and presumably non-GVH-reactive cells in the spleens of endotoxin-treated mice could cause a decrease in the number of GVH-reactive cells contained in any given dose of transplanted cells. As a result, GVH reactivity and associated mortality might decrease. Okunewick et al. (1980) have reported an exponential relationship between the dose of spleen cells transplanted and a fatal GVH response. It is generally considered that postthymic T cells in the donor cell population cause or promote GVH disease in immunosuppressed hosts (Elkins, 1971; Korngold and Sprent, 1978). In our earlier report (Truitt et al., 1978), "dilution" of GVH-reactive cells due to proliferation of non-GVH-reactive cell populations was considered to be untenable with the data reported because there was an inverse relationship between mortality and the dose of endotoxin-treated splenocytes administered for adoptive immunotherapy. For example, in Panel D (Fig. 1), GVH-related mortality decreased as the dose of transplanted spleen cells from endotoxin-treated donors increased. However, as described below, subsequent experiments using several different assays for GVH reactivity in nonleukemic systems did not support these observations.

The GVH reactivity of various doses of untreated and endotoxin-treated DBA/2 donor spleen cells was compared in three different assay systems; (1) Simonsen spleen weight-gain assays in neonatal F_1 mice (Simonsen, 1962); (2) runting in newborn F_1 mice (Russell, 1960); and (3) mortality in lethally irradiated adult mice (Grebe and Streilein, 1976). The results are shown in Table 2. At any equivalent cell dose, the GVH reactivity of endotoxin-treated spleen cells, whether assessed by splenomegaly, change in body weight, or mortality, was lower than that for the identical dose of untreated DBA/2 spleen cells (cf. Groups 1 vs. 5, 2 vs. 6, 3 vs. 7, and 4 vs. 8). Splenocytes from endotoxin-treated DBA/2 donors were reactive in each of the GVH assays used; however, in both the Simonsen and the newborn runting assay a larger number of cells was necessary to produce comparable levels of splenomegaly or runting. For example, whereas inoculation of 4×10^6 untreated DBA/2 spleen cells (Group 2) resulted in a spleen index of

Table 2. Comparison of GVH Reactivity of Graded Doses of Spleen Cells from Untreated or Endotoxin-Treated DBA/2 Donor Mice in Three Different Assay Systems

Group	DBA/2 donor cells		Simonsen spleen weight-gain assays		Newborn runting assays		Adult mortality assays			
	Endotoxin[a]	Dose	No. D2AKF1 mice[b]	Spleen index (S.D.)	No. D2AKF1 mice[c]	Change in body weight	No. AKR mice[d]	MST (days)	% survival on day 30	% survival on day 90
1	–	1×10^6	12	1.15 (0.30)	16	−4.2%	20	12	0	0
2	–	4×10^6	12	1.88 (0.57)	10	−12.0%	20	13	10	5
3	–	16×10^6	11	2.65 (0.70)	15	−22.8%	20	10	20	0
4	–	40×10^6	12	2.54 (0.57)	14	−23.4%	20	9	0	0
5	+	1×10^6	13	1.03 (0.18)	15	+0.6%	20	49	65	20
6	+	4×10^6	14	1.44 (0.35)	14	−1.2%	20	79	75	50
7	+	16×10^6	12	2.11 (0.36)	16	−16.2%	20	89	65	45
8	+	40×10^6	13	1.88 (0.38)	16	−9.6%	20	22	35	10

[a] Endotoxin treatment: 0.4 mg/kg i.p. on days −14, −13, −12; 1 mg/kg on day −9; 2.0 mg/kg on day −7; and 4.0 mg/kg on days −5 and −2. Mice were used as donors on day 0.

[b] Six- to 10-day-old (DBA/2 × AKR)F$_1$ neonates were injected i.p. with the cells indicated. Littermate control mice were given media only ($n = 10$) or 40×10^6 D2AKF1 spleen cells ($n = 15$). Spleen index = (experimental spleen weight/100-g mouse)/(littermate control spleen weight/100-g mouse).

[c] One- to two-day-old D2AKF1 neonates were injected i.p. with cells indicated. Littermate control mice were given media only ($n = 11$) or 1 to 40×10^6 D2AKF1 spleen cells ($n = 15$). All mice were weighed 28 days after injection. Data represent the percent change in the average body weight of experimental mice as compared to the average weight (16.7 g) of control mice.

[d] Number of immunosuppressed (10-Gy TBR) young AKR mice inoculated i.v. with donor cells on day 0.

1.88 in the Simonsen assay and a 12.0% reduction in body weight in the newborn runting assay, almost 16×10^6 cells from endotoxin-treated donors (Group 7) were necessary to produce similar results (2.11 spleen index and -16.2% change in body weight). These results are consistent with the idea that the proportion of GVH-reactive cells was reduced after endotoxin treatment of the donor.

In interpreting the results of adult mortality assays of GVH reactivity, consideration must be given to the amount of hemopoietic support provided to the lethally irradiated host mice. The frequency of spleen colony-forming units (CFU-S) in normal and endotoxin-treated DBA/2 mice was measured using the method of Till and McCulloch (1961). The CFU-S content of spleens from endotoxin-treated DBA/2 donors was measured 2 days after the last injection of endotoxin, i.e., at the same time interval as when they were used as donors in the GVL and GVH experiments. The frequency of CFU-S in endotoxin-treated DBA/2 spleens was 11.3 per 1×10^5 cells; in spleens from untreated DBA/2 donors, the frequency was 0.8 per 1×10^5 cells (unpublished results). There was no significant difference in the transplantation factor (f) [i.e., the proportion of cells homing to the spleen as calculated according to Siminovitch et al. (1963)] between cells from untreated ($f = 8\%$) and endotoxin-treated ($f = 7\%$) donors. Thus, 1×10^6 spleen cells from endotoxin-treated DBA/2 donors contained approximately 1600 CFU-S (113 CFU-S per 1×10^6 cells/0.07), which provided adequate hemopoietic reconstitution (65% survival at 30 days) with negligible acute GVH reactivity (Group 5, Table 2). In contrast, although the GVH reactivity of 1×10^6 untreated donor cells was low in the Simonsen and newborn runting assays (Group 1), only 100 CFU-S (8 CFU-S per 1×10^6 cells/0.08) was given to the lethally irradiated DBA/2 adult hosts. As a consequence, all mice died within 30 days primarily from aplasia rather than GVH disease. In order to provide hemopoietic support comparable to the lowest dose of endotoxin-treated cells used (1×10^6), an inoculum of approximately 16×10^6 untreated splenocytes was necessary. At such doses, the GVH-reactive cell population in untreated donors was substantially higher than in endotoxin-treated donors (cf. Groups 3 and 5).

To verify that the GVH-reactive cell population was simply diluted by proliferation on non-GVH-reactive cells after endotoxin treatment, we used limiting dilution microcytotoxicity assays (lda) to estimate the frequency of anti-AKR-reactive precursor cytolytic T lymphocytes (pCTL) in the spleens of untreated and endotoxin-treated donors. The lda employed was identical to that described by MacDonald et al. (1980). Graded numbers of spleen cells (125 to 4000) from untreated or endotoxin-treated DBA/2 mice (pools of three mice) were cultured in 24 replicate microwells each with an excess (1×10^6) of irradiated AKR stimulator cells for 8 days in the presence of 50% (v/v) T-cell growth factor (TCGF). Lectin-free TCGF was obtained from the supernatant of a secondary MLC (MacDonald *et al.*, 1980). Thereafter, each microculture well was assayed for cytotoxicity against 2.5×10^3 ^{51}Cr-labeled AKR target cells in a 3½-hr cytotoxicity assay. Microculture wells that gave a ^{51}Cr release exceeding the mean spontaneous release from 24 replicate control wells by at least three standard deviations were considered positive. The proportion of negative cultures was plotted against the number of responder cells plated (Fig. 2). Linear regression analysis was applied to the data and the frequency of pCTL calculated according to the Poisson equation (Miller *et al.*, 1977; MacDonald *et al.*, 1980).

The mean frequency of anti-AKR-reactive pCTL from two experiments testing cells from untreated DBA/2 mice was calculated to be 1/1436 as compared to 1/3481 in the spleens of endotoxin-treated DBA/2 mice. Thus, the frequency of antihost-reactive cytolytic T cells was reduced by a factor of 1/2.4. The absolute number of cytolytic cells per spleen did not change since the total number of splenocytes increased 2.4-fold following endotoxin treatment. Using these estimates of the frequency of antihost-reactive pCTL, a dose of 40×10^6 spleen cells from endotoxin-treated DBA/2 donors would contain a dose of pCTL equivalent to 16×10^6 spleen cells from untreated donors (i.e., approximately 11,000 pCTL). At these doses, there was no statistically significant difference in the 30- and 90-day survival rates in adult mortality GVH assays using endotoxin-treated or untreated donor cells (cf. Groups 3 and 8, Table 2). These results support the hypothesis that the decreased GVH reactivity following endotoxin treatment was attributable to a decrease in the number of GVH-reactive cells (pCTL) transplanted.

To test this hypothesis and to evaluate the importance of adequate hemopoietic stem cells on the incidence and intensity of GVH disease, untreated DBA/2 spleen and bone marrow cells were admixed and tested in adult mortality GVH assays. Since mouse bone marrow contains few mature T cells, but an abundance of stem cells, this had the effect of enriching for stem cells without increasing the number of antihost pCTL injected. For comparison, groups of mice were given spleen cells from untreated or endotoxin-treated DBA/2 donors. The results are shown in Table 3. DBA/2 bone marrow alone caused no acute GVH disease in lethally irradiated AKR hosts (Group 2); 92% (11/12) of the mice survived the first 60 days, and 58% survived to 100 days. Mice given 16×10^6 spleen cells from untreated DBA/2 do-

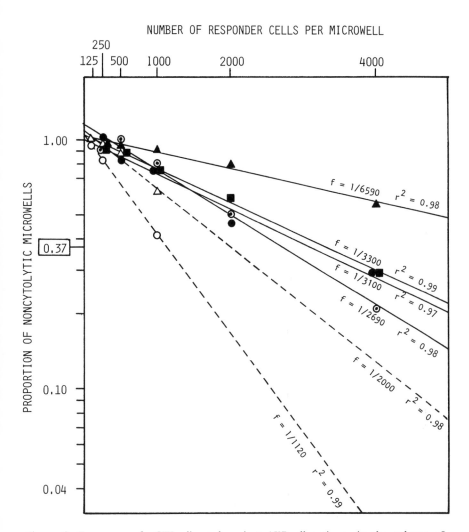

Figure 2. Frequency of pCTL directed against AKR alloantigens in the spleens of untreated (dashed lines) and endotoxin-treated (solid lines) DBA/2 mice estimated by limiting dilution assays. Varying numbers of responder cells were cultured in groups of 24 microwells with 1×10^6 irradiated AKR spleen cells in the presence of TCGF. After 8 days, each microwell was assayed for cytotoxicity against ^{51}Cr-labeled AKR leukemia target cells and the proportion of noncytolytic cultures in each group determined. pCTL frequencies (f) were determined from linear regression analysis of the data.

Table 3. GVH Reactivity of Bone Marrow and/or Spleen Cells from Untreated or Endotoxin-Treated DBA/2 Donor Mice as Measured in Lethally Irradiated AKR Mice

Group	Donor cells ($\times 10^6$)	No. AKR mice transplanted[a]	MST (days)	Percent survival on day		
				30	60	100
1	None	6	10	0	0	0
2	DBA/2 BM (8)	12	> 100	92	92	58
3	DBA/2 SPL (8)	10	9	10	0	0
4	DBA/2 SPL (16)	13	9	8	0	0
5	DBA/2 BM$_x$ (8) + SPL (8)	12	9	8	0	0
6	DBA/2 BM (8) + SPL (8)	25	64	84	52	32
7	ETX-DBA/2[b] SPL (16)	10	68	70	60	30

[a] Lethally irradiated (10 Gy) 6- to 12-week-old AKR mice inoculated i.v. with bone marrow (BM) and/or spleen (SPL) cells from untreated or endotoxin-treated (ETX) DBA/2 donors 3 to 4 hr post-irradiation. BM cells in Group 5 were given 30 Gy of γ-irradiation prior to inoculation.
[b] Endotoxin treatment: 0.4 mg/kg i.p. on days -14, -13, -12; 1.0 mg/kg on day -9; 2.0 mg/kg on day -7; and 4.0 mg/kg on days -5 and -2. Mice were used as donors on day 0.

nors developed acute GVH disease; none of the mice survived beyond 60 days (Group 4). At a dose of 16×10^6 spleen cells, the irradiated AKR host received approximately 11,000 antihost-reactive pCTL and 1600 CFU-S based on the frequency estimates described earlier. In contrast, an identical dose of cells from endotoxin-treated donors would contain only about 4600 antihost pCTL and over 25,000 CFU-S; mice given these cells (Group 7) had an MST of 68 days, and 60% survived more than 60 days posttransplant. A 50% reduction in the number of inoculated antihost pCTL from untreated donor spleens was achieved by cutting the transplant dose in half (Groups 3 and 5); nevertheless, there was no improvement in survival (cf. Groups 3 and 5 to 4). This may be due in part to a concomitant reduction in the number of stem cells transplanted. In contrast, if the antihost pCTL inoculum was reduced by 50% *and* the stem cell inoculum increased by admixing 8×10^6 bone marrow cells with 8×10^6 splenocytes, then 52% of the mice survived to 60 days and 32% to 100 days (Group 6); this was not significantly different from that seen after transplantation of spleen cells from endotoxin-treated donors (cf. Groups 6 and 7). Thus, the phenomenon of mitigation of GVH disease by treating donor mice with bacterial endotoxin could be mimicked using cells from untreated mice by reducing the GVH-reactive (pCTL) population and enriching for hemopoietic stem cells. We cannot say whether the stem cells had an inhibitory effect on the GVH reactivity of donor pCTL.

V. Conclusions

It is apparent from the data described above that modulation of GVH disease by treatment of immunocompetent cell donors with bacterial endotoxin in the system we have applied to the adoptive immunotherapy of AKR leukemia (Truitt *et al.*, 1978) can be explained best in terms of a reduction in the actual number of GVH-reactive cells transplanted. The most compelling evidence for this comes from estimates of the frequency of antihost-reactive pCTL in untreated and endotoxin-treated donor spleens (Fig. 2). Other experimental data are consistent with this hypothesis; for example, splenocytes from endotoxin-treated donors did cause GVH disease in each of three different assay systems treated, but the dose of cells required to produce GVH reactivity comparable in incidence and intensity to that of untreated donor cells was severalfold higher (Table 2). The GVL-reactive cell population also was reduced as evidence by a decrease in the amount of leukemia that could be eliminated from 50% to the AKR hosts.

In leukemic animals, the survival rates following chemoradiotherapy plus spleen and lymph node cell transplantation improved as larger numbers of cells from endotoxin-treated donors were transplanted (Fig. 1D). The increased survival could not be attributed to increased GVL reactivity since the number of cells transplanted exceeded the minimum necessary to eliminate the leukemia (Truitt *et al.*, 1978). These data contrast with those obtained in adult mortality GVH assays using nonleukemic animals in which the survival rate dropped when more than 16×10^6 spleen cells from endotoxin-treated donors were transplanted (cf. Groups 7 and 8, Table 2). Lymphoid cells are preferential targets for GVH reactivity in comparison to other tissues (Elkins, 1971), and the leukemia cells may have provided an "immune distraction" to the transplanted donor cells, which otherwise would have reacted against normal host tissues resulting in an increased level of GVH disease. We have found that GVH-related mortality in lethally irradiated AKR mice given bone marrow and lymph node cells from *H-2*-incompatible SJL/J donors was delayed when leukemia cells were present (unpublished data).

Dilution of the GVH-reactive cell population in the spleens of endotoxin-treated mice was paralleled by expansion of the hemopoietic stem cell compartment. Endotoxin treatment resulted in a 14-fold enrichment in hemopoietic stem cells as compared to untreated mice. As a result, fewer spleen cells from endotoxin-treated donors were necessary to reconstitute hemopoiesis in lethally irradiated hosts. It is quite obvious that if fewer spleen cells are transplanted, then fewer GVH-reactive cells are transplanted, resulting in less GVH disease. Thus, treatment of donor mice with endotoxin achieved *in vivo* the same effect as stem cell separation procedures (Grebe and Streilein, 1976) attempt to achieve *in vitro*, i.e., a reduction in the number of GVH-reactive cells and an enrichment for hemopoietic stem cells in the transplant inoculum.

The hypothesis outlined above does not totally preclude the possibility that other mechanisms might be involved in modulation of GVH disease by treatment of donor mice with bacterial endotoxin. The effect of IgM antibody and/or other serum factors described by Jutila and his associates (Thomson *et al.*, 1978; Rampy and Jutila, 1979) and the ability of adherent spleen cells to inhibit GVH reactivity of nonadherent spleen cells (Thomson and Jutila, 1974), for example, cannot be readily explained by a simple dilution effect. Nor can the results of Chedid (1973) using an *in vitro* treatment procedure. Early data reported by Liacopoulos *et al.* (1967; Liacopoulos and Merchant, 1969) do not appear to be consistent with dilution of GVH-reactive cells, although they did not specifically mention the degree of splenomegaly observed after endotoxin treatment of their donor mice.

A variety of mechanisms may be responsible for the effects of endotoxin on T-cell-mediated GVH reactivity depending on the experimental design and assay systems used. In view of the vast array of biological effects attributed to endotoxin (Morrison and Ryan, 1979), this would not be surprising. The increased application of bone marrow transplantation to a variety of hematologic dyscrasias justifies a thorough investigation of all possible avenues to prevent the complication of GVH disease. Although the contribution of bacterial infections to GVH-associated mortality cannot be disputed, the value of using endotoxin-tolerant donor cells to modulate GVH disease in a clinical setting is open to debate. Other approaches, such as the use of gnotobiotic techniques, may prove to be more valuable and/or practical. More basic information about immunobiology may derive from animal studies such as those described in this chapter than practical applications to man.

ACKNOWLEDGMENTS

The author wishes to express appreciation to Mortimer M. Bortin, Alfred A. Rimm, and Chiu-Yang Shih for their consultations on the experiments and critiques of this manuscript. Special thanks to

Evangeline Reynolds, Mark Freedman, and Deborah Meltzer for technical assistance and to D'Etta Waldoch and Elaine Kobs for preparing the manuscript.

Drugs used in this study were provided by the Drug Synthesis and Chemistry Branch, Division of Cancer Treatment, National Cancer Institute.

This work was supported by Public Health Service Grants CA-21998, CA-18440, and CA-26245 from the National Cancer Institute, the Leukemia Research Foundation, Inc., the Milwaukee Division of the American Cancer Society, and the Board of Trustees, Mount Sinai Medical Center.

The author is a Scholar of the Leukemia Society of America, Inc.

VI. References

Boggs, S. S., Wilson, S. M., and Smith, W. W., 1973, Effects of endotoxin on hematopoiesis in irradiated and nonirradiated W/W^v mice, *Radiat. Res.* **56**:481.

Bortin, M. M., 1974, Graft-versus-leukemia, in: *Clinical Immunobiology*, Vol. 2 (F. H. Bach and R. A. Good, eds.), pp. 287–306, Academic Press, New York.

Bortin, M. M., and Rimm, A. A., 1977, Severe combined immunodeficiency disease: Characterization of the disease and results of transplantation, *J. Am. Med. Assoc.* **238**: 591.

Bortin, M. M., and Rimm, A. A., 1978, Bone marrow transplantation for acute myeloblastic leukemia, *J. Am. Med. Assoc.* **240**:1245.

Bortin, M. M., and Rimm, A. A., 1981, Allogeneic bone marrow transplantation for 144 patients with severe aplastic anemia, *J. Am. Med. Assoc.* **245**:1132.

Bortin, M. M., and Truitt, R. L., 1977, AKR T cell acute lymphoblastic leukemia: A model for human T cell leukemia, *Biomedicine* **26**:315.

Bortin, M. M., Rimm, A. A., and Saltzstein, E. C., 1973, Graft-versus-leukemia: Quantification of adoptive immunotherapy in murine leukemia, *Science* **179**:811.

Bortin, M. M., Truitt, R. L., and Rimm, A. A., 1978, Nonspecific adoptive immunotherapy of T cell acute lymphoblastic leukemia in AKR mice: A model for the treatment of T cell leukemia in man, in: *The Handbook of Cancer Immunology*, Vol. 5, *Immunotherapy* (H. Waters, ed.), pp. 401–429, Garland STPM Press, New York.

Chedid, L., 1973, Possible role of endotoxemia during immunologic imbalance, *J. Infect. Dis.* **128**(Suppl.):112.

Connell, M. St. J., and Wilson, R., 1965, The treatment of x-irradiated germfree CFW and C3H mice with isologous and homologous bone marrow, *Life Sci.* **4**:721.

Damais, C., Lamensans, A., and Chedid, L., 1972, Endotoxines bacteriennes et maladie homologue du nouveau-né, *C. R. Acad. Sci.* **274**:1113.

Elkins, W. L., 1971, Cellular immunology and the pathogenesis of graft-versus-host reactions, *Prog. Allergy* **15**:78.

Gale, R. P., and the UCLA Bone Marrow Transplant Unit, 1979, Current status of bone marrow transplantation in acute leukemia, *Transplant. Proc.* **11**:1920.

Galley, C. B., Walker, R. I., Ledney, G. D., and Gambrill, M. R., 1975, Evaluation of biologic activity of ferric chloride-treated endotoxin in mice, *Exp. Hematol.* **3**:197.

Glucksberg, H., Storb, R., Fefer, A., Buckner, C. D., Neiman, P. E., Clift, R. A., Lerner, K. G., and Thomas, E. D., 1974, Clinical manifestations of graft-versus-host disease in human recipients of marrow from HLA-matched sibling donors, *Transplantation* 18: 295.

Grebe, S. C., and Streilein, J. W., 1976, Graft-versus-host reactions: A review, *Adv. Immunol.* 22:119.

Grohsman, J., and Nowotny, A., 1972, the immune recognition of TA3 tumors, its facilitation by endotoxin, and abrogation by ascites fluid, *J. Immunol.* 109:1090.

Heit, H., Heit, W., Kohne, E., Fleidner, T. M., and Hughes, P., 1977, Allogeneic bone marrow transplantation in conventional mice. I. Effect of antibiotic therapy on long term survival of allogeneic chimeras, *Blut* 35:143.

Jones, J. M., Wilson, R., and Bealmear, P. M., 1971, Mortality and gross pathology of secondary disease in germfree mouse radiation chimeras, *Radiat. Res.* 45:577.

Jutila, J. W., 1973, Etiology of the wasting diseases, *J. Infect. Dis.* 128(Suppl.):99.

Keast, D., 1973, Role of bacterial endotoxin in the graft-vs-host syndrome, *J. Infect. Dis.* 128(Suppl.):104.

Korngold, R., and Sprent, J., 1978, Lethal graft-versus-host disease after bone marrow transplantation across major histocompatibility barriers in mice, *J. Exp. Med.* 148: 1687.

LeFeber, W. P., Truitt, R. L., Rose, W. C., and Bortin, M. M., 1977, Graft-versus-leukemia: Donor selection for adoptive immunotherapy in mice, in: *Experimental Hematology Today—1977* (S. J. Baum and G. D. Ledney, eds.), pp. 239–246, Springer-Verlag, Berlin.

Liacopoulos, P., and Merchant, B., 1969, Effet du traitement des donneurs avec des endotoxines sur la maladie homologue des receveurs adultes irradiés, in: *La structure et les effets biologiques des produits bactériens provenants de germes Gram-négatifs* (L. Chedid, ed.), pp. 341–356, Colloq. Int. sur les Endotoxines, CNRS No. 174, Paris.

Liacopoulos, P., Merchant, B., and Harrel, R. E., 1967, Inhibition de la maladie homologue par un traitement des couris donneuses avec des antigenes bacteriens, *Pathol. Biol.* 15:587.

MacDonald, H. R., Cerottini, J.-C., Ryser, J.-E., Maryanski, J. L., Taswell, C., Widmer, M. B., and Brunner, K. T., 1980, Quantitation and cloning of cytolytic T lymphocytes and their precursors, *Immunol. Rev.* 51:93.

McIntire, R., and Gale, R. P., 1981, Relationship between graft-versus-leukemia and graft-versus-host in man—UCLA experience, in: *Graft-versus-Leukemia in Man and Animal Models* (J. P. OKunewick and R. F. Meredith, eds.), pp. 1–9, CRC Press, Boca Raton, Fla.

Miller, R. G., Teh, H.-S., Harley, E., and Phillips, R. A., 1977, Quantitative studies of the activation of cytotoxic lymphocyte precursor cells, *Immunol. Rev.* 35:38.

Morrison, D. C., and Ryan, J. L., 1979, Bacterial endotoxins and host immune responses, *Adv. Immunol.* 28:293.

Okunewick, J. P., and Meredith, R. F. (eds.), 1981, *Graft-versus-Leukemia in Man and Animal Models*, CRC Press, Boca Raton, Fla.

Okunewick, J. P., Meredith, R. F., Brozovich, B. J., Seeman, P. R., and Magliere, K., 1980, Exponential relationship between spleen cell concentration and fatal graft-versus-host response after transplantation of allogeneic spleen–marrow cell mixtures, *Transplantation* 29:507.

Parr, I., Wheeler, E., and Alexander, P., 1973, Similarities of the anti-tumor actions of endotoxin, lipid A and double-stranded RNA, *Br. J. Cancer* 27:370.

Powles, R. L., Morgenstern, G., Clink, H. M., Hedley, D., Bandini, G., Lumley, H., Watson, J. G., Lawson, D., Spence, D., Barrett, A., Jameson, B., Lawler, S., Kay, H. E.

M., and McElwain, T. J., 1980, The place of bone-marrow transplantation in acute myelogenous leukaemia, *Lancet* **1**:1047.

Rampy, P. A., and Jutila, J. W., 1979, Influence of lipopolysaccharide on graft-versus-host reactivity of lipopolysaccharide-unresponsive C3H/HeJ mice, *Infect. Immun.* **26**: 137.

Reed, L. J., and Muench, H. A., 1938, A simple method of estimating fifty percent endpoints, *Am. J. Hyg.* **27**:493.

Reed, N. D., and Jutila, J. W., 1965, Wasting disease induced with cortisol acetate: Studies in germ-free mice, *Science* **150**:356.

Rose, W. C., Rodey, G. E., Rimm, A. A., Truitt, R. L., and Bortin, M. M., 1976, Mitigation of graft-versus-host disease in mice by treatment of donors with bacterial endotoxin, *Exp. Hematol.* **4**:90.

Russell, P. S., 1960, The weight-gain assay for runt disease in mice, *Ann. N.Y. Acad. Sci.* **87**:445.

Siminovitch, L., McCulloch, E. A., and Till, J. E., 1963, The distribution of colony forming cells among spleen colonies, *J. Cell. Comp. Physiol.* **62**:327.

Simonsen, M., 1962, Graft-versus-host reactions: Their natural history and applicability as tools of research, *Prog. Allergy* **6**:349.

Storb, R., Thomas, E. D., Weiden, P. L., Buckner, C. D., Clift, R. A., Tyler, A., Goodell, B. W., Johnson, F. L., Neiman, P. E., Sanders, J. E., and Singer, J., 1978, One-hundred-ten patients with aplastic anemia (AA) treated by marrow transplantation in Seattle, *Transplant. Proc.* **10**:135.

Thomas, E. D., Storb, R., Clift, R. A., Fefer, A., Johnson, F. L., Neiman, P. E., Lerner, K. G., Glucksberg, H., and Buckner, C. D., 1975, Bone marrow transplantation, *N. Engl. J. Med.* **292**:832, 895.

Thomas, E. D., Buckner, C. D., Banaji, M., Clift, R. A., Fefer, A., Flournoy, N., Goodell, B. W., Hickman, R. O., Lerner, K. G., Neiman, P. E., Sale, G. E., Sanders, J. E., Singer, J., Stevens, M., Storb, R., and Weiden, P. L., 1977, One hundred patients with acute leukemia treated by chemotherapy, total body irradiation, and allogeneic marrow transplantation, *Blood* **49**:511.

Thomson, P. D., and Jutila, J. W., 1974, The suppression of graft-versus-host disease in mice by endotoxin-treated adherent spleen cells, *J. Reticuloendothelial Soc.* **16**:327.

Thomson, P. D., Rampy, P. A., and Jutila, J. W., 1978, A mechanism for the suppression of graft-vs-host disease with endotoxin, *J. Immunol.* **120**:1340.

Till, J. E., and McCulloch, E. A., 1961, A direct measurement of the radiation sensitivity of normal mouse bone marrow cells, *Radiat. Res.* **14**:213.

Truitt, R. L., 1978a, Use of decontamination and a protected environment to prevent secondary disease following adoptive immunotherapy of acute leukemia in mice, in: *Experimental Hematology Today—1978* (S. J. Baum and G. D. Ledney, eds.), pp. 195–201, Springer-Verlag, Berlin.

Truitt, R. L., 1978b, Application of germfree techniques to the treatment of leukemia in AKR mice by allogeneic bone marrow transplantation, in: *The Handbook of Cancer Immunology*, Vol. 5, *Immunotherapy* (H. Waters, ed.), pp. 431–452, Garland STPM Press, New York.

Truitt, R. L., 1979, Successful adoptive immunotherapy of AKR spontaneous T cell leukemia following antibiotic decontamination and protective isolation, in: *Clinical and Experimental Gnotobiotics* (T. Fliedner, H. Heit, D. Niethammer, and H. Pfieger, eds.), pp. 215–220, Gustav Fischer Verlag, Stuttgart.

Truitt, R. L., Rose, W. C., Rimm, A. A., and Bortin, M. M., 1978, Graft-versus-leukemia. VIII. Selective reduction in antihost reactivity without loss of antileukemic reactivity by treatment of donor mice with lipopolysaccharide, *Exp. Hematol.* **6**:488.

van Bekkum, D. W., and de Vries, M. J., (eds.), 1967, *Radiation Chimaeras*, Logos Press, London.

van Bekkum, D. W., and Vos, O., 1961, Treatment of secondary disease in radiation chimaeras, *Int. J. Radiat. Biol.* **3**:173.

van Bekkum, D. W., Roodenburg, J., Heidt, P. J., and van der Waaij, D., 1974, Mitigation of secondary disease of allogeneic mouse radiation chimaeras by modification of the intestinal microflora, *J. Natl. Cancer Inst.* **52**:401.

Walker, R. I., 1978, The contribution of intestinal endotoxin to mortality in hosts with compromised resistance: A review, *Exp. Hematol.* **6**:172.

Walker, R. I., Ledney, G. D., and Galley, C. B., 1975, Aseptic endotoxemia in radiation injury and graft-vs-host disease, *Radiat. Res.* **62**:242.

Weiden, P. L., Flournoy, N., Thomas, E. D., Prentice, R., Fefer, A., Buckner, C. D., and Storb, R., 1979, Antileukemic effect of graft-versus-host disease in human recipients of allogeneic-marrow grafts, *N. Engl. J. Med.* **300**:1068.

Weiden, P. L., Flournoy, N., Sanders, J. E., Sullivan, K. M., and Thomas, E. D., 1981, Antileukemic effect of graft-versus-host disease contributes to improved survival after allogeneic marrow transplantation, *Transplant. Proc.* **13**:248.

Yang, C., and Nowotny, A., 1974, Effect of endotoxin on tumor resistance in mice, *Infect. Immun.* **9**:95.

25

Enhancement of Nonspecific Resistance to Tumor by Endotoxin

R. Christopher Butler

I. Clinical Use of Endotoxin in Tumor Therapy

Since the late 1800s, bacterial products have been used as antitumor agents for the treatment of cancer patients. In spite of the fact that most patients treated with these products have been in advanced stages of the disease, the clinical results obtained have been generally encouraging and occasionally remarkable. Endotoxin has been shown to be one of the most potent antitumor components in these bacterial preparations. Therefore, endotoxin has been extensively studied in animal tumor models to determine the conditions under which it may be used most effectively against tumors.

The clinical use of endotoxin in cancer therapy began with the introduction of Coley's "mixed bacterial toxins" in 1893 (Coley, 1893). This preparation, which is composed of heat-killed *Streptococcus pyogenes* mixed with *Serratia marcescens* toxins, has since been used in the treatment of thousands of cancer patients (Nauts *et al.*, 1953; Fowler, 1969a,b). A review of 1287 cancer patients treated with mixed bacterial toxins demonstrated partial or complete tumor regression in over 50% of the patients (Fowler, 1969). Survival duration for patients responding to this therapy ranged from 5 to 70 years even though most of the tumors had been considered inoperable at the time of initial toxin treatment. Soft tissue sarcomas and malignant lymphomas were most amenable to toxin treatment and demonstrated the highest rates of success.

R. Christopher Butler • Departments of Microbiology and Immunology, The Arlington Hospital, Arlington, Virginia 22205.

A review in 1971 of 52 cases of reticulum cell sarcoma of bone treated by mixed bacterial toxin therapy alone or combined with surgery or radiotherapy found a 5-year survival rate of 64% (Miller and Nicholson, 1971). Permanent cures were achieved in 48% even though a third of these patients had metastases prior to treatment. These results compare favorably against other studies that have found surgery or radiotherapy alone to produce a 5-year survival rate of only 32% in patients without initial metastases (Copeland, 1967).

II. Tumor Hemorrhage and Necrosis

The antitumor effects of mixed bacterial toxins in animals were first observed in 1907 when they were used beneficially in the treatment of lymphosarcoma in dogs (Beebe and Tracy, 1907). By the early 1930s, it was recognized that a gram-negative bacterial toxin was responsible for the antitumor activity in animals. Gratia and Linz (1931) first described the phenomenon of tumor hemorrhage and necrosis leading to tumor regression of transplantable sarcomas in guinea pigs in response to treatment with *E. coli* culture filtrates. Subsequent studies demonstrated that this antitumor effect on sarcomas was a general property of culture filtrates from gram-negative bacteria including *E. coli* (Duran-Reynals, 1935; Shear and Andervont, 1936), *N. meningitidis* (Shwartzman and Michailovsky, 1932), *S. typhosa* (Shwartzman, 1936), and *P. mirabilis* (Fogg, 1936).

In a series of studies, Shear and colleagues attempted to isolate and identify the active component in these gram-negative bacterial filtrates (Felton *et al.*, 1935; Shear and Andervont, 1936; Hartwell *et al.*, 1943). They isolated a toxic lipopolysaccharide (LPS) from *Serratia marcescens* that was highly active in inducing hemorrhage in sarcoma 37 tumors (Hartwell *et al.*, 1943). Subsequent studies further purified this LPS and demonstrated it to be identical to endotoxin (Gardner *et al.*, 1939; Shapiro, 1940). Serologically, LPS was shown to correlate with the bacterial somatic O antigen (Zahl *et al.*, 1942).

Several investigators have studied the relationship between endotoxin structure and tumor hemorrhage activity. Using different series of smooth and rough endotoxin mutants ranging from Ra to Re, it was determined that the Re mutants consisting only of KDO and lipid A were at least as active as the native smooth LPS in inducing tumor hemorrhage (Nowotny *et al.*, 1971; Nigam, 1975; Nakahara *et al.*, 1975). The lipid A moiety has been identified as containing the active site for this particular antitumor effect (Nakahara *et al.*, 1975; Parr *et al.*, 1973). Detoxification of the lipid A moiety by methylation did not significantly reduce its tumor-hemorrhaging potential, thereby indicating that toxicity and antitumor effects may be potentially separable (Nowotny *et al.*, 1971).

In studying the mechanism of endotoxin-induced tumor hemorrhage, O'Malley *et al.* (1962) found that the administration of *Serratia marcescens* LPS to normal mice was followed by the appearance of tumor-necrotizing factors (TNF) in the serum. Serum collected shortly after endotoxin administration was highly active in inducing hemorrhage of sarcoma 37 tumors. Later studies found that preinfection of serum donors with BCG greatly enhanced the production of TNF in the postendotoxin serum (Carswell *et al.*, 1975; Butler and Nowotny, 1976). This TNF was not related to lymphocyte-activating factor (Männel *et al.*, 1980) or to interferon (Ruff and Gifford, 1980) production in response to endotoxin.

III. Enhancement of Nonspecific Resistance to Tumor

A. Tumor Models

The second type of antitumor effect of endotoxin is the enhancement of nonspecific resistance to tumor. This effect is characterized by the inhibition of tumor implantation and growth. The most commonly used tumor model to demonstrate resistance to tumor has been the transplantable ascites tumor which grows as a cell suspension in the peritoneal cavity. Treatment of mice with endotoxin has been shown to inhibit the growth of Ehrlich ascites tumor (Mita, 1972; Hoshi *et al.*, 1973; Nigam, 1975), an ascites form of sarcoma 180 (Prigal, 1968), Rauscher leukemia virus-induced ascites tumor (Braun, 1973), lymphoma L5178Y (Parr *et al.*, 1973), and the TA3-Ha mammary adenocarcinoma ascites tumor (Grohsman and Nowotny, 1972; Yang and Nowotny, 1974; Yang-Ko *et al.*, 1974; Butler and Nowotny, 1976, 1979; Butler *et al.*, 1978).

Endotoxin has also been shown to inhibit the implantation or growth rate of solid subcutaneous tumors such as Sa1 (Tripodi *et al.*, 1970), Ehrlich solid carcinoma (Arai *et al.*, 1975; Mizuno *et al.*, 1968), Moloney virus-induced sarcoma (Strausser and Bober, 1972; Bober *et al.*, 1976), sarcoma 180 (Mizuno *et al.*, 1968), and MT296 mammary carcinoma (Henderson *et al.*, 1976). Thus, endotoxin can inhibit the development of a broad spectrum of tumors in animal models.

The observations by Grohsman and Nowotny (1972) that microgram quantities of endotoxin could enhance survival of mice inoculated with the TA3-Ha ascites tumor and the later work by Yang and

Nowotny (1974) stimulated our investigations on the role of endotoxin in the enhancement of nonspecific resistance to tumor.

B. Optimal Conditions

One of the problems plaguing the clinical use of endotoxin in cancer therapy has been the inconsistency and unpredictability of results. Our early work demonstrated that some of this inconsistency may be due to factors such as dosage, timing of administration, and route of administration (Nowotny and Butler, 1980). For example, doses of 1 to 50 μg of endotoxin could significantly enhance survival from lethal tumor challenge, whereas doses less than 0.1 μg or greater than 100 μg had no protective effect. The widely studied Coley toxins were crude mixtures in which endotoxin concentrations were not quantitated or controlled from one batch to the next. Future clinical evaluations must closely evaluate and standardize endotoxin dosage.

The timing of single or repeated doses of endotoxin may play an even greater role in determining whether endotoxin treatments will be effective or not. In general, prophylactic treatment of mice with endotoxin prior to or together with tumor challenge was more effective than treatment after the time of challenge (Parr et al., 1973; Nowotny and Butler, 1980). However, while a single dose of endotoxin given 1 day or later after tumor challenge was relatively ineffective, repeated treatments over a course of 5 to 14 days produced a moderately strong antitumor effect.

The duration of the enhanced resistance to tumor may be relatively long-lasting. A single injection of 25 μg of endotoxin has been shown to provide significant protection from the TA3-Ha ascites tumor for up to 21 (Nowotny et al., 1975) to 39 days (Nowotny and Butler, 1980). However, at shorter time intervals the endotoxin protection curve became biphasic or multiphasic with peaks of strong protection and nodes of no protection from tumor. This oscillating effect of endotoxin on the host was similar to the oscillating adjuvant effect of endotoxin in promoting SRBC antibody responses in vivo (Behling and Nowotny, 1977). In another tumor system, the administration of a large dose of endotoxin intramuscularly in a mineral oil emulsion enhanced resistance to sarcoma 180 subcutaneous implants 92 days after the endotoxin treatment (Prigal, 1968).

A third important criterion influencing the effectiveness of antitumor therapy is the route of administration. For ascites tumors, local intraperitoneal treatments were significantly more effective than systemic intravenous injections (Yang and Nowotny, 1974). Mizuno et al. (1968) reported that local intradermal injections of endotoxin were more effective than either intraperitoneal or intravenous treatments in inhibiting growth of intradermal solid tumors. In contrast, however, intradermal injections of endotoxin on the abdomen of mice challenged with an ascites tumor were more effective than intraperitoneal doses in enhancing survival. These studies emphasize the need for carefully controlled clinical trials involving comparisons of different endotoxin dosages and the site and timing of administration.

C. Relationship between Endotoxin Structure and Function in the Enhancement of Nonspecific Resistance to Tumor

In contrast to the number of studies performed relating endotoxin structure to function in hemorrhagic necrosis, our studies have been the only ones to systematically study the relationship between endotoxin structure and function in the enhancement of nonspecific resistance to tumor (Ng *et al.*, 1976). The results of these studies have been both unexpected and intriguing.

Studies of a series of *Salmonella minnesota* endotoxin structural mutants ranging from smooth LPS (S1114) to a heptoseless Re glycolipid (R595) indicated that the tumor protection potency decreased with decreasing polysaccharide chain length. This included a rather sharp break in activity between the R5 and the R7 mutants. The only known difference between these two molecules is that the R5 mutant has a glucose residue bound to heptose in the core polysaccharide region. A similar break in activity between R5 and R7 was found in the ability to induce the production of myeloid colony-stimulating factor (CSF) (Butler *et al.*, 1978). A *Salmonella typhimurium* 1102 Re glycolipid produced neither enhanced resistance to tumor nor CSF production.

In contrast to this set of structural mutants, a group of *E. coli* endotoxins demonstrated significantly greater tumor protection and CSF production by three Re glycolipids than by three intact LPS. Studies by Mita (1972) similarly found that an *E. coli* phosphomucolipid (lipid A) effectively inhibited the growth of Ehrlich ascites tumor cells *in vivo*. Chemical studies of our *Salmonella* and *E. coli* heptoseless Re glycolipids demonstrated significant structural differences between each of these glycolipid moieties (Ng *et al.*, 1976), which may be related to the broad range of functional capacity.

These studies might have led us to conclude that the tumor resistance-enhancing active site lies exclusively in the lipid A moiety and is merely modified in expression by polysaccharide side chains, which affect solubility or binding. However, at this time we found several nontoxic polysaccharide-rich (PS) fractions prepared by mild acid hydrolysis of smooth LPS to be active in enhancing tumor resistance (Ta-

502 R. Christopher Butler

Table 1. Enhancement of Resistance to Tumor by the PS Fraction of Endotoxin

Treatment[a]	Percent survival (p)
Saline	30
20 μg S. marcescens LPS	100 (0.001)
20 μg S. marcescens PS	80 (0.01)

[a] Mice received i.p. injections of LPS or PS 1 day before challenge with 1600 TA3-Ha ascites tumor cells.

ble 1). This effect was not due to contamination of the PS with lipid A because chemical analysis found no detectable quantities of fatty acids in these water-soluble products (Nowotny et al., 1975). Furthermore, the PS preparations were devoid of endotoxicity as measured by six different sensitive endotoxin assays. In addition to enhancing tumor resistance, these PS preparations could also perform some of the other beneficial effects of endotoxin including the protection from lethal irradiation (Nowotny et al., 1975), antibody response adjuvancy (Butler and Friedman, 1980b; Behling and Nowotny, 1977), and CSF production (Chang et al., 1974). The treatment of PS with periodate destroyed the ability to induce CSF, thus indicating that this effect was due to PS (Nowotny et al., 1975).

Therefore, while the protective effects of the PS fractions prepared from smooth LPS and the results using S. minnesota and S. typhimurium structural mutants indicate that a significant proportion of the tumor protection activity of these particular endotoxins resides in the PS portion of the molecule, the effectiveness of the E. coli glycolipids indicates that in some endotoxin strains the lipid moiety contains a resistance-stimulating component. Thus, there appears to be a real difference between different endotoxins in both the location of the primary site or sites on the molecule and the relative contribution of each of these sites to the potency in enhancing resistance to tumor.

D. Mechanism of Endotoxin-Induced Resistance to Tumor

1. Cell Types Involved

The mechanism of action of endotoxin in enhancing nonspecific resistance to tumor is complex and appears to involve lymphocytes, macrophages, lymphokines, and monokines. Resistance can be transferred by both cells and serum although serum antibody does not play a role.

Yang and Nowotny (1974) first demonstrated that the adoptive transfer of nucleated splenocytes from endotoxin-pretreated mice could protect recipient mice from a lethal tumor challenge. Depletion of macrophages and other adherent cells from the transferred splenocyte suspensions did not significantly reduce the degree of resistance conferred (Nowotny and Butler, 1980). This transfer of resistance was not due to the transfer of residual endotoxin, which was present at a level of less than 0.001 μg/10^7 cells. Thus, the transfer of nonspecific resistance appears to be a function of activated lymphocytes.

While lymphocytes are probably involved in this resistance mechanism, the relative roles of T and B cells have not been definitively determined. We have found that endotoxin is quite effective in the protection of athymic BALB/c nu/nu mice from tumor, presumably in the complete absence of mature T cells (Nowotny and Butler, 1980). However, even though T cells are not absolutely required, the results of endotoxin treatment of mice in which T cells have been depleted by antithymocyte serum indicated that T cells play a role in maintaining the normal base-level responsiveness against tumors (unpublished observations). It is possible that T-cell activities may augment the endotoxin-induced antitumor activity in normal hosts.

Circumstantial evidence indicated an essential role for the B cell based upon a lack of response of C3H/HeJ mice to the protective effects of endotoxin. Coutinho (1976) demonstrated that the lack of responsiveness of C3H/HeJ mice to endotoxin was due to a specific defect in the B cell rather than to any defect in the response capacity of either macrophages or T cells. The observations that nonadherent cells transferred resistance but that T cells were not required also provides indirect evidence supporting B-cell involvement in this effect.

Even though macrophages may not be essential for the transfer of resistance, there is little doubt that they play a major role as effector cells in nonspecific resistance to tumor. In our studies, the systemic depletion of macrophages with carrageenan prior to treatment with endotoxin prevented an increase in resistance over the level of saline-treated controls (Nowotny and Butler, 1980).

Many other investigators have also demonstrated that endotoxin can stimulate macrophage tumoricidal activity *in vitro* and *in vivo*. Alexander and Evans (1971) demonstrated that endotoxin could stimulate macrophages to inhibit the growth of L5178Y or SL2 lymphoma cells *in vitro*. Removal of macrophages from endotoxin-treated splenocyte suspensions abrogated the cytotoxic activity against Gross virus-induced lymphoma cells *in vitro* (Glaser *et al.*, 1976). Using a human system, peripheral blood monocytes could be activated by endotoxin to kill tumor cells *in vitro* (Cameron and Churchill, 1980). These human macrophages were cytotoxic for the three malignant cell lines tested but had no effect on the three nonmalignant cell lines. Parr *et al.* (1973) demonstrated that mixing peritoneal macrophages with an inoculum of FS6 sarcoma cells inhibited subsequent tumor growth only if the macrophages were mixed with endotoxin.

2. Role of Subcellular Factors in Resistance to Tumor

The relationship between B cells and macrophages may be either unidirectional or reciprocal. The activation of macrophages by endotoxin may involve the production of lymphokines such as macrophage-activating factor (MAF). Wilton *et al.* (1975) found that cultured macrophages could only be activated by endotoxin if B cells were added to the cultures. The addition of T cells had no effect. In addition, cell-free supernatants from endotoxin-stimulated B cells, but not from T cells, could stimulate the macrophages. Therefore, endotoxin could stimulate only B cells to release MAF. Similarly, endotoxin has been shown to induce B cells to produce macrophage migration inhibition factor (MIF) (Yoshida *et al.*, 1973), and monocyte chemotactic factor (CTF) (Wahl, 1974). It may be that a combination of these and other endotoxin-induced B-cell factors is responsible for the *in vivo* activation of macrophages against tumor cells.

We have, in fact, demonstrated that endotoxin-induced soluble factors are capable of mediating the antitumor effects of endotoxin. This role of soluble factors in tumor resistance has recently been discussed in detail (Butler and Nowotny, 1976; Butler *et al.*, 1978). Briefly, we found that postendotoxin serum, containing negligible residual endotoxin, could transfer nonspecific resistance. Post-PS serum also contained factors capable of mediating this effect. The preinfection of serum donors with BCG primed the donor mice to produce more highly active serum in response to LPS or its nontoxic PS component. One lymphokine that is released into the serum in response to both endotoxin and PS is CSF, which can stimulate the proliferation of monocyte and granulocyte precursors *in vivo* (Stanley *et al.*, 1975). There was a consistent relationship between the induction of high CSF levels and the induction of strong resistance to tumor, which suggests a possible role for CSF or CSF-like components in endotoxin-induced tumor resistance (Butler *et al.*, 1978).

In addition, these postendotoxin sera also contained TNF, which was originally described by O'Malley *et al.* (1962) and later by Carswell *et al.* (1975). As with tumor resistance, postendotoxin serum from BCG-infected mice was more potent in producing tumor hemorrhage than the serum from uninfected donors. Perhaps the most significant new finding of this particular study was that the nontoxic PS was as effective as whole endotoxin in stimulating the release of TNF.

From these studies, it appears that postendotoxin serum may contain a wide variety of factors responsible for a variety of effects of endotoxin. We have recently reported that these sera also contain factors capable of enhancing the *in vitro* antibody response to SRBC by mouse splenocytes (Butler *et al.*, 1978; Friedman and Butler, 1979). In this case, we found that the serum antibody response-enhancing factors could be produced *in vitro* by either adherent macrophage cultures or by the transformed P388D$_1$ macrophage cell line in response to either endotoxin or PS. These macrophage-produced helper factors could stimulate antibody production by athymic BALB/c *nu/nu* splenocyte cultures, thus indicating a direct effect upon the B cell. The endotoxin-induced monokines were also effective in restoring the immune response capacity of immunosuppressed splenocytes from Friend leukemia virus (FLV)-infected donors (Butler and Friedman, 1980a,b).

Thus, it is evident that B cells and macrophages interact reciprocally through the production of lymphokines and monokines in response to endotoxin. The elucidation of the precise role of each of these factors in mediating the antitumor effects of endotoxin will depend upon their isolation and purification in quantities sufficient for *in vivo* therapeutic trials.

IV. Antitumor Effects of Endotoxin Combined with Mycobacterial Products

Even though the treatment of mice with endotoxin significantly enhanced their level of resistance to tumor, a substantial proportion of the treated mice were still not protected from tumor. Therefore, we attempted to stimulate the tumor resistance to an even higher level by treating mice with combinations of endotoxin and other bacterial immunostimulatory agents.

Since the preinfection of serum donors with BCG enhanced the potency of postendotoxin serum, this combination of immunostimulants was used to try to directly stimulate resistance to tumor challenge (Table 2.) Preinfection with BCG alone 18 days before tumor challenge significantly increased the level of host resistance. However, the combined treatment with BCG and either endotoxin or PS produced a much stronger protection effect than was achieved using either agent alone.

The mechanism for this additive or synergistic effect could be that the BCG and endotoxin treatments each stimulated different cell types

Table 2. Effect of BCG Infection Followed by Treatment
with LPS or PS on Resistance to TA3-Ha Tumor[a]

Treatment[b]	Percent survival (p)	
None	12	
S. marcescens LPS	65	(0.001)
S. marcescens LPS (BCG preinfected)	91	(0.001)
E. coli O111 PS	35	(0.05)
E. coli O111 PS (BCG preinfected)	90	(0.001)

[a] Mice were challenged with 3000 TA3-Ha cells i.p. on day 0.
[b] Mice were pretreated with 25 μg of LPS or PS on days -4, -3, -2, -1, and 0 with respect to the day of challenge. Some groups were preinfected with approximately 2×10^6 viable BCG 18 days before challenge.

or subgroups of immunologically responsive cells so that when given together there would be a more complete stimulation of the immune system. This hypothesis fits with the observation that BCG is primarily a T-cell or delayed hypersensitivity stimulator (Zbar et al., 1971), whereas most endotoxin effects are mediated by B cells. In addition, however, both BCG and endotoxin (Evans and Alexander, 1972; Pimm and Baldwin, 1974) either directly or indirectly stimulate the activation of macrophages against tumor cells, and it may be this combined effect on macrophages that ultimately mediates the high level of nonspecific resistance.

In an attempt to define the mycobacterial product responsible for the enhanced response to endotoxin, we tested a variety of combinations of BCG components with endotoxin or PS for the enhancement of resistance to tumor. Two water-soluble BCG extracts known to function as adjuvants were tested alone and together with endotoxin. In our system, neither water-soluble adjuvant (WSA) (Chedid et al., 1972) nor muramyl dipeptide (MDP) enhanced resistance to tumor alone and neither enhanced the antitumor activity of endotoxin. However, in a different system (Butler and Friedman, 1980b), MDP with endotoxin or PS functioned synergistically to enhance the antibody response of immunodepressed splenocytes from FLV-infected leukemic mice (Table 3). Therefore, combinations of MDP and endotoxin or the nontoxic PS may potentially yield beneficial effects in an appropriate tumor system.

Two other preparations of mycobacterial components, cord factor and P3, functioned synergistically with endotoxin in enhancing nonspecific resistance to the TA3-Ha ascites tumor (Table 4). P3 is a more purified and less toxic preparation of trehalose dimycolate than cord factor (Ribi et al., 1972). Mineral oil emulsions of P3 and endotoxin

Table 3. Interaction between MDP and LPS or PS on the Antibody Response of Splenocytes from Mice Infected with FLV

Culture treatment[a]	Percent of normal splenocyte response (p)		Percent of FLV-infected splenocyte response (p)	
None	26	(0.0005)	100	
50 μg MDP	32	(0.005)	123	
20 μg LPS	38	(0.0005)	144	(0.025)
20 μg LPS + 50 μg MDP	79	(NS)	303	(0.005)
20 μg PS	54	(0.005)	206	(0.005)
20 μg PS + 50 μg MDP	138	(0.025)	527	(0.0005)

[a] Cultures of 8 × 10^6 normal or FLV-infected splenocytes were treated with LPS, PS, or MDP and immunized *in vitro* with 2 × 10^6 SRBC. The mean numbers of plaque-forming cells per 10^6 cells from eight cultures per group were determined after 5 days of culture.

have also been shown to produce regression of guinea pig line 10 hepatocarcinomas and to reduce metastases (Ribi *et al.*, 1972, 1975, 1976, 1979). In general, our work has supported the work of Ribi and colleagues but has differed substantially in several respects. First, we studied an ascites tumor model, whereas they studied a vascularized subcutaneous tumor. Second, we observed the prevention of tumor take due to enhanced nonspecific resistance, whereas they measured the regression of established tumors. Third, we administered P3 in the form of liposomes in water, whereas they administered P3 only in mineral oil emulsions. In addition, we have made the significant observa-

Table 4. Interaction between P3 (Cord Factor) and LPS or PS in the Enhancement of Nonspecific Resistance to Tumor

Treatment[a]	Percent survival (p)	
Saline	0	
25 μg P3	23	
25 μg *S. minnesota* LPS	28	(0.05)
25 μg *S. minnesota* LPS + 25 μg P3	60	(0.001)
25 μg *E. coli* K12 LPS	30	(0.05)
25 μg *E. coli* K12 LPS + 25 μg P3	70	(0.001)
25 μg *S. marcescens* LPS	25	
25 μg *S. marcescens* LPS + 25 μg P3	65	(0.001)
25 μg *S. marcescens* PS	20	
25 μg *S. marcescens* PS + 25 μg P3	60	(0.001)

[a] Mice received i.p. injections of LPS, PS, or P3 one day before challenge with 3000 TA3-Ha ascites tumor cells.

tion that nontoxic PS preparations could also function synergistically
with P3.

V. Conclusions and Future Prospects

The clinical and experimental studies reviewed here demonstrate
that endotoxin can be a potent antitumor agent for certain types of tumors, especially sarcomas and transplantable ascites tumors. There are
at least two mechanisms for these antitumor effects—tumor hemorrhage and necrosis mediated by TNF and the enhancement of nonspecific tumoricidal responses involving lymphocytes and macrophages.

The principle drawback that has limited the clinical use of endotoxin has been its toxicity. However, our work has demonstrated that
the antitumor activity of the endotoxin molecule can be separated from
the toxic activity either through chemical detoxification (methylation)
or through mild hydrolysis (PS preparation). The PS preparations are
completely nontoxic. Furthermore, we have demonstrated that PS can
perform the following beneficial functions of endotoxin:

1. Enhance nonspecific resistance to a transplantable ascites tumor.
2. Induce the production of subcellular factors including:
   a. Factors mediating nonspecific resistance to tumor.
   b. TNF.
   c. Myeloid CSF.
   d. Antibody response helper factors.
3. Function synergistically with mycobacterial products including
   cord factor and MDP in enhancing tumor resistance and antibody responses.
4. Enhance resistance to lethal irradiation.

Since PS stimulates all of the beneficial effects of endotoxin tested
so far, but does not produce any of the detrimental or toxic effects, this
preparation is the best endotoxin derivative currently available. Without the specter of toxic shock reactions, PS is especially well suited for
clinical trials. Our work also suggests that nontoxic mycobacterial products given together with PS may greatly enhance the effectiveness of
the PS.

ACKNOWLEDGMENT

This work was supported in part by Grant CA27415-02 from the
National Cancer Institute.

VI. References

Alexander, P., and Evans, R., 1971, Endotoxin and double-stranded RNA render macrophages cytotoxic, *Nature New Biol.* **232**:76.

Arai, M., Nakahara, M., Hamano, K., and Okazaki, H., 1975, Isolation and characterization of antitumor lipopolysaccharide from *Proteus mirabilis, Agric. Biol. Chem.* **39**:1813.

Beebe, S. P., and Tracy, M., 1907, The treatment of experimental tumors with bacterial toxins, *J. Am. Med. Assoc.* **49**:1493.

Behling, U. H., and Nowotny, A., 1977, Immune adjuvancy of lipopolysaccharide and a nontoxic hydrolytic product demonstrating oscillating effects with time, *J. Immunol.* **118**:1905.

Bober, L. A., Kranepool, M. J., and Hollander, V. P., 1976, Inhibitory effect of endotoxin on the growth of plasma cell tumor, *Cancer Res.* **36**:927.

Braun, V., 1973, Immunologic and antineoplastic effects of endotoxin: Role of membranes and mediation by cyclic adenosine-3',5'-monophosphate, *J. Infect. Dis.* **128** (Suppl.):118.

Butler, R. C., and Friedman, H., 1980a, Leukemia virus-induced immunosuppression: Reversal by subcellular factors, *Ann. N.Y. Acad. Sci.* **332**:446.

Butler, R. C., and Friedman, H., 1980b, Restoration of leukemia cell immune responses by bacterial products, in: *Current Chemotherapy and Infectious Disease* (J. D. Nelson and C. Grassi, eds.), American Society for Microbiology, Washington, D. C.

Butler, R. C., and Nowotny, A., 1976, Colony stimulating factor (CSF)-containing serum has anti-tumor effects, *IRCS Med. Sci.* **4**:206.

Butler, R. C., and Nowotny, A., 1979, Enhancement of nonspecific resistance to tumor by combinations of immunostimulatory bacterial products, *Cancer Immunol. Immunother.* **6**:255.

Butler, R. C., Nowotny, A., and Friedman, H., 1977, Stimulation of an *in vitro* antibody response by endotoxin-induced soluble factors, *J. Reticuloendothelial Soc.* **22**:28.

Butler, R. C., Abdelnoor, A. M., and Nowotny, A., 1978, Interrelationship between bone marrow colony-stimulating (CSF) and antitumor resistance (TUR) enhancing activities of post-endotoxin sera, *Proc. Nat. Acad. Sci. USA* **75**:2893.

Butler, R. C., Nowotny, A., and Friedman, H., 1980a, Macrophage factors that enhance the antibody response, *Ann. N.Y. Acad. Sci.* **332**:564.

Butler, R. C., Friedman, H., and Nowotny, A., 1980b, Restoration of depressed antibody responses of leukemic splenocytes treated with LPS-induced factors, *Adv. Exp. Med. Biol.* **121A**:315.

Cameron, D. J., and Churchill, W. H., 1980, Cytotoxicity of human macrophages for tumor cells: Enhancement by bacterial lipopolysaccharides (LPS), *J. Immunol.* **124**:408.

Carswell, E. A., Old, L. J., Kassel, R. L., Green, S., Fiore, N., and Williamson, B., 1975, An endotoxin-induced serum factor that causes necrosis of tumors, *Proc. Nat. Acad. Sci. USA* **72**:3666.

Chang, H., Thompson, J. J., and Nowotny, A., 1974, Release of colony stimulating factor (CSF) by non-endotoxic breakdown products of bacterial lipopolysaccharides, *Immunol. Commun.* **3**:402.

Chedid, L., Parent, M., Parent, F., Gustafson, R. H., and Berger, F. M., 1972, Biological study of a nontoxic, water-soluble immunoadjuvant from mycobacterial cell walls, *Proc. Nat. Acad. Sci. USA* **69**:855.

Coley, W. B., 1893, The treatment of malignant tumors by repeated inoculations of erysipelas, with a report of original cases, *Am. J. Med. Sci.* **105**:487.

Copeland, M. M., 1967, Primary malignant tumors of bone: Evaluation of current diagnosis and treatment, *Cancer* **20**:738.

Coutinho, A., 1976, Genetic control of B-cell responses. II. Identification of the spleen B-cell defect in C3H/HeJ mice, *Scand. J. Immunol.* 123:61.

Duran-Reynals, F., 1935, Reaction of spontaneous mouse carcinomas to blood-carried bacterial toxins, *Proc. Soc. Exp. Biol. Med.* 32:1517.

Evans, R., and Alexander, P., 1972, Mechanism of immunologically specific killing of tumor cells by macrophages, *Nature (London)* 236:168.

Felton, L. D., Kauffman, G., and Stahl, H. T., 1935, The precipitation of bacterial polysaccharides with calcium phosphate, *J. Bacteriol.* 29:149.

Fogg, L. C., 1936, Effect of certain bacterial products upon the growth of mouse tumor, *Public Health Rep.* 51:56.

Fowler, G. A., 1969a, Enhancement of Natural Resistance to Malignant Melanoma with Special Reference to the Beneficial Effects of Concurrent Infections and Bacterial Toxin Therapy, Monograph No. 9, New York Cancer Research Institute, New York.

Fowler, G. A., 1969b, Beneficial Effects of Acute Bacterial Infections or Bacterial Toxin Therapy of Cancer of the Colon or Rectum, Monograph No. 10, New York Cancer Research Institute, New York.

Friedman, H., and Butler, R. C., 1979, Immunomodulatory effects of endotoxin-induced factors, in: *Microbiology 1979* (D. Schlessinger, ed.), American Society for Microbiology, Washington, D.C.

Gardner, R. E., Bailey, G. H., and Hyde, R. R., 1939, Hemorrhagic activity of toxic carbohydrate complexes from bacteria on a transplantable rat tumor, *Am. J. Hyg.* 29:1.

Glaser, M., Djeu, J. Y., Kirchner, H., and Herberman, R. B., 1976, Augmentation of cell-mediated cytotoxicity against syngeneic Gross virus-induced lymphoma in rats by phytohemagglutinin and endotoxin, *J. Immunol.* 116:1542.

Gratia, A., and Linz, R., 1931, Le phénomène de Shwartzman dans la sarcome du cobaye, *C. R. Soc. Biol.* 108:427.

Grohsman, J., and Nowotny, A., 1972, The immune recognition of TA3 tumors, its facilitation by endotoxin, and abrogation by ascites fluid, *J. Immunol.* 109:1090.

Hartwell, J. L., Shear, M. J., and Adams, J. R., 1943, Chemical treatment of tumors. VII. Nature of the hemorrhage-producing fraction from *Serratia marcescens (Bacillus prodigiosus)* culture filtrate, *J. Nat. Cancer Inst.* 4:107.

Henderson, J. S., Migliore, R. D., and Berbrayer, D., 1976, Interference by cortisone with endotoxin's adjuvator action on transplantation of a mouse tumour, *J. Cancer Res.* 33: 203.

Hoshi, A., Kanzawa, F., Kuretani, K., Homma, J. Y., and Abe, C., 1972, Anti-tumor activity of protein moiety of the *Pseudomonas aeruginosa* endotoxin, *Gann* 63:503.

Hoshi, A., Kanzawa, F., Kuretani, K., Homma, J. Y., Abe, C., 1973, Anti-tumor activity of constituents of *Pseudomonas aeruginosa, Gann* 64:423.

Männel, D. N., Farrar, J. J., and Mergenhagen, S. E., 1980, Separation of a serum-derived tumoricidal factor from a helper factor for plaque-forming cells, *J. Immunol.* 124: 1106.

Miller, T. R., and Nicholson, J. T., 1971, End results in reticulum cell sarcoma of bone treated by bacterial toxin therapy alone or combined with surgery and/or radiotherapy (47 cases) or with concurrent infection (5 cases), *Cancer* 27:524.

Mita, A., 1972, Immunological evidences for proliferation and migration of lymphocytes induced by administration of potent antitumor phosphomucolipid in tumor-bearing mice, *Rev. Eur. Etud. Clin. Biol.* 17:860.

Mizuno, D., Yoshioka, O., Akamatu, M., and Kataoka, T., 1968, Antitumor effect of intracutaneous injection of bacterial lipopolysaccharide, *Cancer Res.* 28:1531.

Nakahara, M., Kitahara, N., Hamano, K., Arai, M., and Okazaki, H., 1975, Antitumor lipopolysaccharide from heptoseless mutant of *Proteus mirabilis*, *Agric. Biol. Chem.* **39:** 1821.

Nauts, H. C., Fowler, G. A., and Bogatko, F. H., 1953, A review of the influence of bacterial infection and of bacterial products (Coley's toxins) on malignant tumors in man, *Acta Med. Scand.* **145**(Suppl. 276):103.

Ng, A. K., Butler, R. C., Chen, C.-L., and Nowotny, A., 1976, Relationship of structure to function in bacterial endotoxins. IX. Differences in the lipid moiety of endotoxic glycolipids, *J. Bacteriol.* **126:**511.

Nigam, V. N., 1975, Effect of core lipopolysaccharides from *Salmonella minnesota* R mutants on the survival times of mice bearing Ehrlich tumor, *Cancer Res.* **35:**628.

Nowotny, A., and Butler, R. C., 1980, Studies on the endotoxin induced tumor resistance, *Adv. Exp. Med. Biol.* **121B:**455.

Nowotny, A., Golub, S., and Key, B., 1971, Fate and effect of endotoxin derivatives in tumor-bearing mice, *Proc. Soc. Exp. Biol. Med.* **136:**66.

Nowotny, A., Behling, U. H., and Chang, H. L., 1975, Relation of structure to function in bacterial endotoxins. VIII. Biological activities in a polysaccharide-rich fraction, *J. Immunol.* **115:**199.

O'Malley, W. E., Achinstein, B., and Shear, M. J., 1962, Action of bacterial polysaccharide on tumors. II. Damage of sarcoma 37 by serum of mice treated with *Serratia marcescens* polysaccharide, and induced tolerance, *J. Nat. Cancer Inst.* **29:**1169.

Parr, I., Wheeler, E., and Alexander, P., 1973, Similarities of the anti-tumour actions of endotoxin, lipid A and double-stranded RNA, *Br. J. Cancer* **27:**370.

Pimm, M. V., and Baldwin, R. W., 1974, BCG immunotherapy of rat tumours in athymic nude mice, *Nature (London)* **254:**77.

Prigal, S. J., 1968, A new application of mineral oil emulsions: The induction in mice of prolonged resistance to a variety of lethal challenges, *Ann. Allergy* **26:**374.

Ribi, E., Meyer, T. J., Azuma, I., and Abar, B., 1972, Mycobacterial cell wall components in tumor suppression and regression, *Nat. Cancer Inst. Monogr.* **39:**115.

Ribi, E., Granger, D. L., Milner, K. C., and Strain, S. M., 1975, Brief communication: Tumor regression caused by endotoxins and mycobacterial fractions, *J. Nat. Cancer Inst.* **55:**1253.

Ribi, E., Takayama, K., Milner, K., Gray, G., Goren, M., Parker, R., McLaughlin, C., and Kelly, M., 1976, Regression of tumors by an endotoxin combined with trehalose mycolates of differing structure, *Cancer Immunol.* **1:**265.

Ribi, E., Cantrell, J. L., Nowotny, A., Parker, R., Schwartzman, S. C., Von Eschen, K. B., and Wheat, R. W., 1979, Tumor regression caused by endotoxins combined with trehalose dimycolate, *Toxicon* **17:**150.

Ruff, M. R., and Gifford, G. E., 1980, Purification and physico-chemical characterization of rabbit tumor necrosis factor, *J. Immunol.* **125:**1671.

Shapiro, C. J., 1940, The effect of a toxic carbohydrate complex from *S. enteritidis* on transplantable rat tumors in tissue cultures, *Am. J. Hyg.* **31:**114.

Shear, M. J., and Andervont, H. B., 1936, Chemical treatment of tumors. III. Separation of hemorrhage-producing fraction of *E. coli* filtrate, *Proc. Soc. Exp. Biol. Med.* **34:**323.

Shwartzman, G., 1936, Reactivity of malignant neoplasms to bacterial filtrates. II. Relation of mortality to hemorrhagic necrosis and regression elicited by certain bacterial filtrates, *Arch. Pathol.* **21:**509.

Shwartzman, G., and Michailovsky, N., 1932, Phenomenon of local skin reactivity to bacterial filtrates in the treatment of mouse sarcoma 180, *Proc. Soc. Exp. Biol. Med.* **29:** 737.

Stanley, E. R., Hansen, G., Woodcock, J., and Metcalf, D., 1975, Colony stimulating factor and the regulation of granulopoiesis and macrophage production, *Fed. Proc.* **34:** 2272.

Strausser, H. R., and Bober, L. A., 1972, Inhibition of tumor growth and survival of aged mice inoculated with Moloney tumor transplants and treated with endotoxin, *Cancer Res.* **32:**2156.

Tripodi, D., Hollenbeck, L., and Pollack, W., 1970, The effect of endotoxin on the implantation of a mouse sarcoma, *Int. Arch. Allergy* **37:**575.

Wahl, S. M., 1974, Role of macrophages in inducing monocyte (MNL) chemotactic factor (CTF) production by guinea pig T and B lymphocytes, *Fed. Proc.* **33:**744.

Wilton, J. L., Rosenstreich, D. L., and Oppenheim, J. J., 1975, Activation of guinea pig macrophages by bacterial lipopolysaccharide requires bone marrow derived lymphocytes, *J. Immunol.* **114:**388.

Yang, C., and Nowotny, A., 1974, Effect of endotoxin on tumor resistance in mice, *Infect. Immun.* **9:**95.

Yang-Ko, C., Grohsman, J., Rote, N., Jr., Nowotny, A., 1974, Non-specific resistance induced in allogeneic mice to the transplantable TA3-Ha tumor by endotoxin and its derivatives, 8th International Congress on Chemotherapy, Athens, pp. 193–197.

Yoshida, T., Sonozaki, H., and Cohen, S., 1973, The production of migration inhibition factor by B and T cells of the guinea pig, *J. Exp. Med.* **138:**784

Zahl, P. A., Hutner, S. H., Spitz, S., Sugiura, K., and Cooper, F. S., 1942, Action of bacterial toxins on tumors. I. Relationship of tumor-hemorrhagic agent to endotoxin antigens of gram-negative bacteria, *Am. J. Hyg.* **36:**224.

Zbar, B., Bernstein, I. D., and Rapp, H. J., 1972, Suppression of tumor growth at the site of infection with living Bacillus Calmette & Guerin, *J. Natl. Cancer Inst.* **46:**831–839.

26

Induced Enhanced Resistance to Transplantable Tumors in Mice

Paul H. Saluk

I. Historical Perspective

The first systematic studies of bacterial vaccines in the treatment of human cancer were performed by W. B. Coley (1891, 1893, 1898, 1906). Additional reports of cancer treatment with such vaccines appeared although not all were of a positive nature (reviewed by Nauts et al., 1946). The use of Coley's toxins (mixtures of *Streptococcus pyogenes* and *Serratia marcescens*) fell into disfavor because of a lack of consistency, which was due, in part, to a lack of potency and a lack of knowledge regarding dosage, site of administration, and frequency of injections. However, a case study by Nauts et al. (1953) revealed the effectiveness of Coley's toxins when these parameters were taken into consideration. Interest in bacterial products was revived by the experiments of Shwartzman and Michailovsky (1932) and of Duran-Reynals (1935) wherein they showed that bacterial filtrates of gram-negative organisms were effective in producing hemorrhagic necrosis of murine tumors.

Subsequently, Andervont (1936) designed experiments that showed that *E. coli* endotoxin caused the complete regression of primary tumors in mice. The intravenous injections of *E. coli* filtrates resulted in marked hemorrhagic necrosis of dibenzanthracene-induced tumors and, in some instances, a complete regression of the tumors. He noted that mice bearing larger tumors were more susceptible to necrosis than those with smaller tumors. This phenomenon was later shown to be re-

---

Paul H. Saluk ● Department of Microbiology and Immunology, Hahnemann Medical College and Hospital, Philadelphia, Pennsylvania 19102.

lated to the antigenic load of the tumor and to stimulation of the host's immune response (Shear, 1943; Parr *et al.*, 1973; Berendt *et al.*, 1978b,c). Andervont also emphasized the influence of mouse strain in these experiments, a well-known observation today. Shear and Turner (Shear, 1941, 1943; Shear and Turner, 1943) extended these observations and showed that the active agent was a lipopolysaccharide. Shear (1943) produced hemorrhage in primary benzpyrene-induced mouse tumors with 0.1 μg of *S. marcescens* endotoxin. The hemorrhage observed in the primary tumor system was dose dependent in a manner similar to that seen with transplantable tumors (Shear *et al.*, 1943). They also noted that severe hemorrhage was most often seen in larger tumors. Such hemorrhage, produced within a few hours of endotoxin treatment, was not always followed by complete tumor regressions. Few studies on this phenomenon followed. Noteworthy, however, was the observation of Ikawa *et al.* (1952) that small doses of highly purified *E. coli* endotoxin produced a hemorrhagic response followed by regression of the transplantable sarcoma 180.

   This brief historical perspective was selected for the basic observations that stimulated contemporary studies on the enhanced resistance to transplantable tumors afforded by endotoxin administration. The focus of this chapter will be on endotoxin-induced resistance to transplantable tumors in mice with the bulk of the discussions to bear on *in vivo* observations. *In vitro* studies will be presented but will focus on the use of cells obtained from endotoxin-treated mice in an effort to discern mechanisms that may be operative *in vivo*. Those studies that show that small amounts of endotoxin augment the tumoricidal activity of macrophages *in vitro* will not be presented (Hibbs *et al.*, 1977; Russell *et al.*, 1977; Ruco and Meltzer, 1978; Doe and Henson, 1978; Weinberg *et al.*, 1978). Therefore, efforts will be directed towards an evaluation of potential *in vivo* mechanisms.

II. Endotoxin-Induced Resistance *in Vivo*

A. Basic Observations

   Throughout the years, certain basic observations have been made with regard to the induced enhanced resistance of tumors by endotoxin administration in experimental murine systems. With regard to solid tumors, a most constant observation has been the timing of endotoxin administration with respect to tumor growth. Virtually without exception, the intravenous injection of endotoxin was effective when given simultaneously with low doses of tumor cells or during the early growth of primary induced tumors (Andervont, 1936; Shear, 1943; Parr *et al.*,

1973; Berendt *et al.*, 1978b,c). This probably relates to the fact that the host's immune response must be stimulated in order for tumor regression, but not hemorrhagic necrosis, to occur. This is based on the observations that the effects of endotoxin treatment were more potent with highly antigenic tumors (Parr *et al.*, 1973) and that a state of T-cell-dependent concomitant immunity was required (Berendt *et al.*, 1978b,c). In contradistinction to the efficacy of posttumor inoculation of endotoxin for solid tumors, ascites tumors are more affected by injections of endotoxin prior to tumor cell inoculation (Grohsman and Nowotny, 1972; Parr *et al.*, 1973; Yang and Nowotny, 1974; Berendt and Saluk, 1976; Bober *et al.*, 1976; Berendt *et al.*, 1978a; Butler *et al.*, 1978). In one instance (Berendt and Saluk, 1976), intraperitoneal injection of endotoxin as late as 3 days post-tumor cell inoculation was effective in enhancing tumor protection. The difference between the timing of the doses for solid or ascites tumors has been interpreted to represent a distinction in antitumor responses induced by endotoxin in the two systems (Parr *et al.*, 1973) (see Section B).

Generally, low doses of endotoxin (μg amounts) are sufficient to effect tumor protection or regression. However, care must be used in the selection of the correct dose since the amount of tumor burden increases the susceptibility of mice to the lethal effects of endotoxin administration (Shear, 1943; Berendt *et al.*, 1980).

Although Mizuno *et al.* (1968) reported a slight toxic effect of their endotoxin preparation against tumor cells, other investigators have noted a distinct lack of such toxicity (Tripodi *et al.*, 1970; Yang and Nowotny, 1974; Berendt and Saluk, 1976). Thus, endotoxin-induced alterations in the host's system are most likely responsible for antitumor responses.

B. Mechanisms of Endotoxin-Induced Tumor Resistance

1. Solid Tumors

Tripodi *et al.* (1970) were one of the first groups to suggest that endotoxin exerted its tumor resistance by stimulation of host resistance factors. Shortly thereafter, Grohsman and Nowotny (1972) suggested that "the stimulation provided by endotoxin is a sufficient aid to the host to recognize and defeat the lethal allogeneic tumor dose." However, it remained for Parr *et al.* (1973) to directly implicate host lymphocytes with a major role in endotoxin-induced tumor regression. Their work was stimulated by the observations of Alexander and Evans (1973), who showed that macrophages obtained from endotoxin-treated mice could suppress the *in vitro* growth of lymphoma cells. Two major observations from Parr *et al.* (1973) were crucial: (1) antilymphocyte serum treatment of mice abrogated the capacity of endotoxin to regress solid tumors, and (2) the magnitude of endotoxin-induced tumor regression was directly related to tumor antigenicity as determined by standard transplantation tests. Further, they showed that endotoxin-induced tumor hemorrhage but not regression could occur in immunosuppressed mice (e.g., irradiation, ALS, and cortisone). Thus, hemorrhage necrosis seemed necessary but not sufficient to cause tumor regression. This idea was suggested earlier by Tripodi *et al.* (1970), and more recently it has been suggested that necrosis creates intratumor conditions sufficient to allow host effector mechanisms to enter the tumor bed and effect regression (Berendt *et al.*, 1978b,c). In addition, Parr *et al.* (1973) showed that the lipid A component of endotoxin was the active component of endotoxin. Previously, Nowotny *et al.* (1971) demonstrated that tumor hemorrhagic necrosis was independent of the O-antigenic polysaccharide of endotoxin. As mentioned, Parr and others have shown that ascites tumors are relatively refractory to endotoxin treatment when administered post-tumor cell inoculation. Since these tumors are avascular and independent of contributions due to necrosis, the mechanisms by which they are prevented from growing may differ from the regression of solid tumors. Experiments utilizing ascites tumors will be presented below.

Berendt *et al.* (1978b) confirmed the observations of Parr *et al.* (1973) that intravenous administration of endotoxin caused hemorrhagic necrosis of four distinct syngeneic tumors. Only two of these tumors underwent complete regression and these were shown to be antigenic. They further demonstrated that the regression, but not hemorrhagic necrosis, was T cell dependent based on the existence of a longlived immunity to a tumor challenge and possession of T cells capable of passively transferring this immunity to syngeneic, normal recipients. These authors (Berendt *et al.*, 1978c) subsequently demonstrated a relationship between the generation of a state of concomitant immunity and susceptibility to endotoxin-induced tumor regression. They suggest that hemorrhagic necrosis, which occurs following endotoxin administration even in tumors not susceptible to endotoxin-induced regression, creates conditions within the tumor that allow for effector T cells to enter the site and effect tumor regression. However, from a very practical viewpoint, one of the most significant findings was that a state of concomitant immunity was a prerequisite for endotoxin-induced tumor regression in susceptible tumors (i.e., immunogenic tumors). Such a state of concomitant immunity is associated with a highly activated macrophage system (North and Kirstein, 1977). Berendt and co-workers feel that the participation of tumor-necrotizing factor (TNF) (Carswell *et al.*, 1975; Hoffmann *et al.*, 1978) is limited to hemorrhagic necrosis and not to a direct cytotoxic effect on the tumor cells because necrosis occurs in those tumors that are refractory to endotoxin-induced regression. However, host lymphoreticular cells may also be essential for endotoxin-induced hemorrhagic necrosis (Männel *et al.*, 1979). They took advantage of the fact that C3H/HeN mice (LPS-sensitive) and C3H/HeJ mice (LPS-insensitive) are histocompatible. By reconstitution studies of lethally X-irradiated mice, they showed that C3H/HeJ mice, which are incapable of endotoxin-induced tumor necrosis, can be rendered susceptible following reconstitution with bone marrow cells from C3H/HeN mice. Conversely, C3H/HeN mice were rendered insensitive to necrosis following lethal irradiation and reconstitution with bone marrow cells from C3H/HeJ mice.

Therefore, from *in vivo* studies on endotoxin-induced hemorrhagic necrosis and regression of solid tumors, it is obvious that several effector systems are operative. Necrosis does not insure tumor regression but appears to be a necessary component. Regression induced by endotoxin treatment requires immunogenic tumors and is accompanied by the generation of a state of long-live T-cell immunity. However, it is quite possible that additional lymphocytes are required in order to effect the mechanism responsible for tumor necrosis. If it is assumed that endotoxin causes the release of TNF generated during tumor growth by a mechanism similar to the generation of this factor from BCG-treated mice (Green *et al.*, 1976; Carswell *et al.*, 1975; Hoffmann *et al.*, 1978), then this factor may be derived from activated macrophages (Männel *et al.*, 1980).

2. Ascites Tumors

In contrast to the regression of solid tumors, studies that utilize ascites cells may involve different mechanisms of antitumor responses (Parr *et al.*, 1973). In general, for ascites tumors, endotoxin must be administered prior to or shortly after the injection of the tumor cells in order to afford protection. However, since ascites tumors are independent of hemorrhagic necrosis, certain information regarding hemorrhage-independent mechansisms of endotoxin-induced antitumor responses can be discerned. It has been reported that endotoxin can cause a reduction in leukemic lymph nodes and spleens which represents a "real" situation wherein hemorrhage necrosis seemingly did not occur (Kassel *et al.*, 1973).

Grohsman and Nowotny (1972) showed that 1 μg of *Serratia marcescens* endotoxin administered intraperitoneally 1 day prior to the injection of lethal tumor cell doses afforded significant protection against ascites tumor growth. Shortly thereafter, Parr et al. (1973) documented the fact that ascites tumors are prevented from growth by prior injection of endotoxin and attributed this to the development of activated macrophages based on the previous observations of Alexander and Evans (1971) that such macrophages could be harvested from the peritoneum of endotoxin-treated mice. Perhaps the first systematic study of the protective effects of endotoxin against ascites tumors was performed by Yang and Nowotny (1974). They showed that pretreatment of mice with 10 μg of endotoxin 3 to 0 days prior to the inoculation of lethal tumor doses gave optimal protection. Further, direct inoculation of endotoxin into the peritoneum was more effective than intravenous injection. The endotoxin was not directly toxic to the tumor cells. A key observation made during these studies was that spleen cells obtained from endotoxin-injected mice, when transferred into the peritoneum, could passively protect normal mice. These observations stimulated our studies regarding the mechanism(s) by which endotoxin gave tumor protection against ascites tumors (Berendt and Saluk, 1976; Berendt et al., 1978a).

The initial studies showed that 20 μg of S. typhimurium endotoxin administered intraperitoneally 1 day prior to tumor cell inoculation could protect mice from tumor growth. Protection was determined by survival of the mice but direct counts, at various times, on tumor cells in the peritoneum revealed a correlation between the number of tumor cells and survival (unpublished observations). However, the extent of this protection was dependent on the initial tumor burden. As the number of cells in the inoculum increased, the mortality of the mice also increased. This probably represents a "race" between the activation of antitumor effector mechanisms and the growth of tumor cells to an unmanageable level; or, perhaps, to the induction by the tumor of mechanisms that evade the host's response. Of those mice that did succumb to tumor growth after endotoxin treatment, their mean survival times were longer compared to mice not given endotoxin. We also noted no difference in the protection afforded by endotoxin when administered 1 day prior to, simultaneous with, or 3 days post-tumor cell inoculation. Furthermore, the endotoxin was not toxic to the tumor cells since incubation of up to 100 μg of endotoxin did not cause the tumor cells to take up trypan blue nor did it affect the tumorigenicity of the cells as assayed by their capacity to grow and kill mice.

Yang and Nowotny (1974) had demonstrated that spleen cells from mice injected intravenously with endotoxin passively conferred enhanced resistance to ascites tumor cells in normal mice. In order to determine if cells in the peritoneal cavity could passively protect and, thus, imply a role for these cells in the endotoxin-induced ascites tumor protection following peritoneal injections, we transferred peritoneal cells from mice injected 3 days previoulsy with endotoxin. At ratios of 1000 : 1 to 100 : 1 (peritoneal cell : tumor cell), peritoneal cells from endotoxin-treated mice gave enhanced resistance to a lethal ascites tumor cell inoculum. A *Limulus* lysate assay showed that less than 0.01 μg of endotoxin was present in the transferred cells. These experiments confirmed the finding of Yang and Nowotny (1974) and extended them to show that peritoneal cells from endotoxin-treated mice could passively transfer protection to syngeneic, normal mice. Since previous investigators had shown that endotoxin could activate macrophages *in vivo* (Alexander and Evans, 1971; Grant and Alexander, 1974), we performed *in vivo* experiments to test this hypothesis in the above system. When macrophages were removed (by adherence to plastic) from peritoneal cells obtained from endotoxin-treated mice (intraperitoneal injections), the remaining lymphocytes were incapable of conferring passive protection to syngeneic, untreated mice. This strongly implicated the macrophage as having a major role in the endotoxin-enhanced tumor resistance to ascites tumors by totally *in vivo* means (Berendt *et al.*, 1978a). Attempts to recover these macrophages by scraping resulted in a high loss of viability; normal macrophages (i.e., washout cells) were recoverable with a higher degree of viability. However, partial protection was passively transferred when the macrophages from the peritoneal cells of endotoxin-treated mice were removed from plastic by lidocaine treatment and transferred to the peritoneum of untreated mice (Berendt and Saluk, unpublished observations). The incomplete protection given by these macrophages was probably due to damage or alterations of the macrophages by lidocaine. This was shown by their inability to readhere to plastic as avidly as such macrophages taken from the peritoneum.

If the mice were X-irradiated with lethal (660 R) or with sublethal (330 R) doses 2 days prior to tumor inoculation and then injected with normally protective doses of endotoxin, no protection was seen and the mice succumbed to tumor (summarized in Table 1). If, however, immediately following irradiation the mice were reconstituted with macrophage-depleted lymphocytes from the peritoneum and spleen of normal, syngeneic mice, the ability of endotoxin to confer protection was restored to a substantial degree. Thus, endotoxin cannot initiate the events that lead to enhanced tumor protection without the presence of a radiosensitive lymphoid cell population. Since the cells taken from endotoxin-treated mice cannot, in the absence of macrophages, passively transfer protection to normal recipients, they must function in an accessory role in the *in vivo* endotoxin-initiated anti-tumor cell re-

Table 1. Summary of X-Irradiation and Reconstitution Studies[a]

X-Irradiation[b]	Reconstitution[c]	Endotoxin[d]	Antitumor response[e]
−	−	−	No
−	−	+	Yes
+	−	−	No
+	−	+	No
+	+	−	No
+	+	+	Yes

[a] Summary of data from Berendt and Saluk (1976) and Berendt *et al.* (1978a).
[b] Mice were exposed to 330-R sublethal whole-body irradiation 2 days prior to tumor cell inoculation.
[c] Cells used for reconstitution were obtained from untreated syngeneic mice and transferred immediately following irradiation. The recipients were given cells from which macrophages were removed by adherence.
[d] Endotoxin was given intraperitoneally 1 day prior to tumor cell inoculation and 1 day after irradiation and reconstitution.
[e] The antitumor response was evaluated by the survival of mice that received a lethal inoculum of ascites tumor cells.

sponse. Thus, the implied effector function is manifested by macrophages but initiated by cells of the lymphoid population. It is of interest that another function attributable to activated macrophages is also sensitive to sublethal doses of X-irradiation. Using the capacity of macrophages to ingest C3b-coated erythrocytes (Bianco *et al.*, 1975), we have shown that 200 R significantly reduces the capacity of endotoxin to induce peritoneal macrophages to ingest via the C3b receptor (Saluk, Swartz, and Wrigley, unpublished observation) (Table 1).

Although it is assumed that ascites tumors are inhibited *in vivo* by endotoxin injection via a mechanism that differs from the hemorrhagic necrosis and regression of solid tumors, there are certain features in common between the two phenomena. Usually, microgram amounts of endotoxin are sufficient to induce both phenomena; and, more importantly, the regression of solid tumors and the prevention of ascites tumor growth both require lymphocytes. Since hemorrhagic necrosis allows effector cells to be delivered to a solid tumor site and because hemorrhagic necrosis is distinct from the mechanism of tumor regression, it is possible that ascites tumors may be killed via the same effector pathway as solid tumors. The difference in the timing of endotoxin administration required to induce antitumor responses in the ascites vs. solid tumor systems may simply relate to the requirement for hemorrhagic necrosis to occur in the solid tumor system and for the solid tumor to sufficiently stimulate the host's system. Since a state of concomitant immunity is necessary for endotoxin to cause tumor regression (Berendt *et al.*, 1978c), it would be of interest to determine if such a state existed during ascites tumor growth. Further, since concomitant immunity is associated with macrophage activation (Blamey *et al.*, 1969; Nelson and Kearney, 1976; North and Kirstein, 1977), it would also be of interest to determine if such activation occurred early during ascites tumor growth. It is known that activated macrophages can appear in the peritoneum during concomitant immunity to solid tumors (Nelson and Kearny, 1976). Further, macrophages taken from ascites tumors have been shown to be activated by the criterion of enhanced phagocytosis and ingestion via the C3b receptor (Roubin *et al.*, 1981). Thus, a demonstration of either concomitant immunity and/or a highly activated macrophage system early in ascites tumor growth would serve to correlate the mechanism of solid tumor regression and ascites tumor growth prevention mediated by endotoxin administration. However, this remains to be done and is speculative.

Ascites tumors may be susceptible to factors elaborated by the cells discussed above in addition to direct cellular interaction. We have previously shown that an unidentified factor obtained from the peritoneum 3 days postendotoxin injection could, in the absence of cells, passively protect mice from lethal tumor cell growth (Berendt and Saluk, 1976). It was also demonstrated that the fluids containing this factor had less than 0.01 μg of endotoxin. The data in Table 2 show that the cell-free peritoneal fluids obtained as above were not effective unless administered simultaneously with the tumor cells (Berendt and Saluk, unpublished observations). This implies that the factor diffuses rapidly from the peritoneum, is inactivated, or is labile. Table 3 shows that this factor is extremely labile. It loses its activity within 3 hr at 4°C, room temperature, or 37°C. Efforts to stabilize this factor have been unsuccessful. However, the factor may be stabilized by cell membranes or by passage into cells. We were able to mix the cell-free peritoneal fluid with the tumor cells for 45 min on ice; injection of this aliquot

Table 2. The Effect of an Induced Cell-Free Peritoneal Fluid Factor in Suppressing Ascites Tumor Growth

Treatment[a]	No. of mice surviving/total mice	Mean survival time (days)[b]
Saline	0/5	15.4
S-CFPF	0/5	13.2
E-CFPF given right after tumor cells	4/5	(19.0)
E-CFPF given 1 day prior to tumor cells	0/5	14.8

[a] Mice were given cell-free peritoneal fluid (CFPF) from saline-injected mice (S-CFPF) or from endotoxin-injected mice (E-CFPF) at the indicated times.
[b] Value in parentheses is the MST of the mouse that did not survive the treatment.

Table 3. The Lability of an Endotoxin-Induced Factor That Enhances Tumor
Protection of an Ascites Tumor

Treatment[a]	No. of mice surviving/total mice	Mean survival time (days)[b]
E-CFPF incubated at 37°C	1/5	(16.5)
E-CFPF incubated at room temperature	0/5	15.4
E-CFPF incubated at 4°C	1/5	(17.0)

[a] E-CFPF (defined in Table 2) incubated for 3 hr at the indicated temperatures.
[b] Values in parentheses are the MSTs of the mice that did not survive the treatment.

into normal recipients does not protect the mice from tumor cell growth, whereas an aliquot maintained on ice without cells was capable of passively conferring protection. However, the tumor cells that were used to adsorb the factor, when injected in lethal doses, did not grow. Thus, one of the mechanisms operative in endotoxin-induced antitumor activity against ascites tumors may proceed via a factor that is labile in the fluid phase. This lability clearly inhibits characterization efforts and thwarts a complete understanding of the mechanism by which endotoxin confers tumor protection. Such a factor, if operative in the regression of solid tumors, would be extremely difficult to detect experimentally.

Additional factors generated by endotoxin *in vivo* and effective against ascites tumors have been reported. Butler *et al.* (1978) recently described a factor in serum 2 hr after endotoxin administration. This factor, when injected intraperitoneally prior to ascites tumor cells, prevented the growth of the tumor cells. The postendotoxin sera was not directly toxic for the tumor cells. Further, serum obtained following the injection of a polysaccharide-rich preparation of endotoxin obtained by acid hydrolysis also possessed this activity. Since the ascites tumors cannot undergo "hemorrhagic necrosis," it is difficult to determine the relationship between this factor and TNF, especially since the former is not toxic to tumor cells. Most likely, the postendotoxin serum effect on ascites tumors represents substances induced by endotoxin simultaneously with TNF. Only purification and characterization of these substances will answer these questions.

In recent attempts to obtain additional information about the effects of endotoxin on peritoneal cells during the generation of antitumor responses, select experiments were performed using peritoneal exudate cells from endotoxin-injected mice (Saluk *et al.*, 1981). At 3, 5, and 7 days after intraperitoneal endotoxin injection (1–50 μg endotoxin), the macrophages were harvested and assayed for their capacity to inhibit the growth of a murine mastocytoma cell line. Under the days and doses mentioned, these macrophages effectively suppressed the *in vitro* growth of tumor cells. This confirms the observation of Alexander and Evans (1971) and lends credence to the observations obtained totally *in vivo* that macrophages can be activated *in vivo* by endotoxin to mediate antitumor activity (Berendt *et al.*, 1978a). That these macrophages were activated was shown by their enhanced capacity to spread on glass surfaces and to ingest via the C3b receptor (Saluk *et al.*, 1981). Interestingly, the ability of the macrophages, after 24 hr in culture, to inhibit tumor growth was lost whereas these macrophages still ingested C3b-coated erythrocytes. The transient nature of cytotoxic macrophages activated *in vitro* by LPS is well known (Fidler *et al.*, 1976; Hibbs *et al.*, 1977; Russell *et al.*, 1977; Ruco and Meltzer, 1978).

3. Additional Observations

Subsequent experiments have shown that lymphocytes obtained from the peritoneum or spleen of endotoxin-injected mice can activate normal macrophages by the criteria of enhanced spreading and C3b-receptor-mediated ingestion (Wrigley and Saluk, 1981). These experiments were initiated to pursue the notion that endotoxin-activated lymphocytes were obligatory accessory cells in the *in vivo* activation of macrophages (Berendt *et al.*, 1978a). Therefore, the lymphocytes were obtained from endotoxin-injected mice. Although the tumoricidal properties of these macrophages have not yet been analyzed, the data support the observation that endotoxin activates macrophages through its effects on lymphocytes *in vivo*. Using the C3H/HeJ and C3Heb/FeJ strain pair, Nowakowski *et al.* (1980) have also given evidence that lymphocytes are required for endotoxin-induced C3b-receptor-mediated ingestion *in vivo*. Of interest, both T and B cells obtained from the endotoxin-injected mice were necessary to activate the macrophages; either cell population alone was ineffective and both T and B cells had to be obtained from endotoxin-treated mice (Wrigley and Saluk, 1981). This suggests the possibility that endotoxin may activate macrophages for other functional manifestations through its actions on several populations of lymphocytes.

There is also additional evidence of the lymphocyte requirement for endotoxin-induced macrophage function. The activation of guinea pig macrophages by endotoxin (by the criterion of enhanced [^{14}C]glucosamine uptake) was shown to require bone marrow-derived lymphocytes (Wilton *et al.*, 1975). In addition, the enhancement of glucose utilization by murine macrophages following endotoxin treatment was also lymphocyte dependent (Ryan *et al.*, 1979). Lymphocyte cooperation was required for endotoxin to amplify macrophage procoagulant activity (Levy and Edgington, 1980).

III. Discussion

It seems highly likely that the mechanism(s) whereby endotoxin facilitates macrophage activation and tumor cell destruction involves obligatory steps supplied by endotoxin-stimulated lymphocytes. This does not exclude contributions to tumor cell destruction by additional means such as T cells and liberated factors. In fact, all of these systems may be simultaneously operative, illustrating the complexity of endotoxin-induced enhanced resistance to *in vivo* tumor growth. All of these factors must be ultimately considered and point to the care with which studies in this area are to be designed and interpreted. The fact that different lymphocytes stimulated *in vivo* by endotoxin treatment cooperate to augment macrophage function (Wrigley and Saluk, 1981) and that T cells are essential for the induced regression of solid tumors (Berendt *et al.*, 1978b,c) illustrate the need for studies designed to determine the interactions of lymphocyte subsets. Further, macrophage subsets may also be involved. Using lymphocytes obtained from endotoxin-injected mice and normal peritoneal macrophages separated into subsets by Percoll density gradient centrifugation, we have shown that there exist subsets of macrophages that are preferentially induced to ingest via the C3b receptor by endotoxin-activated lymphocytes (Tabor and Saluk, 1981). This may suggest that the availability of macrophages responsive to activation signals from endotoxin-activated lymphocytes will be a consideration in the effective use of endotoxin in experimental systems. For example, the responsive macrophages may be subverted from the area or from being responsive during certain stages of tumor growth.

Additionally, as suggested by Berendt *et al.* (1978b,c), the specific and nonspecific aspects of the host's response to a growing tumor must be evaluated during endotoxin treatment. Another significant aspect of endotoxin-enhanced tumor regression that has generally been ignored is the use of multiple doses of endotoxin. Using a regimen of repeated injections, Bober *et al.* (1976) showed the efficacy of doses as low as 1 ng in the prevention of ascites tumor growth.

It is obvious that much information has been gained regarding the mechanisms by which endotoxin can cause antitumor activity on the part of the host. The early experiments of Coley serve to illustrate the potential practical application of endotoxin in the treatment of human cancer. However, the fact that such treatment was not predictable may be related, in part, to an incomplete understanding of the events as outlined in this chapter. The use of animal models will serve to elucidate how the many host responses activated by endotoxin injection interrelate and provide insight into the efficacious use of endotoxins.

IV. References

Alexander, P., and Evans, R., 1971, Endotoxin and double stranded RNA render macrophages cytotoxic, *Nature New Biol.* **232:**76.

Andervont, H. B., 1936, The reaction of mice and of various mouse tumors to the injection of bacterial products, *Am. J. Cancer* **27:**77.

Berendt, M. J., and Saluk, P. H., 1976, Tumor inhibition in mice by lipopolysaccharide-induced peritoneal cells and an induced soluble factor, *Infect. Immun.* **14:**965.

Berendt, M. J., Mezrow, G. F., and Saluk, P. H., 1978a, Requirement for a radiosensitive lymphoid cell in the generation of lipopolysaccharide-induced rejection of a murine tumor allograft, *Infect. Immun.* **21:**1033.

Berendt, M. J., North, R. J., and Kirstein, D. P., 1978b, The immunological basis of endotoxin-induced tumor regression: Requirement for T-cell-mediated immunity, *J. Exp. Med.* **148:**1550.

Berendt, M. J., North, R. J., and Kirstein, D. P., 1978c, The immunological basis of endotoxin-induced tumor regression: Requirement for a pre-existing state of concomitant anti-tumor immunity, *J. Exp. Med.* **148:**1560.

Berendt, M. J., Newborg, M. F., and North, R. J., 1980, Increased toxicity of endotoxin for tumor-bearing mice and mice responding to bacterial pathogens: Macrophage activation as a common denominator, *Infect. Immun.* **28:**645.

Bianco, C., Griffin, F. M., and Silverstein, S. C., 1975, Studies of the macrophage complement receptor: Alteration of receptor function upon macrophage activation, *J. Exp. Med.* **141:**1278.

Blamey, R. W., Crosby, D. L., and Baker, J. M., 1969, Reticuloendothelial activity during the growth of rat sarcomas, *Cancer Res.* **29:**335.

Bober, L. A., Kranepool, M. J., and Hollander, V. P., 1976, Inhibitory effect of endotoxin on the growth of plasma cell tumor, *Cancer Res.* **36:**927.

Butler, R. C., Abdelnoor, A., and Nowotny, A., 1978, Bone marrow colony-stimulating factor and tumor resistance-enhancing activity of postendotoxin mouse sera, *Proc. Natl. Acad. Sci. USA* **75:**2893.

Carswell, E. A., Old, L. J., Kassel, R. L., Green, S., Fiore, N., and Williamson, B., 1975, An endotoxin-induced serum factor that causes necrosis of tumors, *Proc. Natl. Acad. Sci. USA* **72:**3666.

Coley, W. B., 1891, Contribution to the knowledge of sarcoma, *Ann. Surg.* **14:**199.

Coley, W. B., 1893, The treatment of malignant tumors by repeated inoculations of erysipelas, with a report of original cases, *Am. J. Med. Sci.* **105:**487.

Coley, W. B., 1898, The treatment of inoperable sarcoma with the mixed toxins of erysipelas and *Bacillus prodigiosus*: Immediate and final results in one hundred and forty cases, *J. Am. Med. Assoc.* **31:**389.

Coley, W. B., 1906, Late results of the treatment of inoperable sarcoma by the mixed toxins of erysipelas and *Bacillus prodigiosus*, *Am. J. Med. Sci.* **131:**376.

Doe, W. F., and Henson, P. M., 1978, Macrophage stimulation by bacterial lipopolysaccharides. I. Cytolytic effect on tumor target cells, *J. Exp. Med.* **148:**544.

Duran-Reynals, F., 1935, Reaction of spontaneous mouse carcinomas to blood-carried bacterial toxins, *Proc. Soc. Exp. Biol. Med.* **32:**1517.

Fidler, I. J., Darnell, J. H., and Budman, M. B., 1976, In vitro activation of mouse macrophages by rat lymphocyte mediators, *J. Immunol.* **117:**666.

Grant, C. K., and Alexander, P., 1974, Nonspecific cytotoxicity of spleen cells and the specific cytotoxic action of thymus-derived lymphocytes in vitro, *Cell. Immunol.* **14:**46.

Green, S., Dobrjansky, A., Carswell, E., Kassel, R., Old, L., Fiore, N., and Schwartz, M., 1976, Partial purification of a serum factor that causes necrosis of tumors, *Proc. Natl. Acad. Sci. USA* **73:**381.

Grohsman, J., and Nowotny, A., 1972, The immune recognition of TA3 tumors, its facilitation by endotoxin, and abrogation by ascites fluid, *J. Immunol.* **109:**1090.

Hibbs, J. B., Jr., Taintor, R. R., Chapman, H. A., Jr., and Weinberg, J. B., 1977, Macrophage tumor killing: Influence of the local environment, *Science* **197:**279.

Hoffman, M., Oettgen, H., Old, L., Mittler, R., and Hämmerling, U., 1978, Induction and immunological properties of tumor necrosis factor, *J. Reticuloendothelial Soc.* **23:**307.

Ikawa, M., Koepfli, J. B., Mudd, S. G., and Niemann, C., 1952, An agent from E. coli causing hemorrhage and regression of an experimental mouse tumor. I. Isolation and properties, *J. Natl. Cancer Inst.* **13:**157.

Kassel, R., Old, L., Carswell, E., Fiore, N., and Hardy, W., 1973, Serum-mediated leukemia cell destruction in AKR mice: Role of complement in the phenomenon, *J. Exp. Med.* **138:**925.

Levy, G. A., and Edgington, T. S., 1980, Lymphocyte cooperation is required for amplification of macrophage procoagulant activity, *J. Exp. Med.* **151:**1232.

Männel, D. N., Rosenstreich, D. L., and Mergenhagen, S. E., 1979, Mechanism of lipopolysaccharide-induced tumor necrosis: Requirement for lipopolysaccharide-sensitive lymphoreticular cells, *Infect. Immun.* **24:**573.

Männel, D. N., Moore, R. N., and Mergenhagen, S. E., 1980, Macrophages as a source of tumoricidal activity (tumor necrosis factor), *Infect. Immun.* **30:**523.

Mizuno, D., Yoshioka, O., Akamatu, M., and Kataoka, T., 1968, Antitumor effect of intracutaneous injection of bacterial lipopolysaccharide, *Cancer Res.* **28:**1531.

Nauts, H. C., Swift, W. E., and Coley, B. L., 1946, The treatment of malignant tumors by bacterial toxins as developed by the late William B. Coley, M. D., reviewed in the light of modern research, *Cancer Res.* **6:**205.

Nauts, H. C., Fowler, G. A., and Bogatko, F. H., 1953, A review of the influence of bacterial infection and of bacterial products (Coley's toxins) on malignant tumors in man, *Acta Med. Scand.* **145**(Suppl. 276):103.

Nelson, D. S., and Kearney, R., 1976, Macrophages and lymphoid tissues in mice with concomitant tumour immunity, *Br. J. Cancer* **34:**221.

North, R. J., and Kirstein, D. P., 1977, T-Cell-mediated concomitant immunity to syngeneic tumors. I. Activated macrophages as the expressors of nonspecific immunity to unrelated tumors and bacterial parasites, *J. Exp. Med.* **145:**275.

Nowakowski, M., Edelson, P. J., and Bianco, C., 1980, Activation of C3H/HeJ macrophages by endotoxin, *J. Immunol.* **125:**2189.

Nowotny, A., Golub, S., and Key, B., 1971, Fate and effect of endotoxin derivatives in tumor-bearing mice, *Proc. Soc. Exp. Biol. Med.* **136:**66.

Parr, I., Wheeler, E., and Alexander, P., 1973, Similarities of the anti-tumour actions of endotoxin, lipid A and double-stranded RNA, *Br. J. Cancer* **27:**370.

Roubin, R., Kennard, J., Foley, D., and Zola-Pazner, S., 1981, Markers of macrophage heterogeneity: Altered frequency of macrophage subpopulations after various pathologic stimuli, *J. Reticuloendothelial Soc.* **29:**423.

Ruco, L. P., and Meltzer, M. S., 1978, Macrophage activation for tumor cytotoxicity: Development of macrophage cytotoxic activity requires completion of a sequence of short-lived intermediary reactions, *J. Immunol.* **121:**2035.

Russell, S. W., Doe, W. F., and McIntosh, A. T., 1977, Functional characterization of a stable, noncytolytic stage of macrophage activation in tumors, *J. Exp. Med.* **146:**1511.

Ryan, J. L., Glode, L. M., and Rosenstreich, D. L., 1979, Lack of responsiveness of C3H/HeJ macrophages to lipopolysaccharide: The cellular basis of LPS-stimulated metabolism, *J. Immunol.* **122:**932.

Saluk, P. H., Edmundowicz, S., Wrigley, D. M., and Tabor, D. R., 1981, Co-expression of enhanced spreading, ingestion, and inhibition of tumor cell proliferation by macrophages from lipopolysaccharide-injected mice, *Cancer Immunol. Immunother.* **11:**159.

Shear, M. J., 1941, Effects of a concentrate from *B. prodigiosus* filtrate on subcutaneous primary induced mouse tumors, *Cancer Res.* **1:**731.

Shear, M. J., 1943, Chemical treatment of tumors. IX. Reactions of mice with primary subcutaneous tumors to injection of a hemorrhagic-producing bacterial polysaccharide, *J. Natl. Cancer Inst.* **4:**461.

Shear, M. J., and Turner, F. C., 1943, Chemical treatment of tumors. V. Isolation of the hemorrhage-producing fraction from *Serratia marcescens* (*Bacillus prodigiosus*) culture filtrate, *J. Natl. Cancer Inst.* **4:**81.

Shear, M. J., Perrault, A., and Adams, J. R., 1943, Chemical treatment of tumors. VI. Method employed in determining the potency of hemorrhage-producing bacterial preparations, *J. Natl. Cancer Inst.* **4:**105.

Shwartzman, G., and Michailovsky, N., 1932, Phenomenon of local skin reactivity to bacterial filtrates in the treatment of mouse sarcoma 180, *Proc. Soc. Exp. Biol. Med.* **29:** 737.

Tabor, D. R., and Saluk, P. H., 1981, The functional heterogeneity of murine resident macrophages to a chemotactic stimulus and induction of C3b-receptor-mediated ingestion, *Immunol. Lett.* **3:**371.

Tripodi, D., Hollenbeck, L., and Pollack, W., 1970, The effect of endotoxin on the implantation of a mouse sarcoma, *Int. Arch. Allergy Appl. Immunol.* **37:**575.

Weinberg, J. B., Chapman, H. A., Jr., and Hibbs, J. B., Jr., 1978, Characterization of the effects of endotoxin on macrophage tumor cell killing, *J. Immunol.* **121:**72.

Wilton, J. M., Rosenstreich, D. L., and Oppenheim, J. J., 1975, Activation of guinea pig macrophages by bacterial lipopolysaccharide requires bone-marrow-derived lymphocytes, *J. Immunol.* **114:**388.

Wrigley, D. M., and Saluk, P. H., 1981, Induction of C3b-mediated phagocytosis in macrophages by distinct populations of lipopolysaccharide-stimulated lymphocytes, *Infect. Immun.* **34:**780.

Yang, C., and Nowotny, A., 1974, Effect of endotoxin on tumor resistance in mice, *Infect. Immun.* **9:**95.

27

Enhancement of Antitumor Resistance by Mycobacterial Products and Endotoxin

Edgar Ribi, John L. Cantrell, Kuni Takayama, and Ken-ichi Amano

I. Background History

It has been more than 100 years since the suggestion was made that the control of cancer was mediated by immunologic methods. This hypothesis was based on the observation that tumors either partially or totally regressed in a few patients following an acute bacterial infection. In 1911, William Coley pioneered the study of mixed bacterial vaccines or their product known as "Coley's toxin," for treating cancers, and there is no doubt that these vaccines had some effect in many cases (Nauts *et al.*, 1946). The effective ingredient appeared to be endotoxin, which caused hemorrhagic necrosis of the tumors. Thus, Gratia and Linz discovered in guinea pigs (1931) and Shwartzman and Michailovsky in mice (1932) that when animals with solid tumors are given single i.v. or i.p. inoculations with small doses of endotoxin, their tumors became hemorrhagic within 24 hr. This was originally done by analogy with the local Shwartzman reaction, with the idea that some hypothetical tumor virus might have prepared the site. It appeared that this type of tumor damage was mediated indirectly because little of the injected endotoxin would be likely to make contact with tumor cells,

Edgar Ribi and John L. Cantrell ● Ribi ImmunoChem Research, Inc., Hamilton, Montana 59840. **Kuni Takayama** ● William S. Middleton Memorial Veterans Hospital, Madison, Wisconsin 53705. **Ken-ichi Amano** ● Department of Bacteriology, Hirosaki University School of Medicine, Hirosaki 036, Japan.

and would not exert any direct cytotoxicity in any case (Shapiro, 1940; Brailovsky *et al.*, 1973). Such hemorrhagic necrosis is almost certainly proportional to endotoxic potency, and the Perrault–Shear assay (1949) with sarcoma 37 in mice has proven itself for the evaluation of endotoxicity. Meanwhile, this type of tumor damage was shown to be mediated by an endotoxin-induced serum factor (Carswell *et al.*, 1975; Hoffmann *et al.*, 1976). Endotoxin can also induce macrophages to become tumoricidal (Alexander and Evans, 1971; Hibbs *et al.*, 1977; Chapman and Hibbs, 1977) and induce macrophage chemotaxis (Hausman *et al.*, 1972). Although the antitumor activity of endotoxin has been extensively studied for at least 50 years as a result of the first observations by Coley (Nauts *et al.*, 1946), endotoxin had come to be regarded as of little use for the treatment of tumors (for brief review see Milner *et al.*, 1971), because it rarely led to tumor elimination with concomitant production of systemic tumor immunity but rather to partial regression followed by rapid resumption of growth.

In recent years, much evidence has supported the hypothesis that an aroused immune system, which plays a protective role against an infectious disease, also plays a major role in the host's defense against cancer. The most encouraging experimental and clinical data were reported in the late 1960s and early 1970s from treatment with bacterial preparations other than endotoxins, primarily with an attenuated antituberculosis vaccine consisting of living bacteria of *Mycobacteria tuberculosis* (strain BCG), formalin-killed cells of *Corynebacterium parvum*, alone or admixed with tumor cells. The clinical procedures have been carried out with limited guidance from experimental animal models and hence have been problematical. Improvement of immunological procedures required an experimental animal model that most closely resembled clinical reality. A particularly appropriate model is the line 10 hepatocellular carcinoma, which is syngeneic with its host, the strain 2 guinea pig, and metastases to the draining lymph nodes as early as 4 days after tumor transplantation (Rapp, 1973). With this model, it was demonstrated by Rapp, Zbar, Hanna and co-workers (for review see Zbar *et al.*, 1976) that established tumors and lymph node metastases could be eliminated solely by the immune response following an intratumor injection of viable BCG vaccine. Significantly, during BCG-mediated tumor regression and elimination of regional metastases, there was a development of systemic cell-mediated immunity as demonstrated by a rapid rejection of a second line 10 tumor transplant. This finding indicated progress in a new direction, namely, the development of effective methods to eradicate primary tumors as well as small numbers of disseminated tumor cells that may escape surgery, chemotherapy, or radiation.

II. Mycobacterial Cell Wall Vaccines for Protection against Experimental Tuberculosis and as Cancer Immunotherapeutic Agents

During the time the antitumor efficacy of the viable BCG vaccine was explored, members of the Molecular Biology Section of the Rocky Mountain Laboratory in Hamilton, Montana, were engaged in developing an effective nonliving antituberculosis vaccine consisting of refined cell walls (CW) of mycobacteria attached to minute oil droplets suspended in saline. The CW were as effective against airborne infection with virulent tubercle bacilli in mice or monkeys as the standard viable BCG vaccine. When administered i.v., CW caused a persistent accumulation of activated macrophages that killed the invading tubercle bacille (Ribi et al., 1966b; Barclay et al., 1967; Anacker et al., 1967; Ribi et al., 1971). Subsequently, Zbar et al. (1973) discovered that when CW of BCG was injected into guinea pig tumors, it was as effective as live BCG.

Ribi et al. (1978) fractionated the BCG CW and obtained components necessary for antitumor activity and removed other components that might induce allergic reactions. In this fractionation, the CW was first digested with enzymes to remove most of the protein and then was exhaustively extracted with organic solvents to liberate free lipids. Testing of the fractions in the line 10 tumor model led to the identification of two macromolecules that, when combined, brought about tumor regression and systemic tumor immunity (Azuma et al., 1974). The two required components in the regression of the tumor in the guinea pig model were identified to be trehalose dimycolate (P3) and the cell wall skeleton (CWS). Like the endotoxin, both P3 and CWS possessed adjuvant activity capable of specifically enhancing delayed-type hypersensitivity or cell-mediated immunity to an antigen to which they were admixed (Okuyama et al., 1976; Granger et al., 1976). P3 is nonimmunogenic and does not sensitize against itself (Meyer et al., 1974). CWS, on the other hand, is suggested to function as an adjuvant–antigen combination because it contains esterified mycolic acid, a putative adjuvant, and a polysaccharide–mucopeptide complex, which might act as an antigen (Ribi et al., 1978). In support of this hypothesis, CWS alone, but not P3 alone, afforded regression of tumors. However, when combined, P3 synergistically augmented the efficacy of CWS and produced regression rates equal to those obtained with the original CW (see Table 4). Intratumor injection of combinations of CWS and P3 incorporated into oil droplets has been well tolerated in humans and has produced systemic therapeutic effects in melanoma patients (Richman et al., 1978; Vosika et al., 1979a,b). CWS plus P3 also was used to replace CW in regressing cancer eye in cattle (Kleinschuster *et al.*, 1977; Ward, McLaughlin, Cantrell, and Ribi, unpublished observations). This is important because CW, but not CWS, induces hypersensitivity to tubercular protein (PPD), which is used to test cattle for tuberculosis.

III. Antitumor Properties of Endotoxin Extracts from Re-Mutant Strains of *Salmonella*

During our search for other microbial components, we found that when certain preparations of endotoxin were combined with P3 and oil droplets and injected into established malignant line 10 tumors, a high rate of cures and systemic tumor immunity developed (Ribi *et al.*, 1975, 1976a). This led us to reinvestigate the value of endotoxin as an immunotherapeutic agent. The most powerful endotoxin adjuvants were phenol–water (PW) or chloroform–methanol (CM) extracts from Re (heptoseless)-mutant, gram-negative bacteria. These extracts contain endotoxic glycolipids (Re glycolipids, ReGl) rather than endotoxic

Table 1. Antitumor Effect of Endotoxic Extracts from Wild-Type and Re-Mutant Enterobacteriaceae: No Correlation with Endotoxic Potency

Extract tested	Lethality $CELD_{50}$ $(\mu g)^a$	Tumor regression[b] Cured/total Extract alone (300 μg)		Extract + P3 (300 μg + 150 μg)	
Wild-type *S. minnesota* (LPS)	0.005	0/19	0%	3/19	16%
Re-mutant *S. minnesota* (ReGl-PW)	0.035	1/15	7	13/15	87
Re-mutant *S. minnesota* (ReGl-CM)	0.130	7/50	14	81/94	86
Wild-type *S. typhimurium* (LPS)	0.010	0/10	0	3/18	17
Re-mutant *S. typhimurium* (ReGl-PW)	0.042	21/332	6	403/456	89
Re-mutant *S. typhimurium* (ReGl-CM)	0.028	7/50	14	115/126	91
Wild-type *E. coli* (LPS)	0.003	0/10	0	0/10	0
Re-mutant *E. coli* (ReGl-CM)	0.250	0/10	0	8/10	80
Re-mutant *S. minnesota* (ReGl-PCP)	0.013	—	—	3/10	30

[a] $CELD_{50}$: Dose lethal for 50% of 11-day-old chick embryos when inoculated i.v. in aqueous solution (Milner and Finkelstein, 1966).
[b] Strain 2 guinea pigs bearing 6- or 7-day-old line 10 tumors (8–10 mm in diameter) inoculated intralesionally with doses contained in 0.4-ml volumes. Animals were considered cured when tumors had completely disappeared, metastases were not palpable at the first draining lymph node, and rejected challenge to line 10 tumor transplantation 2 months after treatment (Rapp, 1973).

lipopolysaccharides (LPS), which make up PW extracts from wild-type bacteria (Table 1). Both ReGl and LPS when injected in combination with P3 and oil droplets caused a rapid-developing Shwartzman-like necrotic reaction in the tumors. Following this reaction, the LPS combination led to a mere partial regression of injected tumors, and their growth continued after about 2 weeks. In contrast, injection of the ReGl–P3 combination led to high rates of permanent regression and development of systemic immunity against a challenge with line 10 tumors. Tumor regression with ReGl–P3 occurs more rapidly than with either viable BCG, BCG-CW, or CWS–P3, and older tumors could be treated with great success (Ribi *et al.*, 1975).

Re-mutant bacteria were also extracted with phenol–chloroform–petroleum ether to yield endotoxic glycolipid designated ReGl-PCP. Although the lethality and pyrogenicity of ReGl-PCP were similar to those of ReGl-PW and ReGl-CM, it was relatively ineffective in regressing line 10 tumors (Tables 1 and 2). This indicated that endotoxicity per se was not sufficient to cure tumors and that at least one additional ingredient that is not readily extracted by the above method is needed.

Table 2. Correlation between Amino Acid Content of Endotoxic Extracts from Re-Mutants Strains and Regression of Tumors

Fractions tested	Total amino acids (% of original weight)	CELD$_{50}$ (μg)[a]	FI$_{40}$ (μg)[b]	Tumor regression[c] using 150 μg of each fraction combined with 50 μg P3	
				Cured/total	%
ReGl-PW, *S. typhimurium*	2.47	0.36	0.02	10/10	100
Above after treatment with Triton X-100	0.018	0.110	0.25	1/10	10
ReGl-PCP, *S. minnesota*	0.265	0.013	0.048	3/10	30
ReGl-CM, *S. minnesota*[d]	1.28	0.195	0.415	10/10	100
B4 (= above ReGl-CM refined by chromatography)[d]	0.271	0.167	0.53	1/16	6

[a] See footnote *a* of Table 1 for further information.
[b] Quantity administered to rabbits giving an area under the fever curve (FI) of 40 cm^2 when plotted by a standard method. Determined in aqueous solution (Haskins *et al.*, 1961).
[c] See footnote *b* of Table 1 for further information.
[d] Content (%) of muramic acid, alanine, and glutamine:

	Mur	Ala	Glu
ReGl-CM, *S. minnesota*	0.253	0.136	0.245
B4	not detectable	0.032	0.048

IV. Peptides as Requirement for Immunotherapy of the Guinea Pig Line 10 Tumor with Endotoxin and P3

Treatment of ReGl-CM with Triton X-100, which resulted in the abrogation of tumor regression activity without affecting endotoxicity, was paralleled by a reduction of the amino acid content (Ribi *et al.*, 1976b, 1979a). We have described earlier the removal of peptide-containing fragments of mycobacterial CWS (i.e., precursor or autolysis products of the mycolic acid–arabinogalactan–mucopeptide polymer) from trehalose mycolates by pressure elution chromatography through columns of microparticulate silica gel (Azuma *et al.*, 1974; Ribi *et al.*, 1974). As shown in Table 2, by processing ReGl with the aid of this technique, the resulting fraction, designated B4, was endotoxic but lacked tumor regression potency (Ribi *et al.*, 1979a). We noted muramic acid, alanine, and glutamic acid among the principal nitrogenous components present in the untreated ReGl and that the proportion of each of these components was lowered significantly during the preparation of B4. These are the components that make up the minimal structural entity, N-acetylmuramyl-L-alanyl-D-isoglutamine (MDP), of the peptidoglycan moiety of wax D (Adam *et al.*, 1974) or mycobacterial CW responsible for adjuvant activity (Ellouz *et al.*, 1974; Kotani *et al*, 1975; Merser *et al.*, 1975). Since hydrosoluble peptidoglycan fractions of the CW from *E. coli* possessed adjuvant activity comparable to that of hydrosoluble wax D (Adam *et al.*, 1974), the possibility existed that precursor or autolysis producers of the *Salmonella* CW may have been coextracted with the endotoxic ReGl. Therefore, we tested adjuvant-active MDP and its analogs for their ability to restore line 10 tumor regression potency to the B4–P3 combination (Ribi *et al.*, 1979b). Data in Table 3 show that the addition of the seryl form of MDP (MDP-seryl) did indeed enhance the efficacy of B4–P3. At a dose level of 150 μg of MDP, some animals died of early shock, which will be discussed later. Addition of as little as 0.15 μg of MDP-seryl led to a cure rate of 88%. Even though it appears that MDP-seryl may replace the putative, essential tumor regression ingredient of ReGl, this may not mean that these substances are structurally related.

Data in Table 3 show that the activity could also be restored to refined endotoxic B4 by the addition of a nontoxic lipid side fraction, designated ACP, recovered during the isolation of ReGl and which contained a small amount of peptidic substances (Ribi *et al.*, 1979b), or by the addition of nontoxic Braun's lipoprotein (Braun *et al.*, 1970), also a constituent of the outer membrane and known to contain covalently bound MDP units (Cantrell, Wheat, and Ribi, unpublished findings). Finally, another outer-membrane protein called porin (Nakae,

Table 3. Antitumor Activity of Toxic and Nontoxic Bacterial Compounds Containing Characteristic Cell Wall Peptidoglycan Components (Muramic Acid, Alanine, Serine, and Glutamic Acid)

Fraction tested	CELD$_{50}$ (μg)[a]	Tumor regression[b] using 150 μg of each fraction combined with 50 μg P3	
		Cured/total	%
ReGl-CM, untreated	0.195	10/10	0
B4 (toxic chromatographic fraction of ReGl-CM)	0.167	1/16	6
MDP (synthetic N-acetylmuramyl-L-seryl-D-isoglutamine)	> 10	6/22	27
B4 + MDP without trehalose dimycolate (P3)		0/7	0
B4 + MDP, 150 μg + 0.15 μg		7/8	88
15 μg + 0.15 μg		4/7	57
1.5 μg + 0.15 μg		0/8	0
ACP (nontoxic side fraction of ReGl-CM)	> 5	2/17	12
B4 + ACP		27/38	71
Braun's lipoprotein (contains peptidoglycan units)	8.9	0/12	0
B4 + Braun's lipoprotein		8/22	36
Porin (mixture of protein plus endotoxin free of peptidoglycan units)	0.33	0/6	0
Porin + Braun's lipoprotein		11/14	72

[a] See footnote a of Table 1 for further information.
[b] See footnote b of Table 1 for further information.

1976), which was found contaminated with endotoxic glycolipids but was free of peptidoglycolipids, was inactive alone but, as was to be expected, effectively caused tumor regression when combined with Braun's lipoprotein. In summary, the data listed in Table 3 show that any of the "nontoxic" fractions (ACP, Braun's lipoprotein, or synthetic MDP), when added to the refined endotoxic fraction (B4) restored antitumor activity. They all contained components characteristic of bacterial CW peptidoglycans.

It must be emphasized that highly toxic LPS preparations from wild-type strains, alone or in combination with P3, failed to cure line 10 tumors although initial strong tumor-damaging effects of hemorrhage and necrosis occurred; the tumors eventually continued their fatal course (Ribi $et\ al.$, 1975, 1976b). This is in contrast to the ReGl, which induces similar tumor necrosis initially, followed by complete cure of the dermal tumor and immunity against rechallenge. Wild-type endotoxin, which induces regression of murine tumors with concomitant development of specific systemic tumor immunity, as distinct from hemorrhagic necrosis, is dependent upon the state of T-cell-mediated antitumor immunity that is generated only in response to immunogenic tumors (Berendt *et al.*, 1978). However, the line 10 tumor is weakly immunogenic as classically defined (Rapp, 1973) and thus apparently not immunogenic enough for wild-type endotoxin to evoke the generation of T-cell-mediated immunity. We believe that it is against weakly immunogenic tumors that clinically useful adjuvant immunotherapy must be developed (Alexander, 1977).

V. Synergistic Effect between Endotoxin and Mycobacterial Components

Since ReGl synergistically enhanced the antitumor activity of P3, it was not surprising to find that ReGl likewise enhanced the activity of mycobacterial CWS; both P3 and CWS are mycolic acid esters (Table 4; Ribi *et al.*, 1976b; McLaughlin *et al.*, 1978). Because the tumor regression potency of peptide-free endotoxic B4 could be restored by the addition of MDP-seryl (Table 3), it also was expected that B4, like ReGl, synergistically enhanced the efficacy of CWS containing MDP-units (Table 4). The results support the hypothesis that the presence of peptidic material is a requirement for endotoxin to bring about line 10 tumor regression.

An endotoxin–CWS–P3 combination, which is presently considered to be one of the most effective microbial immunotherapeutic agents, led to rapid destruction of tumors and was successful in treating older tumors (Ribi *et al.*, 1976b). It has an advantage over the endotoxin–MDP–P3 combination inasmuch as this low-molecular-weight MDP, but not the macromolecular CWS, enhanced susceptibility to endotoxic shock (Ribi *et al.*, 1979a). Data in the upper portion of Table 4 show that 150-μg doses of endotoxic ReGl, alone or in combination with P3 or CWS, were not lethal for guinea pigs. Data in the middle portion of Table 4 show that peptide-free endotoxic B4 alone or combined with P3 or CWS also were not lethal. However, when CWS was replaced by an equal quantity of MDP–P3, some of the animals died (18%) within 2 to 5 hr after treatment and all survivors were lethargic for about 24 hr after treatment. Primary tumors of nearly all surviving animals regressed, but only animals that received the MDP–B4 combination to which P3 had been added were also cured (91%) of metastatic disease and resisted contralateral challenge with a lethal dose of line 10 tumor cells. The data show that the use of water-insoluble macromolecular

Table 4. Guinea Pig Line 10 Tumor (Hepatocellular Carcinoma) Regression with BCG Cell Wall Skeleton (CWS) or Adjuvant Peptide (MDP-Seryl) in Combination with Endotoxin (ReGl) or B4 (Purified ReGl) and/or Trehalose Dimycolate (P3)

Materials (150 μg each) attached to oil droplets and injected into solid 6-day-old intradermal tumors[a]	No. of animals tested	Early death (%)	Tumor regression of survivors of early shock (% cured)
ReGl	50	0	14
P3	217	0	0
P3 + ReGl	126	0	91
CWS	25	0	20
CWS + P3	21	0	67
CWS + ReGl	29	0	93
B4	8	0	0
B4 + P3	16	0	11
B4 + CWS	8	0	100
MDP-seryl + B4	8	13[b]	0[c]
MDP-seryl + B4 + P3	18	18[b]	91
MDP-seryl + P3	22	0	27
MDP-seryl	8	0	0
MDP-seryl + B4 dissolved in PBS	9	22[b]	0
MDP-seryl + B4 + P3 dissolved in PBS	9	33[b]	0

[a] See footnote b of Table 1 for further information.
[b] Surviving animals lethargic for at least 24 hr.
[c] Primary tumors of all animals surviving early shock regressed but lymph node metastases proliferated.

CWS in combination with endotoxin may be less toxic than treatments with endotoxin combined with MDP. Data in the lower portion of Table 4 show that intratumor inoculation with MDP and endotoxin dispersed in phosphate-buffered saline (PBS) rather than in an oil-in-water emulsion in the presence or absence of P3 also caused early death. Although the primary tumors of some surviving animals regressed completely, metastatic disease progressed in all animals treated with PBS containing these components. Guinea pigs without tumors were found equally susceptible to the toxic effect of solutions containing MDP and endotoxins.

Results obtained from numerous experiments using graded doses of MDP analogs + P_3 were erratic (Ribi et al., 1981). This finding was also observed within single experiments using varying dosages of the same lot of MDP-seryl and with other analogs of MDP (data not shown). One of several explanations could be that in order for MDP–

P3 to be effective, the participation of a minute quantity of endotoxin is needed, which may be picked up as a contaminant during preparation of the adjuvant emulsion. Studies to test the validity of this and other hypotheses are planned.

VI. Eradication of Residual Tumor Deposits by Systemic Immunity

When a primary line 10 tumor is excised, malignant disease characterized by progressive lethal growth of lymph node metastases develops (Zbar *et al.*, 1971). Injection of BCG-CWS + ReGl–P3 mixtures intradermally between the primary tumor and the draining lymph node and excision of the tumor 1 or 7 days later prevented the development of palpable metastasis in 70% of the animals (Kelly *et al.*, 1978). These results suggested that immunotherapy with nonviable immunostimulants may be a useful adjunct to surgery for eliminating residual tumor deposits.

In a more recent study Zbar *et al.* (1979) demonstrated that BCG-CWS–ReGl–P3 combinations, but not living BCG, were effective when inoculated together with viable line 10 tumor cells contralateral to the surgical site. Injection of sites remote from surgical excisions of the adjuvants alone or tumor cells alone cured no animals. Preliminary data indicate that tumor cells rendered nontumorigenic by X-irradiation when admixed with nonviable microbial components were capable of eradicating microscopic lymph node metastases. Thus, it may be possible to use irradiated autochthonous tumor cells with nonviable adjuvants in human patients with minimal malignant disease. These results provided additional insight into a requirement for postoperative adjuvant therapy to eradicate residual metastatic cancer.

VII. Relating Structure to Biological Activity of Endotoxin

A. Hypothetical Structure of Endotoxin

From a practical point of view, interest has always centered upon the importance of the toxicity and pyrogenicity of endotoxic extracts. These properties are of particular importance since humans are much more susceptible to the toxic effects of endotoxins than are small laboratory animals such as mice and guinea pigs (for review see Milner *et al.*, 1971; Joiner and Wolff, 1981). We and other investigators had previously explored the possibility of using chemical modification techniques to selectively reduce the toxicity and pyrogenicity while retaining adjuvanticity. The aim was to possibly provide chemically defined, nontoxic adjuvants that are capable of enhancing nonspecific resistance to bacterial infectious diseases (Ribi *et al.*, 1959, Notowny, 1963a; Johnson and Nowotny, 1964; Schenck *et al.*, 1969; McIntire *et al.*, 1976) and to synergistically enhance the action of P3, CWS, or MDP (Ribi *et al.*, 1979b). Because these reactions were done with heterogeneous mixtures, the exact structure of any single component of which is unknown, the results were empirical, marginal, and often difficult to reproduce.

The fractionation problem is apparent by examining the general structure of Re-mutant endotoxin as shown in Fig. 1, which is based on the work of several investigators (Lüderitz *et al.*, 1978). Accordingly, the basic unit of endotoxin is the glucosamine disaccharide. The β-hydroxymyristic acid is linked to the amino groups of the sugar. The reducing and nonreducing ends are occupied by the phosphate groups. The *O*-acylated fatty acids (R) of the endotoxin are lauric (C_{12}), myristic (C_{14}), palmitic (C_{16}), and β-hydroxymyristic acids. The nonreducing end is occupied by 2-keto-3-deoxyoctonate (KDO). It is clear that by varying the substituent groups at the reducing end and nonreducing end of the sugar, and by varying the *O*-acyl fatty acids, one can come up with a

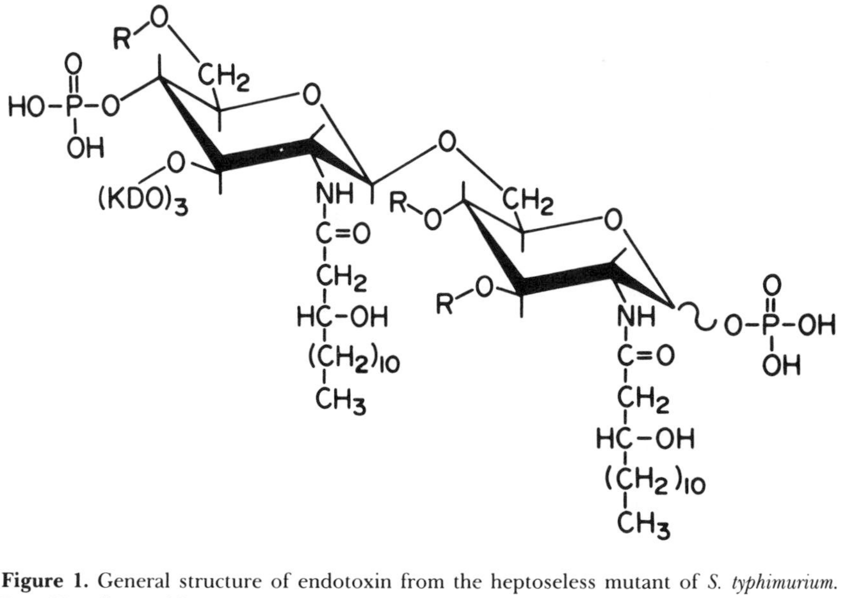

Figure 1. General structure of endotoxin from the heptoseless mutant of *S. typhimurium*. R = H or fatty acid.

large number of possible molecular species. Furthermore, endotoxic glycolipids are known to self-aggregate and exist in polydispersed molecular forms. This makes fractionation a difficult task.

B. Disaggregation of Endotoxin

There are two major factors responsible for the self-aggregating properties of endotoxin: (1) the presence of divalent cations and (2) the strong hydrophobic interaction between molecules. Since most of the structural studies have been done with endotoxic glycolipids liberated from whole organisms with the aid of monophasic solvent consisting of phenol, chloroform, and petroleum ether according to Galanos *et al.* (1969), we chose this method for endotoxin extraction. However, instead of using whole cells, we isolated endotoxin from trypsin-treated, electron microscopically "clean"-appearing CW preparations. In contrast to endotoxin isolated by this procedure from whole cells, endotoxin extracted from CW, when dispersed in water, gave only a slightly opalescent, colloidal solution rather than turbid suspensions. CW extracts also were soluble in chloroform–methanol (86:14) after addition of a minute quantity of water to give a ratio of 86:14:1. Solubility of ReGl in this neutral solvent system facilitated its fractionation by microparticulate gel chromatography, which led to the isolation of B4. Microparticulate silica and unidentified nontoxic material of lower polarity, which were found to contaminate B4, were removed following gel chromatography through a Sephadex LH-20 column. The resulting toxic fraction was designated B5 (Ribi *et al.*, 1982). Clear solutions of B5 were observed in either chloroform or water. To our knowledge, this solubility property has not been reported for any potent endotoxin. We found that phospholipids and divalent cations (Mg^{2+} and Ca^{2+}) were essentially absent in the B5 fraction. Moreover, comparative gel filtration experiments with EDTA-treated ReGl and EDTA-treated B5 using Sephadex LH-60, indicated the absence in B5 of endotoxic particles having a particle weight greater than 10,000. The highly aggregated portion, which was present in ReGl, was eluted from the LH-60 column at the void volume with chloroform–methanol–triethylamine solvent (6:3:1). It amounted to about one-fourth to one-third of the weight of ReGl and was well separated from subsequently eluted endotoxin that was either moderately aggregated or completely disaggregated.

C. Preparation and Properties of Nontoxic Lipid A

It was our working hypothesis that complete solubilization and disaggregation are required before one is able to fractionate amphipathic compounds such as a mixture of endotoxic glycolipids. B5 appears to meet these requirements, and work is in progress to fractionate this multiple component system. Meanwhile, we noted that treatment of B5 with EDTA, which was carried out at pH 4.5 at 70°C, resulted in a reduction of the KDO content. This was not surprising since treatment of endotoxic glycolipids with 0.02 M sodium acetate (pH 4.5) at 100°C for 30 min was reported to cause liberation of its KDO moiety (Rosner et al., 1979). Table 5 shows that the application of two cycles of this treatment to B5 caused the removal of 99% of its KDO moiety without loss of its toxicity. Since the starting material (ReGl) was prepared from polysaccharide-deficient and heptoseless Re-mutant bacteria, the B5 fraction is regarded as a KDO–lipid A complex, and the KDO-free residue (B5-pH 4.5) resulting from treatment at pH 4.5 (residue of dialyzed reaction mixture) meets the definition of lipid A (Lüderitz et al., 1973).

These data show unequivocally that KDO is not essential for toxicity. However, when B5 was treated with 0.1 N HCl for 30 min at 100° C, i.e., according to the classical procedure for preparing lipid A (Westphal and Lüderitz, 1954), the resulting nondialyzable residue (B5-HCl) also was free of KDO. B5-HCl solubilized in 0.5% triethylamine in water was essentially nontoxic ($CELD_{50} > 10 \mu g$) and nonpyrogenic ($FI_{40} = 20 \mu g$) (Table 6). Most important was the finding that the nontoxic B5-HCl, in combination with P3-ACP, retained the degree of tumor regression potency (80% cures) similar to that observed with the toxic B5 (81% cures) (Table 6).

Results of chemical analysis (Table 7) show that the glucosamine and total fatty acid content of the toxic, KDO-depleted, B5-pH 4.5 and the nontoxic B5-HCl were essentially the same, but the B5-HCl was significantly lower in phosphorus content. The molar ratio of glucosamine : phosphorus : fatty acids was 2:2:4 for toxic B5-pH 4.5 and was 2:1:4 for nontoxic B5-HCl. The distribution of normal fatty acids (lauric, myristic, and palmitic acid) and β-hydroxymyristic acid, shown in Table 8, did not indicate a correlation between the content of these components and toxicity.

Table 5. Lethality of KDO-Depleted Endotoxic B5[a]

	Recovery (%)	KDO (μmole/mg)	$CELD_{50}$ (μg)[b]
Untreated B5	—	1.13	0.012
B5-pH 4.5 (B5-lipid A)	41	0.014[c]	0.031

[a] B5 was treated with two cycles of sodium acetate pH 4.5, 100°C, 30 min.
[b] Assay was done in PBS containing 0.5% triethylamine. See Table 1 for further information.
[c] 99% of KDO was removed.

Table 6. Antitumor Effect of Toxic and Nontoxic Glycolipids

Material tested	CELD$_{50}$ (μg)[a]	FI$_{40}$ (μg)[b]	+ P3 Cured/total	%	+ ACP + P3 Cured/total	%
			\<\-\- Line 10 tumor regression[c] \-\-\>			
B5	0.004–0.049 (8)[d]	0.07; 0.029 (2)	1/26 (3)	4	54/67 (8)	81
B5-pH 4.5	0.031	ND	ND		7/8 (1)	88
B5-HCl	5.0; > 10 (3)	5.8; 20 (2)	ND		27/34 (4)	80
ACP	> 10 (1)	> 100 (1)	0/26 (1)	0	—	

[a] See Table 1 for further information.
[b] Determined in PBS containing 0.5% triethylamine. See Table 2 for further information.
[c] Eight or nine animals per lot of sample were tested. See Table 1 for further information.
[d] Number of different lots tested is given in parentheses.

Thin-layer chromatography (TLC) showed that nontoxic B5-HCl contained components of low polarity, which were not present or at least not detected in toxic B5 (Fig. 2). Polar components may have been rendered less polar because of the removal of polar phosphate groups from the acylated glycolipids.

Results of additional experiments support these conclusions (Takayama *et al.*, 1981). We converted ReGl rather than B5 to lipid A by two cycles of treatment with sodium acetate at pH 4.5, 100°C for 30 min, whereby 96% of the KDO content was removed. This preparation was toxic and pyrogenic (CELD$_{50}$ = 0.0546 μg, FI$_{40}$ = 0.64 μg). Lipid A was solubilized in chloroform–methanol 4 : 1 and fractionated on a DEAE-cellulose column (Fig. 3). The first column fraction (III) was eluted with chloroform–methanol 4:1 at the void volume. It consisted of highly aggregated lipid A and was toxic (CELD$_{50}$ = 0.4645 μg). Little material was eluted with more-polar solvents (chloroform–methanol

Table 7. Chemical Composition of ReGl, B5, and Acid-Treated B5

Material tested	Glucosamine	Phosphorus	KDO	Fatty acid
	\<\-\-\-\-\-\-\- nmoles/mg \-\-\-\-\-\-\-\>			
	(moles/2 moles glucosamine)			
ReGl	638 (2)	941 (2.95)	911 (2.86)	1606 (5.03)
B5	932 (2)	1055 (2.26)	1134 (2.43)	1688 (3.62)
B5-pH 4.5	1182 (2)	1122 (1.90)	14.1 (0.02)	2507 (4.24)
B5-HCl	1157 (2)	713 (1.23)	9.8 (0.02)	2340 (4.04)

Table 8. Fatty Acid Composition of ReGl, B5, and Acid-Treated B5

Material tested	Fatty acid (nmoles/mg)			
	C_{12}	C_{14}	C_{16}	OH-C_{14}
ReGl	155.1	40.0	20.1	1290
B5	221.8	159.0	37.7	1269
B5-pH 4.5	116.1	170.4	51.2	2170
B5-HCl	284.6	200.2	47.5	1808

7:3, followed by methanol). When a linear gradient of ammonium acetate (0 to 0.5 M) in methanol–water (99 : 1) was used, an initial sharp peak (fraction IV) appeared. This was followed by a series of poorly resolved peaks (fractions V–VII). Data in Table 9 show that fraction IV differed from the other fractions in that it was relatively nontoxic ($CELD_{50} > 10$ μg) and nonpyrogenic ($FI_{40} = 5.0$ μg). Fraction IV, which was clearly separated from fraction V (see Fig. 3), was about 1000 times less toxic than fractions V and VI. In addition, fraction IV was about 1000 times less pyrogenic than fraction VI. For toxicity and pyrogenicity tests, these fractions were solubilized in 0.5% triethylamine in water. All fractions were active in regressing line 10 tumors when combined with ACP and P3 (78 to 100% cures). These rates were significantly different from controls.

The chemical properties of the lipid A DEAE-column fractions are given in Table 10. The toxic fractions III, V, VI, and VII had total phosphorus values ranging from 1030 to 1150 nmoles/mg. The nontoxic fraction IV contained only 540 nmoles phosphorus/mg. The values for the glucosamine content (500 to 790 nmoles/mg) and ethanolamine content (50 to 90 nmoles/mg, not shown) of the fractions were similar. Normal fatty acid content, which included lauric, myristic, and palmitic acid, ranged from 600 to 1160 nmoles/mg. All fractions, toxic and nontoxic, also had similar hydroxy fatty acid content. These acids, which are typical for endotoxic extracts from Enterobacteriaceae, consisted of both α- and β-hydroxymyristic acids, of which the β-hydroxymyristic acid was the major component (97 to 99 molar %).

A relevant question concerns why fraction IV is nontoxic while all of the other fractions are toxic. The answer may lie in that the toxic fractions still contain the acid-labile sugar-1-phosphate groups, which had not been removed by the moderate hydrolytic condition (pH 4.5 treatment). Alternatively, fraction IV may have constituted a nontoxic portion that was present in the original extract. The TLC patterns (Fig. 4) show that the nontoxic, gradient effluent fraction IV contained low polar components, which were essentially absent in the toxic lipid A

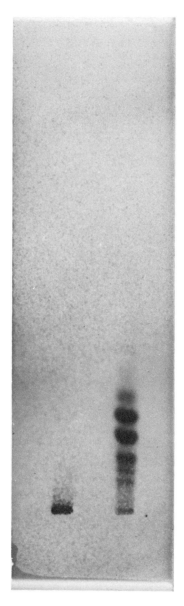

Figure 2. Thin-layer chromatography of B5 (left lane) and nontoxic B5-HC1 prepared from B5 (right lane). Sample size was 25 μg, the silica gel H plate was developed with a solvent system of chloroform–methanol–water–concentrated ammonium hydroxide (50: 25:4:2), and the bands were visualized with dichromate–sulfuric acid reagent followed by charring.

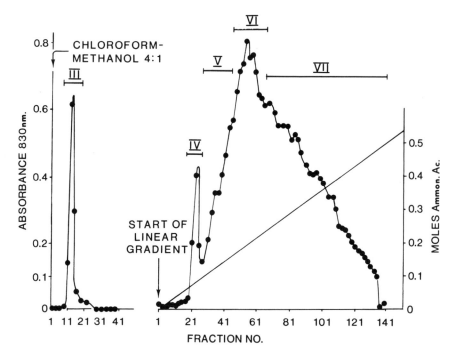

Figure 3. DEAE-cellulose column chromatography of lipid A. Lipid A (315 mg) was dissolved in chloroform–methanol (4 : 1) and applied to a 3.2 × 26-cm DEAE-cellulose column in the acetate form. The order of elution was: 2000 ml of chloroform–methanol (4 : 1), 1000 ml each of chloroform–methanol (7 : 3) and methanol, and 2000 ml of a linear gradient of 0–0.5 M ammonium acetate in methanol–water (99 : 1). Fractions of 12 ml were collected and analyzed for total phosphorus (24).

fractions VI and VII. As speculated previously, polar components may have been rendered less polar by the removal of phosphate groups from the toxic fraction. It is noteworthy that the highly toxic and pyrogenic fractions (VI and VII), by TLC analysis, appear essentially free of phospholipids. The patterns do not reveal bands in the low polar region, which are typical for phospholipids such as phosphatidylethanolamine.

Similar to B5, treatment of one of the KDO-depleted toxic lipid A DEAE-column fractions (VII) with 0.1 N HCl at 100°C for 15 min, resulted in a 200-fold reduction in toxicity with full retention of tumor regression potency (Table 11). The TLC patterns in Fig. 5 indicated a conversion of the mixture of high polar components that make up the toxic fraction VII into the mixture of the lower polar components (higher R_f values) that make up the nontoxic fraction VII.

Table 9. Biological Properties of Fractions Obtained by DEAE-Cellulose
Column Chromatography of Lipid A

Fraction	$CELD_{50}$ $(\mu g)^a$	FI_{40} $(\mu g)^b$	Tumor regression[a] Cured/total	%
III	0.465	—	7/7	100
IV	> 10	5.0	7/9	78
V	0.0417	0.047	8/9	89
VI	0.0398	0.0014	9/9	100
VII	0.0168	0.048	9/9	100

[a] See Table 1 for further information.
[b] See Table 2 for further information.

We have previously shown that MDP dramatically enhanced the lethality of endotoxins when administered i.v. to guinea pigs (Ribi *et al.*, 1979a). Since this was true also for clinically tested immunostimulating agents such as the relatively nontoxic *Pseudomonas* vaccine (Pseudogen), we cautioned the use of combinations of MDP and endotoxin for immunotherapy of cancer. Data in Table 12 show that, at the dose levels tested, Pseudogen, ReGl, and the KDO-free toxic lipid A fractions III, VI, and VII, in combination with MDP caused endotoxic shock in guinea pigs. The addition of as little as 6 μg of MDP to Pseudogen sufficed to cause the death of 4 of 5 animals within 2 to 5 hr. In contrast, adding MDP to the nontoxic lipid A fraction IV or to the B5-HCl did not render these fractions lethal or cause lethargy in the animals.

We have shown that treatment of disaggregated B5 with HCl reproducibly led to complete detoxification ($CELD_{50} > 10$ μg) whereas the similarly treated conventional endotoxic extracts such as ReGl ($CELD_{50} = 0.004$ μg) led to only partial detoxification ($CELD_{50} = 0.25$ μg). It is tempting to speculate that the partial detoxification of this extract was due to steric hindrance; the acid-labile groups being less accessible in the highly aggregated portions of the glycolipid extracts. This may explain the earlier reported results where a mere 10-fold de-

Table 10. Chemical Analysis of the Fractions Obtained by DEAE-Cellulose Column
Chromatography of Lipid A

Fraction	Recovery (mg)	nmoles/mg sample			
		Phosphorous	Glucosamine	Normal fatty acid	Hydroxy fatty acid
III	76.6	1030	720	670	1860
IV	30.4	540	500	810	1440
V	51.9	1030	730	600	1980
VI	88.9	1080	680	1160	1500
VII	142.5	1150	790	870	1660

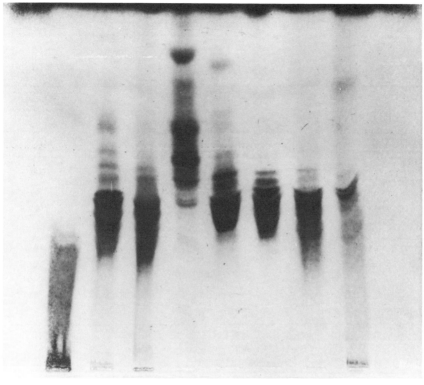

I   II   III   IV   V   VI   VII VIII

Figure 4. Thin-layer chromatography of lipid A fractions from DEAE-cellulose column fractionation. Chromatography was performed as described in Fig. 2 using the solvent system of chloroform–methanol–water–concentrated ammonium hydroxide. Identification from left to right: I, endotoxin (ReGl); II, lipid A; III to VII, column fractions eluted with a linear gradient of 0–0.5 M ammonium acetate in methanol–water (99:1); VIII, a 1.5 M ammonium acetate in methanol–water (99:1) column effluent (24).

crease in toxicity was observed following acid treatment of endotoxic phenol–water extracts (Westphal and Lüderitz, 1954). Acid treatment of aqueous ether extracts from *Salmonella enteritidis*, which were more soluble in water than phenol–water extracts and electron microscopically consisted of minute rodlets rather than of strands of infinite length (Ribi *et al.*, 1966a), produced a fraction that was 1000-fold less toxic (Haskins *et al.*, 1961). It now appears that these results may have been attributed to the presence of less aggregated molecules in this extract. In agreement with this point of view, preparative TLC analysis of reaction mixtures obtained by directly treating ReGl with 0.1 N HCl, resulted in the separation of groups of bands containing lower polar,

Table 11. Detoxification of Lipid A Column Fraction VII

Sample tested	$CELD_{50}$ $(\mu g)^a$	Line 10 tumor regression[a] Sample + ACP + P3 $(150\ \mu g\ +\ 150\ \mu g\ +\ 50\ \mu g)$ Cured/total
ReGl	0.004	5/8 (1/8)[b]
$\downarrow$		
pH 4.5, 100°C, 30 min		
$\downarrow$		
Lipid A fraction VII	0.0168	9/9 (0/9)
$\downarrow$		
0.1 N HCl, 100°C, 15 min		
$\downarrow$		
Lipid A fraction VII-HCl	3.16	7/8
$\downarrow$		
Control: ACP + P3	> 10	0/26

[a] See Table 1 for further information.
[b] Tumor regression in the absence of ACP and P3 is given in parentheses.

VII   VII
HCl

Figure 5. Thin-layer chromatography of toxic column lipid A fraction VII and "nontoxic" fraction VII-HCl prepared from it. Chromatography was performed as described in Fig. 4.

Table 12. Enhancement of Endotoxic Shock by MDP in Guinea Pigs[a]

Material dissolved in PBS and inoculated i.v.	Dose (μg)	Dead/total	CELD$_{50}$ (μg)[b]	FI$_{40}$ (μg)[c]
MDP	150	0/5	> 20	> 100
Pseudogen	150	0/5	4.9	54
MDP + Pseudogen	150 + 150	5/5		
MDP + Pseudogen	30 + 150	5/5		
MDP + Pseudogen	6 + 150	3/5		
ReGl (endotoxin)	150	0/5	0.008	0.155
MDP + ReGl	150 + 150	3/5		
MDP + lipid A fraction III	150 + 150	5/5	0.4645	ND
MDP + lipid A fraction IV	150 + 150	0/5[d]	> 10	5.0
MDP + lipid A fraction VI	150 + 150	3/5	0.0398	0.0014
MDP + lipid A fraction VII	150 + 150	4/5	0.0168	0.048
MDP + B5	150 + 150	3/5	0.167	0.53
MDP + B5-HCl	150 + 150	0/5[d]	> 10	20.0

[a] Groups of five strain 2 guinea pigs (400–500 g) were inoculated i.v. with 0.2 ml PBS solution containing the materials. Lipid A and B5 fractions were dissolved in containing 0.5% triethylamine.
[b] See Table 1 for further information.
[c] See Table 2 for further information.
[d] These animals did not become lethargic. In all other groups where MDP was combined with the lipid A fractions, the surviving animals became lethargic.

nontoxic components from groups of bands consisting of higher polar, toxic components. Components from both groups, in combination with ACP and P3, were capable of causing regression of line 10 tumors (Amano *et al.*, 1982).

The findings discussed here shed light on an old question, namely, whether lipid A per se is toxic or whether the toxicity retained in products of acid hydrolysis of LPS is attributed to residual nondegraded portions of endotoxin (Ribi *et al.*, 1961; Haskins *et al.*, 1961; Nowotny, 1963b; Milner *et al.*, 1971). We now hypothesize that acid treatment produced a mixture of KDO-depleted *O*- and *N*-acylated glucosamine disaccharide (lipid A) molecules, whose toxicity depends on the content of acid-labile phosphate groups, which act as the "toxophor." We may distinguish the "toxic" lipid A diphosphate from the "nontoxic" lipid A monophosphate.

The nontoxic lipid A fractions, B5-HCl, fraction IV, or fraction VII-HCl, are still complex mixtures as are the toxic lipid A fractions from which they were derived. We are presently developing the methodology to fractionate both the toxic and the nontoxic mixtures so that meaningful structural determinations can be made.

VIII. Summary

CW components from mycobacteria and endotoxins from gram-negative bacteria are the most powerful microbial immunopotentiating agents or adjuvants known. As cancer immunotherapeutic agents tested in animal models, they are particularly effective in combinations since mixtures of these adjuvants are more active than components alone. Standard endotoxic extracts from Re-mutant bacteria (ReGl) synergistically enhanced the antitumor activity of mycobacterial CWS or trehalose dimycolate (P3). When ReGl was freed of peptides, phospholipids, and divalent cations, the resulting fraction of disaggregated glycolipids, designated B5, was endotoxic, but lacked tumor regression potency. Precursor or autolysis products of the peptidoglycan moiety of the CWS may have been coextracted with the endotoxic glycolipids. Indeed, tumor regression potency could be restored to the B5–P3 mixture by the addition of synthetic MDP or of a nontoxic, peptide-containing, lipoid side fraction (ACP) recovered during isolation of ReGl.

The toxic compound B5 could be rendered essentially nontoxic ($CELD_{50}$ > 10 μg) and nonpyrogenic (FI_{40} = 20 μg) without loss of tumor regression potency by mild acid treatment. This procedure was previously used to prepare toxic and pyrogenic lipid A from standard endotoxic extracts such as ReGl. It is probably that the incomplete detoxification of ReGl was due to steric hindrance; the acid-labile groups being less accessible in the highly aggregated portions of the glycolipid extract. B5-HCl, as well as other nontoxic lipid A fractions described here (fractions IV and VII-HCl), when combined with P3 and ACP, caused complete regression of line 10 tumors and elimination of regional lymph node metastases in at least 80% of tumor-bearing guinea pigs at a dosage of 150 μg. At this dosage, the toxic fractions led to a comparable cure rate. B5-HCl (8 μg) also synergistically enhanced the tumor regression potency of BCG-CWS (50 μg). All cured animals rejected a rechallenge of a lethal dose of line 10 tumor cells. ReGl or B5, administered (i.v.) at a dose of 150 μg in combination with synthetic MDP, but not with ACP, caused lethal endotoxic shock in 60 to 100% of the guinea pigs. In contrast, when a similar dosage of nontoxic lipid A fractions was combined with MDP, none of the animals died.

These results demonstrate that endotoxic extracts could be selectively detoxified while retaining the ability to synergistically enhance the antitumor properties of mycobacterial adjuvants. The nontoxic products may represent potential candidates for immunotherapy of human cancer. In addition, it will be of interest to determine to what extent nontoxic lipid A components can replace the toxic components in eliciting its numerous other biological activities including those that stimulate the activity of mycobacteria and certain of their components or nonspecifically enhance the resistance against microbial infections.

IX. References

Adam, A., Cirobaru, R., Ellouz, F., Petit, J.-F., and Lederer, E., 1974, Adjuvant activity of monomeric bacterial cell wall peptidoglycans, *Biochem. Biophys. Res. Commun.* **56**:561.

Alexander, P., 1977, Back to the drawing board—The need for more realistic model systems for immunotherapy, *Cancer* **40**:467.

Alexander, P., and Evans, R., 1971, Endotoxin and double-stranded RNA render macrophages cytotoxic, *Nature New Biol.* **232**:76.

Amano, K., Ribi, E., and Cantrell, J. L., 1982, Different structural requirements of endotoxic glycolipid for tumor regression and endotoxic activity, *Biochem. Biophys. Res. Comm.* **106**:667.

Anacker, R. L., Barclay, W. R., Brehmer, W., Larson, C. L., and Ribi, E., 1967, Duration of immunity to tuberculosis in mice vaccinated intravenously with oil-treated cell walls of *Mycobacterium bovis* strain BCG, *J. Immunol.* **98**:1265.

Azuma, I., Ribi, E., Meyer, T. J., and Zbar, B., 1974, Biologically active components from mycobacterial cell walls. I. Isolation and composition of cell wall skeleton and P3, *J. Natl. Cancer Inst.* **52**:95.

Barclay, W. R., Anacker, R. L., Brehmer, W., and Ribi, E., 1967, Effects of oil-treated mycobacterial cell walls on the organs of mice, *J. Bacteriol.* **94**:1736.

Berendt, M. J., North, R. J., and Kirsten, D. P., 1978, The immunological basis of endotoxin-induced tumor regression: Requirement for T-cell-mediated immunity, *J. Exp. Med.* **148**:1550.

Brailovsky, C., Trudel, M., Lallier, R., and Nigam, V. N., 1973, Growth of normal and transformed rat embryo fibroblasts: Effects of glycolipids from *Salmonella minnesota* R mutants, *J. Cell Biol.* **57**:124.

Braun, V., Rehn, K., and Wolff, H., 1970, Supramolecular structure of the rigid layer of the cell wall of *Salmonella, Serratia, Proteus*, and *Pseudomonas fluorescens*: Number of lipoprotein molecules in a membrane layer, *Biochemistry* **9**:5041.

Carswell, E. A., Old, L. J., Kassel, R. L., Green, S., Fiore, N., and Williamson, B., 1975, An endotoxin-induced serum factor that causes necrosis of tumors, *Proc. Natl. Acad. Sci. USA* **72**:3666.

Chapman, H. A., Jr., and Hibbs, J. B., Jr., 1977, Modulation of macrophage tumoricidal capability by components of normal serum: A central role for lipid, *Science* **197**:282.

Ellouz, F., Adam, A., Ciorbaru, R., and Lederer, E., 1974, Minimal structural requirements for adjuvant activity of bacterial peptidoglycan derivatives, *Biochem. Biophys. Res. Commun.* **59**:1317.

Galanos, C., Lüderitz, O., and Westphal, O., 1969, A new method for the extraction of R lipopolysaccharides, *Eur. J., Biochem.* **9**:245.

Granger, D. L., Yamamato, K., and Ribi, E., 1976, Delayed hypersensitivity and granulomatous response after immunization with protein antigens associated with a mycobacterial glycolipid and oil droplets, *J. Immunol.* **116**:482.

Gratia, A., and Linz, R., 1931, Le phénomène de Shwartzman dans la sarcome du cobaye, *C. R. Soc. Biol.* **108**:427.

Haskins, W. T., Landry, M., Milner, K. C., and Ribi, E., 1961, Biological properties of parent endotoxins and lipoid fractions, with a kinetic study of acid-hydrolyzed endotoxin, *J. Exp. Med.* **114:**665.

Hausman, M. S., Snyderman, R., and Mergenhagen, S. E., 1972, Humoral mediators of chemotaxis of mononuclear leukocytes, *J. Infect. Dis.* **125:**595.

Hibbs, J. B., Jr., Taintor, R. R., Chapman, H. A., Jr., and Weinberg, J. B., 1977, Macrophage tumor killing: Influence of the local environment, *Science* **197:**279.

Hoffmann, M. K., Green, S., Old, L. J., and Oettgen, H. F., 1976, Serum containing endotoxin-induced tumour necrosis factor substitutes for helper T cells, *Nature (London)* **263:**416.

Johnson, A. G., and Nowotny, A., 1964, Relationship of structure to function in bacterial O antigens. III. Biological properties of endotoxins, *J. Bacteriol.* **87:**809.

Joiner, K. A., and Wolff, S. M., 1981, The role of endotoxin in human disease and its therapy, in: *Augmenting Agents in Cancer Therapy* (Hersh, E. M., Chirigos, M. A., and M. J. Mastrangelo, eds.), pp. 125–134, Raven Press, New York.

Kelly, M. T., McLaughlin, C. A., and Ribi, E., 1978, Eradication of microscopic lymph node metastases of the guinea pig line 10 tumor after intralesional injection of endotoxin plus mycobacterial components, *Cancer Immunol. Immunother.* **4:**29.

Kleinschuster, S. J., Rapp, H. J., Leuker, D. C., and Kainer, R. A., 1977, Regression of bovine ocular carcinoma by treatment with a mycobacterial vaccine, *J. Natl. Cancer Inst.* **58:**1807.

Kotani, S., Watanabe, Y., Kinoshita, F., Shimono, T., Morisaki, T., Shiba, T., Kusumoto, S., Tarumi, Y., and Ikenaka, K., 1975, Immunoadjuvant activities of synthetic *N*-acetylmuramylpeptides or amino acids, *Biken J.* **18:**105.

Lüderitz, O., Galanos, C., Lehmann, V., Norminen, M., Rietschel, E. T., Rosenfelder, G., Simon, M., and Westphal, O., 1973, Lipid A: Chemical structure and biological activity, *J. Infect. Dis.* **128**(Suppl.):17.

Lüderitz, O., Galanos, C., Lehmann, V., Mayer, H., Rietschel, E. T., and Weckesser, J., 1978, Chemical structure and biological activities of lipid A's from various bacterial families, *Naturwissenschaften* **65:**578.

McIntire, F. C., Hargie, M. P., Schenck, J. R., Finley, R. A., Sievert, H. W., Rietschel, E. T., and Rosenstreich, D. L., 1976, Biologic properties of nontoxic derivatives of lipopolysaccharide from *Escherichia coli* K235, *J. Immunol.* **117:**674.

McLaughlin, C. A., Bickel, W. D., Kyle, J. S., and Ribi, E., 1978, Synergistic tumor-regressive activity observed following treatment of line-10 heptocellular carcinomas with deproteinized BCG cell walls and mutant *Salmonella typhimurium* glycolipid, *Cancer Immunol. Immunother.* **5:**45.

Merser, C., Sinay, P., and Adam, A., 1975, Total synthesis and adjuvant activity of bacterial peptidoglycan derivatives, *Biochem. Biophys. Res. Commun.* **66:**1316.

Meyer, R. J., Ribi, E., Azuma, I., and Zbar, B., 1974, Biologically active components from mycobacterial cell walls. II. Suppression and regression of strain-2 guinea pig hepatoma, *J. Natl. Cancer Inst.* **52:**103.

Milner, K. C., and Finkelstein, R. A., 1966, Bioassay of endotoxin: Correlation between pyrogenicity for rabbits and lethality for chick embryos, *J. Infect. Dis.* **116:**529.

Milner, K. C., Rudbach, J. A., and Ribi, E., 1971, Bacterial endotoxins: General characteristics, in: *Microbial Toxins: A Comprehensive Treatise* (G. Weinbaum, S. Kadis, and S. J. Ajl, eds.), pp. 1–65, Academic Press, New York.

Nakae, T., 1976, Identification of the outer membrane proteins of *E. coli* that produce transmembrane channels in reconstituted vesicle membranes, *Biochem. Biophys. Res. Commun.* **71:**877.

Nauts, H. C., Swift, W. E., and Coley, B. L., 1946, The treatment of malignant tumors by bacterial toxins as developed by the late William B. Coley, M. D., reviewed in the light of modern research, *Cancer Res.* **6:**205.

Nowotny, A., 1963a, Endotoxoid preparations, *Nature (London)* **197:**721.

Nowotny, A., 1963b, Relation of structure to function in bacterial O-antigens. II. Fractionation of lipids present in Boivin-type endotoxin of *Serratia marcescens, J. Bacteriol.* **85:**427.

Okuyama, H., Onoe, K., Takeda, J., and Morikawa, K., 1976, Histological studies on the adjuvanticity of BCG-cell wall, *Recent Adv. RES Res.* **16:**67.

Perrault, A., and Shear, M. J., 1949, The bacterial cell of *S. marcescens* as a source of tumor-necrotising polysaccharide, *Cancer Res.* **9:**626.

Rapp, H. J., 1973, A guinea pig model for tumor immunology: A summary, *Isr. J. Med. Sci.* **9:**366.

Ribi, E., Milner, K. C., and Perrine, T. D., 1959, Endotoxic and antigenic fractions from the cell wall of *Salmonella enteritidis*: Methods for separation and some biologic activities, *J. Immunol.* **82:**75.

Ribi, E., Haskins, W. T., Landy, M., and Milner, K. C., 1961, Preparation and host-reactive properties of endotoxin with low content of nitrogen and lipid, *J. Exp. Med.* **114:**647.

Ribi, E., Anacker, R. L., Brown, R., Haskins, W. T., Malmgren, B., Milner, K. C., and Rudbach, J. A., 1966a, Reaction of endotoxin and surfactants. I. Physical and biological properties of endotoxin treated with sodium deoxycholate, *J. Bacteriol.* **92:**1493.

Ribi, E., Larson, C., Wicht, W., List, R., and Goode, G., 1966b, Effective nonliving vaccine against experimental tuberculosis in mice, *J. Bacteriol.* **91:**975.

Ribi, E., Anacker, R. L., Barclay, W. R., Brehmer, W., Harris, S. C., Leif, W. R., and Simmons, J., 1971, Efficacy of mycobacterial cell walls as a vaccine against airborne tuberculosis in the rhesus monkey, *J. Infect. Dis.* **123:**527.

Ribi, E., Parker, R., and Milner, K. C., 1974, Microparticulate gel chromatography accelerated by centrifugal force and pressure, *Methods Biochem. Anal.* **22:**355.

Ribi, E., Granger, D. L., Milner, K. C., and Strain, S. M., 1975, Brief communication: Tumor regression caused by endotoxins and mycobacterial fractions, *J. Natl. Cancer Inst.* **55:**1253.

Ribi, E., Milner, K. C., Granger, D. L., Kelly, M. T., Yamamoto, K., Brehmer, W., Parker, R., Smith, R. F., and Strain, S. M., 1976a, Immunotherapy with nonviable microbial components, *Ann. N.Y. Acad. Sci.* **277:**228.

Ribi, E., Milner, K. C., Kelly, M. T., Granger, D. L., Yamamoto, K., McLaughlin, C. A., Brehmer, W., Strain, S. M., Smith, R. F., and Parker, R., 1976b, Structural requirements of microbial agents for immunotherapy of the guinea pig line-10 tumor, in: *BCG in Cancer Immunotherapy* (G. Lamoureux, R. Turcotte, and V. Portelance, eds.), pp. 51–61, Grune & Stratton, New York.

Ribi, E., McLaughlin, C. A., Cantrell, J. L., Brehmer, W., Azuma, I., Yamamura, Y., Strain, S. M., Hwang, K. M., and Toubiana, R., 1978, Immunotherapy for tumors with microbial constituents or their synthetic analogues: A review, in: *Immunotherapy of Human Cancer*, pp. 131–154, Raven Press, New York.

Ribi, E., Parker, R., Strain, S. M., Mizuno, Y., Nowotny, A., Von Eschen, K. B., Cantrell, J. L., McLaughlin, C. A., Hwang, K. M., and Goren, M. B., 1979a, Peptides as requirement for immunotherapy of the guinea pig line-10 tumor with endotoxins, *Cancer Immunol. Immunother.* **7:**43.

Ribi, E., Cantrell, J. L., Von Eschen, K., and Schwartzman, S., 1979b, Enhancement of endotoxic shock by adjuvant dipeptide (MDP), *Cancer Res.* **39:**4756.

Ribi, E., Cantrell, J. L., Schwartzman, S. M., and Parker, R., 1981, BCG cell wall skeleton, P3, MDP and other microbial components—Structure activity studies in animals models, in: *Augmenting Agents in Cancer Therapy* (E. M. Hersh, M. A. Chirigos, and M. J. Mastrangelo, eds.), pp. 15–31, Raven Press, New York.

Ribi, E., Amano, K., Cantrell, J., Schwartzman, S., Parker, R., and Takayama, K., 1982,

Preparation and anti-tumor activity of nontoxic lipid A, *Cancer Immunol. Immunother.* **12:**91.

Richman, S. P., Gutterman, J. U., Hersh, E. M., and Ribi, E., 1978, Phase I–II study of intratumor immunotherapy with BCG cell wall skeleton plus P3, *Cancer Immunol. Immunother.* **5:**41.

Rosner, M. R., Tang, J. Y., Barzilay, I., and Khorana, H. G., 1979, Structure of the lipopolysaccharide from *Escherichia coli* heptoseless mutant, *J. Biol. Chem.* **254:**5906.

Schenck, J. R., Hargie, M. P., Brown, M. S., Ebert, D. S., Yoo, A. L., and McIntire, F. C., 1969, The enhancement of antibody formation by *Escherichia coli* lipopolysaccharide and detoxified derivatives, *J. Immunol.* **102:**1411.

Shapiro, C. J., 1940, The effect of a toxic carbohydrate complex from *S. enteritidis* on transplantable rat tumors in tissue culture, *Am. J. Hyg.* **31**(Sect. B):114.

Shwartzman, G., and Michailovsky, N., 1932, Phenomenon of local skin reactivity to bacterial filtrates in the treatment of mouse sarcoma 180, *Proc. Soc. Exp. Biol. Med.* **29:** 737.

Takayama, K., Ribi, E., and Cantrell, J. L., 1981, Isolation of a nontoxic lipid A fraction containing tumor regression activity, *Cancer Res.* **41:**2654.

Vosika, G. J., Schmidtke, J. R., Goldman, A., Ribi, E., Parker, R., and Gray, G. R., 1979a, Intralesional immunotherapy of malignant melanoma with *Mycobacterium smegmatis* cell wall skeleton combined with trehalose dimycolate, *Cancer* **44:**495.

Vosika, G. J., Schmidtke, J. R., Goldman, A., Parker, R., Ribi, E., and Gray, G. R., 1979b, Phase I–II study of intralesional immunotherapy with oil attached *Mycobacterium smegmatis* cell wall skeleton and trehalose dimycolate, *Cancer Immunol. Immunother.* **6:**135.

Westphal, O., and Lüderitz, O., 1954, Chemische Erforschung von Lipopolysacchariden gram negativer Bakterien, *Angew. Chem.* **66:**407.

Zbar, B., Bernstein, I. D., and Rapp, H. J., 1971, Suppression of tumor growth at the site of infection with living Bacillus Calmette-Guérin, *J. Natl. Cancer Inst.* **46:**831.

Zbar, B., Ribi, E., and Rapp, H. J., 1973, An experimental model for immunotherapy of cancer, *Natl. Cancer Inst. Monogr.* **39:**3.

Zbar, B., Ribi, E., Kelly, M., Granger, D., Evans, C., and Rapp, H. J., 1976, Immunologic approaches to the treatment of human cancer based on a guinea pig model, *Cancer Immunol. Immunother.* **1:**127.

Zbar, B., Canti, G., Ashley, M. P., Rapp, H. J., Hunter, J. T., and Ribi, E., 1979, Eradication by immunization with mycobacterial vaccines and tumor cells of microscopic metastases remaining after surgery, *Cancer Res.* **39:**1597.

28

Role of LPS in Recognition and Induced Disease Resistance in Plant Hosts of *Pseudomonas solanacearum*

T. L. Graham

I. Introduction

Nearly all confirmed bacterial plant pathogens are gram-negative organisms, all of which produce some form of cell wall lipopolysaccharide (LPS). Moreover, most successful pathogens or symbionts also produce an extracellular polysaccharide (EPS). With bacterial pathogens (e.g., *Pseudomonas, Xanthomonas, Erwinia*, etc.) the EPS often appear as a loosely associated slime layer, whereas with bacterial symbionts (e.g., *Rhizobium*) the EPS often appear as organized capsules extending some distance from the cell surface. When one considers the most initial events in recognition between the plant cell and a colonizing bacterium, the surface LPS and EPS of the bacterium are obviously tempting molecules to consider as candidates for the initial molecular *triggers* of the cascade of responses between host and microbe that result either in successful invasion (symbiotic or "compatible" pathogenic reactions) or in rejection (nonsymbiotic, nonpathogenic or "incompatible" pathogenic reactions).

It is becoming more and more apparent (Bauer, 1981; Graham, 1981) that either of these overall recognition "phenomena" (success or rejection) is made up of a complex stacking of multiple levels of evolutionarily determined recognition "events," which taken only as a matrix, are specific and *unique* to a particular microbe–plant combination

T. L. Graham ● Monsanto Agricultural Products Co., St. Louis, Missouri 63166.

resulting in the highly specific ramifications one associates with that combination.

Below I review the highly preliminary results emerging from studies on the interaction of *P. solanacearum* and its plant hosts, which suggest a key role of LPS in recognition and the induction of a variety of active host responses. Finally, a "metabolite shunting" model is proposed to possibly help explain the multiple effects of LPS in plant systems.

II. Role of LPS in Attachment of *P. solanacearum* to Host Cells

One of the very first events at the cellular level in the recognition between plant host and microbe is a simple binding or attachment of the microbe within or on the surface of host tissues (or the absence of such a reaction). The potential role of LPS in these simple "binding" recognition events has been reviewed for a number of systems (Bauer, 1981; Carlson, 1981; Dazzo, 1980; Sequeira, 1978). Attachment per se should be viewed as an initial localization event only; whether it alone is a determinant of specificity in any given system is a subject of debate. Nevertheless, attachment is a critical event in that it may contribute to the immobilization of avirulent systemic pathogens (making them more susceptible to active host responses) or alternatively to immobilization of symbionts or virulent tumor-producing bacterial pathogens (allowing them to form the tight cell–cell association needed for further infection).

Early work with *P. solanacearum* actually began by focusing on the identity of the receptor for attachment in the plant cell wall. As had been proposed for *Rhizobium*, plant lectins were considered attractive candidates for this receptor. Indeed, a lectin was isolated from potato tubers and examined for its ability to bind to various *P. solanacearum* isolates (Sequeira and Graham, 1977). Fifty-five virulent isolates and 34 avirulent isolates were tested with the purified lectin from Katahdin potatoes. All avirulent isolates agglutinated with the lectin whereas virulent isolates either failed to agglutinate or agglutinated only at much higher concentrations (3- to 6-fold) of the lectin. Failure of the virulent cells to bind the lectin was correlated to the presence of an extracellular polysaccharide slime (EPS) formed by virulent but not avirulent cells. Agglutination of virulent cells was greatly increased by washing of the slime EPS from the cells, and agglutination of avirulent cells could be prevented by the addition of partially purified slime EPS. Preliminary results suggested that the receptor for the lectin on the bacterial surface might be the LPS. Purified LPS preparations from a single isolate of the virulent bacterium (strain K60) or from the avirulent bacterium spontaneously derived from this virulent strain (strain K60-B1) precipitated with the purified potato lectin; this binding was reversible (as was whole cell binding) by chitin oligomers containing the internal *N*-acetylglucosamine residues reported as the hapten of the potato lectin (Allen and Neuberger, 1973). Binding of the LPS from the avirulent strain was stronger than binding of the LPS from the virulent strain. This suggested that binding sites for the lectin could be in the Rcore–lipid A region since the avirulent K60-B1 isolate appears to be a rough mutant lacking the O antigen present in the parent K60 strain (Graham *et al.*, 1977). This hypothesis was further suggested by the following: (1) deoxycholate-dissociated micelles of either LPS form precipitated much more strongly with the lectin, and (2) isolated lipid A complexed to bovine serum albumin also precipitated in a hapten-reversible manner with the lectin. The hypothesis also seemed reasonable since lipid A contains internal *N*-acetylglucosamine residues.

In summary, then, it was proposed that lectin, LPS, and EPS together might account for the binding or lack thereof of *P. solanacearum* to host cells. Avirulent cells could bind more tightly to the lectin in the host cell wall not only by virtue of a more readily accessible Rcore–lipid A but also due to the lack of EPS. These experiments, however, were obviously highly preliminary and suggested a large number of further studies to test the overall hypothesis.

First of all, although the Rcore is accessible on the bacterial surface, lipid A itself is not generally considered antigenic and thus may not be exposed even in rough bacterial mutants. Second, agglutination and precipitation reactions are at best only semiquantitative and are subject to many complex and nonspecific parameters; a more sensitive and specific quantitative binding assay was needed. Third, these initial studies were performed with potato *tuber* lectin; is this lectin or a similar lectin actually present in the host cell wall of potato or tobacco leaves where infection occurs? And, last, the potato "lectin," although highly purified from other cellular proteins, appeared on gels as a diffuse band that stained in a reddish color with Coomassie blue. These latter characteristics are typical of hydroxyproline-rich glycoproteins of cell wall origin but did not allow a true assessment of the lectin's homogeneity. These various problems are now under systematic investigation in the laboratory of L. Sequeira; some of the emerging results are presented below.

The relative role of Rcore or lipid A in binding and their accessibility on the bacterial surface have not yet been adequately addressed; this is obviously not a simple experimental problem. However, a highly selective phage with putative specificity for components of the O antigen of *P. solanacearum* has been isolated (Whatley *et al.*, 1980) and is now being used to screen for additional possible rough- or smooth-form *P. solanacearum* isolates. These isolates will allow both purification and examination of further LPS molecular species *in vitro* and further whole cell binding studies using a series of bacterial strains with more precisely defined surfaces. Coupled with chemical characterization of the LPS species, this should allow clearer definition of the specific moieties and linkages required for binding.

A highly quantitative and sensitive binding assay for the bacterial cell wall components has been developed utilizing the fact that lectin and lectin–hapten complexes were found to bind to nitrocellulose membranes whereas free hapten did not (Duvick *et al.*, 1979). Using this assay, it has been unambiguously confirmed that both purified LPS and EPS preparations do indeed bind tightly to potato lectin and that the binding of LPS can be quantitatively inhibited by the addition of EPS (Duvick *et al.*, 1979). The assay has also allowed the demonstration that binding of LPS to the lectin is not as sensitive to ionic strength as is binding of the EPS, which is severely depressed by ionic strengths in the range that might be expected in the intercellular fluid of host tissues. This observation raises the possibility that (1) either EPS is not involved in prevention of attachment *in vivo* or (2) that the ionic strength effect could play an additional regulatory role in attachment *in vivo*. Quantitative measurements of ionic strength of intercellular "fluids" are obviously difficult and would be highly subject to the physiological state of the tissue. Parameters such as light intensity (Lozano and Sequeira, 1970; Graham *et al.*, 1977) and temperature (Klement and Goodman, 1967; Ciampi and Sequeira, 1980) are known to dramatically affect the host–parasite interaction between *Pseudomonas* and its hosts; how these parameters could affect ionic strength of intercellular fluids in tobacco has not been explored carefully. In any case, the nitrocellulose membrane assay should allow more extensive studies to be performed especially as more highly chemically characterized LPS and EPS structures are identified from phage-selected isolates.

Concerning the localization of the lectin, preliminary studies have suggested that the potato lectin and a similar lectin in tobacco are localized in the plant cell wall in leaves since it could be extracted by simple vacuum infiltration with buffered saline (Sequeira and Graham, 1977); these initial observations have been further confirmed by staining host tissue cross-sections with FITC-labeled antibody against the lectin (N. Hunter and L. Sequeira, personal communication). Post-staining fluorescence was nearly totally associated with the host cell wall, suggesting that the potato lectin may indeed be accessible for bacterial binding at the host cell wall.

The final problem with the initial bacterial attachment hypothesis was the nature of the plant "lectin" itself. Lectin preparations with very similar properties to those of the potato tuber or leaf lectin were also isolated from tomato callus cultures and from tobacco leaves (T. L. Graham and L. Sequeira, unpublished). These "lectins" were all characterized by gel electrophoresis as a diffuse band that stained reddish with Coomassie blue, suggesting that they might represent similar hydroxyproline-rich glycoprotein fractions. All of the preparations precipitated avirulent forms of *P. solanacearum* and all had some hemagglutinin activity as well. Several observations suggested, however, that the potato "lectin" as isolated by acidic ethanol in the above studies (Sequeira and Graham, 1977) and by Allen and Neuberger (1973) were not identical; these included a sensitivity of the activity of the acidic ethanol-isolated lectin to lyophilization and the requirement of high salt concentrations for the long-term stability of the lectin. An elegant series of comparisons has confirmed this. Further purification of the potato tuber lectin obtained by acidic ethanol has shown that it is *not* a hemagglutinin (J. Leach and L. Sequeira, personal communication). As it is extensively purified, its specific activity for bacterial agglutination increases dramatically as its specific activity as a hemagglutinin decreases markedly. The lectin also differs from that of Allen and Neuberger in its apparent isoelectric point (pI $\sim$ 11.0!), its sedimentation coefficient, and its amino acid analysis. Consistent with its staining on gels, the acidic ethanol lectin is much higher in hydroxyproline and in lysine than the Allen and Neuberger lectin. The extremely high lysine content and isoelectric point are suggestive of a polycationic structure similar to that reported for the apple agglutinin proposed to be involved in *Erwinia amylovora* recognition (Romeiro and Karr, 1980). Indeed, the nitrocellulose membrane binding assay has allowed the demonstration that simple polyanions do inhibit the lectin binding to LPS (J. Duvick and L. Sequeira, unpublished observations). Simple salts do not as strongly inhibit the lectin, suggesting that the "specificity" of the lectin may be for an extended charge distribution. Further hapten studies are needed to confirm this latter possibility. Earlier total inhibition of agglutinin by chitin oligomers (Sequeira and Graham, 1977) could be explained by a combination of a mixed lectin preparation (including the specific Allen and Neuberger hemagglutinin) and a possible contamination of chitin oligomers with *chitosan* oligomers. It should be emphasized that the study of the specificity of lectins has routinely been approached by testing simple available sugars for binding inhibition; these results may not translate well to the true specificities of the lectins for natural polymers. Similar anomalies are known for the soybean seed lectin, which, although "specific" for 2-N-acetylgalactosamine, binds strongly to EPS preparations from *Rhizobium japonicum*, which contain 2-*O*-acetylgalactosamine but not the amino sugar (Mort and Bauer, 1980).

In summary, then, a potato cell wall bacterial *agglutinin* and the bacterial cell wall LPS and EPS of *P. solanacearum* very likely have molecular roles in the attachment recognition of this bacterium, although obviously much work still needs to be pursued to further confirm the overall hypothesis.

III. Active Host Responses to *P. solanacearum* LPS

In addition to a role in simple attachment or immobilization of avirulent *P. solanacearum* in host tissues, the LPS of *P. solanacearum* may be involved in the *induction* of several active host responses. These include an engulfment of the avirulent bacterium by cell wall-like material, an induction of whole tissue resistance to further bacterial colonization (including a protection against the hypersensitive reaction) and, *seemingly* contradictory to the latter, in the induction of the hypersensitive reaction itself. Although at first sight unrelated, these events may be closely related events in a series of responses and may all be triggered (as modeled below) by the same initial LPS–agglutinin interaction.

A. Role of LPS in Induction of Active Wall Immobilization

The *immobilization* of avirulent *P. solanacearum* (Sequeira *et al.*, 1977) is not a simple attachment phenomenon although attachment may certainly be a prerequisite for further host responses. Following attachment, avirulent *P. solanacearum* cells appear to be rapidly surrounded or engulfed by a fibrillar/granular wall-like material apparently of host origin. The engulfing material is itself surrounded by a pellicle-like structure. The engulfment, like attachment, appears to be characteristic of avirulent or nonpathogenic (e.g.,*E.coli*) bacteria and is avoided by those virulent *P. solanacearum* isolates examined. Unlike initial attachment, the engulfment is apparently an active host response. Preliminary studies suggest that the engulfment process itself may be inducible by isolated LPS alone; rough-form LPS from the avirulent isolate K60-B1 when injected into tobacco leaves can be visualized under electron microscopy as a laminated micellar structure attached to the host cell wall. Although only limited observations have been made, in several sections the LPS has been shown to induce a similar invagination of the host plasmalemma, a similar formation of vesicles between the plasmalemma and cell wall, and the LPS appears to be engulfed in a manner very similar to whole K60-B1 cells (Graham *et al.*, 1977). This suggests that bacterial LPS may be involved in the active induction of engulfment as well as in simple bacterial attachment. Further studies on the specificity of the engulfment phenomenon and on the *nature* of the host material involved are obviously needed. Although any number of host wall materials could be involved, polymers such as lignin, β-1,3-glucans, or pectic polymers are attractive candidates due to their known involvement in disease resistance mechanisms.

B. The Induction of Tissue Resistance by LPS

When tobacco tissue is injected with either live compatible (virulent) or incompatible avirulent *P. solanacearum*, a tissue response is observed. With virulent bacteria, the tissue becomes chlorotic and bacteria slowly spread from the initial injection site. With incompatible avirulent bacteria, the host tissue collapses quickly, becoming necrotic within 24 hr and eventually becoming totally desiccated; no bacterial nor symptom spread from the initial injection area is seen. This "hypersensitive" reaction is characteristic of an incompatible avirulent reaction and is considered a form of resistance since pathogen invasion and damage is contained to the few cells that came in direct contact with the invading bacterium.

Initial studies demonstrated that prior injection with heat-killed *P. solanacearum* could totally prevent the response of the tobacco leaf to challenge inoculation with *either* virulent or avirulent forms of the bacterium (Lozano and Sequeira, 1970). Cell-free extracts were also found to possess activity (Sequeira *et al.*, 1972) and the induction of the tissue resistance was later proposed to be at least partially due to lipopolysaccharides from the bacterial cell wall (Graham *et al.*, 1977). Thus, purified LPS preparations, prepared by a variety of conventional procedures from whole cells, isolated cell walls, and culture filtrates of both smooth virulent (K60) and rough avirulent (K60-B1) strains of *P. solanacearum*, induced disease resistance in tobacco at concentrations as low as 50 μg/ml. LPS from nonplant pathogens such as *E. coli* B, *E. coli* K, and *Serratia marcescens* also were active. Preliminary fractionations in the various procedures suggested that cell wall proteins, free phospholipid, and nucleic acids were not necessary for activity. In addition, the activity of LPS from rough forms and the nonspecific activity of LPS from a variety of gram-negative organisms suggested that activity resided in the Rcore–lipid A region of the LPS molecule. If the Rcore–lipid

A linkage was broken with mild acid or the lipid A region deacylated, activity was destroyed. These results, of course, are highly reminiscent of endotoxin activity in animals in that activity may reside in the lipid A region with the Rcore possibly necessary for solubility.

The above results were largely supported by the work of Mazzucchi and Pupillo (1976), who demonstrated the activity of protein–LPS complexes from *Erwinia chrysanthemi* in preventing the hypersensitive reaction in tobacco induced by *Pseudomonas syringae*. Mazzucchi and Pupillo also demonstrated consistent activity in the range of 50 μg/ml.

It should be emphasized that these studies are highly preliminary and obviously require further confirmation and extension. Although they utilize highly purified LPS *preparations* prepared by a variety of procedures, such preparations of a polysaccharide are by nature heterogeneous, and as pointed out by research workers in early work with endotoxin (see, for example, Nowotny, 1969; Ng *et al.*, 1976; Tsang *et al.*, 1974), chemically purified and modified LPS preparations provide useful, but inconclusive, information regarding active LPS components due to the extreme complexity of the molecule and its possible associations with other minor molecular components. The selection and study of a series of rough mutants by the phage selection techniques mentioned above (especially the identification, if possible, of an "Re" mutant, which contains largely only lipid A and KDO) will again greatly facilitate further studies in the structural requirements for activity. Nonetheless, the fact that protection is induced by all heat-killed gram-negative bacteria examined but none of the gram-positive bacteria examined (T. L. Graham and L. Sequeira, unpublished results) and that highly purified LPS preparations from a variety of gram-negative organisms induce resistance, strengthen the proposed activity.

In summary, then, tissue resistance induced by LPS is effective against both virulent and avirulent bacteria and appears to be *evolutionarily* (but not molecularly) a "nonspecific" event. This recognition event and the particular type of resistance induced may represent a very primitive level of recognition established in the coevolution between plant and gram-negative organism. Unfortunately, very little is known of the actual mechanism of the resistance stimulated except that it obviously does *not* result in events *visually* related to the hypersensitive reaction. The total lack of symptoms caused by challenge inoculation suggests a possible inhibition of colonization at a very early stage. Since the *visual symptoms* of the hypersensitive reaction appear to require living bacteria, it is perhaps not surprising that LPS-induced tissue resistance prevents this "resistant" reaction as well. Moreover, as discussed below, whole-plant disease resistance reactions can generally be expected to be a complex series or mixture of independent but highly interrelated events. The injection of intact *living* incompatible bacteria results in the hypersensitive response, while the injection of intact heat-killed incompatible bacteria results only in a tissue resistance response. Obviously, the latter may be a *component* of the former. This is especially suggested by the fact that, as noted above, injected LPS appears to cause many of the same microscopically visible cellular responses (e.g., cell wall engulfment, membrane invagination and vesiculation) as are seen early in the hypersensitive reaction (Graham *et al.*, 1977).

In any case, we now have an additional potential "level" of activity of LPS in host plants in the induction of tissue resistance, leading to the proposal that the *same* recognition event between LPS and host agglutinin that results in attachment and immobilization of *P. solanacearum* may also trigger active host resistance responses (Sequeira and Graham, 1977).

C. Role of LPS in the Induction of the Hypersensitive Reaction

The molecular components involved in the induction of the visual symptom termed the *hypersensitive reaction* (HR) (like attachment and active immobilization, generally considered highly specific to *incompatible* bacteria) have been extremely elusive. Cell-free bacterial fractionation has not successfully led to an "inducer" molecule, suggesting that living bacteria may be needed. In fact, as noted above, heat-killed incompatible bacteria not only do not cause the HR but cause the induction of a resistance *to* the HR. As noted above, this is not totally surprising since resistance to the HR could involve some of the components of the resistance responses induced as part of the HR itself. Tissue necrosis and collapse (the visual symptom) may be a molecularly determined event within the HR that requires a highly specific interaction between the living partners.

Obviously, if living bacteria are needed for observation of the complete HR reaction, a *molecular* study of induction is not simple. However, the availability of *P. solanacearum* strains from the phage selection studies mentioned above is supplying a number of strains for the *genetic* investigation of this phenomenon. A highly selective phage (CH154) has been identified that lyses non-HR-inducing strains but not HR-inducing strains. Preliminary correlations (Whatley *et al.*, 1980) suggest that HR-inducing strains may be rough mutants and that non-HR-inducing strains contain a more complete O antigen, which is the putative binding site for the phage. Whether or not the specificity is defined by the presence or absence of components of the O antigen, an analysis of sugar components of the isolated LPS species by gas chromatography (Whatley *et al.*, 1980) has demonstrated a clear difference in sugar composition between HR-inducing and non-HR-inducing strains (particularly in the relative percentages of glucose, xylose, and rhamnose).

If these results are further confirmed, they would suggest that, once again, components of the same molecular signal potentially involved in attachment, active immobilization, and non-HR tissue resistance may also be required for the induction of the necrosis associated with the HR. Again, it should be emphasized that the HR is an overall reaction and *involves* attachment, active immobilization, and tissue resistance as components. The extension here would be that even the highly specific *necrotic* reaction associated with the HR could require or share components of these early seemingly less specific events.

IV. A Metabolite Shunting Model for Host Plant Resistance Responses

How can the same molecular components be key determinants in attachment, active immobilization, induced tissue resistance, and the induction of the hypersensitive reaction? Or phrased somewhat differently, is there precedence for or at least a logical mechanism by which such seemingly diverse and multiple disease resistance responses could be linked at the *metabolic* level?

The plant possesses an entire array of potential defense mechanisms that it can draw on to defend itself against an invading microbe. These mechanisms are, in fact, often closely related metabolically and often involve multiple branches off the same metabolic pools (e.g., lignins, tannins, isoflavanoid phytoalexins, etc., all originating from a common "phenylpropanoid" pool). However, very little is known on how these potential defenses are actually linked in terms of their induction or their interaction once induced. Recently, however, it was observed that both biotic (natural fungal) and abiotic (chemical) isoflavanoid phytoalexin elicitors may cause the induction of additional nonphytoalexin resistance mechanisms in soybean plants (Lundry *et al.*, 1981). In these studies, plants treated with such diverse phytoalexin elicitors as Triton X-100, 4,4′-bisphenol, $CuCl_2$, and mycolaminaran exhibited a local resistance to invasion by *Phytophthora megasperma* var. *glycinea* (PMG), which generally correlated well with the presence of the phytoalexin glyceollin and involved the typical *phytotoxic* necrotic reaction to phytoalexin elicitation. However, a systemic resistance was also observed in reaction to both biotic *and* abiotic elicitors, which did not correlate to the presence of measurable fungitoxic substances (Lundry *et al.*, 1981). If one examines the metabolism involved (see, e.g, Hahlbrock and Grisebach, 1979), it is apparent that a general stimulation or "loading" of the phenylpropanoid pool could easily lead to a multiple response involving metabolite shunting into both the generally toxic phytoalexins and nontoxic polymeric components such as lignin, which could simply increase the *physical* barriers to infection. These hypotheses are now under further investigation in our laboratory. The key observation, however, was that a series of chemically unrelated abiotic and biotic elicitors (chosen specifically for their ability to elicit glyceollin) routinely elicited other nonphytotoxic and apparently nonfungitoxic defense mechanisms, suggesting a somewhat loose or "leaky" control on the matrix of phenylpropanoid-derived resistance responses.

This concept is again totally consistent with what is known regarding the enzymology of phenylpropanoid metabolism. The actual enzymes involved in isoflavanoid phytoalexin production (and in other metabolic shunts from the phenylpropanoid pool which may contribute to whole-plant disease protection) are under coordinate transcriptional induction control. In parsely cell culture, the enzymes of early phenylpropanoid metabolism (which "load" the phenylpropanoid pool) and of several of the shunts off this pathway are coinduced but in kinetically distinguishable "groups" by several stress phenomena including rapid tissue culture dilution and UV light exposure (Hahlbrock and Grisebach, 1979). Very recently, certain biotic isoflavanoid phytoalexin elicitors have, moreover, been shown to induce representative enzymes of one of these groups in parsley (Hahlbrock *et al.*, 1981). The fact that both biotic and abiotic phytoalexin elicitors induce multiple "levels" of resistance may suggest, then, that the elicitors turn on the early phenylpropanoid enzymes to load the phenylpropanoid pool and then either do not regulate or differentially regulate (depending on the exact reactions observed) several of the pathways leading off phenylpropanoid metabolism resulting in several end products that could contribute to resistance.

It would seem totally reasonable that this type of differential "metabolite shunting" from common pools into a number of end products involved in disease resistance could be involved in the multiple but related resistance responses of tobacco or potato to bacteria as well; for example, attachment and recognition could supply the elicitor-like "signal" that results in the loading of a common metabolic pool(s), which could then be differentially shunted (depending on further signals received from the *actual* interaction) to result in (1) the formation of an induced polymer barrier for immobilization, (2) the formation of selectively bacterial-toxic principles, which could lead to a nonphytotoxic tissue resistance and/or (3) the formation of more generally antibiotic phytoalexin-like molecules, leading to the "suicide" reaction of the visual hypersensitive reaction. The metabolism involved in resistance of solanaceous hosts to invading microbes involves both components of phenylpropanoid and isoprenoid metabolism. Above we discussed primarily phenylpropanoid metabolism. Although very little is currently known of the enzymology or regulatory controls in isoprenoid metabolism in plants, a common feature of both of these pathways is multiple branches off common metabolic pools.

It is thus not difficult at least to rationalize at the *metabolic* level the diversity of responses seen in the *P. solanacearum*/tobacco system. Unfortunately, at this time such a hypothesis is very difficult to evaluate since virtually nothing is known concerning the molecular or metabolic bases for the various levels of resistance described above in the *P. solanacearum*/tobacco system, even though this is one of the "cleanest" host–pathogen systems available for in-depth studies. To enable a realistic evaluation of the role of LPS or any molecule in plant disease recognition and resistance, it is obvious that this largely neglected area of plant biochemistry must be pursued vigorously. We need perhaps to keep a more balanced perspective and focus as much on the actual "nuts and bolts" biochemical mechanisms of resistance as we have in the past on the more romantic aspects of recognition and elements of specificity. Our focus in the biochemistry of plant disease resistance has put us in an awkward position where we have studied the "elicitors" and inducers exhaustively before we were even aware of *what* they were truly eliciting and what *relative* role the proposed end-products have on actual *whole-plant* resistance.

V. Conclusions

It is obvious that much research is needed to evaluate the true role of LPS in recognition and resistance responses of solanaceous plant hosts to *P. solanacearum*. Where possible, I have defined some of the major research needs in the discussions on the individual responses above. However, it is critical that research advance systematically on all fronts or else we can fall into the common trap (wherein fell the phytoalexin concept) where we know a great deal about "events" without knowing their relative importance. The major area of deficiency in plant disease resistance research is that we know very little about the actual matrix of biochemical mechanisms involved and next to nothing on how they work *together* in whole-plant resistance. This is certainly a complex problem, but systems such as the multiple responses observed to LPS offer an exciting starting point.

VI. References

Allen, A. K., and Neuberger, A., 1973, The purification and properties of the lectin from potato tubers, a hydroxy proline-containing glycoprotein, *Biochem. J.* **135**:307.

Bauer, W. D., 1981, Infection of legumes by rhizobia, *Annu. Rev. Plant Physiol.* **32**:407.

Carlson, R. W., 1982, Surface chemistry of *Rhizobium*, in: *Ecology of Nitrogen Fixation* (W. J. Broughton, ed.), Vol. 2, pp. 199–234, Oxford University Press, London.

Ciampi, L., and Sequeira, L., 1980, Influence of temperature on virulence of race 3 strains of *Pseudomonas solanacearum*, *Am. Potato J.* **57**:307.

Dazzo, F. B., 1980, Determinants of host specificity in the *Rhizobium* clover symbiosis, in: *Nitrogen Fixation* (W. E. Newton and W. H. Orme-Johnson, eds.), Vol. 2, pp. 165–187, University Park Press, Baltimore.

Duvick, J. P., Sequeira, L., and Graham, T. L., 1979, Binding of *Pseudomonas solanacearum* polysaccharides to plant lectin *in vitro*, *Plant Physiol.* **63**(Suppl.):134.

Graham, T. L., 1981, Recognition in *Rhizobium*–legume symbioses, *Int. Rev. Cytol.* **13** (Suppl.):127.

Graham, T. L., Sequeira, L., and Huang, T. S. R., 1977, Bacterial lipopolysaccharides as inducers of disease resistance in tobacco, *Appl. Environ. Microbiol.* **34**:424.

Hahlbrock, K., and Grisebach, H., 1979, Enzymic controls in the biosynthesis of lignin and flavanoids, *Annu. Rev. Plant Physiol.* **30**:105.

Hahlbrock, K., Lamb, C. J., Purwin, C., Ebel, J., Fautz, E., and Schäfer, E., 1981, Rapid response of suspension-cultured parsley cells to the elicitor from *Phytophthora megasperma* var. *sojae*, *Plant Physiol.* **67**:768.

Klement, Z., and Goodman, R. N., 1967, The hypersensitive reaction to infection by bacterial plant pathogens, *Annu. Rev. Phytopathol.* **5**:17.

Lozano, J. C., and Sequeira, L., 1970, Prevention of the hypersensitive reaction in tobacco leaves by heat-killed bacterial cells, *Phytopathology* **60**:875.

Lundry, D. R., Bass, J., Castanho, B., and Graham, T. L., 1981, Protection of soybean plants against disease by phytoalexin elicitors, *Plant Physiol.* **67**(Suppl.):75.

Mazzucchi, V., and Pupillo, P., 1976, Prevention of confluent hypersensitive necrosis in tobacco by a bacterial protein–lipopolysaccharide complex, *Physiol. Plant Pathol.* **9**: 101.

Mort, A. J., and Bauer, W. D., 1980, Composition of the capsular and extracellular polysaccharides of *Rhizobium japonicum*: Changes with culture age and correlations with binding of soybean seed lectin to the bacteria, *Plant Physiol.* **66**:158.

Ng, A., Butler, R. C., Chen, C. H., and Nowotny, A., 1976, Relationship of structure to function in bacterial endotoxins. IX. Differences in the lipid moiety of endotoxic glycolipids, *J. Bacteriol.* **126**:511.

Nowotny, A., 1969, Molecular aspects of endotoxic reactions, *Bacteriol. Rev.* **33**:72.

Romeiro, R., and Karr, A., 1980, Isolation of a bacterial agglutination activity for *Erwinia amylovora* from apple, *Plant Physiol.* **65**(Suppl.):135.

Sequeira, L., 1978, Lectins and their role in host–pathogen specificity, *Annu. Rev. Phytopathol.* **16**:453.

Sequeira, L., and Graham, T. L., 1977, Agglutination of avirulent strains of *Pseudomonas solanacearum* by potato lectin, *Physiol. Plant Pathol.* **11**:43.

Sequeira, L., Aist, S., and Ainslie, V., 1972, Prevention of the hypersensitive reaction in tobacco by proteinaceous constituents of *Pseudomonas solanacearum*, *Phytopathology* **62:** 536.

Sequeira, L., Gaard, G., and DeZoeten, G. A., 1977, Attachment of bacteria to host cell walls: Its relation to mechanisms of induced resistance, *Physiol. Plant Pathol.* **10:**43.

Tsang, J. C., Wang, C. S., and Alaupovic, P., 1974, Degradative effect of phenol on endotoxin and lipopolysaccharide preparations from *Serratia marcescens*, *J. Bacteriol.* **117:** 786.

Whatley, M. H., Hunter, N., Cantrell, M. A., Hendrick, C., Keegstra, K., and Sequeira, L., 1980, Lipopolysaccharide composition of the wilt pathogen *Pseudomonas solanacearum*, *Plant Physiol.* **65:**557.

Index